# The Far North: Plant Biodiversity and Ecology of Yakutia

PLANT AND VEGETATION

Volume 3

*Series Editor:* M.J.A. Werger

For other titles published in this series, go to www.springer.com/series/7549

# The Far North: Plant Biodiversity and Ecology of Yakutia

**E.I. Troeva, A.P. Isaev, M.M. Cherosov, and N.S. Karpov**

*Russian Academy of Sciences*
*Siberian Branch*
*Institute for Biological*
*Problems of the Cryolithozone*

Springer

*Editors*
E.I. Troeva
Russian Academy of Sciences
Siberian Branch
Institute for Biological
Problems of the Cryolithozone
41 Lenin Ave.
Yakutsk
Russia 677980
etroeva@mail.ru

A.P. Isaev
Russian Academy of Sciences
Siberian Branch
Institute for Biological
Problems of the Cryolithozone
41 Lenin Ave.
Yakutsk
Russia 677980
forest_forest@ibpc.ysn.ru

M.M. Cherosov
Russian Academy of Sciences
Siberian Branch
Institute for Biological
Problems of the Cryolithozone
41 Lenin Ave.
Yakutsk
Russia 677980
cherosov@mail.ru

N.S. Karpov
Russian Academy of Sciences
Siberian Branch
Institute for Biological
Problems of the Cryolithozone
41 Lenin Ave.
Yakutsk
Russia 677980

ISBN 978-94-007-3193-6 ISBN 978-90-481-3774-9(eBook)
DOI 10.1007/978-90-481-3774-9
Springer Dordrecht Heidelberg London New York



Printed on acid-free paper

Springer is part of Springer Science+Business Media (www.springer.com)

# Contents

# Afterword

Instead of summarizing and concluding the information given in this book, we would like to emphasize on another topic: plans for the future and basic issues requiring further investigations. In other words, what to do next?

There are many interesting themes and issues that are waiting for specialists. The flora of Yakutia is still subject of study, especially the territories of South and South-East Yakutia bordering on the mountainous systems of South Siberia and the Far East. The recent 30 years of investigation have yielded over 400 species of higher vascular plants being first recorded in Yakutia, including its interior regions. And such hard-to-reach places as highlands, especially in the North-East, represent the real botanical "Klondike"; however, it would not be so easy to find a floristic "nugget". This refers both to higher vascular plants and other plant groups.

Despite the low biological diversity (less than 2000 species of higher vascular plants), various territories are characterized by large numbers of endemic species. This is especially true for the north-eastern regions and some ranges in the North. The number of endemic and rare species of Yakutia is probably much higher than presently recorded.

The vegetation of Yakutia features the following interesting phenomena and issues:

- Xerophytization of the vegetation of Central Yakutia. There are still many places in the region to be explored by florists and geobotanists;
- The phytocoenoses of Yakutia are characterized by rather low $\alpha$- and $\beta$-diversities. The reasons for this seem to be clear. However, the level of knowledge is not even throughout the territory of Yakutia due to varying approaches of community description.
- There are unique ecosystems in Yakutia that are more characteristic for more southern latitudes (steppe, tundra-steppe, etc.)
- There are patches of dark coniferous forests in the South, their elements penetrating northwards;
- North-West Yakutia features a specific interrelation between relief and vegetation, particularly where at low altitudes the flat territory is covered by alpine tundra communities;

- The interrelation between the tundra and valley species complexes in the river valley communities in South–East and South Yakutia is intriguing where arctic alpine and alpine species grow at all levels of the floodplain;
- In the Verkhoyansk Range, the link-up of the zonal forest and tundra vegetation at high altitudes needs careful research, when joint boundaries are obliterated.
- The gradual transition between the *Larix* forests and woodlands and *Pinus pumila* shrubberies in the highlands of North-East and South Yakutia is striking and a vegetation continuum is clearly seen. The syntaxonomical delineation of the continuum is of great interest.
- The largest botanical-geographical barrier in North-East Yakutia, the Verkhoyansk Range and other mountainous systems of North-East Russia as a whole, still warrants careful botanical and ecological investigation.

The typology of following vegetation is least studied in Yakutia:

- Bogs;
- Riparian and aquatic vegetation;
- Maritime vegetation;
- Psammophytic vegetation;
- Petrophytic vegetation including that of the slopes of mountain rivers;
- Alpine vegetation.

Other vegetation types are also waiting for more detailed investigation, since the size of Yakutia is very large and provides florists and geobotanists with work for a further hundreds of years.

We hope that foreign specialists, having become interested in the objects and ideas described in this book, will join us to study the Yakutian flora and vegetation in collaboration. For joint projects, please apply to any author:

Institute for biological problems of cryolithozone, 41 Lenin Ave., Yakutsk, 677980, Russia

# Introduction

The flora and vegetation of the Soviet Union have always attracted experts in botany from Western Europe and other countries. Up to the early twentieth century foreign explorers had a better opportunity to study the vegetation of the Russian Empire. After the Great October Socialist Revolution ideological obstacles hampered the development of scientific contacts and joint expeditions for several decades. Presently the situation has finally changed. Important international projects are implemented aiming to study the flora and vegetation of the former USSR countries: Kazakhstan, the Caucasus, Russian territories in the Altai, northern Russia, the Far East, etc. The study of the Yakutian flora and vegetation with international participation has a rather occasional character. However, such episodic cases pave the way for further comprehensive investigations to be conducted in collaboration with foreign colleagues.

Most botanical works on Yakutia have been published in Russia and in Russian. This significantly hampers the distribution of unique and interesting information abroad. And this is a common situation for many Russian regions. The authors of this monograph made an attempt to solve this problem in part.

Yakutia, with an area of over 3 million km$^2$, not only covers one fifth of all Russia. It also features peculiar vegetation growing on perennially frozen grounds, the so-called permafrost or cryolithozone that thaws only several metres deep and allows for a short growing season. How can plants survive under such extreme conditions? What are the adaptation mechanisms that allow them to withstand the cold winters with the lowest temperatures reaching sometimes minus 70°C? How can they successfully grow and propagate during the very short growing season under unfavourable hydrothermal conditions (up to + 30°C in July and 200 mm of annual precipitation)? These are the questions that every botanist would like to have the answers on.

Yakutia contains the vegetation of several natural zones, from the arctic deserts to the middle taiga with the elements of the southern taiga. Biodiversity, biogeography, ecology of flora and vegetation, these are the topics that are interesting to florists and plant sociologists, plant ecologists and ecophysiologists, as well as to other specialists in botany. We expect that this book will be able to fill a gap for foreign specialists on these important issues of nature investigation.

It may appear that the study of less than 2,000 higher vascular plant species and a description of their communities are not worth the trouble. However, after more than a century of investigations, the Yakutian vegetation is generally still an unexplored object, which gradually discloses its mysteries to persistent botanical explorers.

A reader who opens this book should realize that he holds in his hands the first generalization on the flora and vegetation of Yakutia based on long-term investigations by botanists mainly from the joint educational-research laboratory of floristics and phytocoenology of the Faculty of biology and geography (Ammosov Yakut State University) and the Institute for biological problems of the cryolithozone (Siberian Branch of the Russian Academy of Sciences). Most authors of the monograph have had an opportunity to explore various corners of this tremendous region lying in the core of the cryolithozone. They covered hundreds of kilometres by various means of transport: cars, helicopters, air planes, off-highway vehicles, and even on horseback, appreciating every rare opportunity to reach remote places. Many authors were the first to set foot in such lands reigned by wild animals and plants. And it is good that most of the territory of Yakutia still is wild nature.

There is no similar book in a Russian edition. It directly has come out in English, and it is very surprising and exciting that we could make it.

As will be clear from the contents of the book, the flora and vegetation of Yakutia are studied unevenly. This is explained by the history of interest in a certain object, and the presence of persons who initiated and developed research work on those objects.

The flora of Yakutia has been the object of study of many scientists. In the beginning of the twentieth century the Yakutian flora was described by academician V.L. Komarov. However, its structure and principles of spatial organization were revealed in the middle of the twentieth century by Mikhail Nikolaevich Karavaev. He worked at the Yakut State University for a certain period, though most of his life he headed the Herbarium of the M.V. Lomonosov Moscow State University.

Most of Yakutia is covered by forest. The study of forest communities has always been a constituent part of the research activity of the Institute for Biological Problems of the Cryolithozone, Siberian Branch of the Russian Academy of Sciences, as well as of the Yakut State University. Igor Petrovich Scherbakov has long headed the forest school of Yakutia. He initiated the investigation of forests on frozen grounds, which was continued by his followers.

The specialist in meadows, plant ecologist, botanist-geographer Konon Evseyevich Kononov was the founder of the Group of phytosociology at the Faculty of Biology and Geography (Yakut State University). The aim of the Group was to study the syntaxonomy and map the herbaceous vegetation. The main achievements of K. Kononov's research activities were to reveal the main principles of the meadow and steppe vegetation structure and the classification of herb vegetation using the Braun-Blanquet approach. He and his followers and colleagues were the first who familiarized the foreign specialists with the main syntaxa of the Yakutian meadow and steppe vegetation.

The famous geobotanist Valdimir Nikolaevich Andreyev came to Yakutia from Leningrad at a mature age. He studied the tundra communities and trained his followers who still study the flora and vegetation of Yakutia nowadays. His ideas on the principles of the distribution of the tundra vegetation, on the biology and ecology of tundra plants, have determined the general lines of investigation of the tundra flora and vegetation for many years. His organizational abilities and status as a leading botanist of Yakutia have allowed the publication of a number of summarizing works on the flora and vegetation of Yakutia.

We authors would never write this monograph without the works of their predecessors in the twentieth century. The list of those specialists is not limited to the abovementioned recognized leaders of Yakutian botany. Most of our elder colleagues are already no longer with us. However, they brought us up as specialists, and passed on their knowledge, experience and persistency in overcoming obstacles to study the flora and vegetation of Yakutia. We authors express their gratitude and dedicate this book to our teachers.

We authors also acknowledge everyone who took part in the preparation of the monograph, appreciate the patience of the scientific editor and management of the Publishing House. Without all this, the book would never have been published.

Team of authors

# Contributors

**A.P. Arzhakovan** Institute for biological problems of the cryolithozone, 41 Lenin Ave., Yakutsk, 677980, Russia, BNW-07@yandex.ru

**A.I. Averensky** Institute for biological problems of the cryolithozone, 41 Lenin Ave., Yakutsk, 677980, Russia, aaver37@yandex.ru

**N.V. Barashkova** Institute for biological problems of the cryolithozone, 41 Lenin Ave., Yakutsk, 677980, Russia, BNW-07@yandex.ru

**B.Z. Borisov** Institute for biological problems of the cryolithozone, 41 Lenin Ave., Yakutsk, 677980, Russia, bzborisov@mail.ru

**N.P. Bosikov** Institute of permafrost, 36 Merzlotnaya Str., Yakutsk, 677010, Russia, bosik@mpi.ysn.ru

**M.M. Cherosov** Institute for biological problems of the cryolithozone, 41 Lenin Ave., Yakutsk, 677980, Russia, cherosov@mail.ru

**A.P. Chevychelov** Institute for biological problems of the cryolithozone, 41 Lenin Ave., Yakutsk, 677980, Russia, chev.soil@list.ru

**I.I. Chikidov** Institute for biological problems of the cryolithozone, 41 Lenin Ave., Yakutsk, 677980, Russia, chikidov@rambler.ru

**A.P. Efimova** Institute for biological problems of the cryolithozone, 41 Lenin Ave., Yakutsk, 677980, Russia, aytaef@yandex.ru

**A.A. Egorova** Institute for biological problems of the cryolithozone, 41 Lenin Ave., Yakutsk, 677980, Russia, eang@mail.ru

**P.S. Egorova** Institute for biological problems of the cryolithozone, 41 Lenin Ave., Yakutsk, 677980, Russia, yakutgard@mail.ru

**N.B. Ermakov** Central Siberian Botany Garden, 101, Zolotodolinskaya Str., 630090 Novosibirsk, Russia, brunnera@mail.ru

**Y.V. Ermakova** Institute for biological problems of the cryolithozone, 41 Lenin Ave., Yakutsk, 677980, Russia, yermakova68@mail.ru

**I.A. Fedorov** Institute for biological problems of the cryolithozone, 41 Lenin Ave., Yakutsk, 677980, Russia, yakutgard@mail.ru

**V.A. Gabyshev** Institute for biological problems of the cryolithozone, 41 Lenin Ave., Yakutsk, 677980, Russia, v.a.gabyshev@ibpc.ysn.ru

**L.D. Gavrilyeva** Institute for applied ecology of the North, 5 Kalandarishvili Str., Yakutsk, 677025, Russia, adoxa@mail.ru

**P.A. Gogoleva** Yakut State University, Faculty of Biology and Geography, Room 456, 48 Kulakovsky Str., Yakutsk, 677000, Russia, sedum@mail.ru

**A.P. Isaev** Institute for biological problems of the cryolithozone, 41 Lenin Ave., Yakutsk, 677980, Russia, forest_forest@ibpc.ysn.ru

**I.A. Ivanov** Kv. 26, Kirov Str., Yakutsk, 677007, Russia, anere.35@mail.ru

**B.I. Ivanov** Institute for biological problems of the cryolithozone, 41 Lenin Ave., Yakutsk, 677980, Russia, bio@ibpc.ysn.ru

**E.I. Ivanova** Institute for biological problems of the cryolithozone, 41 Lenin Ave., Yakutsk, 677980, Russia, e.i.ivanova@ibpc.ysn.ru

**A.P. Ivanova** Institute for biological problems of the cryolithozone, 41 Lenin Ave., Yakutsk, 677980, Russia, a.p.ivanova@rambler.ru

**N.S. Karpov** Institute for biological problems of the cryolithozone, 41 Lenin Ave., Yakutsk, 677980, Russia, bio@ibpc.ysn.ru

**A.V. Kononov** Institute for biological problems of the cryolithozone, 41 Lenin Ave., Yakutsk, 677980, Russia, a_v_kononov@ibpc.ysn.ru

**L.I. Kopyrina** Institute for biological problems of the cryolithozone, 41 Lenin Ave., Yakutsk, 677980, Russia, l.i.kopyrina@mail.ru

**L.V. Kuznetsova** Institute for biological problems of the cryolithozone, 41 Lenin Ave., Yakutsk, 677980, Russia, kuznetsova90@mail.ru

**L.P. Lytkina** Institute for biological problems of the cryolithozone, 41 Lenin Ave., Yakutsk, 677980, Russia, forest_forest@ibpc.ysn.ru

**T.Ch. Maximov** Institute for biological problems of the cryolithozone, 41 Lenin Ave., Yakutsk, 677980, Russia, t.c.maximov@ibpc.ysn.ru

**L.G. Mikhalyova** Institute for biological problems of the cryolithozone, 41 Lenin Ave., Yakutsk, 677980, Russia, lgmikhalova@rambler.ru

**S.I. Mironova** Institute for applied ecology of the North 5 Kalandarishvili Str., Yakutsk, 677025, Russia, s.i.mironova@ipes.ysn.ru

**N.V. Nikitina** Yakut State Academy of Agriculture, Olyokminsk Branch, 58 Gagarin Str., Olyokminsk, 678100, Russia, nadezhda_nikitin@mail.ru

**A.N. Nikolaev** Institute of permafrost, 36 Merzlotnaya Str., Yakutsk, 677010, Russia, yktnan@mpi.ysn.ru

**E.G. Nikolin** Institute for biological problems of the cryolithozone, 41 Lenin Ave., Yakutsk, 677980, Russia, enikolin@yandex.ru

**P.A. Pavlova** Institute for biological problems of the cryolithozone, 41 Lenin Ave., Yakutsk, 677980, Russia, yakutgard@mail.ru

**B.N. Pestryakov** Yakut State University, Faculty of Biology and Geography, Room 240, 48 Kulakovsky Str., Yakutsk, 677000, Russia, pbn_bot@mail.ru

**K.A. Petrov** Institute for biological problems of the cryolithozone, 41 Lenin Ave., Yakutsk, 677980, Russia, kap_75@bk.ru

**L.N. Poryadina** Institute for biological problems of the cryolithozone, 41 Lenin Ave., Yakutsk, 677980, Russia, lnporyadina@rambler.ru

**A.V. Protopopov** Institute for biological problems of the cryolithozone, 41 Lenin Ave., Yakutsk, 677980, Russia, a.protopopov@mail.ru

**V.V. Protopopova** Institute for biological problems of the cryolithozone, 41 Lenin Ave., Yakutsk, 677980, Russia, a.protopopov@mail.ru

**P.A. Remigailo** Institute for biological problems of the cryolithozone, 41 Lenin Ave., Yakutsk, 677980, Russia, p.a.remigailo@ibpc.ysn.ru

**I.F. Shurduk** Institute for biological problems of the cryolithozone, 41 Lenin Ave., Yakutsk, 677980, Russia, forest_forest@ibpc.ysn.ru

**N.P. Sleptsova** Yakut State University, Faculty of Biology and Geography, Room 240, 48 Kulakovsky Str., Yakutsk, 677000, Russia, endyps@rambler.ru

**R.R. Sofronov** Institute for biological problems of the cryolithozone, 41 Lenin Ave., Yakutsk, 677980, Russia, r.r.sofronov@ibpc.ysn.ru

**E.V. Sofronova** Institute for biological problems of the cryolithozone, 41 Lenin Ave., Yakutsk, 677980, Russia, e.v.sofronova@ibpc.ysn.ru

**N.K. Sosina** Institute for biological problems of the cryolithozone, 41 Lenin Ave., Yakutsk, 677980, Russia, n.k.sosina@ibpc.ysn.ru

**P.A. Timofeyev** Yakut State University, Faculty of Biology and Geography, Room 240, 48 Kulakovsky Str., Yakutsk, 677000, Russia, timpa@inbox.ru

**E.I. Troeva** Institute for biological problems of the cryolithozone, 41 Lenin Ave., Yakutsk, 677980, Russia, etroeva@mail.ru

**I.I. Vasilyeva** Institute for biological problems of the cryolithozone, 41 Lenin Ave., Yakutsk, 677980, Russia, l.i.kopyrina@mail.ru

**V.I. Zakharova** Institute for biological problems of the cryolithozone, 41 Lenin Ave., Yakutsk, 677980, Russia, e.i.ivanova@ibpc.ysn.ru

# Chapter 1
# Natural Conditions

**A.P. Chevychelov and N.P. Bosikov**

The Republic of Sakha (Yakutia) is situated in the north-east of Siberia. The northernmost point of continental Yakutia lies on the Nordwick cape (74°N), while the northernmost island point is situated in the north of the Henrietta Islands (77°N). The southernmost point is on the Stanovoy Range (55° 30′ N) nearly conforming to Moscow's latitude. The westernmost and easternmost points lie on 105° 00′E and 165° E respectively. Thus, the territory of Yakutia stretches for 2000 km from the north to the south, and 2500 km from the west to the east.

The area of Yakutia (3 103 200 $km^2$) makes up one-fifth of the Russian Federation's territory. The Republic is larger than France, Austria, Germany, Italy, Switzerland, England, Finland and Greece put together. Some islands of the Arctic Ocean, including the Novosibirskie Islands, belong to the Republic. Over 40% of Yakutia's territory is beyond the Polar Circle. There are three time zones in its area. The capital of the Republic of Sakha (Yakutia), Yakutsk city, is 8468 km away from Moscow.

In the north, Yakutia borders the Laptev and East-Siberian Seas which are part of the Arctic Ocean. The coastline is approximately 4000 km (Mostakhov et al. 1980; Matveyev et al. 1989).

## 1.1 Geomorphology and Relief

The territory of Yakutia is divided into three large zones: western, eastern and southern Yakutia (Fig. 1.1). According to the natural zonation of the former USSR, the western zone of Yakutia belongs physiographically to the "Middle Siberian plateau" (Korzhuev 1974). The southern part is part of the Aldan-Okhotsk physiographic zone, and the eastern part of the Republic is partly represented by the East-Siberian mountainous country and by vast lowlands of North-eastern Yakutia.

A.P. Chevychelov (✉)
Institute for biological problems of the cryolithozone, 41 Lenin Ave., Yakutsk, 677980, Russia
e-mail: chev.soil@list.ru

E.I. Troeva et al. (eds.), *The Far North: Plant Biodiversity and Ecology of Yakutia*,
Plant and Vegetation 3, DOI 10.1007/978-90-481-3774-9_1,

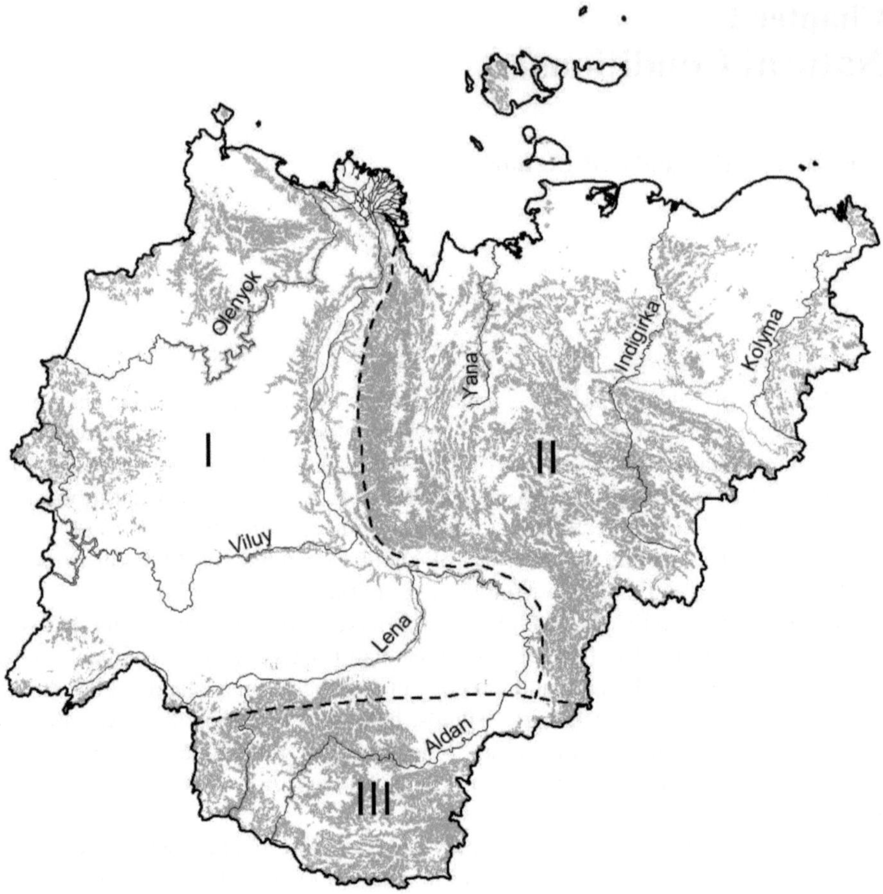


**Fig. 1.1** Physiographical zones of Yakutia (by Korzhuev (1965)) I – West Yakutia, II – East Yakutia, III – South Yakutia

The last two zones are characterized by a complicated latitudinal and altitudinal zonation (Fig. 1.2).

Physiographic zonation (Fig. 1.1) is based on a set of all natural elements though the role of a key element is very important. Thus, key elements for West Yakutia, characterized by plains and plateau, are soil and vegetation, while relief is the major factor for South Yakutia. Relief has an influence upon vegetation and soil patterns, climate and other natural elements and is reflected in vertical (elevational) zonation (Korzhuev 1974).

Geomorphologically the territory of Yakutia consists of two large areas: the plateau and the folded areas. The latter, represented by mountains and high tablelands, covers two thirds of the territory and is largely centered in the north-east, east and south-east of the republic. Tablelands and plateaus are located in West

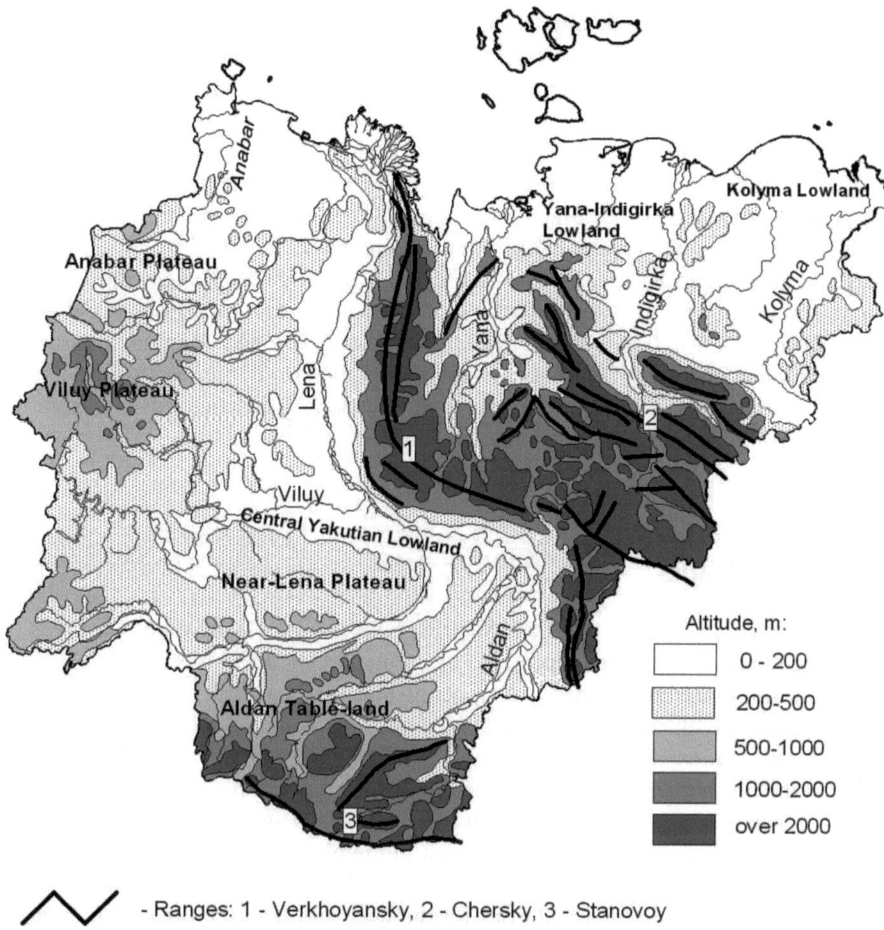


**Fig. 1.2** Orographic scheme of Yakutia (by Korzhuev (1965))

Yakutia, while the northern and central regions are characterized by vast depressions (Korzhuev 1965) including the largest lowlands in Yakutia, the Anabar-Lena and Central Yakutian Lowlands, which lie at altitudes of 50–100 and 60–200 m, respectively. The Central Yakutian Lowland is a colossal depression surrounded nearly from all sides by mountainous systems and high tablelands. It is open only in the north towards the Lena River's lower reaches. The Lena and its major tributaries (the Aldan and Viluy Rivers) divide the depression into the Lena-Viluy and Lena-Aldan parts. These rivers have washed away loose sediments off the plain and formed wide valleys consisting of several terraces of various widths.

The main plateaus of West Yakutia are Lena-Aldan, Anabar-Olenyok, and Olenyok-Viluy.

Most of South Yakutia is situated in the Aldan tableland with an average altitude of 800–1100 m. Eastwards, it merges into the Uchur-Maya upland region and the Aldan-Uchur Range, which reaches as high as 2246 m. The southernmost area of the Aldan tableland borders outlying areas of the Stanovoy Range, the height of which reaches 2000 m in the west and 2412 m in the east.

East Yakutia has an even more complicated surface structure, with mountainous country consisting of vast tablelands, upland regions and mountain ranges. The largest of these are the Verkhoyansk mountain chains (2250–2959 m) located along the Lena and Aldan Rivers. To the east of the Verkhoyansk Range, the Chersky Range is situated, featuring the highest point of the north-east – the Pobeda (Victory) mountain (3147 m). Between the Verkhoyansk and Chersky Ranges vast tablelands are situated: Yana, Oymyakon with the Oymyakon depression in the Upper Indigirka, and Nera. The Yana-Indigirka and Kolyma Lowlands cover the north, while the Yukagir tableland lies in the east-north-east.

## 1.2 Hydrography

Yakutia borders the Laptev and East-Siberian Seas (Fig. 1.1), the shallowest seas of the Arctic Ocean. Their floor represents a continental terrace. The coastline is quite dented forming numerous bays, straits, peninsulas, and capes. The largest bays are the Anabar, Olenyok and Yana Bays. The Yakutian seas are the coldest in the Arctic. Winter temperatures of the water under ice cover range from 0.2° in the river mouths to +2° at the northern borders of the seas. Summer temperatures of the surface water in bays reach +7° to +10°, whereas in the open sea it is only +1° to +3°C. The Laptev and East-Siberian Seas are the iciest in the Northern Hemisphere. For 9–10 months, they stay covered with ice as thick as 1.5–2 m. The severe nature of the Arctic creates unfavourable conditions for the flora and fauna of the seas bordering Yakutia (Mostakhov et al. 1980).

The river network of Yakutia is rather complicated and belongs to the Laptev and East-Siberian Seas basins. Besides the large rivers, a huge number (0.5 million) of small rivers and streams with a total length of over 1.5 million km drain the territory of Yakutia. The Lena River has the largest basin (65% of the territory) and its major tributaries are the Viluy and Aldan Rivers. Other large rivers are the Anabar, Olenyok and Yana Rivers flowing into the Laptev Sea, and the Indigirka and Kolyma Rivers flowing into the East-Siberian Sea. The Aldan and Indigirka Rivers' networks are the most complicated ones while that of the Viluy is poorly developed. Yakutian rivers are mountainous in their upstream parts, transitional mountainous-lowland in their midstream parts, and typical lowland in their down stream parts. As a rule, almost all large rivers form vast deltas. Precipitation (both snow and rain) plays a major role in river feeding. Discharge is uneven over the year – during the warm period melt waters, rain and other sources provide 90–100% of the annual volume, whereas winter drainage does not exceed 10% of the year's total. The greatest water discharge is usually observed in May–June, during spring tides of the rivers.

Maximum water discharge of the mountainous rivers in the southern and north-eastern regions of the republic is often during summer high water (Mozolevskaya 1973).

River water temperature reaches maximum values in July-early August (25–28°). River freezing-over in Yakutia lasts 180–200 days. Spring flood starts during the first ten days of May in the southern regions and in late May in the North. Spring tides are often accompanied by huge ice blockades, yielding a significant increase in water levels which, in turn, leads to floods.

In Central Yakutia rivers get free of ice in the second half of May and freeze in late October – early November (small rivers freeze in the first half of October) (Mozolevskaya 1973).

Lakes are also common in Yakutia, being mostly of thermokarst and flood-plain origin. In the Lena-Viluy Lowland lakes make up 15–30% of the territory. Nidzhily, the largest lake in Yakutia (119 $km^2$), is situated there. Some lakes are as deep as 100 m or even more, though the average depth reaches only 3–10 m.

River drainage rates in the Viluy River basin are much reduced due to water retention by numerous lake depressions in some parts of that region. The same phenomenon is observed in the Lena-Aldan Interfluve, which is known for the large number of lakes that are not drained. Such lakes accumulate moisture and evaporate it intensively during hot periods. As a result, many such lakes gradually dry up.

## 1.3 Climate

In most of the territory of Yakutia, except for its coastline and the highland region in the South, the climate is strongly continental with very low winter and very high summer temperatures, insignificant nebulosity and relatively mild winds, especially in winter.

A major pressure phenomenon is the establishment and growth of the Asian anticyclone system at the very beginning of the winter. This determines the thermal and wind characteristics of the winter period. The high pressure produces extremely steady air with very low surface temperatures, strong surface inversions and low humidity.

The strong continentality and severity of the climate is felt even more rigorous in some parts of Yakutia due to their location in mountain valleys or enclosed tablelands where, in wintertime, cold, heavier air masses stream down from watershed plateaus. Weak circulation and ground radiation make the air stagnant and even colder (Mozolevskaya 1973).

Figures 1.3 and 1.4 depict the standard climate diagrams following the Walter and Lieth (1960) method. The tops of each diagrams list the name of the climate station, its altitude above sea level (m), its average annual temperature (°C), and its average annual precipitation (mm).The diagrams show curves for the average monthly temperatures (1 scale unit = 10°C) and average monthly rainfall (1 scale unit = 20 mm). Months of the year are arranged along the horizontal axis, which

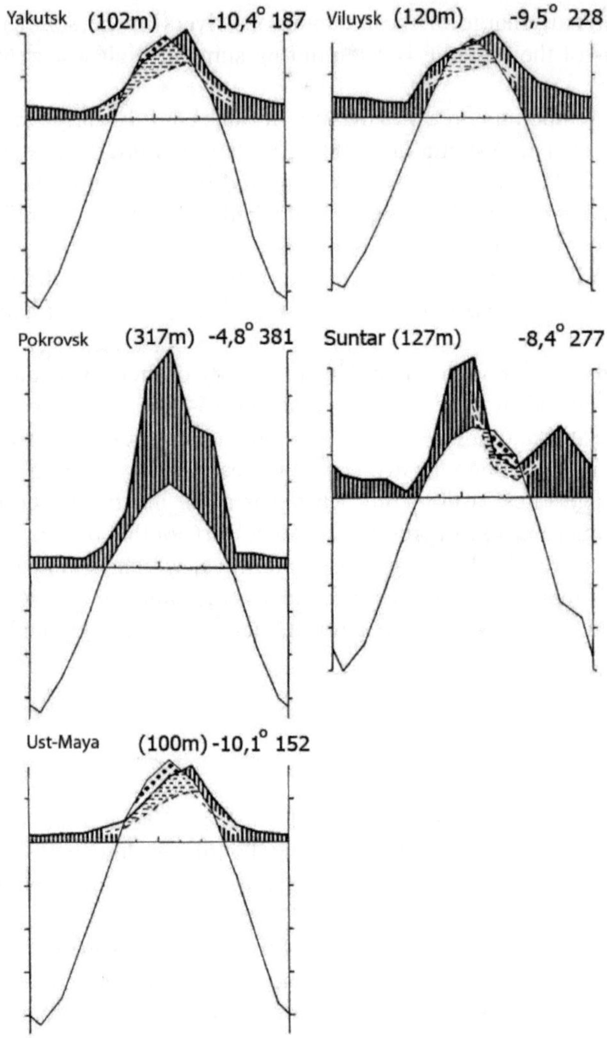


**Fig. 1.3** The standard climate diagrams for Central – South Yakutia

represents 0°C and 0 mm. When the rainfall curve is above the temperature curve the period is considered moist (vertically hatched); when the temperature curve is above the rainfall curve the period is considered dry (dotted). Some climate diagrams show apart from the standard scaling of 10°C against 20 mm of rain an additional monthly rainfall curve for the months of the summer season at a scaling of 10°C against 30 mm of rainfall.

Most of the territory experiences its lowest air temperatures in January, rarely in December or February. In coastal areas, January and February temperatures are almost equal, while on the islands February is the coldest month. From November

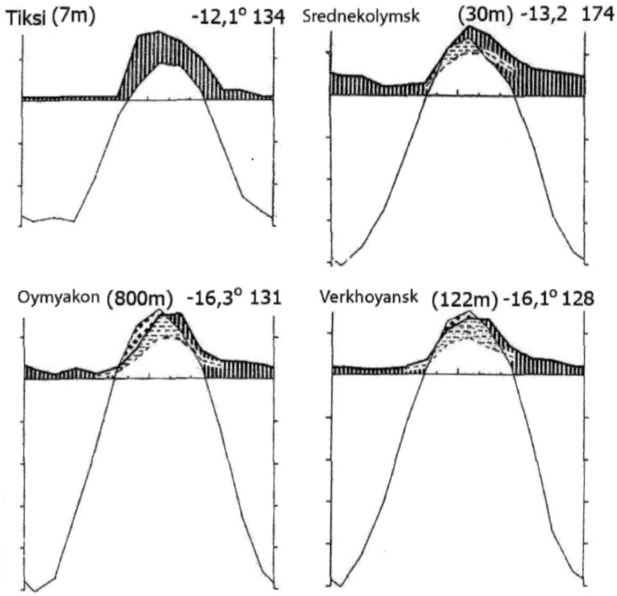


**Fig. 1.4** The standard climate diagrams for North Yakutia

through February, the lowest temperatures are observed in the Oymyakon depression and Yana intermountain trough. Long term winter average air temperatures in Oymyakon and Verkhoyansk are –50°C and –49°C, respectively. Towards the coastal area, the temperature rises to –35°C to –30°C, and to –30°C to –27°C on the islands. Strong surface air inversion (rise in temperature at higher elevations) is a quite ordinary phenomenon during the cold period. They are so constant and intensive, that they have an effect on the mean annual average perennial temperature values at meteorological stations situated in mountains.

Some days, 60° below zero can be observed almost all over the territory. Extremely low temperatures have been recorded in the Oymyakon depression and Yana intermountain trough: –71°C in Oymyakon and –68°C in Verkhoyansk. Minimal temperatures in the South and South-West may reach –58°C to –62°C, and in Central Yakutia –66°C. In coastal areas and on islands, the lowest recorded temperatures are –46°C and –52°C. At a short distance from the shoreline as well as in deep bays, minimal values are considerably less severe.

The duration of the period with temperatures of –50°C or lower varies widely: from 10 h on the Arctic coast to 900 h within the Yana-Indigirka tableland. Yakutia has no analogues both in minimal temperature values and the duration of the period with extremely low temperatures in the Northern Hemisphere (Mostakhov et al. 1980).

A distinct feature of the warm period is the quick rise of the average daily temperatures in spring and their quick drop in autumn. July is the warmest month. The highest temperatures in Yakutia between May to August are observed in the central region. The average July temperature in the central, southern, and south-western parts of the republic is +17 to +19°C, while North of the Viluy River it ranges from +12 to +15°C. Lowest July temperatures (+2 to +5°C) are recorded in coastal areas and on the islands. In highland regions, temperature distributions depend on altitude, relief, and other microclimatic peculiarities. In most of the lowlands maximum temperatures may reach +34 to +38°C, in the coastal areas +29 to +32°C, and on the islands +18 to +24°C.

Characteristic of the thermal conditions of Yakutia are the significant annual amplitudes. Differences between temperatures of the warmest and coldest months and differences between absolute minimum and maximum values in the interior are the world's greatest. Thus, the amplitude of the average monthly temperature varies from 53.0 in Tommot to 61.8 in Ust-Moma with a maximum value of 64.5 in Oymyakon. Amplitudes of absolute minimum and maximum temperatures at those localities reach 99, 102, and 104°C, respectively. Coastal areas and islands are characterized by smaller temperature amplitudes due to their mild winters and cool summers (Mozolevskaya 1973). Sums of temperatures above freezing point vary over a wide range from the coastline towards the lowlands (Table 1.1).

Warm periods with an average daily temperature above zero, in the agricultural zone of the republic, last 150–165 days from the end of April-early May to late September.

The transition to an average daily air temperature above 10°C (which means the start of plant growth) occurs in early July in the tundra, in the first decade of June in the West, and in late May in the central and south-western regions. Back transitions start in the first decade of August in the tundra, in the second and third decades of August in West Yakutia, and in early September in the central and south-western regions (Izyumenko 1966).

The average duration of the period with temperatures above 10°C is 0–50 days in the tundra, 65–90 days in the West, and 90–100 days in Central and South-West Yakutia.

The duration of the frost-free period varies significantly due to the sheer size of the republic and its complicated relief. The longest frost-free period (95–100 days) is recorded in the Middle Lena River valley. In the tundra such period hardly lasts

**Table 1.1** Distribution of sums of positive average daily temperatures in Yakutia in relatively flat areas (°C)

| | Sums of temperatures during periods with temperature above | | | |
|---|---|---|---|---|
| Regions | 0° | 5° | 10° | 15° |
| Sea islands | 50–300 | 0 | | |
| Coastline | 300–500 | 0–300 | 0 | |
| Tundra | 600–900 | 500–800 | 0–500 | 0 |
| Western regions | 1200–1600 | 1100–1500 | 800–1200 | 0–700 |
| Central and south-western regions | 1700–1900 | 1600–1800 | 1400–1600 | 800–1200 |

two months. Sometimes frost may occur all summer long. In highlands, the frost-free period also varies in length. In some years, cold-air surges from the North cause advective radiation frost all over the territory all summer long.

Solar radiation income is very significant in summer thanks to the long duration of daylight and high radiation intensity. The polar day occurs in summer beyond 67°N, while in the rest of the territory there are 18–20 h of day light. Radiation sums for the three summer months in Yakutia total 1590 – 1925 MJ/$m^2$, making up approx. 41–49% of the annual income (Mozolevskaya 1973).

Winter is the clearest season of the year with least nebulosity in February and March. Summer and autumn are characterized by relatively high nebulosity.

The annual precipitation amounts for most of the territory to 200–250 mm, and 350–500 mm in the South and South-West. Precipitation falls unevenly during the year: 15–20% of the total amount occurs during the cold period (November through March), and 4–5 times more (75–80%) during the warm period (April through October). Coastal areas, islands, the Yana and Oymyakon table-lands, the Verkhoyansk and Moma-Selennyakh depressions, as well as the Central Yakutian Lowland are characterized by extremely low amounts of precipitation (150–250 mm per year). In foothills and highland regions, precipitation amounts to 400 mm in the Olyokma and Chara watersheds, the Aldan tableland, and up to 500–700 mm on western slopes of the Aldan-Uchur Range and ridges of the Verkhoyansk Range. Obviously, in the mountains a significant part of the precipitation runs off. During the vegetation period, June is the hardest month for plants when lack of precipitation is observed all over the territory except for its southern and north-western parts. July and August have the highest precipitation values (Izyumenko 1968a).

In most of Yakutia snow cover lasts 225–250 days a year. The shortest period (220–225 days) is observed in the Viluy, Aldan and Middle Lena River valleys, while in coastal areas and on the islands snow cover is longest (260–295 days). Depth of snow cover is insignificant (especially in the North) due to the prevalence of anticyclonic circulation in winter. Deepest snow cover (40–60 cm) is observed in the Upper Aldan River basin, the Kolyma River valley, and in some highland regions. In mountains, snow depth may vary depending on altitude and prevailing wind direction. Snow starts melting in late April, though active solar radiation causes snow evaporating already in March (Izyumenko 1968a).

Being determined by peculiarities in the atmospheric circulation, wind characteristics (direction and speed) vary between winter and summer periods. Winter (September through May) is characterized by prevalence of southern, south-western, and western winds in most of the republic, while in the South-East northern and north-western directions predominate. In summer, winds blow mostly from the North, North-East, North-West and West, and for South-East Yakutia southern winds prevail. Wind speed values are insignificant. In most of the territory, mildest winds (1–2 m/sec) are recorded in January–February. In the enclosed valleys and lowlands (depressions) (Ust-Moma, Oymyakon, Verkhoyansk) wind speed does not exceed 0.2–0.4 m/sec. Increase in cyclonic activity in summer yields wind speeds of up to 5 m/sec. Strong winds are quite uncommon for most parts of the republic (Izyumenko 1968b).

**Table 1.2** Zonal-subzonal change in climate in Yakutia (Izyumenko 1966, 1968a, b)

| Meteostations | True altitude (m) | Climatic indices* $\Sigma$ t>10°C | $C_h$ | $C_c$ | Climate description |
|---|---|---|---|---|---|
| North Yakutia, arctic tundra | | | | | |
| Preobrazhenie Isl. | 6 | 0 | 1.9 | 138 | Arctic, cryogenic, moderately continental, |
| Dunay Isl. | 5 | 0 | 2.6 | - | excessively humid |
| North Yakutia, subarctic tundra | | | | | |
| Ust-Olenyok | 11 | 0 | 1.0 | 176 | Subarctic, long periods of frost, moderately to strongly continental, |
| Kyusyur | 35 | 576 | 0.8 | 220 | semi-humid |
| North Yakutia, northern taiga | | | | | |
| Olenyok | 127 | 846 | 0.6 | 246 | Cold, cryogenic, very to |
| Jarjan | 50 | 950 | 0.7 | 238 | strongly continental, |
| Zhigansk | 53 | 1090 | 0.6 | 256 | semi-arid |
| Central Yakutia, middle taiga | | | | | |
| Viluysk | 110 | 1421 | 0.4 | 251 | Moderately cold, long |
| Yakutsk | 99 | 1565 | 0.3 | 302 | periods of frost, strongly to |
| Ust-Maya | 169 | 1457 | 0.4 | 305 | extremely continental, very |
| Isit | 117 | 1503 | 0.4 | 268 | arid |
| South Yakutia, middle and upper taiga | | | | | |
| Tommot | 283 | 1370 | 1.0 | 272 | Moderately cold to cold, |
| Aldan | 676 | 1281 | 1.3 | 231 | frosty, very to strongly |
| Emeldzhak | 951 | 1062 | 1.9 | 224 | continental, humid to |
| Chulman | 671 | 1158 | 1.3 | 279 | excessively humid |

*$\Sigma$ t>10°C sum of air temperatures above 10°C;, $C_h$ humidity, $C_c$ continentality

Climatic shifts with respect to geographic zonation (i.e. latitudinal and altitudinal aspects) also should be given consideration (Table 1.2). We analyze this here based on integral climatic indices (sum of temperatures above 10°C, precipitation-evaporation ratio ($C_h$), continentality ($C_c$)). Low thermal resources in the Arctic tundra cause low evaporation of precipitation, and thus result in high humidity under the conditions of moderate continentality as observed along the coastline of the northern seas.

In the subarctic tundra an increase in the thermal factor reduces the humidity ($C_h$ = 0.8–1.0) while the continentality hardly rises. In the northern taiga, farther inland, the thermal factor increases ($\Sigma$ t > 10°C = 846–1090°). This means that the continentality increases ($C_c$ = 238–256), and the humidity slightly decreases ($C_h$ = 0.6–0.7). In the middle taiga of Central Yakutia the highest values for the thermal factor are recorded, with the sum of temperatures reaching 1421–1565° while continentality ($C_c$ = 302–305) and aridity are severest ($C_h$ = 0.3–0.4).

Towards the South of Yakutia, mountainous relief makes the climatic indices more complex. There, at higher altitudes (compared to Central Yakutia) in mountain taiga, the thermal factor decreases somewhat ($\Sigma$ t > 10°C = 1062–1370°), which cause the humidity to rise significantly ($C_h$ = 1.0–1.9) while the continentality decreases. It is important to note that depressions and hollows have a profound effect on the abovementioned climatic indices, yielding the well-known phenomenon of thermoaridity in the bottom lands in contrast to surrounding watershed area. Thus, in the Chulman depression that runs into the Aldan upland as far as 500 m, a reduction of 100 m in altitude is accompanied by an increase in continentality (+18) and in sum of temperatures (+43°), and a decrease in annual precipitation (–32 mm).

## 1.4 Permafrost

Permafrost is a basic factor greatly influencing the distribution and functioning of ecosystems, and vegetation in particular (Nekrasov 1984).

The whole territory of Yakutia lies in the zone of perennially frozen soils. The permafrost stratum varies widely in depth within Yakutia. Thus, on above-floodplain terraces, it is as thick as 300–400 m, while in the upper reaches of the Markha River (West Yakutia, south of the Polar Circle) it reaches 1500 m. This is the maximum thickness recorded worldwide. The average annual temperature of perennially frozen grounds at a depth of 15–20 m fluctuates from –1° to –2°C (SW Yakutia) to –10° to –12°C (high watersheds in mountains) (Mostakhov et al. 1980).

The continuous permafrost zone alternates with patches of non-frozen grounds, the so-called *talik*. In Southern Yakutia (the Lena-Aldan and the near-Lena plateaus, as well as upper regions of the Yakokit and Seligar River basins), taliks occupy up to 50% of the area and are concentrated on dry, flat watersheds, under river beds which do not freeze through in winter, under deep lakes, and near permanent sources of underground water. Absence of permafrost is explained from the specific composition of carbonate rocks, their waterproof properties and extensive karst processes, as well as by good infiltration of precipitation, because the average annual temperature of those rocks is as high as 1–2°C due to convective heat transfer by infiltrating precipitation (Petrova 1971).

Firmly frozen, ice-bound strata represent waterproof horizons for precipitation and facilitate to some extent soil wetting by oozing out ground moisture during seasonal periods of thawing (Mozolevskaya 1973).

Ice, as a constituent part of the perennially frozen grounds, determines their heat-transfer and mechanical properties. Ice is a unique solid substance being the lightest rock-forming mineral with the lowest melting temperature. Due to these characteristics, ice is located in the near-surface layers of the earth's crust. Ground, constitutional ice is most widespread. It includes the following types: ice-cement, segregation (migration) ice, and injection (intrusive) ice.

*Ice-cement* is formed from water freezing in between particles of fine-grained rocks without affecting the spatial position of those particles. Due to very small ice crystals, invisible to the naked eye, the frozen rock represents a monolithic, firmly cemented mass.

*Segregation ice* is formed in moist, loamy, micrograined soils under condition of water inflow (migration) to the "freezing front". Moist soil freezing is always accompanied by water inflow from the deeper horizons resulting in the development of ice lenses.

*Injection ice*, like segregation ice, is also formed from underground water, but of larger volumes and under significant pressure. Injection ice represents extensive layers and lenses several tens of meters thick and several hundreds of meters long, very often yielding large frost boils on the surface called *bulgunnyakh* in Siberia and *pingo* in Alaska and Canada.

*Fissure ice (wedge or vein ice)* is the result of water freezing in cracks of rocks as well as in holes and cavities common in carbonate landscapes. *Recurrent-fissure ice* is a very common phenomenon in the lowlands of Northern and Central Yakutia. Its formation process is typical of land surfaces cracking during severely cold weather. Such cracks are called frost fractures, and the whole micro- or nanorelief of the surface is called polygonally-fractured. The ice veins have a wedge, columnar or irregular shape in cross-section (Fig. 1.5).

Various natural processes as well as human activity (such as fires, deforestation, agriculture, use of heavy machinery) induce thermokarst processes. This phenomenon of permafrost thawing leads to land surface destruction resulting in the formation of unique landscapes. The following are the most distinct types of such landscapes.

*Bulgunnyakh (Pingo)* is a domelike hill with an ice core (hydrolaccolith). They occur one by one in depressions of flat tundra, forest-tundra and *alases* (see below) of Central Yakutia. They vary in height from 1 to 30–40 m and continue to grow for up to 1000 years. Landscapes with pingos are sometimes considered as a special type of hillocky tundra.

*Baidjarakh complex (earth mounds, "burial mounds")* represents muddy-peaty, sometimes muddy-loamy or sandy cones of 3–4 m (up to 12 m) high, 3–15 m wide and up to 20 m long (Fig. 1.6). The frost boil landscape is the result of thawing and freezing of recurrent-fissure ice which forms a grid with polygonal cells. The young

**Fig. 1.5** Exposed recurrent-fissure ice veins on eroded slopes of a river bank in the northern taiga zone

**Fig. 1.6** Baidjarakh complex on the slopes of a mature alas

baidjarakhs have a regular, clear shape of an up-side-down bowl, while the old ones are eroded by water and wind, and are covered with vegetation. At the same time, they retain a round or quadrangular shape at their bases. This phenomenon is very common in Central and North Yakutia on above-floodplain terraces, lakes or coastal depressions, and they are grouped in parallel, staggered rows. In the tundra zone the baidjarakhs may cover large areas (up to thousands of hectares).

*Alas* represents a shallow depression, formed by thermokarst, with meadow or steppe vegetation, in the taiga in Central Yakutia (Fig. 1.6). The process of alas formation runs through several stages. Solar radiation induces thawing of buried ice in open spots in the taiga and this leads to the formation of a baidjarakh complex. Gradually, the soil that continues to thaw forms a depression, in which melt water forms a pond or a lake. The lake slowly dries out, giving space to herbaceous vegetation. Later, a new ice lens may form underground, and a pingo may appear in the middle of the alas. There is an important difference between alases and other thermokarst depressions met in Northern Yakutia. The lakes and lake depressions in the lowlands of North-Eastern Siberia develop under conditions of excessive moisture. These lakes are full of water and can be drained only artificially. In Central Yakutia, due to the arid climate, an alas lake desiccates after the resources of underground ice have been depleted, and allows vegetation to grow. (We shall discuss the alas vegetation complex in Chapter 3). Thus, the arid climate is the only factor which leads to the formation of these peculiar landscapes. The particular process of alas evolution, and its distinctive soil and vegetation complex make it worthwhile to treat them as a special alas-type of a landscape in Central Yakutia (Bosikov 1993). Alases represent high-quality hayfields and pastures for livestock and horses.

*Naled (*Icing*)* is very common in Eastern Siberia and the Far East. It develops under the conditions of a severe climate, when deeply frozen ground and ice cover on rivers form a kind of pipe or tunnel for river water. In some places this water is forced to the surface and immediately becomes frozen. This is called a river naled.

Sometimes, this phenomenon occurs far from river valleys caused by underground waters oozing and freezing on the surface (ground naled).

In upland regions thawing of permafrost induces rock slides and solifluction processes on gentle slopes, when thawed, wet soils slide on the perennially frozen substrate.

The perennially frozen grounds are overlain by a soil stratum which thaws in summer (the active layer). In most of Central and North-Western Yakutia, the depth of seasonal thawing varies from dozens of cm to 2–3 m depending on the geographical location, relief, vegetation pattern, soil composition, etc. In Central Yakutia and the western near-Lena regions, soils thaw as deep as 1–4 m. Such values are characteristic for forested watershed areas, whereas waterlogged territories show less deep thawing. In Northern Yakutia (tundra and forest-tundra zones), the depth of the thawing layer is 0.3–0.8 m. In the south and south-west regions with higher amounts of precipitation, soils thaw as deep as 2.5–4.0 m (Mozolevskaya 1973).

Soils start thawing right after the snow cover has melted, and they freeze in autumn as soon as the average daily air temperature comes below 0°C. The freezing process proceeds simultaneously in a top-down and a bottom-top direction, and in January both frozen layers meet. Thawing depth values are not constant and depend on the meteorological features of a year.

Based on the thickness of the seasonally thawed soils, Nekrasov (1984) divides the territory of Yakutia into 5 zones (Table 1.3 and Fig. 1.7) with thaw-layer thickness of sandy loams within 0.5–3.0 m, and of loams within 0.3–1.8 m. The scheme clearly illustrates that the soils of the arctic and subarctic lowland tundra, as well as those of highland areas, i.e. chains of mountain ranges in North-East Yakutia (Verkhoyansk, Chersky, Momsky), thaw to relatively shallow depths (sands 0.5–1.5 m, sandy loams 0.3–1.0 m), while the thawing depths of soils of the strongly thermo-arid Central Yakutian Lowland are much deeper (sands 2.2–3.0 m, sandy loams 1.4–1.8 m).

**Table 1.3** Thawing depths in Yakutia (in m) (see text for explanation)

| | Zone | I | II | III | IV | V |
|---|---|---|---|---|---|---|
| Depth (m) | Sandy | 0.5–1.0 | 1.0–1.5 | 1.5–2.0 | 1.9–2.5 | 2.2–3.0 |
| | Loamy | 0.3–0.7 | 0.7–1.0 | 1.0–1.3 | 1.2–1.6 | 1.4–1.8 |

## 1.5 Soils

Due to the huge area of Yakutia and the greatly different conditions of soil formation, such as landscape, climatic, lithological and geochemical peculiarities, the soil cover is very diverse and comprises a wide range of zonal, azonal, and intrazonal soil types.

Geographically, most of Yakutia (38%) belongs to the middle-taiga subzone of the taiga-cryolithozone region of the boreal belt of Eastern Siberia. 31.5% falls

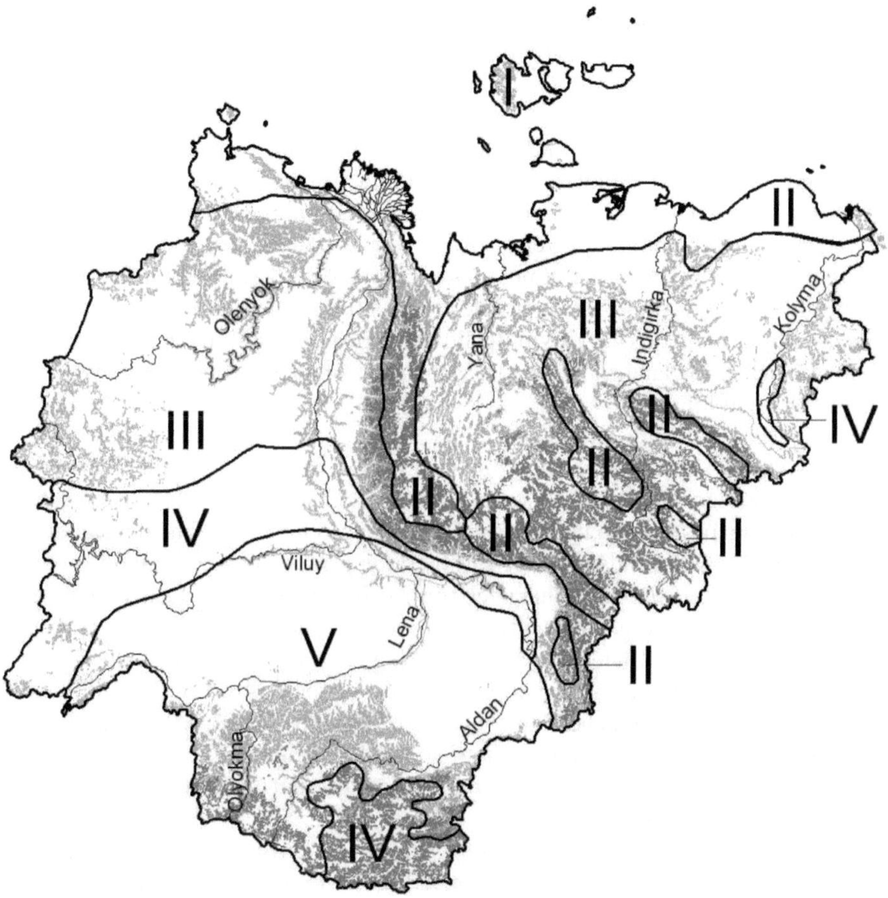


**Fig. 1.7** Thickness of seasonally thawed soil zones in Yakutia

under the northern taiga and 30.5% is tundra. Thus, soils of the tundra and taiga landscapes form the basis of the soil cover structure (SCS) of Yakutia. And within those categories highland landscapes in the tundra and taiga prevail, accounting for 18.3 and 22.4%, respectively (Elovskaya et al. 1979).

## *1.5.1 Tundra Soils*

*Frozen tundra gleyey soils* are formed in watershed landscapes of spotty and small hillock tundra common in the south of the arctic tundra and in the north of the subarctic tundra. These automorphous soils are characteristic for relatively well-drained sites with strongly pronounced, scattered hillock microrelief induced by frost fracturing. The fractured surface consists of flat, fairly well recognizable polygons as large as 70–100 cm in diameter and 15–20 cm in height. The polygon centers are characterized by a patch of scanty vegetation or bare ground. In watershed landscapes of the arctic tundra, these soils, bare or carrying scanty vegetation, make up 50–60%, and in those of the subarctic tundra 15–30% (Elovskaya et al. 1979).

*Frozen tundra humus-gley and humus-gleyey soils* are formed under the same conditions as the previous ones, except for the fact that they develop on polygons covered with vegetation. Such soils develop on various rock types. At a depth of 60–70 cm the whole mineral layer of frozen tundra humus-gley and humus-gleyey soils is underlain by layers of almost pure ice of over 10 cm thick. Most of these loamy soils are of medium-sized particle composition. Frozen tundra humus-gleyey soils develop also on detrital eluvio-dealluvial deposits of prominent surfaces of the Olenyok-Anabar lowland and form just a small part of the microrelief with vegetated patches or small hillocks.

*Frozen tundra humus-peaty-gley soils* are formed in the Primorskaya and Olenyok-Anabar Lowlands occupying low ridges, remnants of the Pleistocene slopes, and polygonal ridges of swamps. They occur together with frozen peatbog soils of earth stripes (the peculiar, permafrost induced landform) and develop under dwarf shrub, moss-lichen, and dwarf shrub-cotton grass tundra communities.

*Frozen tundra podbur* are characteristic for loose, sandy or loamy-sand sediments of the plain tundra, especially in the western part of the Lena River's delta in the Primorskaya and Olenyok-Anabar Lowlands. *Frozen tundra podbur* have a depth of 80–90 cm and cause the horizons above the permafrost to be periodically oversaturated with water and create gleyization. The topsoils are acid or subacid but lower down they gradually change from subacid to neutral. The top layer is also characterized by an accumulation of organic matter, but lower down the organic content sharply decreases.

*Salt-marsh soils* develop in annually flooded areas on saline silty sea sediments which are either bare or carry maritime meadows. The soil profile consists of alternating grayish-brown layers of loamy sand and silt which downwards become wetter and more gleyified. This soil type is subalkaline. The soil-absorbing complex is saturated with bases. The maximal content of organic matter occurs at a depth of 40–60 cm and the highly soluble salt content is 1–2%.

At increasing elevation and diminishing flooding by sea water *marshy soddy gley alkali soils* are formed They characteristically are medium- to subacid, contain insignificant amounts of highly soluble salts, and are of the sulfate-chloride salinization type. On the best drained sites of marshes far from the sea tidal area *marsh soddy gleysols* develop under a cover of vegetation. During summer, such soils thaw as deep as 50–60 cm. They can be either loamy-sand or loamy.

## *1.5.2 Northern taiga Soils*

The northern taiga of Yakutia significantly differs from the tundra (to the North) and middle taiga (to the South) zones by climatic parameters such as humidity, the range of temperature fluctuation, the depth at which permafrost occurs. This results in the development of peculiar soils and vegetation covers.

*Frozen northern taiga typical gleysols* are common for ridges and upper slopes of northern taiga depressions, mainly occurring in the Kolyma Lowland, with a distinct frozen bumpy-fractured microrelief where the bumps and fractures have their own soil profile. Such soils are formed under poor, sparse larch forests. The particle composition is a light or medium loam. Humus horizons are rough, with a high content of decomposed organic matter, fulvoacid, that actively impregnated the entire profile and often accumulates on top of the perennially frozen soil.

*Frozen northern taiga gleyey soils* are common for both banks of the Lena River in the lower parts of the Lower-Lena Lowland. They feature a well-developed mesorelief of ridges. The profiles of such soils are weakly differentiated into genetic horizons, but can be easily divided into two layers by particle composition: an upper (50–60 cm) loamy layer and a lower sandy layer. The upper layer is not saturated with bases. Soils of this type are gleyey and weakly thixotropic.

*Frozen northern taiga humus gleysols* are widely distributed along valleys of large tributaries of the Lena, Indigirka and Kolyma rivers, as well as on lower parts of gentle slopes of ridges and on low ridges. They develop under swamped, sparse larch forests and thus are wet and gleyified. Soils of this type are characterized by a subacid or neutral reaction, a high content of organic matter in the upper organogenic horizons, a relatively high content of humus (no less then 3–5%) in the lower part of the profile, and a medium to heavy loam particle composition.

*Frozen northern taiga podzolized soils* are common for valleys of large rivers, such as the Indigirka, Kolyma, Lena, and occur on emergent parts of the mesorelief (ridges and their gentle slopes). The profiles of such soils can be rather well differentiated into genetic horizons. They develop under influence of both podzolization and gleyization processes. Though at the same time, such soils experience cryoturbation. Soils of this type are characterized by a subacid reaction, different amounts of clay, silt, and total oxides, a medium to light loam particle composition, though sandy varieties also occur (Elovskaya et al. 1979).

*Frozen northern taiga podzolic soils* occupy most of the Lena-Linde interfluve (the Lena basin in the Polar Circle area) situated on the lower and middle levels of a plain with absolute heights of 100–200 m. They are formed on sandy/loamy sandy alluvium under pine forests with or without a lichen cover and herbs. The profile is clear. Soil effervescence is not characteristic for this soil type. The lower soil horizons are gleyey. Frozen northern taiga podzolic soils have an acid reaction, an insignificant absorbing capacity (up to 1.0–1.5 mg-eq), an unsaturated soil absorbing complex, and weak biogenic accumulation of ashy elements.

*Frozen podzolic gleysols* are common in the Lower-Lena and Indigirka Lowlands occupying the upper and middle thirds of ridge slopes. In the landscape, they are situated at lower elevations than northern taiga podzolic soils with which they can form a mosaic. This soil type is characterized by a combination of gleying and podzolizing processes which vary quantitatively depending on particle composition of the frozen podzolic-gley soils. The profile is clearly differentiated into genetic horizons. Permafrost occurs at a depth of 50–70 cm and forms a waterproof layer that impedes intra-soil drainage. Gleying occurs nearly in all horizons with its maximum in the lower part of the profile just above the permafrost.

### 1.5.3 *Soils of the Middle taiga Subzone of Central and Southern Yakutia*

In the Central Yakutian Lowland three soil zones can be distinguished, each with its typical altitudinal level. They are: (1) the steppe zone of frozen chernozems (100–140 m above sea level); (2) the taiga-alas zone of frozen pale solodized and frozen chernozem-meadow soils (140 to 250–270 m a.s.l.); (3) the frozen-taiga, soddy-carbonate typical and podzolized soils of plains and plateaus at 300–550 m a.s.l. (Elovskaya and Konorovsky 1978).

*Frozen chernozems* are formed in the Middle Lena valley, in the most cryo-arid part of the Central Yakutian depression on prominent relief elements (rims, ridges) of young terraces of large rivers above the floodplain, that carry steppe vegetation. The profile of these soils contains humus, carbonate and, sometimes, gypsum horizons. In summer time, frozen chernozems under vegetation cover thaw as deep as 160–170 cm. Being zonal automorphic soils, frozen chernozems are formed in complex with meadow-chernozem and chernozem-meadow soils (see below) as well as with accompanying saline soils (solonetz and solonchak) which develop in depressions of nano- and microrelief.

*Frozen meadow-chernozem soils* develop under meadow-steppe vegetation occupying gentle slopes of terraces above the floodplain. These semi-hydromorphic soils get additional moisture from waters flowing down from the upper parts of the landscape. But the arid climate and the slopes impede the accumulation of water over the permafrost .

*Frozen chernozem-meadow soils* occur in depressions of terraces above the floodplain and in taiga alases under conditions of sufficient moisture supply or even of temporarily excessive moisture under rich meadow forb vegetation. The surface is always covered with polygonal frost fissures. In summer, they thaw to 1.3–1.6 m.

Frozen saline soils are common on terraces above floodplain of the Lena, Viluy, and Amga Rivers within the Central Yakutian Lowland, as well as on alases of Central Yakutia. They are divided into two groups: (1) frozen solonchaks and (2) frozen solonetz. *Frozen solonetz* form complexes with solonetzic chernozems, meadow-chernozem, and chernozem-meadow soils. Frozen solonchak occur in depressions as patches among chernozem-meadow soils occupying small areas (Elovskaya and Konorovsky 1978).

*Frozen pale soils* are common under forest vegetation in the Lena-Viluy and Lena-Amga interfluves of the ancient alluvial plain with varying morphological stages within altitudinal range of 140–270 m. Compared to frozen taiga soils, they are formed under drier conditions under xeric dwarf shrub-herb *Larix* forests, mainly with *Vaccinium vitis-idaea*. In meso-depressions (alas, *charan* (stepped *Betula* forests) landscapes) they alternate with halophytic communities on chernozem-meadow and meadow-chernozem saline and non-saline soils. The peculiar feature of this soil type is the presence of a diagnostic illuvial-carbonate horizon in the profile. *Frozen pale typical* loamy soils usually develop under *Vaccinium vitis-idaea*+herb *Larix* forest of higher growth class and are confined to upper levels of the ancient alluvial plain and north-facing slopes. *Frozen pale solodized soils*, on the contrary, develop at the middle and lower levels (100–170 m) of the plain, and have a solodized horizon A2 in their profile. *Pale grey soils* are formed under sparse *Larix* and *Larix+Betula* forests of high growth class with a herb-dwarf shrub ground layer. *Frozen pale leached soils* occur at high altitudes (250–360 m) in the alluvial depressions of South Yakutia under *Larix* forests. This type represents a transition from pale to frozen taiga soils and is characterized by shallower effervescence (Elovskaya 1987).

*Frozen taiga typical soils* are formed on denudation plains within 300 to 550–600 m altitude on eluvium and eluvio-deluvium of non-carbonate bedrocks under *Larix* and *Pinus sylvestris+Larix* forests. *Frozen taiga podzolized soils* develop under the same conditions except that the bedrock is of lighter granulometric composition (loam sand, sometimes light loam). Rarely they develop on ancient sandy alluvium. The light granulometric composition of melkozem of this soil type results in the absence of permafrost -induced nano-relief. In summer they thaw to a depth of 1.5–2.0 m.

*Frozen soddy-carbonate typical soils* are azonal as, under the conditions of the middle taiga of Central and South Yakutia, they substitute for zonal frozen soils within the Near-Lena and Lena-Aldan plateau surfaces, developing on eluvio-deluvium of the following carbonate bedrocks: Cambrian, Devonian, Silurian limestones and dolomites. The peculiar morphological and physicochemical features of these soils are brown tints in the profile, shallowness and high content of stones and detritus, heavy (heavy loamy-clayey) granulometric composition of melkozem, the neutral-alkalescent reaction of the environment and perfect conditions for forest growth. *Frozen soddy-carbonate podzolized soils* develop on the upper part of gentle slopes of flat watersheds. Compared to typical soils, the forest vegetation on this soil type is characterized by more vigorous growth of *Pinus sylvestris*. Besides, there is a well pronounced podzolized horizon A1A2 in the profile. *Frozen soddy-carbonate leached soils* occur within the carbonate plateaus of South Yakutia on watershed and slope surfaces of so called "table mountains" under mixed *Larix+Picea+Pinus sylvestris* forests of high growth class.

*Frozen soddy carbonate undeveloped soils* occur on steep slopes bordering river valleys. They are shallow and characterized by a shortened soil profile, a high content of detritus and with bedrock close to the surface (Elovskaya 1987).

*Podzolized alfa-humus soils* are formed within the Olyokma-Chara and Aldan plateaus of Southern Yakutia on permafrost -free, thin, eluvio-dealluvial, granite-gneisses which are relatively rich in primary minerals, with free intrasoil drainage. In contrast, podzol formation together with humus accumulation form typical *podzolized soils,* occurring mainly on acid siallitic residual soil material (Chevychelov 2000).

### 1.5.4 Soils of the Mountainous Provinces of Northern, North-Eastern and Southern Yakutia

The soil-vegetation cover of the Yakutian highland provinces, as in every mountainous region, is characterized by vertical zonation. At the same time, the altitudinal positions of these zones and subzones (or belts) are strongly determined by the latitudes of the specific mountain systems. For example, the upper forest border reaches 200–300 m in the northern taiga, 500–800 m in the middle subzone of the northern taiga, 900–1100 m in the middle taiga, and 1350–1750 m in the highest ranges of South Yakutia.

*Barren mountain soils* of fell-fields (bare rock) belt of mountainous systems and ranges of Yakutia do not occur as solid covers but as patches in the mountain tundra zone. *Frozen tundra detrital podburs* develop in the mountain tundra belt under conditions of proper drainage on stony shallow soil eluvium of various dense bedrocks. There are two subtypes of frozen tundra detrital podburs – typical and podzolized. The latter is characterized by a different, podzolized soil profile.

*Mountain tundra gley frozen soils* also develop in the mountain tundra zone occupying flat landscapes with poor drainage and they contain a bit of a detrital layer (flat plateau terraces, table-lands). Humus-gley soils contain both organogenic horizons and an over-permafrost gley layer. The profile is mostly dark-grey in colour with a bluish-grey gley component (Elovskaya et al. 1979). *Mountain tundra thixotropic soils* are found in the Kystyk highlands within the tundra zone of the Chekanovsky and Pronchischev Ranges.

*Frozen northern taiga podzolized gleyey detrital soils* are widely distributed in the northern taiga subzone on the Yukagir Plateau and in parts of the Anyuy Range occupying the upper thirds of slopes and the low flat watershed regions at altitudes of 200–450 m. They develop under the sparse larch forests of the northern taiga with a pronounced dwarf shrub-lichen cover. *Frozen northern taiga typical (thixotropic) detrital soils* are formed on the elevated plains and tablelands of Northern Yakutia at altitudes of 200–500 m. Their distinctive feature is the presence of a thick (50 cm and more) thixotropic horizon in their profile at a depth of 30–40 cm. *Frozen northern taiga disturbed carbonate soils* are observed within the Olenyok-Anabar Plateau at altitudes of 220–500 m. These soils are formed under sparse, low-growing *Larix-Picea obovata* forests on eluvio-dealluvial dolomites and lime-stones and are associated with a distinct, frozen, polygonally-fractured nanorelief.

In contrast with frozen tundra detrital podburs, *frozen taiga detrital podburs* are characterized by a thicker profile of shallow soils (70–80 cm) with thick organogenic horizons (up to 10 cm) and a loamy particle composition.

*Frozen chernozems and frozen chestnut detrital soils* are formed in depressions of upland regions of North-East Yakutia in the mountain taiga belt. They occupy southern slopes with steppe vegetation (grass-forb, sedge, wormwood and other steppe communities). In North-East as well as in Central Yakutia, steppe landscapes with well developed steppe soils in the taiga zone were formed under the specific microclimatic and physiographical conditions of depressions.

*Mountain meadow soils* have a discontinuous distribution below the barren mountain zone in ranges in Southern Yakutia. They are a typical part of a vertical soil zonation of a cyclonic (oceanic) type. These soils are formed under subalpine forb-grass meadows under excessive, non-stagnant moist to wet conditions: lakesides, banks of mountain streams, bottoms of temporary streams, steep slopes under cliffs (Chevychelov and Volotovsky 2001). *Typical burozems,* as a rule, develop in the upper parts of the mountain taiga belt in the south-east of South Yakutia. They occur on steep, leeward slopes under *Betula ermani ssp. lanatai* forests. This soil type is characterized by a well pronounced humus profile and high humus content throughout the mineral layer. Like typical burozems, *illuvial-humic burozems* are also formed in the upper part of the mountain taiga belt though under *Picea ajanensis* forests. From an ecological point of view, they represent a logical transition from burozems to zonal podzols. It is evident that the burozems are typical for a specific height zone in the vertical soil gradient in the upper part of the mountain taiga belt, and they intergrade into the permafrost area (Chevychelov 2000).

### *1.5.5 Intrazonal Floodplain and Bog Soils*

Alluvial soils are formed in floodplains and deltas of the Yakutian rivers. They have a foliated structure. Sometimes they contain residual humus from previous soil formation in the lower parts of their profile in a horizon above the permafrost or in an underlying permafrost layer. All alluvial soils are flooded at spring tides. Soils of the lower floodplains are flooded annually. Soils of the middle floodplains are covered with water every second or third year. Soils of the upper floodplains are flooded every 10–15 years. Besides, the lower floodplain soils are exposed to summer floods as well, resulting in a wide range of saturation levels of soils and underlying ground material. Alluvial landscapes and soils are considered as azonal-intrazonal formations with distinctive features determined by climate and geographical position (Elovskaya 1987).

*Frozen alluvial soddy soils* are formed under mesic and stepped meadows flooded irregularly. The characteristic feature of alluvial soils of the Lena River is the absolute absence of buried humus horizons, while the humus horizon at the surface has a blurred upper border. *Frozen alluvial soddy gley soils* develop on the lower thirds of slopes, on low gentle ridges, and in depressions within the middle and upper floodplains under mesic meadows or herb-rich forests. They are the wettest subtype of alluvial soddy soils. Gleization takes place right under the humus accumulating horizon. *Frozen alluvial soddy humus soils* are formed under floodplain *Larix* forests and mixed *Picea-Betula-Larix* forests. Their distinctive feature is the presence of forest litter and a coarse humus horizon in the soil profile.

*Frozen alluvial chernozem-like soils* occupy the most prominent parts of upper floodplains above typical soddy soils and are restricted to the Yakut steppe soil province (Elovskaya 1987).

*Frozen alluvial peat-gley soils* are common in the middle taiga subzone in excessively wet floodplain depressions under *Carex* and *Carex+Calamagrostis* tussock meadows, as well as under other communities of bog vegetation. These soils are formed under the influence of prolonged stagnation of tidal waters and additional wettening by melt waters flowing down from the upper parts of the landscape.

*Frozen alluvial muddy-peaty-gley soils* are also formed in the middle taiga subzone, but under sedge-grass bogs in the lower floodplain depressions of Central Yakutia, where the level of decomposition and humification is much higher than that of the moss bogs of North and South Yakutia. *Frozen alluvial peat-gley soils* differ from the abovementioned types by a thick peat horizon (20–50 cm) with layers of decomposed organic matter and a shallower seasonally thawed layer and they show stronger gleization. *Frozen alluvial muddy-peat soils* have an even thicker peat horizon (over 50 cm) reaching the permafrost. *Frozen alluvial solonchaks* are formed in small depressions in mesic floodplain meadows on alluvial soddy soils (Elovskaya 1987).

There are two types of bog soils in Yakutia: peat lowland and peat upland soils. *Frozen peat upland soils* are widely distributed especially in South and North Yakutia, being formed in the river and stream valleys surrounded by slopes of non-carbonate rocks. Their formation is favoured by additional drainage of melt and intrasoil water from watersheds, as well as by a colder microclimate and a perennially frozen, icy, waterproof layer. *Frozen peat lowland soils* occur in polygonal-bog landscapes in the tundra zone, under moss-dwarf shrub mires in the northern taiga subzone, and in alas landscapes, taiga river valleys, and around ponds and lakes with *Carex* and grass species in the middle taiga subzone. Unlike peat upland soils, peat lowland soils are characterized by highly humified peat, high ash and exchange base contents, as well as a lower actual and exchange acidity.

## References


Bosikov N P (1993) Evolyutsiya ponyatiya "alas" v merzlotovedenii. In: Voprosy geographii Yakutii. Izd-vo Yakutskogo nauchnogo tsentra, Yakutsk – *Evolution of the term "alas" in permafrost science* (in Russian)

Chevychelov AP (2000) Pochvy tayezhnogo poyasa Tokinskoy kotloviny i khrebta Tokinsky Stanovik. Pochvovedeniye 10: 1187–1196 – *The soils of the taiga belt of the Tokinskaya depression and Tokinsky Stanovik Range* (in Russian)

Chevychelov AP, Volotovsky KA (2001) Pochvy goltsovogo i podgoltsovogo poyasov khrebta Tokinsky Stanovik. Pochvovedeniye 7: 791–797 – *The soils of goltsy and sub-goltsy belts of the Tokinsky Stanovik Range* (in Russian)

Elovskaya LG (1987) Klassifikatsiya i diagnostika merzlotnykh pochv Yakutii. YaF SO AN SSSR, Yakutsk – *Classification and diagnostics of frozen soils of Yakutia* (in Russian)

Elovskaya LG, Konorovsky AK (1978) Rayonirovanie i melioratsiya merzlotnykh pochv Yakutii. Nauka, Novosibirsk – *Regionalization and melioration of frozen soils of Yakutia* (in Russian)

Elovskaya LG, Petrova EI, Teterina LV (1979) Pochvy severnoy Yakutii. Nauka, Novosibirsk – *The soils of Northern Yakutia* (in Russian)

Izyumenko SA (ed) (1966) Spravochnik po klimatu SSSR. Yakutskaya ASSR, Issue 24(2): Temperatura vozdukha i pochvy. Gidrometeoizdat, Leningrad – *Reference book on climate of the USSR. The Yakutian ASSR, Issue 24(2): Air and soil temperature* (in Russian)

Izyumenko SA (ed) (1968a) Spravochnik po klimatu SSSR. Yakutskaya ASSR, Issue 24(4): Vlazhnost vozdukha, atmosphernye osadki, snezhny pokrov. Gidrometeoizdat, Leningrad –

*Reference book on climate of the USSR. The Yakutian ASSR, Issue 24(4): Air humidity, atmospheric precipitations, snow cover* (in Russian)

Izyumenko SA (ed) (1968b) Spravochnik po klimatu SSSR. Yakutskaya ASSR, Issue 24(5): Oblachnost i atmosphernye yavleniya. Gidrometeoizdat, Leningrad – *Reference book on climate of the USSR. The Yakutian ASSR, Issue 24(5): Nebulosity and atmospheric phenomena* (in Russian)

Korzhuev SS (1965) Skhema prirodnogo rayonirovania. In: Gerasimov IP (ed) Yakutia. Nauka, Moscow – *The Scheme of natural zonation* (in Russian)

Korzhuev SS (1974) Morfotektonika i relief zemnoy poverkhnosti (na primere proiskhozhdeniya i vozrasta reliefa Vostochnoy Sibiri). Nauka, Moscow – *Morfotectonics and relief of the Earth's surface (case study of origination and age of the relief of Eastern Siberia)* (in Russian)

Matveev IA et al. (ed) (1989) Atlas selskogo khozyaistva Yakutskoy ASSR. GUGK, Moscow – *Atlas of agriculture of the Yakutian ASSR* (in Russian)

Mostakhov SE, Nekrasov IA, Dmitrieva ZM, Kalmykova AI (1980) Yakutskaya ASSR. Yakutskoye knizhnoe izd-vo, Yakutsk – *The Yakutian ASSR* (in Russian)

Mozolevskaya AK (ed) (1973) Agroclimaticheskie resursy Yakutskoy ASSR. Gidrometeoizdat, Leningrad – *Agroclimatic resources of the Yakutian ASSR* (in Russian)

Nekrasov IA (1984) Vechnaya merzlota Yakutii. Yakutskoye knizhnoe izd-vo, Yakutsk – *Permafrost of Yakutia* (in Russian)

Petrova EI (1971) Pochvy Yuzhnoy Yakutii. Yakutskoye knizhnoe izd-vo, Yakutsk – *The soils of Southern Yakutia* (in Russian)

Walter H, Lieth H (1960) Klimadiagramm Weltatlas. Fischer, Jena.

# Chapter 2
# Flora of Yakutia: Composition and Ecological Structure

**L.V. Kuznetsova, V.I. Zakharova, N.K. Sosina, E.G. Nikolin, E.I. Ivanova, E.V. Sofronova, L.N. Poryadina, L.G. Mikhalyova, I.I. Vasilyeva, P.A. Remigailo, V.A. Gabyshev, A.P. Ivanova, and L.I. Kopyrina**

## 2.1 Floristic Regionalization

**L.V. Kuznetsova**

A floristic regionalization of Yakutia was first made by Karavaev (1958). Presently, a more solid floristic knowledge has been gathered which allows a more accurate delineation of some floristic regions. This formed the basis of the specified and supplemented floristic regionalization (Fig. 2.1) offered by the authors in 2005 (Danilova 2005).

*The Arctic Floristic Region (Arct).* The tundra zone of Yakutia covers 394.3 thousand km$^2$, i.e. about 13% of the territory of the republic. The southern border of the region lies along the right bank of the Kolyma River (as from the Mikhalkino settlement) crossing its valley 20 km to the South of the Pokhodsk settlement, then follows the Stadukhinskaya channel, and in the vicinity of the Andryushkino and Chokurdakh settlements crosses the Indigirka and Alazeya Rivers. At the latitude of the Tit-Ary Island it crosses the Lena River, then runs to the South of the Taimylyr settlement at the Olenyok River, and, finally, to the North of the Saskylakh settlement on the Anabar River. The region includes all the islands of the Arctic Ocean and the zone along the continental coastline of 120–150 km wide. The zone is widest at the Lower Alazeya, Indigirka, Yana and Anabar Rivers (up to 300 km), and narrowest (up to 50 km) at the Kolyma River mouth and near the Buor-Khaya Bay (Gakkel and Korotkevich 1962; Andreyev and Nakhabtseva 1974; Perfilyeva et al. 1991).

The basic climatic indices (air temperature, average precipitation, humidity, etc.) in the tundra zone change in a north–south direction and serve as a determining factor for discerning two subzones: arctic and hypoarctic (Perfilyeva et al. 1991).

In the northernmost islands of the Arctic Ocean the vegetation is represented mostly by arctic deserts and semi-deserts, significant areas of which being covered with rocks and gravel. Along the whole coastline of the Arctic floristic

L.V. Kuznetsova (✉)
Institute for biological problems of the cryolithozone, 41 Lenin Ave., Yakutsk, 677980, Russia
e-mail: kuznetsova90@mail.ru

E.I. Troeva et al. (eds.), *The Far North: Plant Biodiversity and Ecology of Yakutia*, Plant and Vegetation 3, DOI 10.1007/978-90-481-3774-9_2,

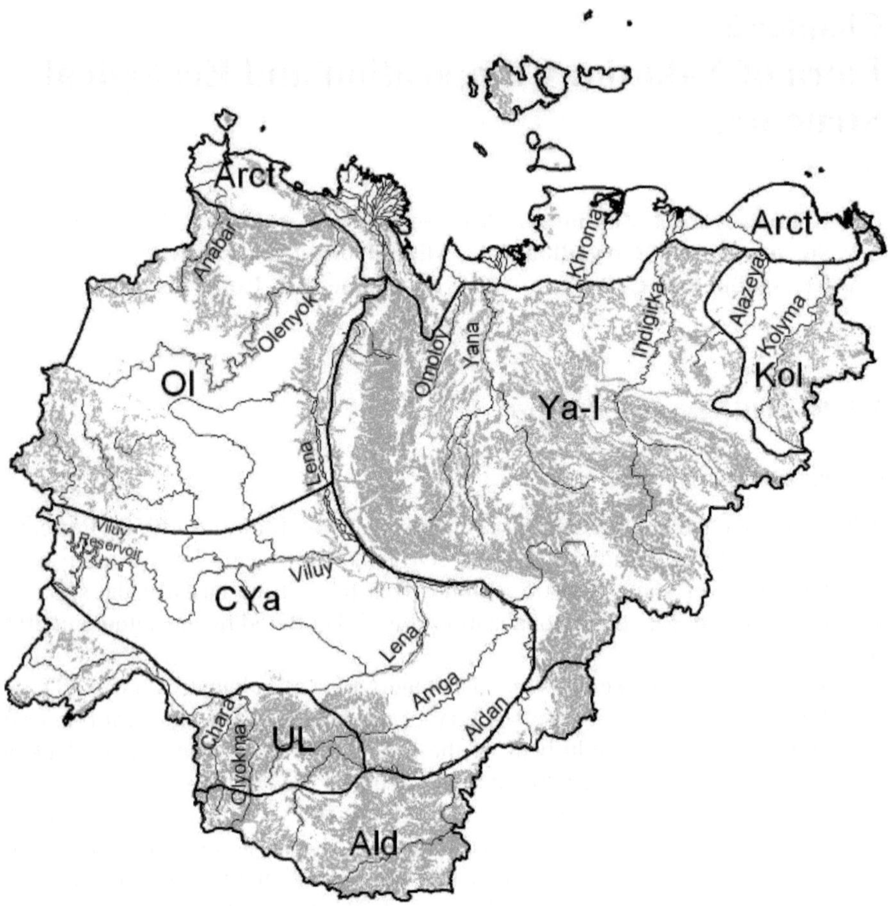


**Fig. 2.1** Florisitic regionalization of Yakutia

region coastal meadows with *Puccinellia phryganodes, Carex subspathacea* and *Calamagrostis deschampsioides* are found, grading into the typical arctic tundra towards the interior. Significant areas in lake depressions, river valleys and deltas are covered by polygonal-ridged tundra landscapes. The Yana-Indigirka Lowland features moss-tussock tundra and bogs. The mountainous part of the tundra zone is occupied by stony deserts characterized by the crustose lichens *Haematomma ventosum, Rhizocarpon geographicum, Umbilicaria hyperborea* and fragments of higher vascular plant groupings (Andreyev et al. 1987; Perfilyeva et al. 1991).

The southern tundra transgrades into forest-tundra, where up to N 72° larches grow as narrow stripes or separate groups along the river valleys (Tikhomirov and Shtepa 1956). The subarctic tundra is characterized by an increase in floristic diversity and a more complicated structure of the vegetation communities.

*The Olenyok Floristic Region (Ol)* occupies the northwestern part of the forest zone of Yakutia covering the Anadar and Olenyok Rivers' basins. The region is situated within the northeastern edge of the Middle Siberian Plateau including the Anabar-Olenyok Plateau and western outskirts of the Chekanovsky Ridge. The southern border of the region almost coincides with the northern border of the distribution areas of *Pinus sylvestris* and *Populus tremula.*

The flora and vegetation of the Olenyok FR is still subject to study. There are only a limited number of publications concerning the investigations of this territory (Komarov 1926; Sochava 1957; Zagrebina 1960; Buks 1963, 1966; Ivanova 1961; Lukicheva 1960, 1963a, b; Vodopyanova 1980; Andreyev et al. 1980; Petrovsky and Plieva 1994; Sosina and Zakharova 2004; Sosina and Sofronov 2004; Zakharova et al. 2004a).

The region is characterized by a continuous distribution of moss and moss-lichen sparse forests with *Larix gmelinii, Betula exilis, Vaccinium uliginosum, V. vitis-idaea* or *Ledum palustre* in the shrub and dwarf shrub layers. Which species predominates depends on the moisture conditions. Small belt communities of *Picea obovata* grow on floodplain edges of large river valleys, though this tree species is more common as a constituent part of larch forests forming the *Picea-Larix* communities. On the left bank of the Lena River, within the Floristic Region (FR), shrubberies of *Pinus pumila* occur on sandy, hilly landforms (Sosina and Zakharova 2004). *Salix* shrubberies are not so common in the region. Pure *Populus suaveolens* communities also occur rarely. More often they grow in a mixture with large *Salix* species *(S. viminalis, S. pseudopentandra, S. pyrolifolia* (Ivanova 1961; Lukicheva 1963a, b).

*Carex* (dwarf shrub-*Carex, Equisetum-Carex*) bogs are quite common among the larch sparse forests. Aquatic vegetation is poor and unvaried, mainly composed of *Nymphaea tetragona, Potamogeton perfoliatus* and *Utricularia vulgaris,* littoral vegetation is represented by *Menyanthes trifoliata* and *Calla palustris.* On elevated landscapes (above 600 m a.s.l.) fell-fields alternate with patches of dwarf shrubs with a dense lichen cover (Ivanova 1961; Sosina and Zakharova 2004).

Meadows of the FR occupy very insignificant areas.There are records of steppe fragments on south-facing slopes with *Festuca kolymensis, Phlox sibirica, Dianthus repens, Euphorbia discolor, Papaver nudicaule, Hedysarum alpinum, Potentilla arenosa,* etc.

*The Yana-Indigirka Floristic Region (Ya-I)* is the largest FR of Yakutia occupying almost the whole basins of the Yana and Indigirka Rivers. It is situated in the eastern part of Yakutia and is strictly separated from the other floristic regions by the Verkhoyansk mountainous system.

The flora and vegetation of the FR have been studied since 1806. At different periods the Region was visited by M. Adams, N.S. Gorokhov, G. Maydell, A.K. Cajander, etc. (Komarov 1926). In the twentieth century the Yana-Indigirka mountainous country was covered in the following works: Sokolov (1923); Birkengof (1932); Yarovoy (1939); Sheludyakova (1938, 1943, 1948a, b, 1957b); Kuvaev (1956); Karavaev (1958); Karavaev and Dobretsova (1964); Prakhov (1957); Yurtsev (1968, 1981); Kildyushevsky (1966); Skryabin (1968); Perfilyeva (1977); Volotovsky (1987); Nikolin (1991, 1992); Nikolin and Petrovsky (1988); Nikolin and Baryshev (1992); Perfilyeva and Shurduk (1999), etc. Its meadow vegetation is reviewed in the monographs "Winter pastures of Northeastern Yakutia" (Andreyev 1974) and "Meadows of Yakutia" (Andreyev et al. 1975).

All over the territory of the FR the mountain northern taiga larch sparse forests prevail composed of *Larix cajanderi* and *L. gmelinii*. At the altitude from 600 to 800 m a.s.l. (in the North) to 120–1,600 m a.s.l. (in the South) these forests are replaced by the shrubberies of *Pinus pumila*, then by mountain tundra, stony and rubbly screes and detritus. Very rarely and only on south-facing rubbly slopes within the forest belt, fragments of *Populus tremula* forests are observed (Andreyev et al. 1987; Kuvaev 1956; Perfilyeva 1977).

There are fragments of relic steppe vegetation in the Yana and Indigirka basins which are confined to south-facing slopes (Sheludyakova 1957b; Karavaev 1958; Karavaev and Dobretsova 1964; Skryabin 1968; Yurstev 1981).

*The Kolyma Floristic Region (Kol).* The region is situated at the northeastern edge of Yakutia, within the basins of the Kolyma and Alazeya Rivers. The most territory (northern and central) of the region is occupied by the Kolyma Lowland. The FR borders on the Alazeya Plateau and the eastern spurs of the Moma Range in the West and the Yukagir Plateau in the Southwest.

The beginning of the floral study in this Region dates back to 1786 when the botanist K. Merk first collected plants in the Kolyma River valley. Later on F.M. Avgustinovich gathered large valuable floristic data from the same territory (Komarov 1926). At present, the flora and vegetation of the Kolyma lowland and the Kolyma valley are most thoroughly investigated, while the mountainous parts of the Region is still subject to study (Trufanova 1972; Permyakova 1973; Andreyev 1974; Perfilyeva and Rykova 1975b; Kozhevnikov 1981; Petrovsky and Zaslavskaya 1981; Khokhryakov 1985; Gogoleva and Cherosov 1987; Zakharova 2006).

The dominating forest species of the FR is *Larix cajanderi.* The sparse near-tundra and northern taiga larch forests alternate with tracks of polygonal-ridged tundra-bogs with forested ridges and a participation of large-hummock tundra with shrubs. As a rule, the pre-tundra forests and sparse forests feature a suppressed, very thin (crown cover not exciding 0.2–0.3) tree layer and a scanty species composition. Southwards, the pre-tundra forests are followed by the zone of northern taiga characterized by an enriched species composition and higher crown cover values (0.3 and more) of trees.

The sparse *Larix* forests of the Moma Range reach 1,000 m a.s.l., but on the north slopes it reaches only up to 400–500 m. Above the forest border, the belt of *Duschekia fruticosa, Betula divaricata* and *Pinus pumila* starts. At higher altitudes the shrubberies get thinner and are replaced by mountain tundra and stony deserts. The mountainous regions are characterized by *chisati*. *Poa-Carex* mountain tundra is also common for the Yukagir Plateau. The south-facing slopes and prominent relief elements of the Labuya Mountain (the Middle Kolyma), contain habitats for steppe fragments.

Various bogs occupy considerable areas of the region, the most widespread of them are tussock bogs *(Calamagrostis langsdorffii, Rubus arcticus, Chamaedaphne calyculata, Vaccinium uliginosum)* and *Carex-Menyanthes* bogs.

*The Central Yakutian Floristic Region (CYa).* The region lies within the vast Central Yakutian Lowland which is composed of four elements: the ancient plateau (300–500 m a.s.l.) of the Cambrian and Jurassic sediments and covered with eluvium of skeletal bedrocks; the ancient alluvial plain (150–280 m a.s.l.) covered with loose sandy-clayey lake-river sediments; the river valley terraces composed of ancient alluvium and representing the lower relief element; and the modern floodplains.

The flora and vegetation of this FR is best studied compared to the rest of the Regions. It was initiated by I.G. Gmelin in 1736, who was followed by numerous explorers. Their routes and publications are thoroughly described by Komarov (1926). The great contributors to a knowledge on the Region's flora in early twentieth century are Cajander (1901), Dolenko (1913, 1916), Abolin (1913, 1929), Drobov (1910, 1914, 1927). Since the middle of the twentieth century the following publications are most important: Povarnitsyn 1932a; Rabotnov 1933a, b, 1935a, b, 1939, 1945; Karavaev 1945, 1955, 1965, 1968; Dobretsova 1959, 1961, 1962; Sheludyakova 1957a, 1958, 1959; Galaktionova and Petrov 1965, 1967; Sheludyakova and Skryabin 1969; Sheludyakova et al. 1954; Karavaev and Skryabin 1971; Ivanova 1967, 1971a, b, 1981; Ivanova and Perfilyeva 1972; Gogoleva 1999; Egorova 2001; Timofeyev 1992, 2003; Efimova and Shurduk 2002; Efimova et al. 2003, Zakharova 2001, 2004a, b; Kuznetsova et al. 2006a, b, Zakharova et al. 2007a, b; including summarized monographs by Karavaev "Conspectus of the flora of Yakutia" (1958), Usanova and Perfilyeva "Identifier of fodder plants of Yakutia" (1966), "Identifier of higher plants of Yakutia" (Tolmachev 1974), as well as by Ivanova "Higher plants of Yakutsk vicinity" (1986) and "Dicotyledons of Yakutsk vicinity" (1990), and "Diversity of the vegetation world of Yakutia" (Danilova 2005).

The Central Yakutian FR greatly differs from the rest of the floristic regions both in soils, climate and flora-vegetation peculiarities. The dominant type is the light coniferous taiga of *Larix cajanderi* with insignificant participation of *Pinus sylvestris.* The role of deciduous tree species in the forest cover is also insignificant. (Scherbakov 1975; Timofeyev et al. 1994).

The forests of *Picea obovata* occur as narrow belt communities confined to the large river valleys and alases. Usually they form their own coenoses, though, sometimes, isolated trees participate in the larch forest communities (Scherbakov 1992; Timofeyev et al. 1994). Original (primary) birch woods of *Betula pendula* occur in wide valleys of the Lena, Aldan, Amga and Viluy Rivers as well as communities surrounding alases and bordering the larch taiga (Timofeyev 1992; Timofeyev et al. 1994). The woods of *Populus suaveolens* are very rare in the Central Yakutian FR covering insignificant areas along the banks of tributaries on the right side of the Lena River. (Timofeyev 2003).

The forest-free areas are partly covered with shrubberies of *Betula exilis* and *B. fruticosa* ("*yerniks*") and other low willow species. The area covered by those yerniks increases northwards.

The steppe communities are confined to elevated parts of the above-floodplain terraces and south-facing slopes of large river valleys and mountains.

Taiga hollows of various origins, either with or without lakes and covered with herb vegetation, represent alas landforms. The most widespread alas types occur on the loamy ancient alluvial sediments covered with halophytic vegetation.

Central Yakutia also features peculiar landscapes (so called "*tukulan*") formed as a result of aeolian processes. They occur in the Lena-Viluy erosional-accumulative plain occupying the above-floodplain terraces of the Viluy, Tyung, Linde, Khoruonka Rivers and the watershed area at 270 m a.s.l. (Korzhuev 1965; Pavlov 1981). The tukulans represent northern sandy deserts scattered in the taiga (Skryabin and Karavaev 1991), and covering a total of 50,000 ha.

*The Upper Lena Floristic Region (UL)* is situated in South–western Yakutia occupying the south–eastern part of the Middle Siberian Plateau and the northern part of the Olyokma-Chara upland. The borders of the region match the distribution area of *Abies sibirica* and *Pinus sibirica.*

The study of the flora and vegetation of the Upper Lena FR is associated with the name of Gmelin (Komarov 1926; Karavaev 1958). However, detailed investigations started as from the middle twentieth century: Kuvaev (1955, 1957); Pozdnyakov (1955); Petrov (1957); Kononov (1971); Scherbakov (1975); Skryabin (1976); Golyakov (1994, 1996, 1997); Kuznetsova (1997, 1998, 1999, 2001); Nikitina and Isaev (1999); Egorova and Ivanova (2001); Kuznetsova and Ivanova (2001); Sofronov (2003); Nikitina (2001, 2004); Egorova (2004, 2006); Volpert (2006), etc.

The region features the most productive forests and the best agroclimatic conditions. The prevailing forest type is a larch forest of *Larix gmelinii* and *L. czekanowskii.* The mesic habitats are covered with larch forests of the *Vaccinium vitis-idaea* type. It is characterized by a sparse undergrowth, prevalence of *Vaccinium vitis-idaea* in the herb-dwarf shrub layer and the absolute absence of a moss-lichen cover. The following species participate in such forests: *Rhododendron dauricum, Duschekia fruticosa, Vaccinium uliginosum, Ledum palustre.* Their degree of dominance determines the community type of the forest (Pozdnyakov 1964; Timofeyev et al. 1994; Scherbakov 1975).

The moist habitats on eluvium of calcareous rocks are covered with *Vaccinium vitis-idaea*-green moss *Larix* forests with participation of *Picea, Pinus sibirica, Abies sibirica* and deciduous trees.

In the Upper Lena FR *Pinus sylvestris* is less strictly confined to the south-facing slopes and also occurs on small ridges, and on east-, west- or even north-facing slopes. Compared to other FRs, pine forest communities growing in mesic habitats are more common in this region.The forests of *Picea obovata* are confined to high floodplain terraces of the large river valleys (the Lena, Olyokma, Chara, Tokko), out of the zone of regular flooding ( Pozdnyakov 1955; Scherbakov 1975; Timofeyev et al. 1994; Koropachinsky 1996).

The forests of *Pinus sibirica* occur strictly within the Upper Lena FR. These forests do not represent pure stands of this tree species but always have an admixture of *Larix* and *Picea* with *Vaccinium vitis-idaea* and green moss. This is the basic type of Yakutian *Pinus sibirica* forests, growing on slopes and summits as high as 600–800 m a.s.l. (Timofeyev et al. 1994; Nikitina and Isaev 1999; Nikitina 2001, 2004).

Like *Pinus sibirica, Abies sibirica* does not form pure stands but occurs as an admixture to *Picea obovata, Larix* and *Pinus sibirica* forests (Ivanova and Krivoshapkin 1998; Krivoshapkin 1998b; Kuznetsova and Ivanova 2001).

Another unique forest community in the Lena-Aldan Plateau is the large grass-dwarf shrub-green moss dark coniferous-larch forest. The tree stand is very rich in species: *Abies sibirica, Pinus sylvestris, Larix gmelinii, L. czecanowskii, L. sibirica, Betula pendula, Populus tremula, Picea obovata.* In the Upper Lena the *Populus tremula – Vaccinium vitis-idaea* woods of post-fire origin are found (Scherbakov 1975).

The upper forest border lies at 1,000–1,250 m a.s.l. and is composed of *Larix* and *Picea* woodlands and fragments of *Betula ermannii ssp. lanata* communities. The subalpine belt consists of communities of *Pinus pumila* and *Rhododendron aureum* and fragments of mountainous *Calamagrostis*-forb meadows. Stony places with fragments of shrubby-lichen tundra and *Carex*-sphagnum bogs are common for watersheds. Calcareous rocks are the habitats for *Dryas-Kobresia*-lichen tundra.

The Nyuya and Peleduy Rivers' banks are covered with meadows dominated by *Brachypodium pinnatum* with participation of *Agrostis gigantea* and *Hordeum brevisubulatum.* In dryer places *Trifolium pratense* and *Vicia cracca* may join. The Upper Lena is also characterized by meadows with *Agrostis gigantea* confined to annually flooded channels (Kuvaev 1957, Andreyev et al. 1975; Andreyev et al. 1987). In the vicinity of Olyokminsk town the slopes of the Lena valley are covered with steppe vegetation including numerous rare species: *Sibbaldianthe adpressa, Oxytropis pilosa,* etc. The valleys of the Olyokma, Chara and Tokko are characterized by petrophytic steppes with *Polygala sibirica* and *Phlojodicarpus sibiricus* (Andreyev et al. 1987).

*The Aldan Floristic Region (Ald)* is situated in southern Yakutia. The northern border of the region lies above the Middle Aldan and then crosses the Middle Olyokma and Tokko. In the South it is limited by the Stanovoy Range and in the North–east by the Sette-Daban Range.

Before the October Revolution of 1917 the Region was explored by E. Laxman, G.W. Steller, G.I. Langsdorf, A.F. Middendorf, Yu.I. Shtubendorf, D.V. Domracheyev, V.F. Boydush, M.I. Gubelman, V.N. Sukachev, V.P. Drobov, etc. The history of exploration is described in detail by Komarov (1926) and Karavaev (1958). The results of the investigations into the FR's flora and vegetation are covered in: Elenevsky (1933); Povarnitsyn (1932b, 1933); Rabotnov (1934,1936a, b); Poyarkova (1953); Tyulina (1956, 1957, 1959, 1962); Pozdnyakov and Gortinsky (1960); Pozdnyakov (1961b); Galaktionova and Permyakova (1964); Scherbakov (1964); Volotovsky (1992, 1994, 1996); Volotovsky and Chevychelov (1991); Volotovsky and Kuznetsova (1998); Golyakov (1996); Kuznetsova (1997, 1998, 2001, 2003a, b), etc.

The main edificatory species of the region is *Larix cajanderi.* It is accompanied by *Picea obovata.,* and by *P. ajanensis* and *Pinus sylvestris* in the South–east, sometimes by *Abies sibirica* and *Pinus sibirica* in the West.

The native bedrock has great influence on the species composition. On calcareous bedrocks (mainly lime-stones and dolomites) larch and pine forests prevail, featuring a high productivity and floristic diversity. The mixed forests develop on moist soils featuring 6–7 tree species and a high floral diversity. The dominating trees are *Larix gmelinii* and *Picea obovata. Pinus sylvestris* plays a significant role in the community too (Tyulina 1957; Pozdnyakov 1961b; Volotovsky 1992).

On the steep lime-stone outcrops of the Middle Aldan so called "tundra-steppes" occur characterized by a combination of typical petrophytic steppe plants and arctic-alpine xerophytes. Above the tree-line fragments of forb-moss communities of *Pinus pimula* grow on leached soils, as well as *Dryas* tundra on dry stony places. Almost the whole surface of calcareous peneplains above the tree-line is occupied with the *Dryas-Kobresia*-lichen tundra. All the abovementioned communities are characteristic for calcareous rocks (Andreyev et al. 1987; Volotovsky 1992). The great majority (over 120 species) of endemic and subendemic plants of the FR are obligate calciphils: *Asplenium ruta-muraria, A. viride, Botrychium robustum, B. virginianum, Juniperus davurica, Chrysosplenium saxatile, Callianthemum isopyroides, Braya siliquosa, Petasites sibiricus* and some other species (Volotovsky 1992).

The plant communities growing on acid soils are characterized by a poor floristic composition and a lower typological diversity. The acid metamorphic or igneous rocks are occupied with pure larch forests (sometimes with an admixture of *Pinus sylvestris)* with a well developed moss-lichen cover. The tree-line communities on crystalline native bedrock contain *Picea ajanensis* (Sokolov 1923; Povarnitsyn 1932b, 1933; Tyulina 1956, 1962; Scherbakov 1975; Volotovsky 1994).

The *Betula ermanii ssp. lanata* forests on the Aldan upland and Stanovoy Range have an uneven distribution. They are rather common for the highest ranges of the subalpine zone (up to 1,800 m a.s.l.), particularly in the Tokinsky Stanovik and Aldan-Uchur Ranges consisting mainly of Archaean crystalline bedrock. (Rabotnov 1936b; Tyulina 1956, 1959, 1962; Volotovsky and Chevychelov 1991, Volotovsky 1994).

The *Betula* forests usually have a secondary origin. *Betula pendula* seldom forms original stands, more often occurring as an admixture in other forest communities. mostly larch forests of the *Vaccinium vitis-idaea, Ledum palustre* or *Vaccimium uliginosum* types. Belts of original birch forests with thin herbage represent regular riparian communities in the Aldan River valley and along oxbow banks (Scherbakov 1975).

*Populus tremula* forests are formed as a stage of secondary, pyrogenic, succession. They are common on calcareous slopes of various aspects with sufficient moisture supply. Significant areas along the banks of the large rivers and on islands are occupied with *Salix viminalis* shrubberies, as well as with *Populus suaveolens* and *Chosenia arbutifolia* stands representing initial successional stages of forest vegetation in river valleys. Riparian vegetation of small rivers is represented by *Salix schwerinii, Spiraea salicifolia* and *Betula fruticosa.*

Shrub communities contain *Pinus pumila* and *Betula divaricata* in mountains while *B. exilis* is confined to rather wide valleys with poor drainage.

The meadow vegetation of the Aldan River basin has an uneven distribution. Large meadow tracts are often represented by very wet tussock communities with *Calamagrostis langsdorffii* and *Carex juncella* in the lower, annually flooded floodplains of the Aldan River and its tributaries. The interridge hollows of the upper floodplains are occupied with *Carex* meadows, almost always invaded with shrubs, occur as small fragments among birch forests and large willow shrubberies (Povarnitsyn 1932b; Andreyev et al. 1975; Andreyev et al. 1987).

The water bodies of the FR are dominated by plants of the *Potamogetonaceae, Ranunculaceae* and *Sparganiaceae* (Skryabin and Karavaev 1991).

The mountain areas of the FR feature a clear belt zonation. Mountain forests and woodlands appear at 650–800 m a.s.l. Higher, in the subalpine belt, communities with *Pinus pumila* and bare screes prevail. The ranges with moist conditions (increased snow accumulation) are the habitats for communities of *Rhododendron aureum* (*Phyllodoce*-lichen and *Phyllodoce*-moss types). Various *Salix* shrubberies (*Salix krylovii* or *S. saxatilis*) grow in trough valleys with sufficient moisture supply and along the banks of mountain springs.

Subalpine meadow communities cover insignificant areas from several tens to over 200 $m^2$ on steep slopes or as narrow belts stretching along erosional hollows.

At 1,500–1,800 m a.s.l. the alpine belt of rock debris is represented by stony deserts with epilithic lichen vegetation and mountainous, mainly poly-dominant dwarf shrub and sedge tundra (Andreyev et al. 1987; Volotovsky 1992; Kuznetsova 2003a, b). The most widespread dwarf shrub-lichen tundra develops on prominent stony landforms and is very monotonous and scanty in species composition.

The largest areas are occupied with the mesotrophic dwarf shrub-*Sphagnum* and *Carex-Sphagnum* bogs of a transitional type. They are not diverse and are very poor in floral composition. The lowland eutrophic marshes do not cover significant areas though are very common in landscapes with lakes (oxbow, thermokarst or moraine).

All altitudinal belts in the mountains bear widespread rock debris (so called "*kurumy*"). Their areas increase with altitude. The largest areas of fell-fields are found in the alpine belt (the most elevated altitudinal belt is represented by fell-fields) where they prevail over tundra landscapes. They are characterized by a specific epilithic lichen vegetation including *Umbilicaria proboscidea, U. hyperborea, Parmelia centrifuga, P. stygia, Rhizocarpon geographicum,* etc.

## 2.2 Higher Vascular Plants

**L.V. Kuznetsova, V.I. Zakharova, N.K. Sosina, and E.G. Nikolin**

The list of higher vascular plants of Yakutia is composed based on the following fundamental works: "Identifier of higher vascular plants of Yakutia" (Tolmachev 1974); "Arctic flora of the USSR"(Tolmachev 1960, 1963, 1964, 1966a, b, 1971, 1975; Tolmachev and Yurtsev 1980, 1983; Yurtsev 1984, 1986, 1987), "Flora of Siberia" (Krasnoborov 1988, 1997; Krasnoborov and Malyshev 1992; Malyshev 1997; Malyshev and Peshkova 1987, 1990a, b, 1993, 1994; Polozhy and Malyshev

1988, 1994; Peshkova 1996; Polozhy and Peshkova 1996; Malyshev et al. 2003), "Vascular plants of the Far East" (Kharkevich 1985; 1987–1989, 1991, 1992,1995), "Conspectus of the flora of Siberia" (Baikov 2005), "Diversity of the vegetation world of Yakutia" (Danilova 2005), as well as on the data of the Herbarium (SASY) of the Institute for Biological Problems of Cryolithozone, Siberian Branch Russian Academy of Sciences. The taxa names are given according to Cherepanov (1995) and Danilova (2005). The taxonomical and ecological-geographical analyses of the flora of Yakutia were conducted based on the revised and supplemented list of the species and subspecies of vascular plants (Danilova 2005). The inventory of life forms and ecological groups of the vascular plants of Yakutia was made based on the classification scheme by Sekretaryova (2004) considering the abovementioned references as well as Bezdelev and Bezdeleva (2006).

The beginning of floristic investigations in the territory of Yakutia dates back to 1733, when I.G. Gmelin participated in the Great Northern Expedition headed by V. Bering. Great contributions to the knowledge of the Yakutian flora and vegetation was made by numerous explorers. Komarov (1926, 1927) prepared the most detailed summary of the investigations that took place up to 1920s. Starting from the mid twentieth century the most important results of Yakutian floral investigations have been published in a number of key monographs: Galaktionova et al. (1962); Tolmachev (1974); Andreyev et al. (1975); Scherbakov (1975); Andreyev et al. (1987); Egorova et al. (1991); Timofeyev et al. (1994); Ivanov (2003, 2005); Dolinin et al. (2000); Danilova (2005); Cherosov (2005), etc.

Yakutia attracted many specialists from various institutes of the former USSR interested in the flora and vegetation of the region (Petrov 1930; Poyarkova 1953; Tyrtikov 1955, 1958; Gorodkov 1956; Pivnik 1958; Yurtsev 1959, 1961, 1962, 1964, 1968, 1981; Tikhomirov et al. 1966; Tikhomirov and Shtepa 1956; Tyulina 1956, 1957, 1959, 1962; Kildyushevsky 1964, 1966; Utkin 1959, 1961, 1965; Pozdnyakov 1952, 1961a, b, 1969; Pozdnyakov and Gortinsky 1960; Alexandrova 1960–1963, 1970; Lukicheva 1960, 1963a, b; Buks 1963, 1966; Korobkov 1981; Tsvelev and Yurtsev 1984; Sumina 1986, etc.)

The flora of Yakutia numbers 1,984 species (including 84 subspecies) of higher vascular plants belonging to 509 genera and 111 families. The leading families are *Asteraceae* with 214 species (10.79% of total number of species), *Poaceae* 209 (10.28%), *Cyperaceae* 161 (8.11%), *Ranunculaceae* 112 (5.65%), *Brassicaceae* 105 (5.29%), *Fabaceae* 104 (5.24%), *Rosaceae* 101 (5.10%), *Caryophyllaceae* 90 (4.54%), *Salicaceae* 60 (3.01%), *Scrophulariaceae* 57 (2.88%), *Polygonaceae* 50 (2.52%), *Lamiaceae* 48 (2.41%), *Saxifragaceae* 48 (2.41%), *Apiaceae, Boraginaceae* and *Ericaceae* with 36 species (1.82%), *Chenopodiaceae* 34 (1.71%), *Juncaceae* 31 (1.56%), *Orchidaceae* 27 (1.36%) and *Primulaceae* 22 (1.11%). The first ten families comprise 61.13% of the flora of Yakutia. Monospecific families (24) concern the species seldom occurring in southern regions: *Onocleaceae, Thelypteridaceae, Sinopteridaceae, Hypolepidaceae, Polypodiaceae, Najadaceae, Scheuchzeriaceae, Commelinaceae, Hemerocallidaceae, Paeoniaceae, Oxalidaceae, Monotropaceae* and *Lobeliaceae,* but also those that recently penetrated into the territory of the republic (*Cannabaceae, Amaranthaceae,*

*Convolvulaceae, Hydrophyllaceae, Cucurbitaceae),* as well as *Trapaceae* having only one fossil record. Species of *Ceratophyllaceae, Empetraceae, Cornaceae, Adoxaceae, Diapensiaceae* are widely distributed in Yakutia. Since the last inventory that took place in 1974, new families have been revealed: *Isoetaceae* and *Commelinaceae.*

The following genera are largest as regards species and subspecies numbers: *Carex* (125), *Salix* (57), *Potentilla* (49), *Artemisia* and *Saxifraga* (43 each), *Oxytropis* (38), *Poa* and *Draba* (33 each), *Taraxacum* (32), *Ranunculus* (31), *Stellaria* and *Pedicularis* (27 each), *Astragalus* (22), *Rumex* and *Papaver* (21 each), *Juncus* (20), *Festuca* (18), *Saussurea* and *Viola* (17 each), *Calamagrostis* and *Thymus* (15 each). The leading 10 genera contain 511 species and subspecies comprising 25.76% of the whole flora. Recent studies enriched the Yakutian flora by new records of genera, usually monospecific: *Isoetes, Commelina, Acelidanthus, Clintonia, Neottianthe, Echinochloa, Phalaris, Ptilagrostis, Amaranthus, Clematis, Paraquilegia, Subularia, Agrimonia, Alchemilla, Sibbaldia, Anthriscus, Eryngium, Ferulopsis, Pastinaca, Phacelia, Anoplocaryum, Verbascum, Lagopsis* and *Anthemis.* Altogether 411 species were added to the flora, and that amounts to 23% of the flora of Yakutia.

An analysis of the longitudinal patterns shows that the majority of the species and subspecies have circumpolar distributions (456 species or 23.98% of the flora). Eurasian species are rather numerous as well (292 or 14.72%), being characteristic mainly for the southern and central parts of Yakutia. The Asian-American element contains 205 species (10.33%), and increases in number north- and eastwards. The species and subspecies of the Asian distribution type are also important in Yakutia: North Asian (10.03%), East Asian (6.80%), all-Asian (2.37%), Central Asian (1.56%). The first two elements show decreasing trends in their numbers of species from the South northwards, the highest number of the North Asian species being concentrated in the mountainous southern and north–eastern regions, while the rest of the Asian elements are concentrated in the southern and central parts of Yakutia. The species of South Siberian (6.10%), Far Eastern (5.54%) and Euro-Siberian distribution (4.44%) occur mainly in the southern and eastern mountainous regions. Participation of the European element in Yakutian flora is very insignificant (0.55%), most of them being adventive species occurring in South–West Yakutia. The group of species with restricted distribution ranges (9.17%) includes the plants with all-Siberian (2.11%), East Siberian (2.22%), and North-Asian (0.71%) distribution areas, and also a few endemics of the Far East (0.30%). The species that are endemic to Yakutia, comprise 3.83% and grow mainly in the Arctic zone and mountainous systems of the Northeast.

For an analysis of the latitudinal distribution areas the flora was divided into the following complexes: boreal (dark coniferous forest, light coniferous forest, pre-boreal), azonal (aquatic, aquatic-bog, riparian, meadow), steppe (forest-steppe, mountainous steppe, steppe, desert-steppe), arctic (arctic, hypoarctic), alpine (alpine, alpine tundra, general mountain (i.e. common for all altitudinal belts), hypoarctic-montane). The territory of Yakutia is covered for 80% with forest, which explains the prevalence of the boreal complex in the flora (23.64%), the importance of which increases from the North southwards. The species of the light taiga are the dominating group (17.24%) with the dark coniferous forest species (3.48%) and preboreal species (2.92%) being less numerous. The latter two groups are characteristic only for southern regions. The group of alpine tundra (10.33%) and hypoarctic-montane (2.62%) species predominates in the Arctic and in the mountains of the Northeast. The alpine (6.05%) and general mountain (4.13%) species occur in South Yakutia. The third largest group is taken by the azonal complex (22.48%) with considerable prevalence of meadow (10.53%) and aquatic-bog (6.45%) species, and notable presence of aquatic (2.92%) and riparian (2.57%) species. The steppe complex of species and subspecies is rather numerous (19.05%). The groups of forest-steppe (6.55%), mountain-steppe (6.86%), steppe proper (5.04%), and desert-steppe (0.60%) taxa are characteristic for the central and southern regions of Yakutia and much scarcer in the Arctic and Subarctic zones. The arctic complex includes 11.69% of the flora with 7.76% of arctic species and 3.93% of hypoarctic ones. The proportion of naturally introduced adventive plants comprises 4.69% of the entire flora.

Within Yakutia, in the ecological groups, based on response to moisture conditions, xerophytes prevail (21.02%). They are followed by mesophytic (18.25%) and hygrophytic (10.48%) groups. Hydrophytes (2.92%), hydatophytes (1.82%) and eurytopic species and subspecies (1.46%) are rather few. In the mountainous eastern regions and the Arctic xerophytes are common, whereas the central and southern regions carry more mesophytes. Mesohygrophytic plants (13.81%) prevail in the flat western and eastern regions.

As regards life forms the Yakutian flora contains predominantly arboreous and herbaceous plants. The leading position is taken by polycarpic herbs (87.25%) dominated by short-rhizome (18.65%), long-rhizome (14.87%) and taproot (13.56%) species and subspecies. The group of corm plants is not numerous, comprising only 2.72% of the total species number. The flora of herbaceous monocarpic species comprises 11.14%, half of which (4.69%) are weeds. There are four parasitic species recorded in Yakutia. They are the representatives of *Cuscutaceae* and *Orobanchaceae* families. *Boschniakia rossica* is common all over Yakutia, while *Cuscuta europaea, C. lupuliformis* and *Orobanche coerulescens* prefer warmer southern regions. This is also true for two insectivorous species *Drosera anglica* and *D. rotundifolia.* The group of woody plants (10.53%) includes trees (1.41%), shrubs (4.08%), dwarf shrubs (3.18%) and dwarf semishrubs (1.87%). The most common and dominating forms are shrubs and dwarf shrubs.

The Red Data Book of the Republic of Sakha (Yakutia) (Dolinin et al. 2000) includes 337 species and subspecies of higher vascular plants of 180 genera and 64 families, making up about 19% of the whole flora of Yakutia. According to the classification of rare and endangered plant species adopted by the IUCN (the World Conservation Union), all the rare species are grouped into four categories of rarity. Category 0 is represented by *Trapa natans* recorded only in fossil state. Category I includes 8 species of low abundance and known only from 1 to 3 sites. Category II is designated for vulnerable plants and contains 23 species characterized by a steady decrease of their distribution area and population numbers due to anthropogenic impact. Category III is the largest one (273 species and subspecies) and is divided into 4 groups. Group IIIa contains rare endemic species of Yakutia (17 species). Group IIIb consists of the species endemic in North–Eastern Siberia (38). Group IIIc (74) includes the species with rather vast distribution areas though rare throughout their areas. Group IIId includes the species having rarity status only within Yakutia (144). Category IV (32) comprises species with questionable records within Yakutia.

*The Arctic Floristic Region (Arct).* The Arctic flora amounts to 789 species and subspecies of higher vascular plants of 220 genera and 70 families totalling 39.77% of the whole floristic diversity of Yakuita. The leading families are *Poaceae* 96 (12.17%), *Asteraceae* 78 (9.89%), *Cyperaceae* 64 (8.11%), *Brassicaceae* 61 (7.73%), *Caryophyllaceae* 54 (6.84%), *Ranunculaceae* 53 (6.72%), *Fabaceae* 43 (5.45%), *Rosaceae* 40 (5.07%), *Salicaceae* 29 (3.68%), *Saxifragaceae* 27 (3.42%), *Scrophulariaceae* 25 (3.17%), *Polygonaceae* 22 (2.79%), *Juncaceae* 15 (1.90%), *Ericaceae* 14 (1.77%), *Papaveraceae* 11 (1.39%) and *Primulaceae* 9 (1.14%). The first 10 families include 69.07% of total Arctic flora. The monospecific families (18) include widespread species in Yakutia: *Empetraceae, Adoxaceae, Diapensiaceae*, as well as *Botrychiaceae, Dryopteridaceae, Cryptogrammaceae, Juncaginaceae, Lemnaceae, Liliaceae, Convallariaceae, Iridaceae, Fumariaceae, Linaceae, Limoniaceae, Menyanthaceae, Orobanchaceae* and *Caprifoliaceae*. Several monospecific families or families with few species (*Lycopodiaceae, Orchidaceae, Pyrolaceae, Plantaginaceae, Rubiaceae*) occur in the boreal zone and are mostly not characteristic for the tundra zone.

The first ten leading genera include *Carex* (42), *Salix, Draba* and *Potentilla* (27 each), *Saxifraga, Ranunculus* and *Poa* (23 each), *Taraxacum* (21), *Stellaria* (16) and *Pedicularis* (14). They contain 243 species making up 30.8% of total Arctic flora. *Dupontia, Koenigia, Honckenya, Jurtsevia* and *Raphanus* are characteristic genera for the Arctic, of them only *Dupontia* contains three species, while the rest is monospecific. Most species of *Draba, Potentilla, Taraxacum* and *Saxifraga* grow in mountainous arctic and subarctic semi-sodded stony tundra landscapes.

The Arctic flora predominantly contains species and subspecies with circumpolar (30.94%) and Asian-American (19.27%) distribution areas and they are mainly tundra (17.49%) and alpine tundra (22.69%) plants.

The boreal complex of the Region's flora contains 114 species and subspecies or 14.45% with predomination of the light coniferous group (11.53%), mainly of circumpolar (3.80%), North Asian (2.79%) and Eurasian (1.77%) distribution. Most of those species grow within the Subarctic zone, and only a small part of the boreal complex penetrates into the Arctic itself through the large river valleys (the Lane, Indigirka, Kolyma, Olenyok, Anabar, Yana Rivers).

The majority of species and subspecies of the Asian (24.33%) and Far Eastern (1.27%) complexes are widely distributed in the north–eastern part of the Region. The flora of the north–western part features Eurasian (10.39%) and Euro-Siberian (1.77%) elements. There are records of adventive *Melandrium album* and *Batrachium peltatum* (with European distribution areas) in large human habitation areas in the Arctic zone.

The Region is characterized by 68 endemic and semi-endemic species and subspecies. This makes up 8.62% of the total Arctic flora. Of these 3.8% are Yakutian endemics, 2.54% Siberian and 2.16% North–East Russia endemics. Species and subspecies such as *Poa trautvetteri, Oxytropis sordida ssp. arctolenensis, Myosotis asiatica ssp. kolymensis, M. czekanowskii, Artemisia henriettae, A. triniana, Taraxacum lenense* and *T. semitubulosum* are characteristic exclusively for this Region, while *Alopecurus roshevitzianus, Hyalopoa lanatiflora, Poa pratensis ssp. skrjabinii, Minuartia jacutica, Stellaria jacutica, Papaver angustifolium, P. czerskii, Saxifraga lactea, Astragalus vallicola, Hedysarum vicioides, Oxytropis adamsiana ssp. janensis, O. middendorfii ssp. orulganica, Androsace gorodkovii* occur sporadically in other regions of Yakutia.

Concerning their response to moisture conditions, the largest group of the Region's flora are xero- and mesoxerophytes including 169 (21.42%) and 124 (15.72%) taxa, respectively. A significant amount of mesohygrophytes (18.38%) and hygrophytes (11.41%) reflects the abundant wetlands. Life form analysis shows the prevalence of herbaceous polycarpics (657–83.27%) characterized by short-rhizome (257–32.57%), taproot (159–20.15%) and long-rhizome (145–18.38%) perennial herbs.

Sixty two species and subspecies growing in the Arctic Floristic Region are listed in the Red Data Book of Yakutia. The Arctic zone represents the only habitat for *Dendranthema arcticum ssp. polare, Taraxacum lenense, T. semitubulosum, Artemisia henriettae, A. triniana, Parnassia kotzebuei, Astragalus kolymensis, Caragana jubata, Lathyrus maritimus, Braya pilosa, Draba arctogena, D. groenlandica, D. eschscholtzii, D. pohlei, Ranunculus spitzbergensis, Papaver anjuicum, P. paucistaminum, Poa abbreviata, P. pseudoabbreviata, P. trautvetteri, Festuca hyperborea, Potentilla pulchella, P. egedii.*

The Arctic FR contains the State Sanctuary "Ust-Lensky", covering almost the whole area of the Lena River delta. According to Egorova et al. (1991) 402 higher vascular plants of 150 genera and 43 families including 26 rare and endangered species and subspecies entered into the Red Data Book of Yakutia, occur there.

*The Olenyok Floristic Region (Ol).* The species composition of this region is rather depauperate, with the highest floristic diversity recorded in young floodplains, on alluvium, as well as on screes and rock outcrops (Lukicheva 1963a, b; Petrovsky and Plieva 1994). The flora consists of 701 species and subspecies of higher vascular plants in 203 genera and 73 families (35.33%). The leading families are *Poaceae* 75 (10.70%), *Cyperaceae* 74 (10.56%), *Asteraceae* 68 (9.70%), *Brassicaceae* 45 (6.42%), *Ranunculaceae* 43 (6.13%), *Caryophyllaceae* 39 (5.56%), *Salicaceae* 34 (4.85%), *Rosaceae* 29 (4.14%), *Saxifragaceae* 23 (3.28%), *Polygonaceae* and *Fabaceae* 22 (3.14%) each, *Scrophulariaceae* 19 (2.71%), *Juncaceae* 17 (2.43%), *Ericaceae* 14 (1.10%), *Primulaceae* 12 (1.71%) and *Lamiaceae* 10 (1.43%). The ten leading families comprise 67.62% of the whole flora of the Region. The Region is characterized by 19 monospecific families. Besides *Empetraceae, Adoxaceae* and *Diapensiaceae* that are widely spread all over Yakutia, they include those with the northern border of their distribution area in this Region: *Alismataceae,*

*Araceae, Santalaceae, Portulacaceae, Droseraceae, Balsaminiaceae* and *Cornaceae*.

The ten leading genera are *Carex* (54), *Salix* (31), *Saxifraga* (19), *Draba* (16), *Ranunculus* (15), *Potentilla* and *Poa* (13 taxa each), *Artemisia* and *Taraxacum* (12 each), *Thymus* (10). In total they contain 195 species and subspecies making up 27.82% of total Region's flora. There are 121 monospecific genera in the Region, with *Stevenia* recorded solely for this FR.

Many species in the Olenyok Floristic Region have a circumpolar distribution area (33.81%) mainly those of the alpine tundra (8.42%) and light coniferous (5.85%) zonal groups. The next largest group are the species and subspecies of the North Asian (14.83%) and Asian-American (14.01%) elements. The North-Asian taxa are mainly light coniferous (3.42%) and alpine tundra (3.14%) plants, while the Asian-American taxa are maily alpine tundra plants (4.99%) but also some hypoarctic-montane species and subspecies (1.10%). The Eurasian element is rather large too (13.70%) and is dominated by light coniferous (2.57%) species. As a whole, the boreal complex comprises 161 taxa (18.40%) represented mainly by the light coniferous group (17.56%). The important alpine geoelement (26.68%) in the Region is dominated by the alpine tundra (20.26%) zone group. The considerable azonal complex (18.40%) includes mainly aquatic-bog (6.70%) and meadow (6.13%) plants. Of the 82 (11.70%) species and subspecies belonging to the steppe complex 41 taxa belong to the mountain steppe group. There are 45 (6.42%) endemic plants of which 16 are endemic for Yakutia: *Elytrigia villosa, Poa trautvetteri, Papaver angustifolium, P. pulvinatum ssp. lenaense, Hedysarum vicioides, Androsace gorodkovii, Thymus karavaevii, T. oxyodontus, T. verchojanicus, Artemisia subarctica, Taraxacum amgense,* as well as *Carex macrostigmatica, Oxytropis czekanowskii, O. karavaevii* and *Thymus sergievskajae* being specific for this Region.

As regards moisture conditions the most important group is represented by meso-hygrophytic plants (130 taxa or 18.55% of the total species composition of the Region). The next largest groups are xerophytes (18.11%) and mesoxerophytes (15.69%).

The prevailing life form is perennial herbs, including 556 taxa (79.32%) with short-rhizome (226–32.24%), long-rhizome (159–22.68%) and taproot polycarpics (116–16.55%) being most common.

15 rare species are listed in the Red Data Book of Yakutia (Dolinin et al. 2000). Of these *Petasites radiatus, Oxytropis czekanowskii, O. karavaevii,* and *Carex macrostigmatica* are found exclusively within the Olenyok Floristic Region.

*The Yana-Indigirka Floristic Region (Ya-I).* The flora of this Region contains 981 species and subspecies of higher vascular plants belonging to 297 genera and 77 families. The leading families are *Poaceae* 103 (10.50%), *Asteraceae* 101 (10.30%), *Cyperaceae* 97 (9.89%), *Brassicaceae* 58 (5.91%), *Caryophyllaceae* 57 (5.81%), *Rosaceae* 53 (5.40%), *Ranunculaceae* 52 (5.30%), *Salicaceae* 45 (5.59%), *Fabaceae* 43 (4.38%), *Scrophulariaceae* 31 (3.16%), *Polygonaceae* and *Saxifragaceae* 30 (3.06%) each, *Juncaceae* 20 (2.04%), *Ericaceae* 19 (1.94%), *Apiaceae* 16 (1.63%), *Lamiaceae* 14 (1.43%), *Chenopodiaceae* and *Primulaceae* 12

(1.22%) each, *Boraginaceae* 11 (1.12%), *Gentianaceae* 10 (1.02%). The first ten families include 640 taxa comprising 65.24% of total flora of the Region. There are 16 monospecific families of which two have species in the central part of the Verkhoyansk Range and are of special interest. They are *Isoëtes echinospora (Isoetaceae),* and the North-Asian mountain steppe species *Ephedra monosperma (Ephedraceae)* which is more characteristic for southern regions of Yakutia. The largest genera are *Carex* (75), *Salix* (40), *Potentilla* (27), *Artemisia* (23), *Saxifraga* (22), *Poa* (19), *Ranunculus* (18), *Stellaria, Draba* and *Taraxacum* (16 each). The leading 10 genera comprise 272 species and subspecies making up 27.73% of total Region's flora. There are 146 monospecific genera of which *Ermania* occurs only in this FR.

Like the previous Regions, in the Yana-Indigirka Floristic Region taxa with circumpolar distributions (273–27.83%) also are most numerous. They are followed by the Asian-American (14.37%) and North-Asian (13.56%) groups. Rather common are also the Eurasian (11.62%) element confined to the western part of the Region, and Northeast-Asian (6.42%) element which is best represented in the East. The flora also contains species and subspecies of South-Siberian origin (4.89%). The Region has the highest number of endemic plants of Yakutia (36). Of these *Hyalopoa lanatiflora ssp. momica, Eremogone jacutorum, Papaver indigirkense, Corydalis gorodkovii, Potentilla tollii, Oxytropis czerskii, O. incana, O. darpirensis, O. middendorfii ssp. jarovoji, Castilleja tenella,* and *Salix darpirensis* occur only in this Region. The total number of endemic species and subspecies in the Region is 76 (7.75%) of which most have a broader endemic status.

The alpine complex is characterized by the highest species numbers and comprises 338 taxa (34.45%) including alpine tundra (16.11%), alpine (7.34%), hypoarctic-montane (6.52%) and general mountain (4.49%) species and subspecies. The boreal complex contains 21.20% of the total Region's flora, most of which belong to the light coniferous (16.92%) zonal group. The azonal complex (20.49%) is rather diverse and is represented mostly by meadow (8.05%) and aquatic-bog (6.83%) plant groups. It is composed of the arctic (4.99%) and hypoarctic (3.36%) plant groups penetrating from the bordering Arctic zone. The steppe complex comprises 15.49% of the total floristic diversity of the Region, including 7.23% of mountain steppe, 4.08% of forest steppe, 3.97% of steppe plants as well as two taxa of desert-steppe groups. Most mountain steppe species are the same as for the Central Yakutian Floristic Region, though there are species typical solely for Ya-I: *Stellaria jacutica, Potentilla tollii, Helictotrichon krylovii,* etc.

The ecological analysis of the flora shows prevalence of the xerophytic group, including xerophytes (21.81%) and mesoxerophytes (16.31%). The groups of mesohygrophytes (17.33%) and mesophytes (15.09%) are rather numerous too. The dominating life form in the Region is polycarpic herbs, represented by 775 (79.00%) species and subspecies of which short-rhizome (32–32.82%), long-rhizome (195–19.88%) and taproot (172–17.53%) perennial herbs are most abundant. The occurrence of annual and biennial herbs is higher then in the previous Region (8.26%).

Presently the number of species listed in the Red Data Book of the Republic of Sakha (Yakutia) totals 79 taxa. For Yakutia, a number of rare species, such as *Ermania parryoides, Potentilla tollii, Oxytropis czerskii, O. incana, O. darpiensis, Salix darpirensis, S. erythrocarpa, S. rotundifolia, Hyalopoa lanatiflora ssp. momica, Papaver indigirkense, Cardamine conferta, Draba eriopoda, Claytonia sarmentosa* are observed only within the Ya-I.

The flora of this unique and interesting though hard-to-reach Region is still to be studied properly. No doubt further investigations will reveal more species, including specific ones for this Region.

*The Kolyma Floristic Region (Kol).* The flora of the Region amounts to 685 (34.53%) species and subspecies of higher vascular plants of 226 genera and 73 families with *Poaceae* 77 (11.24%), *Cyperaceae* 72 (10.51%), *Asteraceae* 66 (9.64%), *Ranunculaceae* 44 (6.42%), *Brassicaceae* 39 (5.69%), *Caryophyllaceae* and *Rosaceae* 32 (4.67%) each, *Salicaceae* 30 (4.38%), *Fabaceae* 29 (4.23%), *Scrophulariaceae* 23 (3.36%), *Polygonaceae* 20 (2.92%), *Saxifragaceae* 19 (2.77%), *Juncaceae* and *Ericaceae* 17 (2.48%) each as the most important families. The first 10 leading families include 444 taxa, or 64.82% of total flora composition of the Region. The families *Lycopodiaceae, Huperziaceae, Botrychiaceae, Alismataceae, Liliaceae, Convallariaceae, Ceratophyllaceae, Geraniaceae, Urticaceae, Euphorbiaceae, Violaceae* and some others contain one species each within the Region. The Kolyma FR harbours a small number of genera of 10 or more species: *Carex* (54), *Salix* (25), *Potentilla* (18), *Saxifraga* and *Ranunculus* 16 species each, *Potamogeton* and *Artemisia* 15 each, *Draba* (13), *Pedicularis* (12), *Astragalus, Stellaria* and *Taraxacum* 10 taxa each. All abovementioned 12 genera include 214 taxa or 31.64% of the total flora composition. There are records of 116 monospecific genera. The Region has no endemic genera specific for the Region.

The geographical analysis shows that the prevailing longitudinal element is circumpolar, containing 220 taxa (32.12% of local flora). The second largest group is the Asian-American element (119–17.37%). The North-Asian (12.99%) and Eurasian (11.68%) geoelements are reasonably large too. Endemic species comprise 31 taxa (4.53%), mainly those that are endemic in the North–East of Russia and Yakutia: *Poa anadyrica, Festuca kolymense, F. auriculata, Vicia macrantha, Astragalus kolymensis, Erigeron muirii,* etc. The alpine complex is most significant in the Region as regards the number of species and subspecies. It includes 35.03% of total local flora, 18.54% of these belonging to the alpine tundra zonal group. The second place is taken by the azonal complex of species (25.26%), represented by the meadow and aquatic-bog groups and containing 55 taxa (8.03%) each, especially aquatic (5.26%) and riparian (3.94%) plants. The boreal element (18.54%) is composed mainly of the light coniferous group (14.60%). The Arctic complex (8.91%) penetrates into the Region from higher latitudes. The southern outskirts of the Region feature a well pronounced steppe complex of species and subspecies (over 10%), including the mountain steppe (5.99%), forest steppe (3.07%) and steppe proper (1.90%) groups.

According to the ecological analysis, the mesophytic (18.39%) and xerophytic (16.50%) groups are most important in the Region, though the number of mesoxerophytic, mesophytic (91 taxa or 13.28% each) and hygrophytic (12.70%) components is significant too. Just as in the previous Regions, the prevailing life form in the Kolyma FR is polycarpic herbs, including 534 species and subspecies (77.96%) with mainly short-rhizome (218–31.82%), long-rhizome (137–20.00%) and taproot plants (106–15.47%).

20 species are listed in the Red Data Book of Yakutia. Most of those grow at the north–eastern or western limits of their distribution areas. The flora of the Region is greatly influenced by the flora of Chukotka.

*The Central Yakutian Floristic Region (CYa).* The flora of the Region contains 1,032 taxa of higher vascular plants belonging to 381 genera of 97 families, that is 52.02% of the total flora of Yakutia. The leading families are *Poaceae* 103 (9.98%), *Asteraceae* 97 (9.40%), *Cyperaceae* 91 (8.82%), *Ranunculaceae* 61 (5.91%), *Rosaceae* 53 (5.13%), *Fabaceae* 47 (4.55%), *Caryophyllaceae* and *Brassicaceae* 44 (4.26%) each, *Polygonaceae* 32 (3.09%), *Apiaceae* 27 (2.61%), as well as *Salicaceae* 24 (2.32%), *Chenopodiaceae* 23 (2.22%), *Scrophulariaceae* 22 (2.12%), *Lamiaceae* 19 (1.83%), *Boraginaceae* 17 (1.64%), *Primulaceae* and *Orchidaceae* 16 (1.54%) each, *Juncaceae* 14 (1.35%), etc. The first 10 families contain 599 species and subspecies, i.e. 58.09% of the total local flora. 29 monospecific families occur in the territory of the Central Yakutian FR, four of them found solely in the Region. They include *Trapa natans* (*Trapaceae)* found in fossil state, *Caulinia flexilis* (*Najadaceae),* and weed species (*Commelina communis (Commelinaceae)* and *Phacelia tanacetifolia (Hydrophyllaceae)).* Several species from the monospecific *Aspleniaceae, Polypodiaceae, Trilliaceae, Cannabaceae, Amaranthaceae, Hypericaceae, Convolvulaceae, Sambucaceae, Paeoniaceae* and *Cucurbitaceae* grow at the northern border of their distribution areas.

The largest genera are *Carex* (65), *Salix* (22), *Potentilla* (20), *Artemisia* (18), *Ranunculus* (17), *Potamogeton* (16), *Rumex* (13), *Stellaria* (12), *Juncus, Astragalus* and *Elymus* (11 each). These comprise 216 taxa or 20.93% of total the Region's flora. The following genera with one or a few species are recorded exclusively within the territory of the Region: *Caulinia, Blysmus, Bolboschoenus, Phacelia, Ferulopsis, Gagea, Agrimonia, Commelina, Verbascum, Spiranthes, Vaccaria, Redowskia, Thellungiella, Alchemilla, Thermopsis, Cenolophium, Eryngium, Peucedanum, Scabiosa* and *Phalaris*.

Species and subspecies with circumpolar distributions prevail. They consist of 261 taxa (25.29% of the local flora). Eurasian are more (19.57%) and North Asian (10.85%) and Asian-American (7.46%) species are less frequent. The South Siberian element is present (6.89%) with a predominance of the mountain steppe and forest steppe zonal groups. The list of endemic species includes 59 species (5.72%), 2.33% of them being endemic for Yakutia, such as *Artemisia karavaevii, Koeleria skrjabinii, K. karavaevii, Festuca karavaevii, F. skrjabinii, Redowskia sophiifolia, Thermopsis jacutica, Dracocephalum jacutense* and *Thymus sergievskajae*.

The core of the Central Yakutian FR is composed of azonal (30.14%), boreal (29.26%) and steppe (over 25%) zonal complexes. The azonal complex contains mainly meadow (14.44%) and aquatic-bog (8.53%) plant groups. The boreal complex is represented mainly by the light coniferous (22.67%) group and the steppe complex by almost equal shares of forest steppe (9.50%), steppe proper (7.85%) and mountain steppe (7.66%) elements. The relatively small alpine complex (13.08%) contains mainly hypoarctic-montane (4.75%) and alpine tundra (4.55%) species.

As regards moisture response mesophytes (20.64%) followed by xerophytes (16.96%), xeromesophytes (14.54%), and mesoxerophytes (13.47%) are the most important. The number of species growing under excessive or variable moisture conditions is rather high: 12.79% mesohygrophytes and 10.85% hygrophytes. The prevalent life form is herbaceous polycarpics containing 758 species and subspecies (73.45% of the local flora), mainly consisting of short-rhizome (327–31.69%) and long-rhizome (219–21.22%) plants. The Region's flora features a high number of annual and biennial herbaceous plants: 161 taxa (15.06%), the great majority being weeds.

There are 79 Red Data Book species and subspecies in the Region, 14 of these exclusive for this territory (Dolinin et al. 2000). First of all, they are local endemics *Koeleria skrjabinii, K. karavaevii, Festuca karavaevii, F. skrjabinii, Redowskia sophiifolia* and *Thermopsis jacutica,* but also records of the species made in $18^{th}$-early $20^{th}$ centuries that still need to be proved: *Oxytropis glabra, Veronica scutellata* and *Listera cordata.* The Region also harbours two *Gagea* species: *G. pauciflora* and *G. provisa* (the latter is a relic Pleistocene species). *Trapa natans* was found fossil in peatlands of the Mamontova Gora (Mammoth Mountain) location in Central Yakutia. The rest of the rare species represent plants growing at the border of their distribution areas (like *Convolvulus arvensis,* an adventive weed, and *Mertensia davurica).*

Despite the fact that the flora of the Central Yakutian FR is the best studied, investigations of the last 10 years have yielded a number of new records for Yakutia: *Bromopsis karavajevii, Eriochloa villosa, Leymotrigia wiluica, Leymus buriaticus, Phalaris canariensis, Carex diplasiocarpa, Rumex rossicus, R. ucranicus, Suaeda corniculata ssp. erecta, Sedum pallescens, Agrimonia pilosa, Alchemilla murbeckiana, Potentilla altaica, Bupleurum scorzonerifolium, Carum buriaticum, Eryngium planum, Linaria vulgaris, Verbascum nigrum, Phacelia tanacetifolia, Cirsium esculenthum.*

The flora of certain communities of the CYa deserve special attention due to their peculiarity. The flora of the loose or semi-sodded ancient eolian sands ("tukulans", pronounced as *too-koo-lany*) occurs only in Central Yakutia. It contains 70 species of higher vascular plants, 60% of them being of Eurasian or circumpolar distribution: *Elytrigia repens, Poa pratensis, Calamagrostis epigeios, Festuca rubra, F. ovina, Arenaria stenophylla, Antennaria dioica,* etc. The Asian complex is represented by *Campanula langsdorffiana, Dianthus versicolor, Silene polaris, Euphorbia discolor, Thymus mongolicus.* The Siberian, Daurian-Mongolian and Manchurian-Siberian groups contain 25 species: *Selaginella sibirica, Koeleria seminuda, Festuca lenensis, Polygonum angustifolium, P. laxmannii, Arabis septentrionalis, Minuartia laricina, Corispermum crassifolia, Eritrichium sericeum, Saxifraga spinulosa,* etc. The following species of the alpine tundra zonal group are also abundant on sandy substrate: *Rumex graminifolius, Minuartia verna, Dianthus repens, Tanacetum bipinnatum,* etc. The endemic psammophytes are *Artemisia karavaevii, Koeleria skrjabinii, K. karavaevii, Festuca karavaevii, F. skrjabinii* and *Thymus sergievskajae* (Dobretsova 1962; Skryabin and Karavaev 1991; Andreyev et al. 1987; Dolinin et al. 2000).

The steppe flora along with the Eurasian forest steppe species (*Poa angustifolia, Helictotrichon schellianum, Cerastium arvense, Eremogone saxatilis, Silene repens, Veronica incana, Sedum purpureum,* etc.), Siberian mountain steppe (*Vicia multicaulis, Carex supina, C. pediformis, Allium strictum, Oxytropis uralensis, Castilleja pallida,* etc.) and Daurian-Mongolian steppe species (*Stipa krylovii, S. capillata, Agropyron cristatum, Cleistogenes squarrosa, Carex duriuscula,* etc.), contain plants endemic for Yakutia and East Siberia (*Krascheninnikovia lenensis, Dracocephalum jacutense, Thermopsis jacutica, Elytrigia jacutorum, Artemisia pubescens, Carex lanceolata, Isatis jacutensis*). The unique Natural Park "Lena Pillars" comprises 464 species and subspecies of higher vascular plants of 276 genera and 81 families. The flora of the Park is composed of steppe and forest-steppe plants (about 26%). Rock ledges and crevices are the habitats for *Redowskia sophiifolia,* the stenolocal endemic species of the Lena and Sinyaya Pillars (natural monuments, resulting from weathering of carbonate rocks along river banks). It is considered a Neogene- or Pleistocene-aged relic plant (Dolinin et al. 2000; Sosina and Isaev 2003).

*The Upper Lena Floristic Region (UL).* The flora includes 1,085 species and subspecies of 402 genera and 101 families. The following families are richest in species and subspecies: *Asteraceae* 104 (9.59%), *Poaceae* 96 (8.85%), *Cyperaceae* 95 (8.76%), *Ranunculaceae* 65 (5.99%), *Rosaceae* 50 (4.61%), *Fabaceae* 45 (4.15%), *Caryophyllaceae* and *Brassicaceae* 37 (3.41%) each, *Polygonaceae* 29 (2.67%), *Lamiaceae* 27 (2.49%), *Apiaceae* and *Salicaceae* 26 (2.40%) each, *Orchidaceae* 23 (2.12%), *Chenopodiaceae* and *Scrophulariaceae* 19 (1.75%) each, *Boraginaceae* and *Ericaceae* 17 (1.57%) each, *Primulaceae* 16 (1.48%), *Gentianaceae* 12 (1.01%), *Juncaceae* 11 (1.01%), etc. The first ten families consist of 585 species and subspecies, 53.92% of the total Region's flora. The flora of the Region contains 27 monospecific families, including *Sinopteridaceae* and *Hemerocallidaceae* which are specific for this Region. The following genera are largest in number of species and subspecies: *Carex* (75), *Salix* (21), *Potentilla* (16), *Artemisia* and *Ranunculus* (15 each), *Potamogeton* (14), *Viola* (13), *Astragalus* (12), *Allium* (11), *Saussurea* and *Elymus* 8 taxa each. All these 208 species make up 19.17% of local flora. The territory of the Region harbours 212 monospecific genera, of which *Aleuritopteris, Pteridium, Apera, Milium, Panicum, Epipactis, Epipogium, Neottianthe, Salsola, Coptis, Neslia, Sibbaldianthe, Ferulopsis, Pastinaca, Nymphoides, Pulmonaria, Amethystea, Betonica, Glechoma, Origanum, Veronicastrum, Anthemis, Helichrysum, Tragopogon, Tripolium* and *Tussilago* solely occur in this FR.

The Upper Lena FR is characterized by prevalence of plants with circumpolar distributions (270 taxa or 24.88% of local flora). The Eurasian element contains about 20 taxa (19.26%); the North Asian 115 (10.60%). The Asian-American and South Siberian are almost equally strong: 6.64% and 6.54% respectively.

The endemic group includes 35 species and subspecies (3.23%), eight (0.74%) of them are endemic for Yakutia. *Eritrichium aldanensis* is observed strictly within boundaries of the FR, while *Papaver jacuticum, P. nudicaule ssp. gracile, Dryas viscosa, Euphorbia maackii, Eritrichium jacuticum, Taraxacum kuvajevii* and *Krascheninnikovia lenensis* were recorded in other Floristic Regions of the republic as well.

The boreal zonal complex leads among the latitudinal geoelements of the Region, containing 362 species and subspecies (33.36%). It is dominated by the light coniferous group (24.79%). The azonal (27.37%) and steppe (21.38%) complexes are also numerous. The azonal complex is dominated by meadow (13.46%) plants, and the steppe complex by forest steppe (8.39%) and mountain steppe (6.82%) elements.

As regards moisture conditions the Region contains mainly mesophytes (23.41%) and xerophytes (15.39%), though xeromesophytes (13.27%) and mesoxerophytes (12.81%) are also important. The Upper Lena FR has the highest number of hydrophytes (3.69%). The number of mesohygrophytes (11.17%) and hygrophytes (10.05%) is lower compared to the previous FR. The leading life forms are perennial herbs (776–71.52%), mainly short-rhizome (355–32.72%) and long-rhizome (224–20.65%) plants, as well as herbaceous monocarps, including 151 taxa (13.91%).

The records of rare and endangered species of the Upper Lena FR number 136 taxa, most of them growing at the border of their distribution area. The territory of the Region is the exclusive habitat for 36 species, such as *Paeonia anomala, Viola uniflora, Pteridium aquilinum, Sibbaldianthe adpressa, Oxytropis pilosa, Hedysarum gmelinii, Potentilla acaulis, Festuca pseudosulcata, Tussilago farfara, Origanum vulgare, Mentha dahurica, Lycopus angustus, Milium effusum, Helichrysum arenarium, Pulmonaria mollis, Nymphoides peltata, Anagallidium dichotomum, Persicaria hydropiper,* etc. The Region has most *Orchidaceae* of Yakutia: 23 of the 27 known species. Two of these, *Epipactis helleborine* and *Epipogium aphyllum,* found in upper reaches of the Lena River, are listed in the Red Data Book of Yakutia. The last years of investigations yielded two more species of this family: *Platanthera biflora* and *Neottianthe cucullata.* Moreover, recent studies resulted in numerous new records for Yakutia: *Cardamine amara, Tragopogon sibiricus, Ferulopsis hystrix, Pastinaca sylvestris, Anthemis subtinctoria, Sparganium gramineum, S. stoloniferum, Agrostis bodaibensis, Festuca pseudosulcata, F. sibirica, Melica altissima, Carex dioica, C. nanella, C. nigra, C. viridula, C. vulpina, Polygonum propinquum, P. rigidum, Chenopodium urbicum, Thalictrum appendiculatum, Helichrysum setigerum, Lathyrus vernus, Euphorbia humifusa, Mentha canadensis, Galium aparine, G. palustre, Sambucus manshurica, Adenophora coronopifolia, Hieracium pseudofariniramum, Ptarmica acuminata, Scorzonera glabra.*

According to Golyakov (1994), the Olyokminsky Sanctuary, situated in the Upper Lena FR, contains 654 species and subspecies of higher vascular plants belonging to 292 genera and 80 families. 35 taxa of 16 genera and 12 families are spore plants, making up 5.35% of the floral diversity of the Region. The family spectrum of the Sanctuary has many features in common with the boreal zone of the Northern Hemisphere though the leading family is *Cyperaceae* with 64 (9.8%) species and subspecies. *Orchidaceae* is also well represented with 18 (2.75%) taxa. The genus spectrum has a typical boreal character and is dominated by *Artemisia, Potentilla, Pedicularis*, followed by *Aconogonon (Polygonaceae)* and to a lesser extent by *Veronica.* The flora of the Sanctuary contains many species with distribution areas exceeding the territory of Asia (426 taxa or 65.13%). There are cosmopolites (26 or 3.97%), and Holarctic species (190–29.0%). More than half of the species and subspecies in the Sanctuary also occur in Europe (378 or 57.79%) and North America (264 or 40.36%). Within the Asian geographical elements (34.86%) almost half belong to the Far East group (113 species and subspecies), 35 are Asian proper, and 80 Siberian. Endemics and subendemics number 25 (3.8%), all of them being endemics of Yakutia: *Rumex jacutensis, Gastrolychnis gracilis, Minuartia jacutica, Eritrichium jacuticum, Euphrasia jacutica, Saussurea hypargyrea, Aconogonon amgense, Anemonastrum calvum, Dryas viscosa,* etc. 74 species in the Sanctuary are listed in the Red Data Books of the Russian Federation (Golovanov et al. 1988) and the Republic of Sakha (Yakutia) (Dolinin et al. 2000): *Cypripedium macranthon, Cypripedium calceolus, C. guttatum, Orchis militaris, Aconogonon amgense, Calypso bulbosa, Iris laevigata, Parietaria micrantha, Spiraea elegans, Papaver setosum, Trollius uncinatus, Clematis fusca, Carex pseudocyperus,* etc.

*The Aldan Floristic Region (Ald).* The Region contains 1,166 taxa of 402 genera and 97 families, 58.77% of the total flora of Yakutia. The richest families are *Cyperaceae* 106 (9.09%), *Asteraceae* 103 (8.83%), *Poaceae* 98 (8.40%), *Ranunculaceae* 78 (6.69%), *Rosaceae* 61 (5.23%), *Caryophyllaceae* 48 (4.12%), *Salicaceae* and *Brassicaceae* 45 (3.86%) each, *Fabaceae* 41 (3.52%), *Polygonaceae* 34 (2.92%), *Scrophulariaceae* 30 (2.57%), *Saxifragaceae* 29 (2.49%), *Apiaceae* and *Ericaceae* 24 (2.06%) each, *Juncaceae* 22 (1.89%), *Orchidaceae* and *Gentianaceae* 19 (1.63%) each, *Lamiaceae* 18 (1.54%), *Boraginaceae* 17 (1.46%), *Primulaceae* 16 (1.37%), etc. The first ten families contain 659 species and subspecies (56.52% of the local flora). The largest genera are *Carex* (77), *Salix* (40), *Saxifraga* (24), *Potentilla* (20), *Ranunculus* (18), *Artemisia* (17), *Viola, Juncus* and *Poa* (15 each), *Pedicularis* (14); together 255 species or 21.87% of the total floral diversity of the Region.

Species and subspecies with circumpolar distributions are the most frequent (299–25.64%). Species with Eurasian and North Asian distributions comprise 15.78% and 11.84%, respectively. The proportion of East Asian (8.49%) and Asian-American (8.32%) elements is almost equal. Species characteristic for the Far East (7.55%) and South Siberian (5.58%) floras penetrate into the Region, e.g. *Aquilegia glandulosa, Bergenia crassifolia, Paraquilegia microfilla, Salix cardiophylla, Saxifraga stellerana, Primula cuneifolia, Isotes asiatica, Picea ajanensis, Pedicularis kuznetzovii, Viola kusnezowiana, V. patrinii, Clintonia udensis, Acelidantus anticleoides, Saussurea baicalensis*, etc.

58 species and subspecies (4.97%) are endemic, including 1.37% endemic for Yakutia. There are 5 local endemics: *Aconogonon amgense, Sorbocotoneaster pozdnjakovii, Anoplocaryum helenae, Adenophora jacutica* and *Saussurea hypargyrea.*

The most important geographic element of the Region is the boreal complex, containing 371 species and subspecies (31.82%) with a prevalence of the light coniferous group (22.56%). The Region features the highest proportion of the dark coniferous group (68 taxa or 5.49% of the local flora). The alpine complex is strong as well (27.62%) with mainly species of the alpine tundra (9.01%) group. The azonal complex (23.07%) contains mainly meadow plants (10.98%), and the steppe complex (15.95%) forest steppe (6.52%) and mountain steppe (5.66%) species and subspecies.

There are 206 monospecific genera, 23 of these occurring only in the Aldan FR: *Phegopteris, Schizachne, Acelidanthus, Clintonia, Habenaria, Dichodon, Clematis, Paraquilegia, Bunias, Subularia, Bergenia, Sieversia, Sorbocotoneaster, Phyllodoce, Calathiana, Halenia, Clinopodium, Lagopsis, Lobelia, Ligularia.*

Mesophytic plants (22.38%) predominate in the Region, represented by forest and meadow species and subspecies.The mountainous topology of the Region, widespread stony and rubbly sodded and semi-sodded slopes and placers, explain the rather high contribution of xerophytes (14.92%) and mesoxerophytes (15.52%). Most of the mesohygrophytic (15.09%) and hygrophytic (11.49%) plants are concentrated in the middle and lower reaches of large river valleys and are abundant in lakes, mainly of the oxbow type.

The prevailing life form is the herbaceous polycarp with 878 taxa (75.30%), including short-rhizome (407–34.91%), long-rhizome (20.33%) and taproot (162–13.89%) species. The Region leads in arboreous species diversity (152–13.04%). There are 15 tree taxa (17 in all of Yakutia), 76 taxa of shrubs (89 in all of Yakutia) and 49 taxa of dwarf shrubs (67 in Yakutia). Only in the Yana-Indigirka FR there are more species of dwarf semishrubs (19) than in the Aldan FR (12) (and 35 for Yakutia as a whole).

For the Aldan FR there are 152 rare and endangered species listed in the Red Data Book of Yakutia, 68 of these found strictly within the boundaries of the Region. A number of species, including rare ones, are characteristic for the Far East and South Siberian floras. Most of them grow in the Region at the border of their distribution areas: *Aquilegia glandulosa, Bergenia crassifolia, Paraquilegia microphylla, Salix cardiophylla, Saxifraga stellerana, Primula cuneifolia, Isoëtes asiatica, Pedicularis kuznetzovii, Viola kusnezowiana, V. patrinii, Clintonia udensis, Acelidanthus anticleoides, Saussurea baicalensis, Pulsatilla ajanensis, Rhodiola rosea, Habenaria linearifolia, Paris hexaphylla, Cinna latifolia,* etc. The stenolocal endemic and spontaneously interspecific hybrid *Sorbocotoneaster pozdnjakovii* and the recently recorded *Anoplocarium helenae* are restricted to the Aldan River valley.

In the southeastern outskirts of the Region, bordering on the Khabarovsk Territory, the Resource Reserve "Bolshoye Tokko" is situated. It exceeds all the known protected areas of the republic in numbers of rare and endemic species. For instance, of the 542 recorded higher vascular species 86 are rare, many of them growing at the border of their distribution area: *Saxifraga staminosa, Phyllodoce caerulea, Driopteris expansa, Ligularia sibirica, Taraxacum zhukovae, Saussurea soczavae, Mertensia rivularis, Dicentra peregrina, Subularia aquatica, Aquilegia amurensis, Picea ajanensis, Salix divaricata, Pulsatilla ajanensis, Cinna latifolia* etc. Some are endemics of the North–East: *Cardamine conferta, Papaver nivale, Chrysosplenium saxatile, Eritrichium jacuticum, Saxifraga tilingiana.* Species such as *Braya siliquosa, Gastrolychnis saxatilis, Salix dshugdshurica, S. cardiophylla, Saxifraga davurica, S. stelleriana, Claytonia eschscholtzii, Callianthemum isopyroides, Listera savatieri* are rare throughout their distribution areas. New records for Yakutia are: *Acelidanthus anticleoides, Botrychium boreale, Stellaria calycantha, Pinguicula algida, Allium maakii, Saxifraga staminosa, Ptilagrostis alpina, Senecio subfrigidus, Trollius uniflorus, Festuca chionobia, Carex kreczetoviczii, C. malyschevii,* etc. Species such as *Potentilla biflora, Subularia aquatica, Viola kusnezowiana, Isoetes asiatica, Sieversia pusilla, Saxifraga tilingiana* are known solely for the Tokinsky Stanovik Range.

Thus, the Aldan FR contains a significant amount of the sub-oceanic Far East species, which is not typical for Yakutia as a whole. The Region features the highest floral diversity of Yakutia and harbours communities that are unique both for Yakutia and North Asia.

As mentioned before, Yakutia is first of all an administrative unit of Russia. And allocation of floristic regions here in is rather relative. This first of all refers to the frontier territories. For instance, the Kolyma FR, situated in the far north–east of Yakutia, has many plants (including endemics) in common with the Chukotka flora. The flora of the north–eastern part of the Aldan FR is very similar to that of Okhotia, while its south–eastern part contains the taxa typical for the Zeya Floristic Region of the Far East, etc.

Based on the abovementioned analysis of the taxonomical and ecological-geographical structure of the flora of Yakutia and its Floristic Regions the following conclusions can be made. The flora of Yakutia numbers 1,900 species and 84 subspecies. Considering that the level of spatial diversity (Arrhenius equation) for this region is 0.1 (Malyshev 1972), the floristic abundance of Yakutia is 1,407 species and subspecies per 100 thousand sq. km. The Aldan Floristic Region has the richest species diversity in Yakutia. This can be explained both by the southern location of the Region (i.e. more favourable climatic conditions) and the penetration of synanthropic plants due to active industrial development of the Region. The poorest floras of the Olenyok and Kolyma FRs are the result of their rather high latitudinal location and monotonous plains of over-wetted landforms. Another possible reason of the low diversity is underestimation of flora due to lack of knowledge on the Regions.

## 2.3 Mosses

### *2.3.1 Leafy Mosses*

**E.I. Ivanova**

The story of investigations of the leafy mosses of Yakutia dates back to the end of the nineteenth century, when H.V. Arnell processed an enormous collection of H. Nilson-Ehle, a member of the Swedish Polar Expedition of 1898. The collection included the moss samples collected from Irkutsk vicinity and the Lena River valley (from Verkholensk through the river mouth). He identified 14 sphagnum and 233 bryum moss species. At that time collections where made by botanists studying vascular plants, amateurs, and local lore students. So the samples where sent for identification to the specialists, mostly to the Finnish bryologist V.F. Brotherus. He recorded 28 moss species from the Novosibirskie Islands collected by A. Byalynitski-Biruli, and 50 species from Bolshoye Toko Lake area collected by O.I. Kuzeneva and N.I. Prokhorov. At the Botanical Museum in Helsinki harbours the moss collection made by the Finnish botanist A. K. Cajander in 1901 during the complex expedition along the Lena River. Those mosses were also identified mostly by V. Brotherus. The collection comprises 97 leafy mosses of 56 genera and 26 families. Establishment of the Research Base of the Academy of Sciences of the USSR (AS USSR) in Yakutsk in 1947 resulted in the switch from random studies to the planned complex investigation of the vegetation cover of Yakutia. The moss collections made by T.A. Rabotnov, L.N. Tyulina, V.B. Kuvayev, L.A. Dobretsova, V.I. Ivanova (Perfilyeva), etc. made up the basis of the bryological herbarium founded in 1949 at the Yakut Branch of the AS USSR. At that time the moss species where identified by bryologists of the Komarov Botanical Institute (KBI AS USSR, Leningrad): I.I. Abramov, A.L. Abramova, L.I. Savich-Lyubitskaya, Z.M. Smirnova, O.M. Afonina, etc. In the 1970s the first qualified bryologist of Yakutia, Nelli Ayusheevna Stepanova, started her activities. She made a great contribution to establishing the bryological research of the Republic (Danilova 2005).

The present collection of the Institute of Biological Problems of Cryolithozone, Russian Academy of Sciences (SASY), numbers about 30,000 samples of leafy mosses. The basis of the bryological herbarium is represented by the collection of Stepanova (1970–1984, the tundra zone of Yakutia), Nikolin (1985–1992, Central Verkhoyanye), and Ivanova (1991–1995, Southern Yakutia). There are also the duplicates (a kind gift of the colleagues from KBI RAS) from the collections of H. Nilson-Ehle, V. Broterus, A.E. Katenin, B.A. Yurtsev, I.D. Kildyushevsky, O.M. Afonina, etc. representing mostly the Arctic and north–eastern part of Yakutia. There is also the duplicate material collected by Martti Ohenoja (Botanical Museum in Oulu, Finland) in the Kolyma River basin in 2003. Nowadays, mosses are actively sampled in all regions of Yakutia. It has already yielded a large number of rare and interesting species. The duplicates of Yakutian mosses are stored at numerous Russian and foreign herbariums (H – Helsinki, LE – Saint-Petersburg,

MNA – Moscow, IRK – Irkutsk, PZV – Petrozavodsk, NSK – Novosibirsk, etc.) (Danilova 2005). In 2007 the "Exicates of bryophytes of Siberia" (Ivanova and Pisarenko 2007) was published in collaboration with the bryological herbarium of the Central Siberian Botanical Garden, Siberian Branch of the Russian Academy of Sciences (NSK).

An annotated list of leafy mosses of Yakutia based on herbarium samples (SASY, LE, MHA, H) and supplemented by reference data is given in the monograph "Diversity of the vegetation world of Yakutia" (Danilova 2005). The taxonomical position of *BRYOPHYTA* species in the article are given according to the "List of mosses of Eastern Europe and Northern Asia" (Ignatov et al. 2006).

The flora of leafy mosses of Yakutia comprises 523 species (537 taxa including varieties) belonging to 173 genera, 51 families, 18 orders, and 5 classes of the *BRYOPHYTA* division. The class *Sphagnopsida* comprises 35 species of 1 genus, 1 family, and 1 order. The class *Andreaeopsida* – 4 species and 1 variety of 1 genus, 1 family, and 1 order. The class *Polytrichopsida* – 19 species and 2 varieties of 7genera, 1 family, and 1 order. The class *Tetraphidopsida* – 2 species, 2 genera, 1 family, and 1 order. The class *Bryopsida* – 463 species and 11 varieties of 162 genera, 47 families, 11 orders (Table 2.1). Compared to other regions of Russia, the flora of leafy mosses of Yakutia is rather rich and relatively properly studied. For instance, there are 477 species of leafy mosses recorded in Altai, 530 species in the Ural region, 450 species in Karelia, 450 species in the Primorsk Territory, 431 species in the Republic of Komi, 406 species in the Leningrad Oblast, 352 species in the Tver Oblast, and 467 species in Chukotka. Quantitatively, the moss flora of Yakutia makes up 82% of the Siberian moss flora (Danilova 2005) and 40% of that of Eastern Europe and Northern Asia (the former USSR) (Ignatov et al. 2006). The ratio of the number of Yakutian leafy moss species to higher vascular plants of Yakutia is 1 : 3.8, or 26.5%. This indicates the significant role of mosses in the vegetation cover and its close connection with the northern regions.

The taxonomical analysis of the leafy moss flora of Yakutia proves its pecularity. There are 15 leading families containing 51–13 species each (Table 2.2) or 393 species in total which is 75.1% of the moss flora of Yakutia.

Thirteen families are monospecific, and the rest of the families are represented by less that 11 species. There are 173 genera containing 3 species on average. Twenty of them are quite large – 37 to 6 species each (Table 2.3). They comprise 243 species or 46.5% of the total moss flora of Yakutia. Eighty five genera consist of 1 species. A rather large number of monospecific genera and families is a characteristic feature for many northern bryofloras. The regional peculiarities are expressed by the leading families *Pottiaceae* and *Grimmiaceae* (92 species, or 17.6%), as well as the genera *Schistidium, Grimmia, Encalypta, Didymodon* and *Tortula* (58 species, or 11%) indicating mountainous regions with widespread rock outcrops and stones. The leading families *Amblystegiaceae, Sphagnaceae, Mniaceae, Brachytheciaceae, Calliergonaceae* (135 species, or 25.8%) and genera *Sphagnum, Brachythecium, Plagiomnium, Mnium, Warnstorfia* and *Drepanocladus* (72 species, or 13.8%) indicate the wide distribution of wet and waterlogged landscapes, especially in the North of Yakutia. As a whole, the family and genera spectra define the Yakut mosses as belonging to the mountain taiga flora.

**Table 2.1** Taxonomical composition of the leafy moss flora of Yakutia

| Classes, orders, families | Number of genera | Number of species and varieties |
|---|---|---|
| ***Class SPHAGNOPSIDA*** | | |
| ***Order Sphagnales*** | | |
| *Sphagnaceae* | *1* | *35* |
| ***Class ANDREAOPSIDA*** | | |
| ***Order Andreaeales*** | | |
| *Andreaeaceae* | *1* | *4+1* |
| ***Class POLYTRICHOPSIDA*** | | |
| ***Order Polytrichales*** | | |
| *Polytrichaceae* | *7* | *19+2* |
| ***Class TETRAPHIDOPSIDA*** | | |
| ***Order Tetraphidales*** | | |
| *Tetraphidaceae* | *2* | *2* |
| ***Class BRYOPSIDA*** | | |
| ***OrderBuxbaumiales*** | | |
| *Buxbaumiaceae* | *1* | *1* |
| ***Order Timmiales*** | | |
| *Timmiaceae* | *1* | *6* |
| ***Order Catoscopiaceae*** | | |
| *Catoscopiaceae* | *1* | *1* |
| ***Order Funariaceae*** | | |
| *Funariaceae* | *3* | *4* |
| ***Order Encalyptaceae*** | | |
| *Encalyptaceae* | *2* | *10* |
| ***Order Bryoxiphiales*** | | |
| *Bryoxiphiaceae* | *1* | *1* |
| ***Order Scouleriaceae*** | | |
| *Scouleria* | *1* | *1+1* |
| ***Order Grimmiaceae*** | | |
| *Grimmiaceae* | *9* | *41+1* |
| *Seligeriaceae* | *2* | *4* |
| ***Order Dicranales*** | | |
| *Leucobryaceae* | *1* | *1* |
| *Dicranaceae* | *3* | *28+1* |
| *Rhabdoweisiaceae* | *9* | *17* |
| *Ditrichaceae* | *4* | *11* |
| *Pottiaceae* | *17* | *51+3* |
| *Fissidentaceae* | *1* | *5* |
| *Schistostegaceae* | *1* | *1* |
| ***Order Splachnales*** | | |
| *Meesiaceae* | *4* | *7* |
| *Splachnaceae* | *4* | *14* |
| ***Order Orthotrichaceae*** | | |
| *Orthotrichaceae* | *3* | *10* |
| ***Order Hedwigiaceae*** | | |
| *Hedwigiaceae* | *1* | *1* |
| ***Order Bryales*** | | |
| *Bryaceae* | *4* | *40* |
| *Mielichhoferiaceae* | *2* | *14* |

**Table 2.1** (continued)

| Classes, orders, families | Number of genera | Number of species and varieties |
|---|---|---|
| *Mniaceae* | *7* | *26* |
| *Bartramiaceae* | *4* | *9* |
| *Aulacomniaceae* | *1* | *3* |
| ***Order Hypnales*** | | |
| *Fontinalaceae* | *2* | *3+1* |
| *Fabroniaceae* | *1* | *1* |
| *Plagiotheciaceae* | *7* | *17* |
| *Pterigynandraceae* | *1* | *1* |
| *Leucodontaceae* | *1* | *2* |
| *Hypnaceae* | *2* | *3* |
| *Entodontaceae* | *1* | *2* |
| *Pylaisiadelphaceae* | *1* | *1* |
| *Pseudoleskeaceae* | *1* | *1* |
| *Anomodontaceae* | *2* | *2* |
| *Neckeraceae* | *2* | *2+1* |
| *Climaciaceae* | *1* | *1* |
| *Hylocomiacae* | *5* | *6+1* |
| *Brachytheciaceae* | *8* | *24* |
| *Calliergonaceae* | *5* | *13* |
| *Scorpidiaceae* | *4* | *9* |
| *Pylaisiaceae* | *6* | *17+1* |
| *Rhytidiaceae* | *1* | *1* |
| *Pseudoleskeellaceae* | *1* | *5* |
| *Leskeaceae* | *3* | *3* |
| *Thuidiaceae* | *3* | *5* |
| *Amblystegiaceae* | *17* | *37+1* |
| In total: 5 classes, 18 orders, 51 families | 173 | 523+14 |

The system of geographic elements was used for a geographic analysis of the moss flora of Yakutia. This system was developed by Lazarenko (1956), and later was supplemented and specified by Shlyakov (1961), Bardunov (1974), and Ignatov (1996). Following Ignatov (1996), the reliability of the analysis requires exclusion of the species with a questionable geographic details. In our case we selected 118 such species (22.6% of the total amount of species) including taxa with wide (pluri-zone) distributions and those preferring different natural zones in various regions of the Globe (that is not characteristic for vascular plants). The rest of the species can be arranged into the following zone elements: arctic, boreal, hemiboreal, nemoral, and arid.

The arctic element comprises 47.8% of the species considered and is represented by two sub-elements. The arctic sub-element includes *Aplodon wormskioldii, Ceratodon heterophyllus, Cinclidium subrotundum, Funaria arctica, Hennediella heimii, Myuroclada rotundifolia, Sanionia georgicouncinata, Seligeria polaris*, etc. (8.1% of total amount of analyzed species). These species are

**Table 2.2** Leading families of leafy mosses of Yakutia

| Families | Yakutia | Number of species and their rank places in floristic regions | | | | | | |
|---|---|---|---|---|---|---|---|---|
| | | Arct | Ol | Ya-I | Kol | CYa | UL | Ald |
| *Pottiaceae* | 51 (1) | 29 (2) | 13 (3) | 39 (1) | 8 (3–1) | 28 (1) | 25 (2) | 18 (3) |
| *Grimmiaceae* | 41 (2) | 25 (4–1) | 5 (8–1) | 26 (3–1) | 4 (7–1) | 17 (5) | 18 (5) | 21 (2–1) |
| *Bryaceae* | 40 (3) | 26 (3) | 7 (6–1) | 26 (3–2) | 6 (5–1) | 12 (8–1) | 8 (10–1) | 12 (6–1) |
| *Amblystegiaceae* | 37 (4) | 25 (4–1) | 18 (2) | 25 (4) | 16 (1) | 22 (3) | 29 (1–1) | 17 (4–1) |
| *Sphagnaceae* | 35 (5) | 30 (1) | 22 (1) | 28 (2) | 8 (3–2) | 23 (2) | 29 (1–2) | 26 (1) |
| *Dicranaceae* | 28 (6) | 18 (5) | 12 (4) | 23 (5) | 8 (3–3) | 16 (6–1) | 23 (3) | 21 (2–2) |
| *Mniaceae* | 26 (7) | 14 (7) | 7 (6–2) | 20 (6) | 4 (7–2) | 19 (4) | 21 (4) | 17 (4–2) |
| *Brachytheciaceae* | 24 (8) | 13 (8–1) | 7 (6–3) | 12 (11) | 9 (2) | 16 (6–2) | 15 (6–1) | 12 (6–2) |
| *Polytrichaceae* | 19 (9) | 15 (6) | 7 (6–4) | 13 (10) | 6 (5–2) | 9 (9) | 13 (7–1) | 11 (7–1) |
| *Rhabdoweisiaceae* | 17 (10–1) | 13 (8–2) | 6 (7–1) | 15 (8) | – | – | 10 (8) | 14 (5) |
| *Plagiotheciaceae* | 17 (10–2) | 13 (8–3) | 7 (6–5) | 16 (7–1) | 6 (5–3) | 13 (7) | 13 (7–2) | 12 (6–3) |
| *Pylaisiaceae* | 17 (10–3) | 11 (10–1) | 8 (5–1) | 16 (7–2) | 7 (4–1) | 12 (8–2) | 15 (6–2) | 11 (7–2) |
| *Mielichhoferiaceae* | 14 (11–1) | 13 (8–4) | – | 14 (9) | 4 (7–3) | 5 (10–1) | – | 8 (9–1) |
| *Splachnaceae* | 14 (11–2) | 12 (9) | 5 (8–2) | – | 5 (6) | – | – | – |
| *Calliergonaceae* | 13 (12) | 11 (10–2) | 8 (5–2) | – | 8 (3–4) | 7 (9–1) | 9 (9) | 9 (8–1) |
| *Ditrichaceae* | – | – | 5 (8–3) | – | 7 (4–2) | 5 (10–2) | – | 8 (9–2) |
| *Encalyptaceae* | – | – | – | – | – | – | – | – |
| *Orthotrichaceae* | – | – | – | – | – | 5 (10–3) | – | 7 (10) |
| *Bartramiaceae* | – | – | 6 (7–2) | – | 4 (7–4) | 5 (10–4) | – | 9 (8–2) |
| *Scorpidiaceae* | – | – | 7 (6–6) | – | 4 (7–5) | 7 (9–2) | 8 (10–2) | 8 (9–3) |
| In total: | 393 | 268 | 150 | 263 | 114 | 221 | 236 | 231 |
| % of total species number | 75,1 | 71,1 | 79,4 | 68,3 | 82,6 | 80,1 | 72,1 | 77,0 |

**Table 2.3** Leading genera of leafy mosses of Yakutia

| | | Number of species and their rank places in floristic regions | | | | | | |
|---|---|---|---|---|---|---|---|---|
| Genera | Yakutia | Arct | Ol | Ya-I | Kol | CYa | UL | Ald |
| *Bryum* | 37 (1) | 24 (2) | 7 (3) | 25 (2) | 6 (2–1) | 11 (3) | 7 (4–1) | 11 (3) |
| *Sphagnum* | 35 (2) | 30 (1) | 22 (1) | 28 (1) | 8 (1) | 23 (1) | 29 (1) | 26 (1) |
| *Dicranum* | 19 (3) | 14 (4) | 12 (2) | 17 (3) | 6 (2–2) | 14 (2) | 17 (2) | 16 (2) |
| *Schistidium* | 18 (4) | 15 (3) | – | 8 (7–1) | – | 6 (6–1) | 8 (3–1) | 7 (5–1) |
| *Grimmia* | 13 (5–1) | – | – | 12 (5) | – | 8 (5) | 5 (6–1) | 8 (4–1) |
| *Pohlia* | 13 (5–2) | 13 (5) | – | 13 (4) | 4 (3–1) | 5 (8–1) | 7 (4–2) | 8 (4–2) |
| *Brachythecium* | 10 (6) | 6 (9–1) | 5 (4–1) | 9 (6–1) | 6 (2–3) | 9 (4) | 6 (5–1) | 7 (5–2) |
| *Encalypta* | 9 (7–1) | 9 (6) | – | 6 (9–1) | – | – | 4 (7–1) | 4 (8–1) |
| *Didymodon* | 9 (7–2) | 7 (8–1) | – | 8 (7–2) | – | 5 (8–2) | 6 (5–2) | 5 (7–1) |
| *Tortula* | 9 (7–3) | 6 (9–2) | 4 (5–1) | 6 (9–2) | – | – | – | – |
| *Stereodon* | 9 (7–4) | 8 (7) | 4 (5–2) | 9 (6–2) | 4 (3–2) | 7 (6–2) | 8 (3–2) | 7 (5–3) |
| *Dicranella* | 8 (8–1) | – | – | 6 (9–3) | – | – | 5 (6–2) | 4 (8–2) |
| *Orthotrichum* | 8 (8–2) | – | – | 7 (8–’1) | – | 4 (9–1) | – | 5 (7–2) |
| *Plagiomnium* | 8 (8–3) | – | – | 6 (9–4) | – | 7 (6–3) | 6 (5–3) | 5 (7–3) |
| *Polytrichum* | 7 (9–1) | 7 (8–2) | 5 (4–2) | 6 (9–5) | 6 (2–4) | 6 (7–1) | 6 (5–4) | 6 (6) |
| *Mnium* | 7 (9–2) | 6 (9–3) | – | 7 (8–2) | – | 6 (7–2) | 6 (5–5) | 5 (7–4) |
| *Timmia* | 6 (10–1) | 5 (10–1) | 5 (4–3) | 5 (10–1) | – | 6 (7–3) | 6 (5–6) | 4 (8–3) |
| *Sciuro-hypnum* | 6 (10–2) | – | – | – | – | – | 4 (7–2) | – |
| *Warnstorfia* | 6 (10–3) | 5 (10–2) | – | – | 4 (3–3) | – | 4 (7–3) | 4 (8–4) |
| *Drepanocladus* | 6 (10–4) | – | – | 5 (10–2) | 4 (3–4) | 4 (9–2) | 4 (7–4) | – |
| *Calliergon* | – | – | – | – | – | – | – | – |
| In total: | 243 | 155 | 64 | 183 | 48 | 121 | 108 | 122 |
| % of total species number | 46,5 | 41,1 | 33,9 | 47,5 | 34,8 | 43,8 | 33,0 | 40,6 |

common for the Arctic region or the northern mountain systems adjacent to the Arctic. The rest of the species belong to the arctic-alpine sub-element. They are *Andreaea rupestris, Brachythecium turgidum, Aulacomnuim turgidum, Conostomum tetragonum, Dicranum elongatum, Encalypta affine, Grimmia mollis, Isopterygiopsis alpicola, Orthothecium chryseon, Paludella squarrosa, Pohlia crudoides, Polytrichastrum alpinum, Pseudocalliergon trifarium, Psilopilum laevigatum, Racomitrium lanuginosum, Rhizomnium pseudopunctatum, Scorpidium scorpioides, Stegonia latifolia, Stereodon bambergeri,* etc. These species form the basis of the Arctic moss flora and of the mountain regions of moderate latitudes. Following Ignatov (1996) we also included here the taxa forming the complex of mineral bogs which are widespread in the southern forest-tundra and northern taiga. Previously, these species belonged to the hypoarctic (subarctic) element.

The boreal element (31.1%) includes the taxa that are widespread in the boreal zone of the Holarctic. They are most species of *Sphagnum, Dicranum* as well as *Aulacomnium palustre, Calliergon giganteum, Climacium dendroides, Helodium blandowii, Hylocomium splendens, Pleurozium schreberi, Pohlia nutans, Polytrichum commune, P. strictum, Plagiomnium ellipticum, Plagiothecium laetum, Rhytidiadelphus triquetrus, Tetraphis pellucida, Tomentypnum nitens, Warnstorfia exannulata,* etc. The mosses of this element play a significant phytocoenotic role in various forest and shrub communities, waterlogged sparse forests, mires, etc. Previously, the boreal element included also those taxa with questionable geographic position which where omitted from this analysis (118 species).

The hemiboreal element (10.4%) comprises the species with a distribution that is limited by the zone of coniferous-broad-leaved forests and the southern part of the boreal zone (Ignatov 1996). Previously, these taxa belonged either to the nemoral or boreal elements. They form the basis of the taiga epiphyte moss flora, though epixylous mosses are also present in this element. We included the species of *Orthotrichum,* as well as *Amblystegium serpens, Brachythecium rivulare, B. salebrosum, Calliergonella cuspidata, Campylidium sommerfeltii, Hygroamblystegium varium, Mnium stellare, Neckera pennata, Plagiomnium confertidens, Pylaisia polyantha, Thuidium assimile, Zygodon sibiricus,* etc.

The nemoral element (6.2%) includes the moss species common for broad-leaved forests of the moderate zone of the Northern Hemisphere. The following species occur in the Yakutian flora, especially in Southern Yakutia: *Brachythecium buchananii, Bryhnia scabrida, Claopodium pellucinerve, Dicranella heteromalla, Homalia trichomanoides, Leucodon pendulus, L. scuiroides, Mnium spinulosum, Oxystegus tenuirostris, Oxyrrhynchium hians, Platygyrium repens, Pylaisia selwynii, Taxiphyllum wissgrillii, Trachycystis ussuriensis, Thuidium delicatulum,* etc. In Central Yakutia they are located within the territory of the Natural Park "Lena Pillars" where they grow in numerous relatively well protected habitats on rocks (shady ravines, deep crevices, various microniches). The nemoral complex there has rather a relic character (Solomonov et al. 2001).

The arid element (4.4%) includes *Aloina brevirostris, A. rigida, Didymodon johansenii, Grimmia anodon, G. tergestina, Jaffueliobryum latifolium,*

*Molendoa sendtneriana, Indusiella thianschanica, Pseudocrossidium obtusum, Pterygoneurum ovatum, P. kozlovii, P. subsessile, Syntrichia laevipila, Tortula acaulon*, etc. characterized by their xeromorphic structure. Unlike other elements, the arid element in Yakutia is extrazonal. It penetrates the forest zone occurring in habitats with rather dry conditions – south-facing slopes with steppe vegetation, dry rocks, anthropogenically modified landscapes, etc.

The most bryophytes of Yakutia have a wide circumpolar distribution, while the rest of the taxa are of the following distribution types: Eurasian (*Hamatocaulis lapponicus, Pylaisia selwynii*); Asian (*Iwatsukiella leucotricha, Plagiomnium confertidens, Trachycystis ussuriensis*); Asian-American (*Grimmia jacutica, Hypnum subimponens, Olygotrichum falcatum, Lyellia aspera, Myuroclada maximoviczii, Timmia sibirica,* etc.); Eurasian-American (*Didymodon asperifolium var. gorodkovii, Encalypta brevicollis, E. brevipes, Grimmia torquata, Myurella sibirica, Trichostomum arcticum, Schistidium agassizii, S. andreaeopsis,* etc.); and amphi-oceanic and oceanic (*Bryoxiphium norvegicum, Orthotrichum sordidum*). Records of these species in the territory of Yakutia are evidence of floristic connections with Northern America, Eastern Asia, mountain systems of Siberia and Central Asia.

Endemism in the bryophytes of Siberia, like in Russia generally, is still to be studied. The proportion of endemic moss species is always less than that of other higher plants. Bardunov (1992) listed 19 endemic moss species for Siberia. Later, a detailed taxonomical revision of mosses from the Asian part of Russia by Ignatov (1996) yielded 4 endemic genera and 10 species including the following Yakutian species: *Barbula jacutica Ignatova, Myrinia rotundifolia (Arnell) Broth.* (Yakutia, Taimyr Peninsula), *Myurella acuminata Lindb. & Arnell* (Yakutia, the Lower Yenisey River, Taimyr Peninsula, Buryatia), *Bardunovia baicalensis Ignatov & Ochyra* (Yakutia, Baikal Lake region) and *Didymodon hedysariformis Otnyukova* (Yakutia, Tyva).

The distributional typification of mosses in Yakutia, like the other plant groups, is based on the floristic regionalization offered by Karavaev (1958). Its modern version, updated according to recent floristic investigations is depicted in Fig. 2.1. The distribution of the bryoflora over the floristic zones is uneven which can be explained by the geographic peculiarities of the regions and insufficient data.

*The Arctic Floristic Region (Arct).* At present, the flora of leafy mosses of this region counts 385 taxa (377 species and 8 varieties) belonging to 39 families and 120 genera. The Arctic flora is special because of its taxonomical structure and high species saturation of a number of families (Tables 2.2 and 2.3). The families with the highest species number are as follows: *Sphagnaceae*, *Pottiaceae*, *Bryaceae, Grimmiaceae*, *Amblystegiaceae*, *Dicranaceae*, *Polytrichaceae*, *Mniaceae*, *Brachytheciaceae, Rhabdoweisiaceae, Plagiotheciaceae*, *Mielichhoferiaceae, Splachnaceae, Pylaisiaceae* and *Calliergonaceae*. They comprise 268 species of the 377 of the Arctic flora (71.1%). The rest of the families consist of less than 11, and four more families include 1 species. There are 120 genera with 3.1 species each on average. The leading genera are *Sphagnum*, *Bryum, Schistidium, Dicranum, Pohlia, Encalypta, Stereodon, Didymodon, Polytrichum, Brachythecium, Tortula, Mnium, Timmia, Warnstorfia.* These taxa include 30–5 species each comprising in total of 155 species, i.e. 41.1% of the entire Arctic leafy moss flora. Forty genera consist of 1 species. The large number of monospecific genera is a characteristic feature of the Arctic flora. Species of the following families *Sphagnaceae, Bryaceae, Amblystegiaceae, Dicranaceae, Polytrichaceae, Mniaceae, Brachytheciaceae, Mielichhoferiaceae, Splachnaceae, Pylaisiaceae, Calliergonaceae* and genera *Sphagnum, Bryum, Dicranum, Pohlia, Polytrichum, Brachythecium, Mnium, Warnstorfia* are typical of the Arctic flora and characteristic for typical tundra-bog complexes. The commonness of rubbly mountain habitats (ranges Ulakhan-Tas, Kular, Polousny, Kharaulakhsky, Kondakovskoe upland, the Novosibirskie Islands, etc.) explains the frequent occurrence of the moss families *Pottiaceae, Grimmiaceae, Plagiotheciaceae* and genera *Schistidium, Didymodon, Tortula, Encalypta, Stereodon, Timmia.* There are 31 species found strictly within this floristic region. They include purely Arctic moss species, such as *Sphagnum arcticum, S. steerei, Funaria arctica, Seligeria polaris, Hennediella heimii, Aplodon wormskjoldii,* etc. The presence of *Didymodon giganteus* in this region is of the great interest. There is only one record of this species in Russia made by H. Nilson-Ehle in 1898 – Bulunsky region, the Lower Lena River, the Kumakh-Surt locality. It was given the "0" rarity category in the new edition of the Red Data book of Russia (Trutnev 2008), i.e. probably extinct, no more existing in the moss flora of Russia.

*The Olenyok Floristic Region (Ol)* is one of the most poorly studied areas in bryological aspect. At present, 190 bryophyte taxa are recorded (189 species and 1 variety) belonging to 34 families and 91 genera. There are 17 leading families comprising 150 species (79.4%) (Tables 2.2 and 2.3). The large amount of species of the families *Sphagnaceae*, *Amblystegiaceae*, *Dicranaceae*, *Mniaceae*, *Pylaisiaceae, Calliergonaceae*, *Bryaceae*, *Brachytheciaceae, Polytrichaceae, Plagiotheciaceae, Splachnaceae, Mielichhoferiaceae, Scorpidiaceae* is typical for the extensive larch sparse forests and bog complexes in which river- and lake-side vegetation is common. A survey of the Muna River basin featuring cliffy limestone outcrops yielded the following leading families: *Pottiaceae*, *Grimmiaceae, Rhabdoweisiaceae, Bartramiaceae,* and genera: *Sphagnum, Dicranum, Bryum, Tortula, Stereodon, Brachythecium, Polytrichum, Timmia.* They include 64 species (33.9%) with *Fissidens adianthoides* and *Tortula obtusifolia* as characteristic species for this region.

*The Yana-Indigirka Floristic Region (Ya-I)* is one of the most interesting regions in bryofloristic respect. There are 385 species and 5 varieties of 134 genera and 43 families. Thirteen leading families include 263 species (68.3%) (Tables 2.2 and 2.3). The regional flora is characterized by a rich species diversity of the families *Pottiaceae*, *Grimmiaceae*, *Plagiotheciaceae*, *Pylaisiaceae, Rhabdoweisiaceae* and the genera *Grimmia, Didymodon, Schistidium, Tortula, Encalypta, Dicranella, Stereodon, Timmia,* growing in mountainous stony tundra and bedrock outcrops. Besides, the species richness of the *Pottiaceae* family can be also explained by the presence of relic steppe vegetation on slopes of southern aspect in the Indigirka River basin. Prevalence of the families *Dicranaceae*, *Mniaceae*, *Amblystegiaceae, Sphagnaceae*, *Bryaceae*, *Polytrichaceae*, *Mielichhoferiaceae, Brachytheciaceae* and the genera *Sphagnum, Bryum, Dicranum, Pohlia, Plagiomnium, Polytrichum,*

*Brachythecium, Mnium, Drepanocladus* indicates the wide distribution of mountainous northern taiga larch sparse forests, abundance of mountain rivers and streams, and waterlogged mountain tundra. It is interesting that *Orthotrichum* is one of the leading genera which can probably be explained by its occurrence in mountain river valleys, and concave elements of mountain topology especially in the southern part of the Verkhoyansk Range. The most interesting moss species in this region are the following:

*Encalypta brevipes* – entered into the new edition of the Red Data Book of Russia (Trutnev 2008) as a rare arctic-alpine species with a naturally low population number. In Yakutia it was recorded in the lower Lena River basin, the Selennyakh Range spurs (Dolinin et al. 2000), and the Indigirka River basin, the Suntar-Khayata Range (sampled by N. Ermakov);

*Indusiella thianschanica* – entered into the new edition of the Red Data Book of Russia (Trutnev 2008) as a rare species with a disjunctive distribution area and naturally low population numbers. Records in Yakutia: the Suntar-Khayata Range, the Yudoma River basin (Ignatov et al. 2001), the Lena Pillars (Solomonov et al. 2001);

*Barbula jacutica* – a species new for science, found in the Yudoma-Maya upland, Tarbagannakh spring. The latest record was made in the Taimyr Peninsula (Ignatova 2001; Fedosov 2006);

*Tortella alpicola* – a rare species found simultaneously at several locations in Russia. Records in Yakutia: the Orulgan Range, the Undyulyung River basin (sampled by E. Sofronova); the Yudoma-Maya upland, the Yudoma River basin (Otnyukova et al. 2004);

*Pterygoneurum kozlovii* – a world-wide rare species. Records in Yakutia: the Indigirka River basin, Yakutsk vicinity, Tabaga and Kangalassy Capes (Afonina et al. 1979; Pisarenko 2006);

*Mielichhoferia macrocarpa* – entered into the new edition of the Red Data Book of Russia (Trutnev 2008) as a rare species with a strongly disjunctive distribution area and naturally low population numbers all over the area. There are only two records in Russia, one of them being in Yakutia (the Suntar-Khayata Range, sampled by V. Zolotov). The second record was made in the Republic of Buryatia (Eastern Sayan, the Tunkin Range).

*The Kolyma Floristic Region (Kol)*, like the OlenyokFR, is very poorly studied in bryofloristic respect. There are 138 species of leafy mosses belonging to 33 families and 72 genera. The leading families, such as *Amblystegiaceae, Brachytheciaceae, Sphagnaceae, Dicranaceae, Calliergonaceae, Mniaceae, Bryaceae, Ditrichaceae, Polytrichaceae, Scorpidiaceae,* etc. indicate the abundance of lakes and wetlands in the region with prevalence of sparse pre-tundra and northern taiga larch forests alternating with tundra bog tracts in the Kolyma lowland (Tables 2.2 and 2.3). *Pottiaceae, Grimmiaceae, Bartramiaceae, Plagiotheciaceae* and *Pylaisiaceae* are leading families and reflect the presence of landscapes with rocky substrates in the Upper and Middle Kolyma River basin (the Yukagir tableland spurs). This region is the only place of occurrence of *Warnstorfia trichophyll,* collected in 1966 on the right bank of the Kolyma River, on thermokarst lake sides 70 and 100 km downstream from Srednekolymsk (sampled by E. Trufanova). *Myrinia rotundifolia* is of special interest too. It was entered into the Red Data Book of Russia (Trutnev 2008) as a stenolocal endemic species with few populations. Its records in Yakutia: the Lower Lena River basin, Kumakh Surt locality; the Middle Kolyma River, Lobuya locality (Dolinin et al. 2000), and the Berezovka River basin (sampled by E. Ohenoja, identified by Martti Ohenoja). It was considered as an endemic species of Yakuta until it was later recorded in the Taimyr Peninsula (Fedosov 2006).

*The Central Yakutian Floristic Region (CYa)* contains 279 taxa (276 species and 3 varieties) of leafy mosses belonging to 43 families and 121 genera. The leading families (*Sphagnaceae, Amblystegiaceae, Mniaceae, Calliergonaceae, Scorpidiaceae*) and genera (*Sphagnum, Brachythecium, Plagiomnium, Mnium, Drepanocladus, Plagiomnium)* include 221 species or 80.1% (Tables 2.2 and 2.3). The bryoflora of the Region is dominated by *Dicranaceae, Brachytheciaceae, Ditrichaceae, Pylaisiaceae, Bryaceae, Polytrichiceae, Mielichhoferiaceae,* i.e. the families typical of the taiga zone. Cliffy limestone outcrops as well as steppe landscapes in the region explain the leading status of the following families *Pottiaceae, Grimmiaceae, Plagiotheciaceae, Bartramiaceae*, and genera *Schistidium, Grimmia, Didymodon* and *Stereodon.* The species found in the territory of the Natural Park "Lena Pillars" are of special interest: *Trachycystis ussuriense, Leucodon sciuroides, Taxiphyllum wissgrillii, Entodon schleicheri, Oxyrrhynchium hians,* etc. They belong to the nemoral complex and probably remain as relics under the arid climatic conditions of Central Yakutia (Solomonov et al. 2001). *Bryoxiphium norvegicum* also deserves special attention. It was collected by K. Krivoshapkin in the Middle Aldan River basin, in the vicinity of the Kyuptsy settlement. There is one more record of this species in Yakutia – the Udokan Range (sampled by L. Kuznetsova). Russia-wide, this nemoral species with an amphi-oceanic distribution occurs only in Yakutia and Chukotka (Ignatov et al. 2006).

Like the Yana-Indigirka FR, the Central Yakutian Region features a large amount of species of the arid complex, growing in the relic steppe communities of the Middle Lena River basin: *Aloina rigida, Jaffueliobryum latifolium, Molendoa sendtneriana, Indusiella thianschanica, Pterygoneurum ovatum, P. kozlovii, P. subsessile,* and *Syntrichia laevipila.*

*The Upper Lena Floristic Region (UL)* counts 327 species and 6 varieties in 45 families and 132 genera. There are 14 leading families including 236 species (72.1%) and 18 leading genera (108 species – 40.6%) (Tables 2.2 and 2.3). The bryoflora of this region has much in common with that of the Aldan FR. It is also characterized by species growing under warmer and more humid climatic conditions including those occurring on stony and rubbly slopes. The species endemic for this region are: *Amblyodon dealbatus, Campylidium calcareum, Campylophyllum halleri, Pallustriella commutata, P. desipiens, Rhizomnium magnifolium, Rhodobryum roseum, Rhynchostegium riparioides, Schistidium duprettii, Sciurohypnum populeum,* etc.

*The Aldan Floristic Region (Ald).* The leafy moss flora counts 300 species and 6 varieties belonging to 49 families and 132 genera. The leading families of the region include 231 species (77%), and the leading genera 122 species (40.6%) (Tables 2.2 and 2.3). Like the Upper Lena FR, the bryoflora of the

Aldan Floristic Region reflects a humid climate and mountainous landscapes. Many eastern species occurring in this region grow at the northern border of their distribution areas (*Haplohymenium triste, Leucodon pendulus).* There are interesting species such as *Andreaea obovata, Tetrodontium brownianum, Dicranum montanum, Paraleucobryum longifolium, Plagiomnium drummondii, Bartramia subulata, Lescuraea saxicola, Ochyraea mollis,* etc.

As a whole, the flora of leafy mosses of Yakutia is considered as a mountain taiga flora. The presence of the Arctic and arid species as well as penetration of species from the South are evidence of the heterogeneity and diversity of Yakutian natural conditions.

The ecological-coenotic characteristics of Yakutian bryoflora are covered in the works of Afonina et al. (1980), Stepanova (1986), Krivoshapkin (1998a, b, 2001), Krivoshapkin and Ivanova (2000), Egorova and Ivanova (2001), Ivanova (2001, 2005), Kuznetsova and Ivanova (2001), etc. Mosses play a significant role in the vegetation cover of Yakutia. In the taiga zone, including mountain regions, the bryoflora makes up 20–40% of the total flora of higher plants and in the North over 40%. In the tundra mosses cover 80–90% of the ground.

*Bryoflora of the Arctic deserts and tundra.* The Arctic deserts are characterized by a predominance of mosses and lichens. The pioneer moss species are *Bryum arcticum, B. rutilans, Bryoerythrophyllum recurvirostre, Ceratodon purpureus, Psilopilum cavifolium, P. laevigatum.* Later *Cinclidium arcticum, Distichium hagenii, Drepanocladus polygamus. Aulacomnium turgidum, Drepanocladus sendtneri,* and *Polytrichastrum alpinum var. fragile* become very common. Waterlogged hollows are dominated by *Hamatocaulis lapponicus, H. vernicosus* and *Plagiomnium curvatulum.* Raised patches are abundantly covered with *Bartramia ithyphylla, Orthothecium strictum,* as well as species of *Bryum, Calliergon, Campylium, Drepanocladus. Ceratodon purpureus, Distichium capillaceum, Leptobryum pyriforme* and *Pohlia cruda* predominate on baidjarakh slopes. The small-hillock tundra of the Novosibirskie Islands is characterized by *Aulacomnium turgidum, A. palustre, Dicranum elongatum, D. flexicaule, Ditrichum flexicaule, Polytrichastrum alpinum, Sanionia uncinata, Tomentypnum nitens.* Swamped landscapes of flat lowlands and river valleys are covered with *Hamatocaulis vernicosus, Meesia triquetra, Polytrichum jensenii, Warnstorfia sarmentosum,* while higher points carry *Aulacomnium palustre, Ditrichum flexicaule, Tomentypnum nitens,* and sometimes polsters of *Sphagnum fimbriatum, S. obtusu.*

In small-hillock spotty tundra mosses cover 30–50% of the total area. *Aulacomnium turgidum, Dicranum flexicaule, D. elongatum, D. spadiceum, Hylocomium splendens var. obtusifolium, Polytrichum juniperinum, Racomitrium lanuginosum* grow on hillocks, while *Hymenoloma crispulum, Oncophorus wahlenbergii, Sphagnum fimbriatum* are characteristic for concave parts. The polygonal-fractured waterlogged arctic tundra is covered mostly with herbs, sedges and cotton-grass. Mosses on higher parts of the microrelief support *Aulacomnium turgidum, Dicranum laevidens, Hylocomium splendens var. obtusifolium, Polytrichum strictum, Tomenypnum nitens*, while *Calliergon cordifolium, Sphagnum aongstroemii, S. fimbriatum* grow in fractures. The northern subarctic tundra is located to the South of the Arctic subzone and consists mainly of hillocky, poorly spotted relief with a prevalence of *Dryas, Vaccinium vitis-idaea, V. uliginosum*, dwarf willows and tussocky *Eriophorum* communities. The content of mosses and lichens in both dryad-dwarf willow and forb-dryad associations is insignificant (about 10%). However, the moss layer of green moss-dwarf willow, dryad-dwarf willow-green moss associations is well developed and represented by *Aulacomnium turgidum, Dicranum elongatum, Hylocomium splendens var. obtusifolium, Polytrichum strictum.* On tussocky tundra *Aulacomnium palustre, Brachythecium mildeanum, Cinclidium arcticum, Dicranum laevidens, Drepanocladus sendtneri, Hamatocaulis vernicosus, Meesia triquetra, Sanionia uncinata, Sphagnum balticum, S. girgensohnii, S. fimbriatum, S. orientale*, etc. predominate. The southern subarctic tundra lays to the South of the northern subarctic subzone, its southern border coinciding with the northern border of pre-tundra forests. Tussocky tundra landscapes are common there all over the area. Sphagnum tussock bog (glade) communities are dominated by *Sphagnum girgensohnii, S. obtusum, S. squarrosum, S. warnstorfi*, etc., and green moss bogs support *Aulacomnium turgidum, A. palustre, Dicranum elongatum, Hylocomium splendens var. obtusifolium*. Polygonal-ridged tundra bogs have an abundant moss cover. In the centre of waterlogged hollows *Scorpidium revolvens* forms a continuous "carpet" with participation of *Cinclidium subrotundum*. On ridges *Aulacomnium palustre, A. turgidum, Hylocomium splendens var. obtusifolium, Tomentypnum nitens, Sphagnum balticum, S. warnstorfii* form thick sods. *Brachythecium mildeanum* with a touch of *B. turgidum, Drepanocladus polygamus, Sanionia uncinata* occur in willow communities in the river valleys. *Myrinia pulvinata and Sanionia uncinata* grow on willow twigs.

*The bryoflora of high-moutain landscapes of Yakutia* is poorly studied. The mountains of Arctic Yakutia are characterized by lichen, dryad, dwarf shrub-green moss tundra communities with a moss cover represented by *Abietinella abietina, Aulacomnium turgidum, Dicranum flexicaule, D. elongatum, D. laevidens, Ditrichum flexicaule, Hylocomium splendens var. obtusifolium, Polytrichum piliferum, Rhytidium rugosum, Tomentypnum nitens*. The microdepressions are occupied by *Sanionia uncinata, Sphagnum aongstremii, S. balticum, S. compactum, S. girgensohnii, S. lenense, S. warnstorfii, Racomitrium lanuginosum,* etc.

In mountain tundra landscapes of South and North–East Yakutia species of the following families predominate: *Amblystegiaceae, Bryaceae, Dicranaceae, Hypnaceae, Grimmiaceae, Mniaceae, Polytrichaceae, Pottiaceae, Sphagnaceae.* The following species are observed: *Anomobryum julaceum, Bryum cyclophyllum, B. wrightii, Catoscopium nigritum, Conostomum tetragonum, Dicranum leioneuron, D. spadiceum, Didymodon asperifolius, Grimmia mollis, Hamatocaulis lapponicus, Hylocomiastrum pyrenaicum, Hylocomium splendens var. obtusifolium, Mnium marginatum, Myurella sibirica, Oligotrichum falcatum, Pohlia crudoides, Polytrichastrum alpinum var. septentrionale, P. formosum, P. sexangulare,* etc.

In high-mountain regions stone deserts, rock debris (kurumy) and cliff outcrops on mountain river banks occupy vast territories. *Abietinella abietina, Andreaea blyttii, A. obovata, A. rupestris, Arctoa fulvella, Blindia acuta,* as well as the species of *Grimmia* and *Schistidium* grow on the bare surfaces of stones and boulders.

However, most of the species prefer stony substrate covered with melkozem-humus material: *Anomobryum julaceum, Didymodon rigidulus, Kiaeria blyttii, Myurella sibirica, Ulota curvifolia,* etc. Species of *Polytrichastrum, Amphidium lapponicum, Blindia acuta, Conostomum tetragonum, Hymenostylium recurvirostrum, Lyellia aspera, Oligotrichum falcatum, Rhabdoweisia crispata, Seligeria campylopoda,* etc. are common at damp cliff niches and between stones. Very rarely, *Schistostega pennata* can be found in small caves where light does not penetrate.

A survey of alpine meadows of Southern Yakutia yielded records of *Aulacomnium, Bartramia, Brachythecium, Dicranum, Polytrichum, while Bartramia ithyphylla, Olygotrichum falcatum, Paraleucobryum longifolium, Pohlia drummondii, Polytrichastrum alpinum var. septentrionale, Sciuro-hypnum reflexum*, etc. growing in subalpine meadows. Nival meadows are characterized by *Dicranum majus, D. spadiceum, Bartramia subulata, Hylocomiastrum pyrenaicum, Hylocomium splendens var. obtusifolium, Grimmia mollis, Niphotrichum panschii, Pohlia drummondii, Polytrichastrum alpinum var. septentrionale, Sciuro-hypnum reflexum*, etc.

*The forest bryoflora* is rich and diverse comprising over 200 species of leafy mosses. Very few species dominate, though they often form a thick and solid carpet, especially in moist and moderately moist forests. Having a strong effect on water and temperature conditions of soils, the moss layer is an important factor in the summer soil heating dynamics impeding permafrost thawing. A dense moss cover has a negative influence on seed germination and the growth of tree sprouts and herbs which results in hampering forest renewal. At the same time, the moss cover effectively protects plants during the winter period.

The moss cover of *sparse pre-tundra forests* is well developed, featuring the following dominant species: *Aulacomnium turgidum, A. palustre, Dicranum elongatum, Hylocomium splendens var. obtusifolium, Pleurozium schreberi, Polytrichum juniperinum, Tomentypnum nitens*. Common species are *Brachythecium mildeanum, Dicranum laevidens, Polytrichum jensenii, Rhytidium rugosum, Sanionia uncinata, Sphagnum squarrosum*. The damper landscapes are abundantly covered with *Cinclidium arcticum, Hamatocaulis lapponicus, Sphagnum warnstofii, Straminergon stramineum, Warnstorfia fluitans*.

The forests in dry environments are mainly *Pinus sylvestris*, sometimes *Larix* forests. The herb and the moss-lichen layers of these forests are poorly developed. The dominating moss species are *Abietinella abietina, Ceratodon purpureus, Dicranum spadiceum, D. undulatum, Polytrichum juniperinum, P. piliferum, Pleurozium schreberi, Rhytidium rugosum.* The moss layer of mesic larch forests is dominated by *Aulacomnium acuminatum, Dicranum fuscescens, D. polysetum, D. undulatum, Hylocomium splendens, Pleurozium schreberi* with co-dominants *Aulacomnium palustre, A. turgidum, Dicranum bonjeanii, D. flexicaule, Polytrichum strictum, Sanionia uncinata, Tomentypnum nitens* etc. The vegetation cover of bryophytes of moist forests makes up 80%. Due to the increased moisture and rather diverse environment in such forests they feature a maximal moss species diversity. The most characteristic species are *Aulacomnium palustre, Calliergon giganteum, Calliergonella lindbergii, Dicranum fuscescens, D. polysetum, D. spadiceum,*

*Hylocomium splendens, Plagiomnium ellipticum, Polytrichum strictum, Ptilium crista-castrensis, Rhytidiadelphus triquetrus, Tomentypnum nitens, Sanionia uncinata.* Sufficient moisture favours growth of *Sphagnum angustifolium, S. capillifolium, S. girgensohnii, S. russowii, S. squarrosum.*

Waterlogged and sphagnum larch forests are common in tracts of taiga in forest depressions or mountainous habitats. Increased moisture, sometimes reaching bog conditions, favours abundant growth of sphagnum mosses covering up to 90% of an area. They are *Sphagnum angustifolium, S. balticum, S. capillifolium, S. girgensohnii, S. magellanicum, S. russowii, S. rubellum, S. warnstorfii* etc. Common green moss species are *Aulacomnium acuminatum, A. palustre, Campylium stellatum, Dicranum flexicaule, D. undulatum, Polytrichum commune, P. jensenii, Tomentypnum nitens* etc.

The diversity of moss species composition in *Picea* forests is rather high, including the dominants *Hylocomium splendens, Rhytidiadelphus triquetrus, Pleurozium schreberi, Aulacomnium palustre, Tomentypnum nitens, Ptilium crista-castrensis.* On the prominent parts of the relief *Aulacomnium turgidum* and *Sanionia uncinata* are added, while dryer habitats are characterized by insignificant participation of *Abietinella abietina* and *Rhytidium rugosum.* Epiphytes occupy the lower twigs. *Sphagnum* species grow in watered depressions. *Polytrichum piliferum* and *Ceratodon purpureus* are recorded on bare substrate. In herb *Picea* forests various proportions of *Rhytidiadelphus triquetrus, Hylocomium splendens* form a background with participation of *Pleurozium schreberi* and isolated patches of *Aulacomnium palustre, A. turgidum, Tomentypnum nitens, Ptilium crista-castrensis*. The depressions are characterized by *Climacium dendroides*, while the promontories carry *Abietinella abietina.* The species diversity in *Picea* forests is based on the epiphyte and epixylous mosses (see below). As the quantity of mosses in the forests of *Picea ajanenis* growing in Southern Yakutia increases the abundance of species of *Brachytheciacea, Dicranaceae, Polytrichaceae*, and *Sphagnacea* decreases. Together with the common species the following mosses are found there: *Brachythecium latifolium, Dicranodontium denudatum, Dicranum scoparium, Cynodontium asperifolium, Homalia trichomanoides, Polytrichastrum alpinum var. septentrionale, Ulota curvifolia*, etc. Only in the *Picea ajanenis* forest *Tetrodontium brownianum* was recorded.

The moss layer of deciduous forests is usually poorly developed due to limiting effects of litter. Sometimes mosses may grow on decaying birch bark or the bottom part of poplars (rarely birch or aspen) trunks. Ground mosses are *Rhytidium rugosum, Abietinella abietina*, and sometimes species of *Amblystegiaceae, Brachytheciaceae, Pylaisiaceae,* and *Plagiotheciaceae.* Pioneer mosses are found in alluvial landscapes.

Epiphyte moss species are not numerous. Their distribution is uneven though a preference for humid air conditions is noted. They are found mostly on north-facing slopes with high precipitation or in over-wetted forests in the river valleys. Any strict relation to tree species was not found. The most common epiphyte is *Orthotrichum speciosum.* Not so common species are *Pylaisia polyantha, Orthotrichum obtusifolium* and *Ulota curvifolia;* while *Orthotrichum laevigatum,*

*Pylaisia selwyni, P. steerei*, and *Zygodon sibiricus* are even more rare. The lower parts of tree trunks, lower branches, and sometimes exposed roots are occupied by *Brachythecium salebrosum, Campylium stellatum, Cynodontium strumiferum, Dicranum fuscescens, D. polysetum, Myurella julacea, Neckera pennata, Plagiothecium laetum, Pohlia cruda, Ptilium crista-castrensis, Thuidium assimile, Sanionia uncinata*, etc.

The similarity of epiphytic and epilithic bryofloras has long been noted in literature. Epiphytes switching to stony substrate is explained by unfavourable climatic conditions. Almost all the Siberian epiphytes may be at the same time epilithophytes, for example, *Orthotrichum speciosum, Pylaisia polyantha*, etc.

The species composition of the epixylous bryoflora is not specifically related to forest type but only to substrate and moisture level. Mosses occupy decaying trees in the following way: trees at the initial stage of decomposition that still have a bark are occupied with epiphytes (*Orthotrichum speciosum, O. obtusifolium, Pylaisia polyantha*, etc.). Later, as the decay process goes on, epiphytes almost disappear, except for those growing on trunk basements, and mosses characteristic for rotten wood start growing (*Oncophorus wahlenbergii, Dicranum fragilifolium*, etc.). Strongly decomposed wood and fallen trees are occupied by *Tetraphis pellucida* and sometimes even by ground mosses such as *Dicranum, Aulacomnium turgidum, Hylocomium splendens, Pleurozium schreberi, Ptilium crista-castrensis, Rhytidium rugosum, Sanionia uncinata*, and *Tomentypnum nitens*.

*The bryoflora of shrub communities* is relatively diverse. In shrubberies with *Betula exilis* (so called *yerniks*) the moss layer is often represented by *Aulacomnium palustre, Bryum pseudotriquetrum, Calliergon giganteum, Dicranum acutifolium, D. fuscescens, D. spadiceum, D. undulatum, Paludella squarrosa, Pleurozium schreberi, Polytrichum commune, P. strictum, Scorpidium cossonii, Sphagnum angustifolium, S. balticum, S. capillifolium, S. fimbriatum, S. fuscum, S. girgensohnii, S. magellanicum, S. rubellum, Straminergon stramineum, Tomentypnum nitens, Warnstorfia exannulata*. The moss layer of *Betula fruticosa* communities is well developed covering up to 50%. The most common species are *Aulacomnium palustre, A. turgidum, Campylium stellatum, Hylocomium splendens, Paludella squarrosa, Pleurozium schreberi, Sanionia uncinata, Scorpidium revolvens, Sphagnum warnstorfii, S. capillifolium, Tomentypnum nitens*.

Willow communities in the large river valleys constantly undergo spring floods and this results in scanty herbaceous and moss covers. The common moss species are *Brachythecium mildeanum, Bryum pseudotriquetrum, Calliergon giganteum, Calliergonella cuspidata, C. lindbergii, Climacium dendroides, Leptobryum pyriforme, Niphotrichum canescens, Pohlia nutans, P. wahlenbergii, Sanionia uncinata, Warnstorfia exannulata*. Willow trunk bases are occupied by *Amblystegium serpens, Brachythecium salebrosum, Campylium stellatum, Eurhynchiastrum pulchellum, Leskea polycarpa, Platydictya jungermannioides, Pylaisia polyantha, Sanionia uncinata, Timmia megapolitana*.

*Pinus pumila* communities are characterized by dense moss-lichen covers dominated by *Aulacomnium turgidum, Dicranum elongatum, D. undulatum, Pogonatum dentatum, Hylocomium splendens var. obtusifolium, Pleurozium schreberi,*

*Polytrichum juniperinum, P. piliferum, Ptilium crista-castrensis, Sanionia uncinata.* In over-wetted microdepressions the following species grow: *Aulacomnium palustre, Scorpidium revolvens, Sphagnum fimbriatum, S. girgensohnii, S. russowii, S. warnstorfii. Andreaea rupestrtis* grows on isolated stones, while stony substrate is the habitat for *Niphotrichum canescens, Schistidium agassizii*, as well as common pioneer species *Ceratodon purpureus, Pohlia cruda, P. nutans.*

*The bog bryoflora of Yakutia* still requires further investigation. The moss layer of dwarf shrub-sedge-green moss bogs is, as a rule, well developed and covers up to 40%. The prevailing species are *Aulacomnium palustre, Bryum pseudotriquetrum, Calliergon giganteum, C. cordifolium, Campylium stellatum, Drepanocladus sendtneri, Plagiomnium ellipticum, Tomentypnum nitens.* The moist areas and wet hollows are dominated by *Calliergonella lindbergii, Pseudocalliergon brevifolius, Scorpidium scorpioides, Warnstorfia exannulata.* Sphagnum mosses (*Sphagnum capillifolium, S. fuscum, S. warnstorfi)* are not so common. Such bogs are found on over-floodplain terraces, lake-side depressions and depressions in watershed areas. Sphagnum bogs are also common in watershed areas and river valleys. Sphagnum species often form a solid carpet with predomination of *Sphagnum angustifolium, S. fuscum, S. capillifolium, S. compactum, S. girgensohnii, S. lenense, S. magellanicum, S. orientale* and insignificant participation of the green mosses *Warnstorfia exannulata, Aulacomnium palustre,* as well as species of *Dicranum* and *Calliergon.*

*The meadow bryoflora.* It is hard to discern dominating moss species in meadow communities. The species composition of different meadow types varies insignificantly, but their abundance increases with overwetting resulting in a deterioration of the aeration of the soil and this impedes the proper development of herbs. This, in turn, leads to waterlogging and a decrease in meadow productivity. The alas meadows show a zonation of vegetation belts related to the moisture level in the soil. Lakes are very common for alases. Their waterlogged sides are covered with *Aulacomnium turgidum, A. palustre, Dicranum undulatum, Drepanoclaadus aduncus, Sphagnum balticum, S. fuscum, S. russowii, S. squarrosum, Warnstorfia fluitans.* Loamy soils are the habitat for *Calliergon giganteum, Campylium stellatum, Climacium dendroides, Plagiomnium ellipticum, Warnstorfia exannulata, Tomentypnum nitens,* etc. On the transect from the lake to edge of an alas the moss composition changes. Typical for the sparse moss cover are *Abietinella abietina, Ceratodon purpureus, Pleurozium shreberi, Pohlia nutans* and *Rhytidium rugosum.*

Dry meadows are situated on heights, often high ridges of river floodplains. They are characterized by a lower diversity of bryophytes: *Aulacomnium turgidum, Ceratodon purpureus, Distichium capillaceum, Pohlia nutans, Rhytidium rugosum.* Meadow communities of the small rivers and streams of the taiga are, as a rule, properly saturated with moisture and possess a well-developed moss cover (up to 40%). The neighbourhood of forests enriches the species composition with forest species (*Hylocomium splendens, Pleurozium schreberi).* Swamping of most of these meadows indicates their transitional character and shows their relation to riparian shore-line vegetation. This explains the presence of numerous hygro- and hydrophyllous mosses: *Calliergon giganteum, C. cordifolium, Calliergonella lindbergii, Drepanocladus aduncus, Hygrohypnum luridum, Scorpidium cossonii, Sphagnum balticum, S. capillifolium, S. fuscum, S. riparium, S. warnstorfii, Warnstorfia exannulata.*

Grass and sedge meadows occupying floodplains of large rivers are regularly flooded with tide waters. Characterised by a high productivity they are intensively used in agriculture which results in a very poor moss layer (cover less than 10%). The most common species are *Drepanocladus sendtneri, D. aduncus, Hamatocaulis lapponicus, Pohlia cruda, Scorpidium revolvens, Tomentypnum nitens* and *Warnstorfia fluitans.*

*The bryoflora of ponds and streams.* It consists of mosses growing in water as well as in littoral and riparian landscapes. Such mosses as *Dichelyma falcatum, Fontinalis anthipyretica, F. hypnoides, Hygrohypnella ochrace, H. polare, Ochyraea duriuscula, Schistidium agassizii* and *Warnstorfia exannulata,* which grow under water or in swift rivers and streams.

The rare underwater species *Scouleria aquatica* of the Far East and Eastern Siberia is peculiar because of its ability to survive the periods with little or no water. *Scouleria aquatica* forms tough thick carpets on half-submerged boulders and stones in the Lena, Olyokma, Aldan, Viluy, Kolyma and other rivers. Strips of sand-pebble alluvium are characteristic for banks of large rivers and often undergo flooding. On these patches mosses may occur in pure carpets though usually they are disturbed by frequent floods and the mechanical action of ice. The most characteristic species for these communities are *Bartramia pomiformis, Bryum pseudotriquetrum, Calliergon giganteum, C. lindbergii, Calliergonella cuspidata, Ceratodon purpureus, Funaria hygrometrica, Leptodyctium riparium, Niphotrichum canescens, Pohlia wahlenbergii, Plagiomnium ellipticum, Pseudocalliergon brevifolium, Warnstorfia exannulata, W. fluitans.* Very peculiar moss communities occupy rather wide strips of 30–100 cm along the banks of mountain rivers and streams with swift currents. They are also exposed to frequent flooding. Such moss communities are rich in species and form thick cushions of 20–30 cm high. Dominants and co-dominants are *Aulacomnium palustre, Bryum pseudotriquetrum, Calliergon cordifolium, C. giganteum, Calliergonella lindbergii, Campylium stellatum, Catoscopium nigritum, Climacium dendroides, Cratoneuron filicinum, Drepanocladus sendtneri, Helodium blandowii, Pohlia wahlenbergii, Plagiomnium ellipticum, Philonotis fontana, Sanionia uncinata, Scorpidium cossonii, Timmia bavarica, Tomentypnum nitens, Warnstorfia exannulata.* Such communities are absent on the banks of large rivers.

*Mosses of sedge, herb-moss communities.* Moving away from water bodies, the significance of vascular plants in vegetation communities increases. Their moss flora composition depends on the adjacent vegetation types. Such communities are characterized by a high species diversity and a outspoken mosaic structure of the moss cover. The mosaic elements are monospecific, or seldom mixed, hummocks of *Aulacomnium turgidum, Brachythecium campestre, B. salebrosum, Hylocomium splendens, Pleurozium schreberii, Ptilium crista-castrensis, Rhytidiadelphus triquetrus*, as well as by *Sphagnum, Polytrichum, Dicranum* species. Sometimes it is hard to delimit moss and herb-moss groupings and sometimes they alternate.

*The bryoflora of icings (naled).* Icings are formed usually in river valleys due to action of permanent sources of underground waters in winter time. In summer icings may thaw or remain unchanged (like "Bulus" icing, which is well-known in Yakutia). There are three zones distinguished on the characteristics of the icing formation and thawing period: the icing alluvium zone; the low shrubs zone; the zone of high shrubs and isolated trees. The icing alluvium is characterized by a few species, probably because these sites stay under ice for a prolonged period of time. The most characteristic species for the alluvium zone are *Catoscopium nigritum, Cratoneuron filicinum, Scorpidium revolvens* and *S. scorpioides.* The zone of low shrubs features a rather large number of mosses: *Bryum pseudotriquetrum, Campylium stellatum, Catoscopium nigritum, Cratoneuron filicinum, Distichium inclinatum, Sanionia uncinata, Scorpidium revolvens, Tomentypnum nitens, Trichostomum arcticum.* The relatively quick ice melting, the presence of streams and ground water as well as developed soil horizons in this zone favour the development of a proper moss cover. The zone of high shrubs and isolated trees contains abundant mosses but with a relatively low species diversity. Common species are *Aulacomnium palustre, Campylium stellatum, Dicranum undulatum, Hylocomium splendens, Pleurozium schreberi, Sanionia uncinata, Tomentypnum nitens.* Sphagnum mosses are recorded only in this zone since the prolonged duration of the ice cover and abundant mineral salts in the other two zones most likely impede the growth of sphagnums. The icings are unique habitats for a whole number of interesting and rare moss species: *Amblyodon dealbatus, Bryobrittonia longipes, Catoscopium nigritum, Dicranella grevilleana, Orthothecium intricatum, Pseudocalliergon turgescens* and *Trichostomum arcticum.*

*The bryoflora of steppe communities* of Yakutia is poorly studied. There are some data, mainly collected in the Middle Indigirka basin (Afonina et al. 1980) and in the vicinity of Yakutsk. Unlike lichens, which are common in steppe communities, mosses play a subordinate role. Mosses grow in small patches on open, unsodded places, their diversity being limited to 5–8 species. The characteristic and constant species of steppe communities of Yakutia are relatively widespread species of dry habitats, such as *Abietinella abietina, Bryoerythrophyllum recurvirostrum, Bryum caespiticium, Didymodon icmadophilus, Rhytidium rugosum, Syntrichia ruralis*, and those specific only for steppe vegetation, such as *Tortula acaulon, Pseudocrossidium obtusulum, Pterygoneurum ovatum, P. kozlovii, P. subsessile.* The most interesting and rare species is *Pterygoneurum kozlovii* which earlier on was only recorded from Russia (Pisarenko 2006). The specific steppe epilithic species *Grimmia tergestina* and *Syntrichia laevipila* grow in petrophyte communities on detrited stony substrates. They belong to the arid geographical element and prevail in the bryoflora of the steppe and forest-steppe regions of Europe and Asia and thus have an extrazonal character in Yakutia. The bryoflora of forest-steppe communities of the Middle Indigirka basin is supplemented by the species growing in larch forests and shrub communities on soils (species of *Polytrichum, Dicranum* genera, *Aulacomnium turgidum, Abietinella abietina, Hylocomium splendens, Rhytidium rugosum, Sanionia uncinata*), as well as on trunks and twigs (*Amblystegium serpens var. juratzkanum, Orthotrichum speciosum, Pylaisia polyantha).* Besides the specific steppe species, the meadow-steppe communities in the vicinity of Yakutsk contain mosses characteristic for meadow vegetation (*Campylium, Drepanocladus* genera). Steppe communities disturbed due to overgrazing or fire feature abundant *Ceratodon purpureus, Bryum argenteum, B.caespiticium*, and sometimes *Aloina brevirostris* and *A. rigida* on bare ground.

*The bryoflora of rock landscapes.* Rock outcrops and stony debris are found on slopes and gentle summits, as well as on river and stream banks. Development of epilithic mosses is greatly determined by humidity, substrate moisture level, light, presence of humus-melkozem soils, and bedrock composition. Rock vegetation has a fragmentary distribution and can be either xeric or mesic depending on moisture conditions. The characteristics of the moss layer vary due to location and moisture conditions, though its species composition is always rich and peculiar. Exposed stones usually lack any vegetation except for crustose lichens and some epilithic mosses which prepare the substrate for further occupation by other plants. Dry or moderately moist stones, cliffs, and large boulders are the place for growth of the following species: *Grimmia, Schistidium, Abietinella abietina, Andreaea rupestris, Hedwigia ciliata.* The species growing on dry rocks usually found on south-facing bank slopes are: *Abietinella abietina, Ditrichum flexicaule, Hypnum cupressiforme, Grimmia longirostris, Rhytidium rugosum, Schistidium plathyphyllum, Stereodon vaucheri.* Moist cliffs of north-facing slopes of river banks and near ground water sources are characterized by sufficiant moisture, mineral salt content and shade and thus represent perfect conditions for moss cover development (about 80%). Mosses form a continuous carpet covering crevices, soil, and stones. The most characteristic species for this habitat type are *Anomodon minor ssp. integerrimus, Hypnum cupressiforme, Encalypta rhaptocarpa, Neckera pennata, Orthotrichum anomalum, Pylaisia polyantha, Timmia austriaca.* Near ground water sources the following hydrophytes and hygrophites grow: *Calliergon cordifolium, C. giganteum, Calliergonella lindbergii, Cratoneuron filicinum, Philonotis fontana, Plagiomnium ellipticum.* Cavities, small ledges, crevices, where shallow soils may accumulate, are occupied by *Aulacomnium palustre, A. turgidum, Dicranum fuscescens, D. undulatum, Distichium capillaceum, Drepanocladus aduncus, Entodon concinnus, Hylocomium splendens, Hymenostylium recurvirostrum, Grimmia jacutica, Plagiomnium ellipticum, Pleurozium schreberii, Polytrichum strictum, Thuidium assimile, Trichostomum crispulum, Sanionia uncinata,* etc.

Carbonate rocks are occupied by numerous obligate and facultative calciphilous species: *Abietinella abietina, Anomodon minor ssp. integerrimus, Bryobryttonia longipes, Campylidium calcareum, Campylium stellatum, Cynodontium tenellum, Cyrtomnium hymenophylloides, Encalypta alpina, E. rhaptocarpa, Hymenostylium recurvirostrum, Hypnum cupressiforme, Molendoa sendtneriana, Myurella julacea, Orthothecium chryseon, O. strictum, Trichostomum crispulum, Syntrichia ruralis, Stereodon vaucheri.*

Stony debris are found on summits and steep slopes. Crustose or sometimes fruticose lichens are common there. Mosses playing a minor role in such communities usually occupy spaces between stones. Such species as *Dicranum flexicaule, D. fuscescens, D. undulatum, Hylocomium splendens, Pleurozium schreberi,*

*Ptilium crista-castrensis* grow together with fruticose lichens and dwarf shrubs in shallow microrelief depressions, while *Andreaea rupestris, Grimmia longirostris, Drepanium recurvatum, Schistidium plathyphyllum, S. pulchrum,* etc. are found on stone surfaces.

Slopes may subside forming “stone rivers” or screes: accumulation of blocks (kurumy) sliding down the slopes. These large stone fell-fields are covered with mosses and later forest vegetation gradually establishes. The stones covered with melkozem are occupied by *Bartramia subulata, Cynodontium strumiferum, Ditrichum flexicaule, Niphotrichum canescens, N. panschii, Pogonatum urnigerum, Polytrichum piliferum, Rhytidium rugosum, Tortella fragilis, T. tortuosa, Stereodon vaucheri.* Soils or shallow hollows are covered with *Hylocomium splendens, Plagiomnium ellipticum, Sanionia uncinata,* etc.

It is hard to distinguish forest moss species and those growing on isolated boulders and stones in mountain forests. The fact is that most mosses growing on this substrate are not forest species but species of rock landscapes. Exposed boulders and stones in *Larix spp., Pinus sylvestris, Picea obovata* or *Betula lanata* forests are occupied by species of *Grimmia, Schistidium,* as well as *Andreaea rupestris* and *Kiaeria starkei.* Large numbers of species growing on stones can be observed in pine forests on carbonate rocks: *Abietinella abietina, Anomodon minor ssp. integerrimus, Didymodon ferrugineus, Haplohymenium triste, Myurella julacea, Neckera pennata var. tenera, Orthotrichum anomalum, Pseudoleskeella catenulata. Pohlia wahlenbergii* was recorded on floodplain pebble alluvium in willow communities.

*The bryoflora of disturbed habitats.* Nowadays, the problem of anthropogenic impact on bryophytes development and natural moss cover recovery in disturbed landscapes has become of actual importance. There are numerous deposits of valuable minerals in Yakutia. The mining industry and agriculture are being vigorously developed. An inevitable increase of anthropogenic pressure may yield depauperization of biological diversity. This, in turn, may result in species extinctions that have not yet been documented.

Bryophytes form the pioneer vegetation in disturbed landscapes in habitats that are non-suitable for other plants and they promote humus and moisture accumulation. The substrate (soil, wood, stones) and moisture conditions are of primary significance for moss development though the surrounding vegetation also plays an important role. The moss flora of disturbed habitats comprises the whole spectrum of ecological groups from hydrophytes to xerophytes, as well as epigeious, epixylous, epilithic, and coprophilous species.

The bryophytes that often settle in anthropogenically transformed landscapes are: *Polytrichum juniperinum, P piliferum, P. strictum, Ceratodon purpureus, Funaria hygrometrica, Tetraplodon mnioides, Bryum argenteum, B. pseudotriquetrum, Leptobryum pyriforme, Pohlia cruda, P. nutans.*

In forest cuttings (for power lines) the moss layer recovers by means of pioneer species (*Polytrichum piliferum, Ceratodon purpureus, Leptobryum pyriforme, Pohlia cruda,* etc.) and forms hummocks. Roads and paths are overgrown both with pioneer mosses and other species from surrounding natural communities:

*Aulacomnium turgidum, Sanionia uncinata, Ptilium cricta-castrensis,* and species of *Polytrichum* and *Dicranum spp.*. Camp-fire places are usually covered with abundant *Funaria hygrometrica, Ceratodon purpureus,* their hummocks occupying the whole area of a camp-fire. The year after a forest fire *Funaria hygrometrica, Leptobryum pyriforme, Ceratodon purpureus* and *Polytrichum piliferum* appear and are gradually substituted by ordinary forest species. Clearings are overgrown with *Polytrichum strictum, Ceratodon purpureus, Pohlia nutans, Aulacomnium turgidum, Sanionia uncinata.*

The bryoflora of regions in South Yakutia disturbed by mining activities (5–7% of the territory) is well studied (Ivanova 2001). Overgrowing rates of technogenic terraces depend on the mechanical composition of ground material and soil moisture level. The process starts from over-wetted places in shallow depressions of microrelief. After 5–6 years floodplain willow shrubs develop on bare pebble alluvium. The moss layer is represented basically by pioneer species of relatively low species diversity: *Ceratodon purpureus, Pohlia cruda, P. nutans, Bryum pseudotriquetrum.* As the willow communities grow older the pioneer mosses are replaced by permanent mesohygrophytes and species number increases. The willow shrubs are followed by poplar and aspen forests containing epiphytes. Terraces of pebbly or large waste stones and bedrock remain lifeless for numerous decades. Later solitary pioneer mosses appear there in small hollows or places close to water: *Polytrichum piliferum, Ceratodon purpureus.* On barren ground, patches of sodded soil remain in the form of hills of 50–80 cm high, occupied by a mixture of pioneer and forest moss species. The edges of stone terraces have embankments (probably as a result of stripping) on which an increased moss diversity is observed. The recovery takes place there due to the accretion of some forest species that may have remained or that are widespread in adjacent intact forest communities: *Aulacomnium palustre, A. turgidum, Hylocomium splendens, Sanionia uncinata.* The moss layer does not develop in the areas undergoing technical recultivation where meadow communities are formed.

The phytocoenotic role of bryophytes in Yakutia, as in all of Siberia, is very significant. Mosses play an important role in the structure and composition of all major vegetation types. However, they acquire maximal significance in tundra, bog and many forest type vegetation. There they represent the dominant and co-dominant species of the ground cover. Mosses are also the key factor in the recovery of natural vegetation in technogenic landscapes.

## 2.3.2 Liverworts

**E.V. Sofronova**

The first records of liverworts of Yakutia date back to the beginning of the twentieth century (Arnell 1913). Further investigations of liverworts were continued in 1950s. The mosses were sampled by geobotanists who would send the collections for identification basically to bryologists of the Komarov Botanical Institute

(Academy of Sciences of the USSR): I.I. Abramov, A.L. Abramova, A.L. Zhukova, K.I. Ladyzhenskaya, and of the Polar-Alpine Botanical Garden: R.N. Shlyakov. Sometimes the mosses were sent to the Czech hepaticologist J. Duda and other specialists (Sofronova 2005).

The analysis of Yakutian liverworts is based on the annotated list compiled from herbarium and reference data (Sofronova 2005, 2008a, b; herbarium collection of the author) and follows the system accepted for the list of *MARCHANTIOPHYTA (=HEPATICAE)* and *ANTHOCEROTOPHYTA (=ANTHOCEROTAE)* of the former USSR (Konstantinova et al. 1992). Genera composition is given according to the slightly modified system of Grolle (1983).

At present, the liverwort flora of Yakutia consists of 199 species and 19 varieties belonging to 63 genera, 34 families, 5 orders – *Calobryales, Metzgeriales, Treubiales, Jungermanniales, Marchantiales* and 2 classes – *Jungermanniopsida* and *Marchantiopsida* of the *MARCHANTIOPHYTA* division.

The class *Jungermanniopsida* includes 179 species of 51 genera, 29 families and 4 orders. The class *Marchantiopsida* is represented by 20 species of 12 genera, 5 families and 1 order *Marchantiales*. Species of the division *ANTHOCEROTOPHYTA*, which belongs to the liverworts, were not recorded in Yakutia, though they occur in the Primorsky Krai (Gambaryan 1992). Comparing to the liverwort flora of other regions, the flora of Yakutia is rather well studied. For example, the areas of Chukotka (Afonina and Duda 1993), Karelia (Bakalin 1999) and Kamchatka (Bakalin 2007), of which exist rather most complete inventories, 175, 174 and 218 species were recorded, respectively. Taking into account the vast territories and diverse landscapes of the Republic of Sakha (Yakutia) it may be assumed that the real species number is 15–30% higher.

There are three leading liverwort families in all the floristic zones of Yakutia – *Lophoziaceae, Scapaniaceae* and *Jungermanniaceae* (Table 2.4), which is characteristic for the whole liverwort flora of the Northern Holarctic.

By number of species the family *Cephaloziaceae* is fourth among the leading families of liverworts in Yakutia. As a rule, its species are a characteristic feature of the floras of flat landscapes. At the same time, the widespread mountainous landscapes in Yakutia determine the presence of the arctic-montane species: *Pleurocladula albescens* (Hook.) Grolle, *Cephalozia ambigua* C. Massal. and *Odontoschisma elongatum* (Lindb.) A. Evans, *O. macounii* (Austin) Underw. The *Cephaloziellaceae* is the fifth family, occurring, as far as we presently know, mainly in the mountainous regions. This family is very well represented in the Arctic, Yana-Indigirka, and Aldan Floristic Regions. The sixth place is taken by *Gymnomitriaceae*. It was recorded only in the Arctic FR and in mountainous regions (the Yana-Indigirka and Aldan FRs) and all the representatives of this family occurring in this area belong to arctic-montane element and generally inhabit acid bedrock. The species of the *Aytoniaceae* (seventh in the spectrum) prefer various bedrock outcrops. Such outcrops are very common in Yakutia which explains the rather high species diversity of the family. *Geocalycaceae*, *Frullaniaceae* and *Calypogeiaceae* take the places 8–10 in species numbers. The first is characteristic of the boreal floras of the lowlands. The representatives of *Frullaniaceae* are

**Table 2.4** Status of the leading families of the liverwort flora of Yakutia and its floristic regions

| Rank position in Yakutian flora | Leading families | Species number in Yakutian flora | Arc | | Ol | | Ya-I | | Kol | | CYa | | UL | | Ald | |
|---|---|---|---|---|---|---|---|---|---|---|---|---|---|---|---|---|
| | | | n[a] | N[b] | n | N | n | N | n | N | n | N | n | N | n | N |
| 1 | *Lophoziaceae* | 47 | 32 | 1 | 14 | 1 | 40 | 1 | 7 | 1 | 15 | 1 | 21 | 1 | 34 | 1 |
| 2 | *Scapaniaceae* | 27 | 19 | 2 | 6 | 3 | 18 | 2 | 3 | 2–3 | 7 | 2–3 | 7 | 2 | 20 | 2 |
| 3 | *Jungermanniaceae* | 23 | 16 | 3 | 7 | 2 | 13 | 3 | 1 | 5–6 | 7 | 2–3 | 6 | 3 | 14 | 3 |
| 4 | *Cephaloziaceae* | 11 | 8 | 4–5 | 1 | 7–8 | 4 | 7–8 | 3 | 2–3 | 2 | 7–8 | 4 | 4 | 6 | 4–5 |
| 5 | *Cephaloziellaceae* | 10 | 8 | 4–5 | 2 | 4–6 | 8 | 4 | 2 | 4 | 2 | 7–8 | 3 | 5–7 | 5 | 6–7 |
| 6 | *Gymnomitriaceae* | 8 | 3 | 6–8 | – | – | 5 | 6 | – | – | – | – | – | – | 5 | 6–7 |
| 7 | *Aytoniaceae* | 7 | 2 | 9 | 1 | 7–8 | 6 | 5 | – | – | 4 | 4–5 | 3 | 5–7 | 3 | 10 |
| 8–10 | *Geocalycaceae* | 6 | 3 | 6–8 | 2 | 4–6 | 2 | 9 | – | – | 4 | 4–5 | 3 | 5–7 | 4 | 8–9 |
| 8–10 | *Calypogeiaceae* | 6 | 3 | 6–8 | 2 | 4–6 | 4 | 7–8 | 1 | 5–6 | 3 | 6 | 2 | 8 | 4 | 8–9 |
| 8–10 | *Jubulaceae* | 6 | 1 | 10 | – | – | 1 | 10 | – | – | 1 | 9 | 1 | 9 | 6 | 4–5 |

[a]– absolute number of species; [b]– rank position in the flora of the given floristic region

most common in the regions with a humid climate. This, probably, explains why the species are recorded only within the Aldan floristic region featuring the northern border of their distribution area. The *Calypogeiaceae* play a more significant role in plain, water-logged regions. Since most of the territory of Yakutia consists of mountains and lies within the sharply continental climate area, the species diversity of these three families is rather low compared to the earlier mentioned families.

A geographical analysis of the liverworts is based on the geographical-genetic classification by Konstantinova (2000). The distribution of species which were not included in that paper was taken from Shlyakov (1979a, b–1982); Potemkin (2001); Bakalin (2005); Schuster (1966, 1969, 1974, 1992ab), Paton (1999); Damsholt (2002). *Apotreubia* sp., *Fossombronia* sp., *Lophozia lantratovae* Bakalin, *Riccia fluitans* L., *R. frostii* Austin were omitted from the analysis since it is not clear to which geographical element they belong. *Scapania nemorea* (L.) Grolle with an amphi-atlantic distribution was also excluded since its occurrence in the territory of the Republic is very questionable. The geographical position of a number of species was recently revised (Sofronova 2003).

We distinguish the following geographical elements in the liverwort flora of Yakutia: Arctic-montane – common for the Arctic, Subarctic, as well as the highlands of moderate latitudes; Arctic-boreal-montane – from the northern tundra to southern borders of taiga, as well as in highland at low altitudes; Boreal – species confined to the coniferous forests of the Holarctic; Montane – mountain species of the temperate part of the Northern Hemisphere; Arctic – the tundra zone and polar deserts of the Arctic; Nemoral – species common in broad-leaved forests; Holarctic – common for the Holarctic.

The arctic-montane and arctic-boreal-montane species (Table 2.5) provide the basis of the liverwort flora of Yakutia. The boreal and montane species are a little less in number. The role of the arctic and nemoral species in theYakutian liverwort flora is rather insignificant. The Holarctic element is represented by 2 species: *Blasia pusilla* L. and *Cephalozia bicuspidata* (L.) Dumort. The cosmopolitan group includes 4 liverworts: *Aneura pinguis* (L.) Dumort., *Marchantia polymorpha* L.

**Table 2.5** Number and proportion of liverwort species of various geographical elements in Yakutia and its floristic regions

| | Yakutia | | Arc | | Ol | | Ya-I | | Kol | | C-Ya | | U-L | | Ald | |
|---|---|---|---|---|---|---|---|---|---|---|---|---|---|---|---|---|
| Geographical element | n[a] | %[b] | n | % | n | % | n | % | n | % | N | % | n | % | n | % |
| Arctic-montane | 55 | 28 | 44 | 36 | 12 | 23 | 40 | 30 | – | – | 12 | 19 | 10 | 16 | 32 | 26 |
| Arctic-boreal-montane | 49 | 25 | 37 | 31 | 22 | 42 | 39 | 29 | 12 | 57 | 27 | 42 | 31 | 48 | 39 | 31 |
| Boreal | 32 | 16 | 11 | 9 | 11 | 21 | 18 | 14 | 5 | 24 | 16 | 25 | 16 | 25 | 25 | 20 |
| Montane | 24 | 12 | 8 | 7 | 1 | 2 | 12 | 9 | 1 | 5 | 3 | 5 | 3 | 5 | 14 | 11 |
| Arctic | 16 | 7.5 | 15 | 12 | 2 | 4 | 11 | 8 | – | – | – | – | – | – | 6 | 5 |
| Nemoral | 11 | 6 | 5 | 4 | 2 | 4 | 3 | 2 | – | – | 2 | 3 | 1 | 1.5 | 5 | 4 |
| Holarctic | 2 | 1 | 2 | 2 | 1 | 2 | 2 | 1.5 | 1 | 5 | 1 | 1.5 | 2 | 3 | 2 | 2 |

[a]– species number; [b]– % of total liverwort flora of Yakutia or of the given floristic region.

s.l., *Reboulia hemisphaerica* (L.) Raddi and *Ricciocarpus natans* (L.) Corda. The species of the latter two groups are widespread in the Holarctic. This allowed omitting them in the further analysis.

The species not concerned in Konstantinova's work (2000) were studied as far as possible using the "focal point" principle offered by Yurtsev (1968). Thus, *Scapania nemorea* was excluded (see above) as well as 14 more species with disjunctive or unclear distributions. The analysis revealed a predominance of liverwort species with circumpolar distributions (146 or 73%). This is characteristic for both the Yakutian and other Holartcic liverwort floras. The next biggest group is that with species of mainly "eastern distributions", i.e. those with none or isolated records in Europe. It comprises 13 species (6.5% of the entire Yakutian liverwort flora) including 2 species with amphi-Pacific distribution: *Scapania microdonta* (Mitt.) Müll. Frib. and *S. plicata* (Lindb.) Potemkin. There are 5 (2.5%) species with Eurasian-Western American, Eurasian-Greenland and Eurasian distributions in Yakutia, and 4 species (*Herbertus sakuraii* (Warnst.) S. Hatt., *Frullania dilatata* (L.) Dumort., *F. parvistipula* Steph., and *Nardia japonica* Steph.) occurring basically in the Asian part of the continent or Beringia. *Prasanthus suecicus* (Gottsche) Lindb. is more common in Europe though its numerous records in the Yamal and Chukotka peninsulas, as well as in Yakutia, suggests that it is rather frequent in Siberia. Liverworts with mainly a "western distribution" have not been recorded. Thus, there is no floristic transition from western to eastern elements. The liverwort group of 13 (6.5%) species with amphi- and near-ocean distributions occurs mainly in the Arctic and Aldan Floristic Regions indicating both the more humid climate there compared to the Yana-Indigirka FR and the connection of the Aldan FR with the near-Pacific floras.

The overwhelming majority of Yakutian liverworts prefer substrates with adequate moisture supplies from either stagnant or permanently running water. With the exception of *Ricciocarpus natans*, however, none of the liverworts is strictly confined to underwater substrates. Typical xerophytic liverworts are absent in Yakutia though some liverwort species are found in xeric landscapes including several types of mountain tundra, shrubs of *Pinus pumila*, dry cliff crevices, upper parts of barren rock slopes, etc. Unlike leafy mosses, there are very few liverwort species with an arid ecology. However, many liverworts grow in what, at first sight, seem dry habitats. This is explained by the effects of permafrost; its thawing provides permanent, sometimes rather small but sufficient, moisture for the development of liverworts.

Many liverwort species are not strictly confined to a specific substrate type. The majority of recorded Yakutian liverworts prefer soils. The rest of the species preferring other substrate types represent mostly rare (1–3 records in the studied region) plants. The mountainous relief of most of Yakutia and the widespread bedrock outcrops predetermine the prevalence of liverwort species growing on various rocky substrates, such as bare rock, stones covered with humus or melkozem. Humus-covered stones feature numerous calciphilous and basidiphilous species, such as *Arnellia fennica* (Gottsche) Lindb., *Bucegia romanica* Radian, *Leiocolea badensis* (Gottsche ex Rabenh.) Jørg., *Reboulia hemisphaerica* (L.) Raddi, *Sauteria alpina*

(Nees) Nees and species of *Asterella* and *Mannia*, etc. These species occur mostly in mountainous regions and in the Central Yakutian Floristic Region. Acidophilous species prefer, as a rule, shallow melkozem soils (*Anastrophyllum saxicola* (Schrad.) R.M. Schust., *Scapania microdonta, S. rufidula* Warnst., *S. sphaerifera* H. Buch et Tuom., *Tetralophozia setiformis* (Ehrh.) Schljakov, and some other species). These liverworts are characteristic only for the Yana-Indigirka and Aldan FRs. The species of *Frullania, Jungermannia* as well as *Porella platyphylla* (L.) Pfeiff. are found sometimes on bare rock. In the forests the availability of sufficient moisture determines that most liverworts grow on decayed wood, sphagnum cushions, and bark. All liverworts preferring bark grow on trunk bases, exposed roots, and sometimes on the lower parts of trunks. The composition of these groups is determined by species growing in surrounding ecotopes, such as soils and decaying wood. That is why these species can not be considered as epiphytes, and they should not be compared with the epiphytes growing in broad-leaved forests in the European part of Russia or in the South of the Far East. There are only two species found at the height of 20–30 cm above ground – *Ptilidium pulcherrimum* (G. Web.) Vain. and *Radula complanata* (L.) Dumort., and at the height of several meters *Frullania bolanderi* Austin. These species are also recorded on rotten wood and/or stones. There is no strong correlation between tree species and "epiphyte" species. Species characteristic for organic substrate (sphagnum mosses, decayed wood, bark, etc.) are *Calypogeia sphagnicola* (Arnell et J. Perss.) Warnst. et Loeske, *Mylia anomala* (Hook.) Gray (usually on sphagnums), *Anastrophyllum michauxii* (F. Web.) H. Buch, *Scapania apiculata* Spruce, *S. glaucocephala* (Taylor) Austin (strictly on decomposed wood), *Anastrophyllum hellerianum* (Nees ex Lindenb.) R.M. Schust., *Chiloscyphus minor* (Nees) J.J. Engel et R.M. Schust., *Ch. profundus* (Nees) J.J. Engel et R.M. Schust., *Lepidozia reptans* (L.) Dumort., *Lophozia longidens* (Lindb.) Macoun, *Ptilidium pulcherrimum, Tritomaria exsectiformis* (Breidl.) Schiffn. ex Loeske (usually on decayed wood), *Frullania bolanderi* (usually on the bark of *Picea ajanensis, Betula lanata, Picea abies, Populus sp.*, etc.), etc. Many species growing on decaying wood and/or bark are more common in the plains in the southern floristic regions: the Upper Lena and Aldan.

As a whole, soils and rocky substrates are inhabited by arctic-boreal-montane and arctic-montane species, as well as by the species of the arctic and montane elements, while organic ecotopes are the substrate for both arctic-boreal-montane and boreal species.

The occurrence of liverworts, like other plants, is correlated to the mechanical and chemical characteristics of the substrate. Thus, the Olenyok Floristic Region mainly covers the vast territory of Cambrian carbonate bedrock, while the Yana-Indigirka FR represents ranges of acid shale-sandstone sediments of the so called Verkhoyansk sedimentary complex (the Middle Carbon – Upper Jura). Calcium bearing bedrocks in this zone are very scarce and limited in area. These two regions, lying practically at the same latitude, possess 14 and 25 calciphilous species, respectively. However, the proportion of calciphilous liverworts is significantly higher (27%) in the first FR than in the second one, despite the vast territory of the latter (19%) (Table 2.5). It is noteworthy that in the Yana-Indigirka FR the calciphilous species are basically arctic and arctic-montane species which are absent in the Olenyok Floristic Region. In contrast, the Olenyok FR is characterized by boreal calciphils being absent in the Yana-Indigirka region.

Unlike leafy mosses, liverworts do not play a significant role in the development of the vegetation cover, though they are remarkable in biogeocoenoses of the tundra zone, highlands, and landscapes characterized by the absence of dense vegetation cover and well-developed soils. Characteristic of such landscapes are *Anastrophyllum minutum* (Schreb.) R.M. Schust., *Barbilophozia barbata* (Schmidel ex Schreb.) Loeske, *Blepharostoma trichophyllum* (L.) Dumort., *Lophozia excisa* (Dicks.) Dumort., *Marchantia polymorpha* L. subsp. *ruderalis* Bischl. et Boisselier, *Plagiochila porelloides* (Torrey ex Nees) Lindenb., *Ptilidium ciliare* (L.) Hampe, *Schistochilopsis incisa* (Schrad.) Konstantinova, *Tritomaria quinquedentata* (Huds.) H. Buch and some other species. They are very common and abundant in almost all Floristic Regions.

Clearer patterns were analysed following the generally accepted floristic regionalization of Yakutia (Fig. 2.1). Presently, the liverwort flora of the Arctic, Yana-Indigirka, and Aldan Floristic Regions of the Republic is best studied. The Kolyma FR is omitted from the analysis due to lack of data.

*The Arctic Floristic Region (Arct).* Unlike other plant groups, the list of liverworts of this floristic region includes the species growing in the tundra zone of Yakutia following the work of Stepanova (1986).

There are records of 121 species and 10 varieties of liverworts of 49 genera and 27 families in this region. Sixteen species and 5 varieties were found only in this FR, one of them, *Diplophyllum albicans* (L.) Dumort., being rare for Siberia.

In the spectrum of the leading families of the region the first 3 places are traditionally taken by *Lophoziaceae, Scapaniaceae*, and *Jungermanniaceae*, though their species diversity is enriched by Arctic species such as *Anastrophyllum cavifolium, Cryptocolea imbricata, Tritomaria heterophylla, Scapania obcordata* being absent in the regions further south except for the Yana-Indigirka FR, and by species of amphi-oceanic distribution: *Jungermannia caespiticia* Lindenb., *J. exsertifolia* Steph., etc. (Table 2.4). The families *Cephaloziaceae, Calypogeiaceae*, and *Gymnomitriaceae* also occupy a rather high rank, due to the vast territories occupied by various tundra types and tundra-bog complexes, as well as by cliff outcrops along the Kharaulakh Range spurs with a properly developed nival morphostructure. The *Aytoniaceae* takes only the 9$^{th}$ place due to the limited distribution of calcium-bearing bedrocks, since this family is represented mainly by calciphils. The *Frullaniaceae* on the 10$^{th}$ place contributes only one arctic-montane species: *Frullania nisquallensis* Sull.

In the Arctic FR the arctic-montane species provide the basis of the flora with a smaller participation of the arctic-boreal-montane species. (Table 2.5). The arctic element here is more numerous than the boreal element. The proportion of nemoral species in the flora of the Arctic FR is also significant.

*The Olenyok Floristic Region (Ol).* There are scant data on liverworts for this region. Presently, 52 species and 3 varieties of 32 genera and 21 families have been recorded. The single record of *Moerckia hibernica* (Hook.) Gottsche was made in this region (Kildyushevsky 1964). This species is rather rare in Siberia as it is strictly confined to carbonate soils.

The leading families in the region are those with widespread species. The importance of the *Calypogeiaceae, Geocalycaceae*, and *Aytoniaceae* (Table 2.4) reflects the typically plain territory in this part of the boreal zone and the wide distribution of various calcium-bearing bedrock outcrops. The absence of the *Gymnomitriaceae* is probably due to the limited distribution of habitats with a regular and prolonged snow cover and the presence of calcium-bearing bedrock outcrops on the north–eastern edge of the Middle Siberian pateau, whereas the species of this family are acidophilous.

About half of the liverworts known form the Olenyok FR are arctic-boreal-montane (Table 2.5). The arctic-montane and boreal species are much less numerous. The Arctic element (*Plagiochila arctica, Scapania hyperborea*) makes up only 4%, as does the nemoral element (*Jamesoniella autumnalis, Riccia cavernosa*).

*The Yana-Indigirka Floristic Region (Ya-I)* is the largest one with 132 species and 8 varieties of liverworts of 53 genera and 31 families. 17 species and 2 varieties occur only in this region, 6 of them being rare in Russia and the whole world: *Apotreubia sp., Cephaloziella polystratosa* (R.M. Schust. et Damsh.) Konstantinova, *Eocalypogeia schusteriana, Fossombronia sp., Lophozia decolorans* (Limpr.) Steph., in Siberia *Scapania plicata*, and in the continental part of Siberia *Prasanthus suecicus*.

The first three leading families (Table 2.4) comprise the same Arctic species as mentioned for the Arctic FR and in addition *Schistochilopsis elegans* and *Scapania zemliae*. There are also arctic-montane and montane species recorded only in this FR, for instance *Lophozia decolorans* and *Scapania plicata*.

Rather important are the *Gymnomitriaceae* (5$^{th}$ place) and *Aytoniaceae* (6$^{th}$ place). This is to be expected since the FR has a mountainous topology with a wide distribution of rocky landscapes inhabited by the species of these families. The genus *Gymnomitrion* is important, as well as *Prasanthus* which occurs only in this region. While the Aldan FR features more species of *Marsupella*, there are only several records of *Marsupella emarginata* in the Yana-Indigirka FR. Probably, *Marsupella* avoids regions with a low relative air humidity and a sharply continental climate such as in the mountains of north–eastern Yakutia. The *Geocalycaceae* is 9$^{th}$. Presently, this family is recorded only for the South of this Floristic Region where it borders with the Aldan FR, and for the Middle Indigirka valley. Thorough investigations of the Orulgan, Suntar-Khayata and Ulakhan-Chistay Ranges didn't yield species of this family. Like in the flora of the Arctic FR, the *Frullaniaceae* is 10$^{th}$ and contains the same arctic-montane species.

In the Yana-Indigirka FR the core of the liverwort flora is represented by the arctic-montane and arctic-boreal-montane species (Table 2.5). The boreal species number is only half of that and is insignificant in the Yana-Indigirka FR, even compared to the Olenyok FR.

Boreal species were recorded only at the boundary with the Aldan and Central Yakutian FRs: *Anastrophyllum hellerianum, Chiloscyphus ssp., Conocephalum conicum* (L.) Underw., *Lepidozia reptans, Ptilidium pulcherrimum.* Apparently the

Verkhoyansk Range is a significant barrier, impeding penetration of the boreal flora northward in north–eastern Yakutia, whereas Arctic species are found far in the South following the Verkhoyansk Range's chains. Accordingly, they make up 8% of the Yana-Indigirka FR, which is a little less than in the Arctic FR and much more than in the Olenyok FR. Notebly, Arctic species such as *Anastrophyllum cavifolium, Plagiochila arctica* Bryhn et Kaal., *Pseudolepicolea fryei* (Perss.) Grolle et Ando, *Scapania zemliae, Tritomaria heterophylla* are often found in the Orulgan and Yulakhan-Chistay Ranges, while in the Suntar-Khayata Range they become relatively rare or are absent. The 3 nemoral species are *Lejeunea cavifolia* (Ehrh.) Lindb., *Jamesoniella autumnalis* (DC.) Steph. and *Scapania nemorea.* But their occurrence in Yakutia has to be confirmed.

*The Kolyma Floristic Region (Kol)* is one of the most poorly studied regions concerning liverworts. At present, there are only 21 species and 1 variety of 15 genera and 10 families recorded.

*The Central Yakutian Floristic Region (CYa).* There are records of 64 species and 1 variety of liverworts of 35 genera and 23 families. The region is the only habitat for *Riccia rhenana.*

The first three leading families are also *Lophoziaceae, Scapaniaceae* and *Jungermanniaceae* (Table 2.4) containing also such boreal species as *Anastrophyllum hellerianum, Scapania glaucocephala.* These species are characteristic for boreal floras and are absent in the northern floristic regions.

The increased role of the *Aytoniaceae,* comprising a large number of calciphilous species in the Central Yakutian region, corresponds to the widespread occurrence of outcrops of calcium-bearing sedimentrary bedrocks. Compared to northern floras of the Arctic and Kolyma FRs and to mountainous floras of the Yana-Indigirka and AldanFRs, the *Geocalycaceae* has a higher rank. This is characteristic for the boreal liverwort floras of the plains. For the same reasons as in the Olenyok FR, the *Gymnomitriaceae* are also absent in the CYa. As in the other FRs, the *Frullaniaceae* also here occupies the last place, containing the same arctic-montane species.

Arctic-boreal-montane species are the core of the liverwort flora of the Central Yakutian FR (Table 2.5). The participation of boreal species is considerable. There is a large number of calciphils among the arctic-montane and montane species: *Arnellia fennica, Asterella saccata, Athalamia hyalina, Leiocolea badensis, Scapania gymnostomophila,* etc. The nemoral element includes 2 mosses: *Porella platyphylla* and *Riccia rhenana,* being rare in the territory of Yakutia. The nearest record of *Porella platyphylla* is in the Upper Aldan River, in the Uchur River basin (Sofronova 2005), while that of the *Riccia rhenana* is in the Kuznetsk Alatau (Konstantinova et al. 1992). The presence of these species in this floristic zone has most likely a relic character. Species of the Arctic element are absent.

*The Upper Lena Floristic Region (UL)* totals 64 species and 1 variety of liverworts of 32 genera and 20 families. It has no species specific to this FR.

The first three leading families are the same (Table 2.4) as for the rest of the floristic regions. They contain boreal species occuring also in the Central Yakutian and/or Aldan FRs: *Anastrophyllum hellerianum, Nardia japonica, Scapania apiculata, S. glaucocephala.*

The *Cephaloziaceae* is 4th, as its species prefer plains, as do the species of the *Geocalycaceae* (5–7th postion). In this floristic region the overwhelming majority of the species of the *Aytoniaceae* occur along the Lena banks on cliff outcrops of calcium-bearing bedrocks – limestones and dolomites. The absence of the *Gymnomitriaceae* is probably due to the absence or scarcity of places with prolonged snow cover and acid bedrock outcrops. The *Frullaniaceae* is represented by one boreal species, *Frullania bolanderi*.

Arctic-boreal-montane species provide the core of the liverwort flora of the Upper Lena FR (Table 2.5). Boreal and arctic-montane species are considerably less. Montane species make up only 5% of the liverwort flora of the region. It should be noted that all the arctic-montane and montane species found in this FR grow in spurs on the Patom Plateau. The plain landscapes feature only calciphilous species. Arctic species are absent and there is only one nemoral species, *Jamesoniella autumnalis*.

*The Aldan Floristic Region (Ald).* There are 125 species and 8 varieties of 48 genera and 28 families known for this region. 12 species and 3 varieties of liverworts comprising numerous nemoral and boreal species occur only in the Aldan FR and represent in Yakutia probably the northern border of their distribution area: *Frullania crispiplicata* Yuzawa et S. Hatt., *F. dilatata, F. parvistipula, Herbertus aduncus* (Dicks.) Gray, etc. Rare liverworts of Russia and Siberia include *Diplophyllum obtusatum* (R.M. Schust.) R.M. Schust., *Frullania crispiplicata, F. inflata, Lophozia savicziae* Schljakov. V.A. Bakalin described the new species *Lophozia lantratovae* recorded in the Udokan Range (Bakalin 2003).

The *Lophoziaceae, Scapaniaceae* and *Jungermanniaceae* again take the first three places (Table 2.4). Though here they are joined by both arctic (*Anastrophyllum sphenoloboides* R.M. Schust., *Barbilophozia hyperborea* (R.M. Schust.) Stotler ex Potemkin, *Lophozia polaris,* etc.), arctic-montane (*Jungermannia polaris* Lindb., *Lophozia propagulifera* (Gottsche) Steph., *L. silvicoloides* N. Kitag., *Scapania kaurinii*, etc.) or montane (*Scapania microdonta, S. rufidula, S. sphaerifera,* etc.) species, occurring also in the Arctic and Yana-Indigirka FRs, and boreal species (*Scapania apiculata, S. glaucocephala*) recorded only in the Central Yakutian and Upper Lena FRs.

The 4–5th position of the *Frullaniaceae*, above the nemoral and boreal species, is noteworthy. The *Cephaloziaceae* takes a similarly high position, due to the appearance of the boreal species that are more common for South Siberia: *Cephalozia lunulifolia* (Dumort.) Dumort., *Cladopodiella fluitans* (Nees) H. Buch. The considerable species diversity of the *Gymnomitriaceae* is explained by the mountainous character of the zone. Because the *Frullaniaceae* and *Gymnomitriaceae* are richer in species in this FR the relatively position of the *Geocalycaceae* and *Calypogeiaceae* is lower but their species diversity is not. The *Aytoniaceae* is 10th probably due to the prevalence of acid bedrock in the region.

The core of the flora of the Aldan FR is represented by arctic-boreal-montane, arctic-montane and boreal species (Table 2.5). The participation of montane species is considerable. In spite of the fact that a number of calciphilous montane species is not recorded, like *Leiocolea badensis*, *Asterella saccata,* etc., the species diversity of the montane element is higher than that of the Yana-Indigirka FR due to presence of the species of amphi-oceanic and near-oceanic distributions: *Bazzania tricrenata, Frullania inflata, Herbertus aduncus, Nardia scalaris* Gray as well as of *Anastrophyllum michauxii, Diplophyllum obtusatum* prefering, probably, regions with humid climate.

It is notable that the liverwort flora of the Aldan FR contains arctic species. These species are possibly distributed far southward thanks to the Verkhoyansk Range and South Yakutian mountains. The nemoral element is represented by *Frullania ssp.* and *Porella platyphylla*, the genera of which most species are common in tropical and sub-tropical regions.

Thus, comparison of the spectra of leading families of the floristic regions of Yakutia shows that only three leading families, *Lophoziaceae, Scapaniaceae* and *Jungermanniaceae*, are the most important in all regions. This pattern is characteristic for many liverwort floras of the Holarctic North. The *Cephaloziaceae* has a rather high position in the Arctic, Kolyma and Upper Lena Floristic Regions which is natural for the floras of the plains. In the Aldan FR this family also takes a prominent first place (4–5) though boreal species (*Cephalozia lunulifolia, Cladopodiella fluitans*) are more common in South Siberia. In the Yana-Indigirka and Aldan FRs a high position is taken by the *Gymnomitriaceae* comprising arctic-montane species and this is related to the mountainous topology of the region. The increased role of the *Aytoniaceae,* composed of calciphilous species, in the Yana-Indigirka, Central Yakutian and Upper Lena regions is explained by the rather widespread occurrence of limestones rather then by the presence of highlands in these regions. Compared to the northern floras of the Arctic and Kolyma Floristic Regions as well as the mountain floras of the Yana-Indigirka and Aldan FRs, the *Geocalycaceae* has a stronger position in the floras of Central and West Yakutia (the Olenyok, Central Yakutian, Upper Lena FRs), which is primarily characteristic for the boreal floras. The *Calypogeiaceae* plays a more considerable role in the plains and marshlands of the Olenyok, Kolyma, and partly in the Arctic and Cenrtal Yakutian FRs. Position of the *Frullaniaceae* family is indicates the occurrence of boreal (*Frullania bolanderi*), nemoral (*F. dilatata, F. crispiplicata, F. parvistipula*) as well as montane liverworts (*Frullania inflate*) of mainly near-oceanic distribution. This family is not important in all the Floristic Regions except for the Southern Aldan FR ($4^{th}$ – $5^{th}$ position).

The geographical spectra of floristic regions characterized by prevailing plains (the Olenyok, Central Yakutian, Upper Lena FRs) are rather similar (Table 2.5). Thus, nearly half of their floras are composed of arctic-boreal-montane species, boreal and arctic-montane species having less importance. The liverworts of the arctic element are absent in the Central Yakutian and Upper Lena FRs and make up to 4% in the Olenyok FR. Arctic-montane species provide the core of the Arctic FR with smaller participation of arctic-boreal-montane plants (Table 2.5). The arctic element plays more significant role there than the boreal element. The portion of arctic-boreal-montane species is also very considerable in the mountainous Yana-Indigirka and Aldan FRs (Table 2.5). The arctic-montane species are fewer or equal there, while the proportion of boreal species there is significantly smaller than in the plain taiga FRs. Boreal species are considerably fewer in the Yana-Indigirka FR than in the Aldan or Olenyok FRs. This can be explained by the siginificant role of the Verkhoyansk Range as a natural barrier preventing the penetration of the boreal flora northwards into the North–East of Yakutia, whereas the arctic species move far to the South along the chains of the Verkhoyansk Range and South Yakutian mountains. Their proportion in the liverwort flora of the Yana-Indigirka floristic region makes up 9%, in the Aldan FR 5%, and in the Olenyok FR only 4%.

*The liverwort flora of Yakutian forests* is richest. The coniferous forests feature the highest species diversity due to the widespread larch forests in the region and the ecological flexibility of larch (*Larix cayanderi* and *L. gmelinii*). In spite of the vast territory they occupy, the larch forests are characterized by a non-specific liverwort flora in terms of presence of rare species. Rare species occurring in larch forests are *Cryptocolea imbricata* R.M. Schust., *Scapania obcordata* (Berggr.) S.W. Arnell and *S. plicata*. The forests of *Picea obovata* and *P. ajanensis,* though having a limited distribution in Yakutia, feature a rather rich liverwort flora compared to other forest types. The most common species for these forests are *Anastrophylum michauxii, Scapania apiculata, S. glaucocephala*, though they occur seldom in larch forests. The forests of *Pinus sylvestris* and *Pinus sibirica* occupy mainly dry habitats with sufficient insolation. Until now, no liverworts have been recorded in the pine forests of the Central Yakutian FR. In the Upper Lena FR the liverworts grow only on bark or decayed wood: *Ptilidium pulcherrimum, Barbilophozia attenuata* (Mart.) Loeske. Forests of *Abies sibirica* are very rare in Yakutia, usually forming mixtures with other forest types. Hence, their liverwort flora is very diverse and is composed purely of boreal species rather rare for Yakutia: *Anastrophyllum hellerianum, Frullania bolanderi, Ptilidium pulcherrimum, Radula complanata, Scapania apiculata*, etc. Deciduous forests of Yakutia occupy insignificant areas. Most interesting of these are communities of *Betula ermanii ssp. lanata* characterized by a diverse liverwort flora. Such forests occupy interslope depressions in mountainous regions as well as habitats with thick snow covers, moisture-saturated soils, and high air humidity. All this represents favourable conditions for liverwort growth resulting in a high species diversity. Only this vegetation type is the place of occurrence of *Bazzania tricrenata* (Wahlenb.) Lindb. As a rule, only *Ptilidium ciliare* plays a significant role in the species composition of the ground cover of deciduous forests. Rarely, in forests of *Betula ermanii ssp. lanata* in the southern part of the Verkhoyansk mountainous system, a solid liverwort-green moss ground cover is developed, sometimes with prevalence of liverworts. Such species as *Calycularia laxa* Lindb. et Arnell, *Schistochilopsis incisa, Lophozia ventricosa* (Dicks.) Dumort., *Tritomaria quinquedentata* etc. form a ground cover of variable density. This phenomenon has never been recorded in forests of *Betula ermanii ssp. lanata* in Southern Yakutia characterized by a dense herbaceous cover. Due to weak competitiveness, liverworts are out-competed by vascular plants or green mosses and grow only on decayed wood, bark or stones. This is a very common phenomenon for the Upper-Lena and Aldan floristic regions. In other forest types the liverworts are not a key element of ground cover composition.

*The liverwort species diversity of shrubberies* of Yakutia is very high as well, especially in *yerniks* of *Betula exilis* or *Betula fruticosa* being widespread in both plain and mountainous floristic zones, river valleys and mountain slopes. The rich liverwort flora of *yerniks* is characterized by a low specificity: The vegetation cover is composed of such common species as *Ptilidium ciliare, Barbilophozia kunzeana* (Huebener) Müll. Frib. and some others. The flora of willow communities is very scanty despite proper moisture conditions and is represented by pioneer species (*Blasia pusilla, Marchantia polymorpha* subsp. *ruderalis*, etc.) or species of wide ecological amplitude *(Barbilophozia barbata, Tritomaria quinquedentata*, etc.). Willow shrubberies on calcium-bearing bedrocks feature a little more diverse liverwort flora at the expense of calciphils: *Leiocolea badensis*, *Pellia endiviifolia* (Dicks.) Dumort., *Schistochilopsis elegans* (R.M. Schust.) Schljakov. As a whole, the liverworts never dominate in the vegetation cover of willow communities. *Pinus pumila* shrubberies form the subalpine shrubby altitudinal belt in mountains (this zone is situated under the fellfields belt, the so called *"goltsy"*) at a height of over 1,000 m above sea level on south-facing slopes and crests. Low moisture levels, absence of micro-depressions and niches make this landscape unfavourable for liverwort growth. Thus, very small amounts of common liverwort species grow their. The ground cover is represented by the species tolerant to xeric conditions – *Anastrophyllum saxicola, Scapania microdonta, Tetralophozia setiformis*.

*The liverwort flora of bog communities* is rather diverse due to proper moisture levels and the wide distribution in both plains and mountainous landforms, river valleys and peneplain summits. The liverworts are abundant only in sphagnum bogs though they do not significantly participate in the development of the ground cover, except for *Ptilidium ciliare*, which sometimes forms up to 80% of the cover. Degenerative, sometimes living sphagnums are often covered with a peculiar "crust" of a mixture of various species with *Anastrophyllum minutum, Barbilophozia binsteadii* (Kaal.) Loeske, *B. kunzeana, Calypogeia sphagnicola*, *Mylia anomala* as the most important species. Though the common species are most abundant in this vegetation type, the bog flora is very specific, especially in mountain bogs. The following interesting records have been made there: *Jamesoniella undulifolia* (Nees) Müll. Frib., *Radula prolifera* Arnell, *Schistochilopsis grandiretis* (Lindb. ex Kaal.) Konstantinova.

The liverworts of the tundra zone of Yakutia have not been studied yet, so there are no records on liverwort flora peculiarities in the lowland tundra. All information on this region represents the analysis of the reference data.

*The liverwort flora of the mountain tundra* is rather rich having a large number of species widespread in mountainous landscapes: *Anastrophyllum saxicola, Blepharostoma trichophyllum, Lophozia excisa, Ptilidium ciliare, Tetralophozia setiformis*, etc. Unlike in other vegetation formations, the phytocoenotic role of liverworts in the mountain tundra is very high. The following species play remarkable coenotic roles in the ground cover: *Anastrophyllum saxicola, Barbilophozia barbata, Blepharostoma trichophyllum, Cephalozia pleniceps* (Austin) Lindb., *Lophozia excisa, Ptilidium ciliare, Scapania simmonsii* Bryhn et Kaal., *Tetralophozia setiformis*, *Tritomaria quinquedentata,* as well as *Preissia quadrata* (Scop.) Nees growing under conditions of sufficient moisture and calcium bearing bedrocks. Rare species found in the mountain tundra of Yakutia are *Lophozoia perssonii* H. Buch et S.W. Arnell, *Prasanthus suecicus, Radula prolifera,*

*Scapania zemliae* S.W. Arnell. In mountains of the Yana-Indigirka Floristic Region (Orulgan and Ulakhan-Chistai Ranges) the following Arctic species are prominent in the vegetation cover: *Anastrophyllum cavifolium* (H. Buch et S.W. Arnell) Lammes, *Scapania zemliae, Tritomaria heterophylla* R.M. Schust.

*The liverwort flora of riparian and littoral communities* (vegetation growing in water and on the banks of large and small rivers, mountain streams, floodplain lake-sides and other near-water habitats) is very diverse and highly specific. The liverworts vigourously overgrow the banks of ponds, rivers, and especially of mountain streams. Some species expand on disturbed lake-sides, river and stream banks, occupying areas as big as several square metres: *Blasia pusilla, Mesoptychia sahlbergii* (Lindb. et Arnell) A. Evans, *Pellia neesiana* (Gottsche) Limpr., *Plagiochila porelloides, Preissia quadrata, Riccia ssp., Scapania crassiretis* Bryhn, *Schistochilopsis incisa*, etc. Stones in small rivers and stream beds are the substrate for *Jungermannia ssp., Marsupella emarginata* (Ehrh.) Dumort., etc. Rare species in this region are *Apotreubia sp., Bucegia romanica, Chiloscyphus fragilis* (Roth) Schiffn., *Jungermannia confertissima* Nees, *J. subelliptica* (Lindb. ex Kaal.) Levier, *Lophozia polaris* (R.M. Schust.) R.M. Schust. et Damsh., etc.

Rock debris, cliff outcrops and screes (rocky and stony habitats) occupy significant areas both in mountainous and level landscapes. *The liverwort flora of the rocky and stony substrates* is very peculiar. Cliff outcrops feature the highest species diversity, while screes the lowest. The common petrophytic species of various bedrock compositions are *Asterella gracilis* (F. Web.) Underw., *A. saccata* (Wahlenb.) A. Evans, *Athalamia hyalina* (Sommerf.) S. Hatt., *Diplophyllum taxifolium* (Wahlenb.) Dumort., *Gymnomitrion ssp., Mannia ssp., Reboulia hemisphaerica, Sauteria alpina*, etc. *Barbilophozia barbata, Tritomaria quinquedentata* and some other species form solid covers on screes. Rock debris is densely covered with *Anastrophyllum saxicola, Blepharostoma trichophyllum, Scapania microdonta, Tetralophozia setiformis*. The following species are common and abundant for cliff outcrops of various bedrock compositions: *Apometzgeria pubescens* (Schrank) Kuwah., *Jungermannia borealis* Damsh. et Váňa, *Leiocolea badensis, Mannia ssp., Preissia quadrata, Scapania gymnostomophila* Kaal., *Tetralophozia setiformis, Tritomaria quinquedentata*, etc. There are interesting records of *Bucegia romanica, Eocalypogeia schusteriana* (S. Hatt. et Mizut.) R.M. Schust., *Frullania inflata* Gottsche, *Scapania kaurinii* Ryan, *S. sphaerifera*, etc.

*Anthropogenic landscapes* in the region occupy insignificant areas compared to the European part of Russia or Europe. Liverworts do not occur there as a rule, except for *Marchantia polymorpha* ssp. *ruderalis*. However, there are disturbed habitats where the liverworts may form a solid cover – deserted forest roads, winter roads (used only in winter time when the snow cover protects the liverworts), and landscapes with partly destroyed vegetation and soil cover. The most common species there are *Aneura pinguis, Blasia pusilla, Marchantia polymorpha* ssp. *ruderalis*. There are several records of *Calycularia laxa* in overgrazed reindeer pastures. The following rare species occur in such disturbed habitats: *Fossombronia sp., Lophozia perssonii, Nardia geoscyphus* (De Not.) Lindb., *N. japonica, Tritomaria heterophylla.*

Thus, the liverwort flora of Yakutia may be characterized as typically Holarctic. Though the presence of the Arctic species (*Anastrophyllum cavifolium, A. sphenoloboides, Barbilophozia hyperborea, Cryptocolea imbricata, Lophozia polaris, L. savicziae, Pseudolepicolea fryei, Scapania hyperborea, S. obcordata, S. zemliae, Schistochilopsis elegans, Tritomaria heterophylla*) in the Ya-I and Ald, nemoral species (*Jamesoniella autumnalis, Lejeunea cavifolia, Porella platyphylla, Riccia cavernosa, R. glauca, R. rhenana*) in the Arct, Ol, CYa, and the species of amphi-oceanic, near-oceanic or Pacific distributions in the continental part of Yakutia (*Bazzania tricrenata, Nardia japonica, N. scalaris, Scapania plicata, Schistochilopsis laxa, etc.*) point to the significant peculiarity of the liverwort flora of Yakutia. The pattern of leading families and species distribution patterns reflect regional peculiarities in the Yakutian liverwort flora, namely its position, mainly in the taiga zone, and mountainous landscapes. There is no evidence of floristic migration in the region from the west eastwards. Mountainous relief also predetermines the prevalence of liverwort species of riparian and littoral habitats thanks to their multiplicity on mountain stream banks and rocky and stony landscapes. The species composition of these liverwort floras is very specific. The soil-inhabiting liverwort flora is richest. The mountainous relief as well as abundant rock outcrops in level areas explain the high species diversity of liverworts growing on rocky and stony substrate. The prevailing forest vegetation type and, as a rule, proper moisture conditions of vegetation formations favour active colonization of decomposed wood by liverworts.

**Acknowledgments** E. Sofronova expresses her thanks to A.D. Potemkin (KBI RAS, Saint-Petersburg) for valuable advices during working on this paper. The work was partly supported by the Project "Liverworts and hornworts of Russia" of the Fundamental Researches Program of the RAS Presidium "Biodiversity and gene pool dynamics" as well as by the Russian Foundation for Basic Research № 07-04-00325a.

## 2.4 Lichens

### L.N. Poryadina

The lichens of Yakutia have become object of investigation over 100 years ago. Data on lichens are given in the geobotanical publications where vegetation characteristics include only dominant or widespread lichen species. Yakutian lichens were also studied as a fodder base for reindeer-breeding. Lichen biodiversity was studied by A.N. Oksner, Yu.V. Rykova, I.I. Makarova, M.P. Andreyev, N.N. Fesko, M.P. Zhurbenko, L.N. Poryadina. The lichen herbarium of the Institute for biological problems of cryolithozone (SASY) totals over 6,000 samples collected during various geobotanical expeditions and identified by V.I. Perfilyeva, V.P. Savich, E.K. Stukenberg, M.P. Tomin, I.I. Makarova, Yu.V. Rykova, L.N. Poryadina, T. Ahti, etc. The lichen collection from the Arctic zone of Yakutia (collected by V.I. Perfilyeva, O.I. Sumina, E.G. Nikolin, etc. and identified by I.I. Makarova, M.P. Andreyev, M.P. Zhurbenko) is deposited in the Herbarium of the Komarov Botanical Institute of the Russian Academy of Sciences in Saint-Petersburg (LE) as well as the lichens from North–East Yakutia collected and identified by M.P. Zhurbenko.

Following the system of D.L. Hawksworth (Hawksworth et al. 1995), the lichen flora of Yakutia comprises 703 species and 10 varieties belonging to 173 genera, 57 families and 13 orders of two classes, *Ascomycotina* and *Basidiomycotina.*

The leading order *Lecanorales* includes 534 species which is 76.0% of the total lichen species number. The largest families are *Parmeliaceae* (101 species – 14.4%), *Physciaceae* (53–7.5%), *Lecanoraceae* (51–7.3%), *Cladoniaceae* (50–7.1%), *Hymeneliaceae* (35–4.9%); and the largest genera *Cladonia* (50–7.1%), *Lecanora* (28–3.9%), *Aspicilia* (28–3.9%), *Rhizocarpon* (27–3.8%), *Caloplaca* (23–3.2%). The families *Coniocybaceae, Mycocaliciaceae, Protothelenellaceae, Thelocarpaceae, Thrombiaceae* as well as the genera *Lepraria, Leptorhaphis* and *Pseudosagedia* are of vague taxonomical position. There are 3 orders, 10 families and 82 genera composed of one species.

The order *Arthoniales* includes two monogeneric families: *Arthoniaceae* (4 species) and *Chrysothricaceae* (2). The orders *Gyalectales* and *Pezizales* are monospecific. The order *Leothiales* consists of the families *Baeomycetaceae* (3 species of 1 genus) and *Icmadophilaceae* (4 species of 4 genera); the order *Lichinales* comprises one family *Lichinaceae* (2 species of 2 genera); the order *Ostropales* numbers 3 monospecific families *Gomphillaceae, Graphidaceae, Stictidaceae* and *Thelotremataceae* having 3 species of 1 genus; the order *Patellariales* includes two species of one genus.

The order *Peltigerales* comprises 39 species (5.5% of the total species number) belonging to 7 genera of 4 families: *Lobariaceae* (5 species, 2 genera), *Nephromataceae* (8, 1), *Peltigeraceae* (24, 2), *Placynthiaceae* (2, 2). The order *Pertusariales* includes 26 species (3.7%) of 3 genera and 2 families: *Megasporaceae* (1, 1) and *Pertusariaceae* (25, 3). The order *Teloschistales* comprises 36 species (5.1%) of 6 genera and 2 families: *Fuscideaceae* (4, 3) and *Teloschistaceae* (32, 3). The order *Verrucariales* numbers 33 species (4.4%) of 10 genera of the familiy *Verrucariaceae.*

The *Lecanorales* provides the core of the lichen flora of Yakutia including 534 species, 121 genera and 30 families. The leading families are *Parmeliaceae* (101 species, 14.4%), *Physciaceae* (53, 7.5%), *Lecanoraceae* (51, 7.3%), *Cladoniaceae* (50, 7.1%), *Hymeneliaceae* (35, 4.9%), *Rhizocarpaceae* (28, 3.9%), *Stereocaulaceae* (23, 3.2%), *Lecideaceae* (20, 2.8%), *Porpidiaceae* (20, 2.8%), *Umbilicariaceae* (20, 2.8%), *Collemataceae* (19, 2.7%), *Acarosporaceae* (18, 2.6%), *Bacidiaceae* (17, 2.4%), *Catillariaceae* (14, 1.9%), *Psoraceae* (14, 1.9%). Genera that are large by species number are *Cladonia* (50, 7.1%), *Lecanora* (28, 3.9%), *Aspicilia* (28, 3.9%), *Rhizocarpon* (27, 3.8%), *Caloplaca* (23, 3.2%), *Stereocaulon* (21, 3.0%), *Lecidea* (19, 2.7%), *Rinodina* (15, 2.1%), *Melanelia* (14, 1.9%), *Buellia* (14, 1.9%), *Toninia* (12, 1.7%), *Acarospora* (11, 1.6%), *Collema* (11, 1.6%), *Porpidia* (10, 1.4%), *Bryoria* (9, 1.3%), *Cetraria* (9, 1.3%), *Hypogymnia* (8, 1.1%), *Physcia* (6, 0.9%). There are 52 genera and 3 families of one species each.

Basidial lichens are represented by *Lichenomphalina hudsoniana* and *L. umbellifera.*

The ratio of number of lichen species over number of vascular species is the lichen coefficient (Mattick 1953). The value of the lichen coefficient for the temperate zone of the Holarctic varies from 0.3 to 0.6, the value for the Salair Range is 0.6, for Mongolia 0.35 (Sedelnikova 1993; Golubkova 1983). According to the data of SASY, the flora of vascular plants of Yakutia numbers 1,984 species and varieties, thus, the lichen coefficient for the region is 0.35. This value is considered to be somewhat understated due to the unequal level of knowledge for the region. For example, the mountainous part of South Yakutia (Tokinsky Stanovik Range; Fesko 1990b) is studied only at the level of macrolichens. The lichen flora of the Stanovoy and Chersky Ranges is poorly investigated, as well as the central part of the Verkhoyansk Range.

The lichen flora composition of Yakutia is similar to that of Holarctic floras and is composed of a number of geographical elements. The boreal character of the flora is clear from the presence of the following families: *Parmeliaceae, Cladoniaceae, Lecanoraceae, Peltigeraceae* and genera: *Cladonia, Bryoria, Hypogymnia, Peltigera.* The families *Physciaceae, Rhizocarpaceae, Lecideaceae, Porpidiaceae, Parmeliaceae, Hymeneliaceae, Umbilicariaceae, Stereocaulaceae* and genera *Lecidea, Porpidia, Umbilicaria, Stereocaulon* are characteristic for mountainous lichen floras. The role of *Polyblastia* is significant (14–15$^{th}$ position in the flora by species number), the representatives of which are common in Yakutia, mainly in the arctic tundra.

The geographical analysis was conducted following the classification of geographical elements of Oksner modified by Sedelnikova (1985, 1990). The geographical elements represent species clustered according to the confinement of their distribution areas or development centres to floristic or botanical-geographical dominions, areas or regions (Bykov 1973). There are 8 elements in the lichen flora of Yakutia. The distribution area types are determined according to the latitudinal distribution of a species within each element.

The boreal element includes the species growing in the Holarctic coniferous forest belt. This is the largest group of species which is explained by the diverse microecological conditions for lichen growth on plains as well as in the forest and subalpine mountain belts with low hypsometric altitudes of the mountains (up to 2,000 m). The boreal element includes the following distribution area types: pluriregional, Holarctic, Eurasio-American, Holarctic-Notarctic, Eurasian, Asian.

The montane element is in the second largest comprising lichens growing in mountain forests of the temperate zone of the Holarctic, often occurring in foothills, sometimes in level areas and very rarely in the high-mountain belt. The montane lichens include 10 distribution area types: pluriregional, Holarctic, Eurasio-American, Holarctic-Notarctic, Eurasian, American-Asian, Eurasio-African, American-Asian-African, Eurasio-Greenland, and Submediterranean. The pluriregional and Holarctic lichens play the most significant role in this element.

The arctic-alpine element of the lichen flora is also considerable in Yakutia. The species of this element grow in the Arctic. Further south they have a disjunctive distribution area and occur in the high-mountain belt of the temperate Holarctic (the distribution area types are pluriregional, Holarctic, Eurasio-American, Holarctic-Notarctic, Eurasian, Asian, Asian-North American). The arctic-alpine lichens, like in the previous elements, are predominantly the species of wide distribution: Holarctic, Eurasio-American, Holarctic-Notarctic and pluriregional. There was a connection between the Arctic and the highlands of the temperate Holarctic during the ice ages of the Pleistocene when floral migration and mixing took place.

The hypoarctic-montane element comprises the species common for the Hypoarctic (those that do not reach high arctic latitudes) as well as for the middle and upper mountain belts of the temperate Holarctic. The Yakutian flora of this element is represented by lichens of pluriregional, Holarctic, Eurasio-American, Holarctic-Notarctic and Eurasian distributions.

The steppe element includes the species widespread in the steppe zone and mountain steppes of the Holarctic. The lichens of the steppe element belong to 8 distribution area types (pluriregional, Eurasio-American, Asian, Eurasio-African, Central Asian, African-Asian-North American, Holarctic, Eurasian, the latter two being best represented).

There are strong connections between the steppe floras of Mongolia and Central Asia. The Mediterranean region has also played a certain role in the formation of the steppe flora of Yakutia.

The nemoral element includes the species growing in the zone of broad-leaved forests of the Holarctic of the European and Chinese-Japanese regions (Oksner 1940b; Sedelnikova 1985). The nemoral lichens of Yakutia have pluriregional, Holarctic, Eurasio-American, Holarctic-Notarctic and Eurasian distributions. The Holarctic, pluriregional and Eurasio-American lichens are best represented. Most nemoral lichens are most likely of Turgai origin. Many of them still occur as relics in the Aldan-Indigirka Interfluve in *Chosenia-Populus suaveolens* forests.

The alpine element comprises the lichens of highlands of the Holarctic. Origination of these species mainly goes back to highlands of Asia, and their insignificant representation in the lichen flora of Yakutia is explained by the lack of records from the Yakutian mountain region.

The Arctic element includes the lichen species common both for the continental and island zones of the Arctic having a Holarctic distribution.

The lichen flora of Yakutia consists of the species of various origins that are of great interest from a florogenetic point of view. The core of the lichen flora is represented by species with pluriregional, Holarctic and Eurasio-American distributions.

The Yakutian lichen life-form classification follows the systems of Oksner (1974), Golubkova (1983) and Pristyazhnyuk (1996) with habit-physiognomic characteristics taken as a criterion. The following taxonomical units are used: division (based on substrate), type (based on direction of thallus growth), class (morphological types), group, subgroup (detailed morphological characteristics of lichen thallus) and life-forms themselves. The following life-forms of Yakutian lichens are recognized.

**Division I** Endogenic lichens developing inside the substrate.

**Type 1.** Plagiotropic lichens with thallus aligned horizontally to the substrate.

**Class 1.1.** Crustose lichens.

Group 1.1.1 Endoxylous species of *Xylographa, Graphis.*

Group 1.1.2 Endolithic species (*Petractis, Sarcogyne, Verrucaria* spp., etc.).

**Division II** Epigenic lichens with thallus developing on the substrate's surface.

**Type 1.** Plagiotropic lichens with thallus aligned horizontally to the substrate.

**Class 1.1.** Crustose lichens tightly adherent to the substrate with their lower surface.

Group 1.1.1. Homogenous crustose lichens with thallus having one and the same structure both in the centre and in the periphery.

Group 1.1.2. Dimorphic species having two morphologically different components: crustose thallus in the centre and leaf-shaped lobes along the edges.

Group 1.1.3. Squamulose lichens with thallus composed of scattered or dense scales forming continuous crust, sometimes with lobes along the edges. Epiliths, epigeoids, epiphleoids.

Group 1.1.4. Polymorphic crustose lichens. These crustose lichens comprise species having various life-forms (as in the above-mentioned groups) under various ecological conditions. Epixylous, epiphleoid, polysubstrate lichens.

**Class 1.2.** Umbilicate lichens fixed to stony substrates by means of gompha: *Lasallia, Umbilicaria spp*.

**Class 1.3.** Foliose lichens.

Group 1.3.1. Lobed rhizoidal lichens.

Group 1.3.2. Dissected lobed rhizoidal lichens.

Group 1.3.3. Inflated lobed non-rhizoidal lichens with thallus as a form of hollow lobes due to core destruction. Epiphleoid, epilithic, polysubstrate species: *Brodoa intestiniformis*, *Hypogymnia* genus.

Group 1.3.4. Dissected largely lobed non-rhizoidal lichens.

**Type 2.** Plagio-orthotropic lichens with primary thallus as horizontal scales or warty outgrowths and secondary thallus composed of vertical excrescences.

**Class 2.1.** Warty- or squamulose (scaly)-fruticose lichens

Group 2.1.1. Subulate and scyphoid lichens. Epigeoid, epiphleoid, polysubstrate species.

Group 2.1.2. Fruticose-branched lichens. Epigeoids, epiliths.

**Type 3.** Orthotropic lichens vertically oriented in space.

**Class 3.1.** Fruticose lichens.

Group 3.1.1. Platy-lobed lichens inhabiting soil, wood substrate, stones: *Allocetraria cucullata, Cetraria ericetirum, C. islandica, C. laevigata, Cetrariella delisei, Evernia esorediosa, E. mesomorpha, Vulpicida juniperinus, V. tilesii, Ramalina capitata, R. polymorpha*, etc.;

Group 3.1.2. Filamentous- and angular lobed lichens. Epilithic, epiphleoid, epixylous, epigeoid species.

Group 3.1.3. Fruticose-branched lichens with thallus rounded in cross section. Epigeoid, epilithic, epixylous species.

Group 3.1.4. Inflated thallus lichens. Thallus has a radial section and wide central cavity. The species of *Dactylina* and *Thamnolia vermicularis* grow on soil; *Ramalina dilacerate* grows on wood.

The slow growth of their thallus impedes the lichens to compete with fast growing vascular plants or mosses under more or less favourable conditions. Therefore the lichens inhabit ecological niches which are not suitable for other plants. Although lichens grow on a great variety of substrates, most species are selective and inhabit only a few or even only one substrate type (Golubkova 1977). As regards substrate type the lichen flora of Yakutia comprises 19 ecological groups.

The lichens inhabiting the substrate's surface belong to episubstrate species which comprise several groups. The epiliths grow on stony substrates (stones, boulders, detritus). This group is predominated by crustose lichens of the following genera: *Acarospora, Arctoparmelia, Aspicilia, Buellia, Carbonea, Caloplaca, Cetraria, Fuscidea, Lecanora, Lecidea, Lecidella, Lasallia, Melanelia, Ochrolechia, Pertusaria, Physcia, Physconia, Porpidia, Pseudephebe, Ramalina, Rinodina, Rhizocarpon, Stereocaulon, Tephromela, Thelidium, Xanthoria*, etc. Stony substrates often represent the habitat for rare and relic species (Sedelnikova 1990; Sedelnikova and Laschinsky 1990; Muchnik 1997) such as: nemoral relics *Arthonia arthonioides, Candelaria concolor, Physcia dimidiata, Physconia detersa, Xanthoria fallax;* periglacial steppe relics *Acarospora glaucocarpa, Aspisilia desertorum, A. obscurata, Lasallia rossica, Lecanora campestris, Lecidea lurida,*

*Lobothallia alphoplaca, Melanelia tominii, Neofuscelia pulla, Phaeophyscia sciastra, Psorotichia schaereri, Ramalina capitata, R. polymorpha, Verrucaria muralis;* glacial relics *Lecanora intricata, Pilophorus cereolus, Rhysoplaca chrysoleuca, Umbilicaria vellea. Asahinea scholanderi* growing on stony substrates is included into the Red Data Book of the USSR (Borodin et al. 1984) and the Russian Soviet Federative Socialist Republic (Golovanov et al. 1988).

The epigeous lichens that inhabit soil include: *Cladonia, Collema, Leptogium, Allocetraria, Bryocaulon, Cetraria, Dactylina, Vulpicida, Xanthoparmelia, Psora, Stereocaulon, Trapeliopsis, Baeomyces, Peltigera, Catapyrenium, Endocarpon spp.*, etc.

Epiphleoids grow on bark of trees and shrubs. *Larix* is inhabited by *Calicium, Bryoria, Hypocenomyce, Hypogymnia, Lecidea, Melanelia, Nephroma, Rinodina, Tuckermannopsis spp. Populus* and *Chosenia* are home for *Caloplaca holocarpa, Phaeophyscia ciliata, Physcia adscendens, Physconia grisea, Pseudosagedia aenea, Rinodina septebtrionalis; Populus and Chosenia* as well as their fallen twigs are inhabited by *Caloplaca ferruginea, C. haematites;* bark of *Chosenia* and *Larix* is the substrate for *Candelariella lutella, C. xanthostigma;* bark of *Dushekia fruticosa* for *Biatora sphaeroides, Chaenotheca ferruginea, Lecanora orae-frigidae, Ramalina dilacerata; Juniperus* and *Pinus pumila* for *Vulpicida juniperinus; Dushekia* and *Betula divaricata* for *Rinodina exiguella; Dushekia* and fallen twigs for *Lecanora fuscescens*. Fallen trees form the only substrate for *Bryoria bicolor, B. furcellata, Buellia badia, Cladonia sulphurina, Lecidea hypopta, L. turgidula, Micarea prasina*. Branches of *Pinus pumila* are inhabited by *Lecanora argentata, L. vogulorum*. The lichens growing on more than two tree species, inhabiting shrub, bark and fallen trees, are called polyepiphleoids and include 15 species of the genera *Biatora, Buellia, Evernia, Hypogymnia, Lecanora, Melanelia, Physcia, Rinodina, Tuckermannopsis*.

The epixylous lichens growing on decaying wood include *Arthonia aspersella, Buellia disciformis, B. griseovirens, B. insignis, Lecanora hagenii, L. saligna, Xylographa vitiligo*, etc. Epibryophytes grow on moss tussocks: *Lecanora frustulosa, Xanthoparmelia subramigera, Icmadophila ericetorum*, etc. Epiphytorelic lichens inhabit higher plant remains: *Basidia bagliettoana, Micarea assimilate*, etc. The epilichenophyte *Thelocarpon epibolum* grows on thallus of *Aspicilia sp.* The coprofilous species *Punctelia subrudecta* was recorded on old excrements of animals.

Besides the species confined to one type of substrate (so called obligate epiphleoids, epiliths, epigeoids, etc.) there are lichen groups occurring on two or more substrate types (following Skye 1968; Muchnik 1991, 1997). Epiphleoid-epigeoids inhabiting both fallen trees and soil include *Cladonia amaurocraea, C. cariosa, C. cenotea, C. cryptochlorophaea, C. decorticata, C. macilenta, C. parasitica, Lecidea vernalis, Bryoria nitidula, Saccomorpha uliginosa*, etc. Epiphleoid-epiliths are represented by *Arctoparmelia separata, Brodoa intestiniformis, Candelariella vitellina, Chrysothrix candelaris, Evernia mesomorpha, Mycobilimbia hypnorum, Pertusaria leptophora, Vulpicida pinastri.* Epiphleoid-epixylous species grow on both bark of living trees and shrubs and on decaying wood: *Bryoria simplicior, Calicium abietinum, Lecanora symmicta, Parmeliopsia ambigua, Rinodina archaea, R. exigua.* Epiphleoid-epiphytorelic lichens are represented by *Lecidella elaeochroma*, inhabiting bark of *Larix, Chosenia, Populus, Pinus pumila*, fallen trees and plant remains. The epiphleoid-epibryophyte *Micarea melaena* was recorded on larch bark, fallen trees and mosses. Epilith-epigeoids inhabiting stones and soil are represented by *Alectoria ochroleuca, A. nigricans, Arctoparmelia centrifuga, Asahinea chrysantha, Parmelia saxatilis, Psora rubiformis, Stereocaulon alpinum, S. tomentosum, Xanthoparmelia somloensis.* Epibryophyte-epiphytorelic lichens grow on mosses and plant remains: *Caloplaca saxifragarum, Lecanora epibryon, Ochrolechia upsaliensis.* Epibryophyte-epigeoids grow on mosses and soil: *Diploschistes muscorum, Phaeophyscia constipata, Trapeliopsis granulosa.*

Lichen species occurring on more than two substrate types make up the group of polysubstrate lichens. Mosses, plant remains and tree bark are inhabited by *Amandinea punctata, Caloplaca cerina, C. sinapisperma;* mosses, plant remains and soil by *Buellia geophila, Caloplaca jungermanniae, Rinodina turfacea;* mosses, wood and bark by *Mycoblastus sangiunarius.* Stones, soil and fallen trees represent the habitat for *Parmelia omphalodes*; stones, soils, fallen and living trees for *Parmelia sulcata, Vulpicida pinastri*; stones, tree bark and wood for *Imshaugia aleurites.* Mosses, wood and chosenia bark are inhabited by *Hypogymnia subobscura*; mosses, stones, fallen trees, bark of *Larix, Chosenia* and *Pinus pumila* by *Hypogymnia physodes;* soil, decaying wood and bark of fallen trees are the habitat for *Cladonia botrytes. Lepraria incana* was recorded practically on all substrate types.

Besides the abovementioned groups of episubstrate lichens there are records of endosubstrate lichens in Yakutia. Thanks to biochemical reactions, the endosubstrate lichens form microscopic cavities inside the substrate where crustose thallus developes, while fruitbodies (the apothecium or only a pore of the perithecium) develop on the substrate's surface. The growth of epilithic endosubstrate lichens is favoured by physio-chemical bedrock weathering processes. Such lichens are represented by endoliths (*Petractis, Sarcogyne, Verrucaria spp.*) and endoxylous plants (*Xylographa, Graphisspp.*).

*The Arctic Floristic Region (Arct).* The Arctic zone of Yakutia belongs to the East Siberian sector of the Russian Arctic (Andreyev et al. 1996a, b) and includes 4 areas. According to the data of Andreyev et al. (1996a, b) the lichens of the *Anabar-Olenyok area* make up 183 species of 68 genera and 32 families. They were studied by Almquist (1883); Malme (1932), Sochava (1933, 1934), Makarova (1985), Makarova and Perfilyeva (1984, 1985), Perfilyeva et al. (1981), Rykova (1978). The lichen flora of the *Kharaulakh area* (154 species, 98 genera, 39 families) was the subject of study of Cajander (1903), Oksner (1939, 1940a, b, c), Rykova (1978), Makarova (1989, 1996, 1998), Makarova and Perfilyeva (1989), Zhurbenko et al. (2002). The works of Oksner (1939, 1940a, b, c), Perfilyeva and Rykova (1975a), Andreyev (1983, 1984) were dedicated to the lichens of the *Yana-Kolyma area* (156 species, 67 genera, 29 families). The lichens of the *Novosibirskie Islands* (228 species, 87 genera, 34 families) were studied by Elenkin (1909), Gorodkov

(1956), Sumina (1986), Makarova et al. (1988), Zhurbenko and Hansen (1992), Samarsky et al. (1997). The lichen flora of the Novosibirskie Islands is characterized as arcto-alpine.

The lichen flora of the tundra zone of Yakutia was summarized by Fesko (Egorova et al. 1991) who lists 403 species of 97 genera and 33 families belonging to 6 orders (following the system of Poelt 1973). They comprise 262 species of 75 genera, 28 families of 5 orders for the arctic tundra and 288 species, 78 genera, 31 families of 4 orders for the hypoarctic tundra.

The core of the tundra flora of Yakutia is provided by lichens of the order *Lecanorales* including 93.1% of the total species number (375 species). The order *Lecanorales* is represented by 27 families and 86 genera. The average species number in a family is 12.2, while the average genus number is 2.9. There are 12 families with a higher species diversity level than the abovementioned value: *Lecideaceae* (65 species), *Parmeliaceae* (46), *Cladoniaceae* (37), *Physciaceae* (35), *Lecanoraceae* (29), *Teloschistaceae* (2I), *Aspiciliaceae* (19), *Pertusariaceae* (19), *Verrucariaceae* (18), *Peltigeraceae* (16), *Stereocaulaceae* (I4), *Usneaceae* (13). These families include 332 species which is 82.4% of the total lichen flora. Ten leading families comprise 305 species (75.6%). About the same families have the largest number of dominating species. The rest, 23 families, include 98 species (24.3%). Monospecific families are *Arthoniaceae, Microglaenaceae, Mycocaliciaceae, Gyalectaceae, Placynthiaceae, Arctomiaceae, Agyriaceae*. The leading families in genera number are *Lecideaceae* (17 genera), *Physciaceae* (9), *Parmeliaceae* (7), *Usneaceae* (7), *Verrucariaceae* (6). One third of the families is monogeneric. The average number of species in a genus is 4.2. There are 24 genera with more than the average number of species. Ten leading genera include 180 species or 44.6% of the total lichen flora. The number of monospecific genera is rather large (41), including widespread *Bryocaulon, Thamnolia*, and rather rare *Arthonia, Staurothele, Thelidium, Thelopsis, Psoroma, Arctomia, Carbonea.* Rare arctic lichens are *Lecania alpivaga, Lecanora torrida, Caloplaca friesii* (Makarova 1985).

The following species play a major role in the formation of ground lichen synusiae: *Cladonia rangiferina, C. arbuscula, C. amaurocraea, C. coccifers, C. gracilis, C. macroceras, C. pyxidata, Cetraria cucullata, C. delisei, C. islandica, C. laevigata, C. nivalis, Dactylina arctica, D. ramulosa, Alectoria nigricans, A. ochroleuca, Bryocaulon divergens, Asahinea chrysantha, Thamnolia vermicularis, Ochrolechia frigida, O.upsaliensis, Sphaerophorus globosus, Stereocaulon alpinum.*

Rare lichen species of the tundra zone in Yakutia are *Cetrelia alaskana, Parmelia teretiuscula, Cetraria andrejevii, C.microphylla, C.inermis, C.nigricascen, Asahinea scholanderi, Coelocaulon aculeatum, Umbilicaria krascheninnikovii.* Besides these, there is a number of species being rare for the Yakutian tundra though common in other regions of Russia: *Peltigera venosa, Stereocaulon vesuvianum, S. condensatum, S. glareosum, Leptogium saturninum, L. lichenoides, Parmelia infumata, Nephroma parile, Umbilicaria cylindrica, U. decussata, U. muehlenbergii, Cladonia acuminata, C. decorticata, C. furcata, C. cyanipes.*

The lichen flora of the arctic tundra subzone is represented by 262 species belonging to 5 orders, 28 families and 75 genera. Ten leading families make up 76.0% of the total species number of the subzone. *Lecideaceae, Parmeliaceae, Cladoniaceae, Physciaceae, Aspiciliaceae, Verrucariaceae, Lecanoraceae* characteristically contribute many species to the arctic and hypoarctic floras (Piyn 1984; Andreyev 1984; Makarova 1983). The lichen flora of the arctic tundra of Yakutia differs from that of the hypoarctic tundra by presence of the families *Microglaenaceae, Gyalectaceae, the* genera *Microglaena, Gyalecta, Collema, Carbonea, Catillaria, Psora, Toninia, Usnea, Pilophonis, Fulgensia, Catolechia, Orphniospora, Phaeorrhiza, Rinodinella* and by species with a wide distrbution range of families *Verrucariaceae* and *Aspiciliaceae.* Of the 403 lichen species of the tundra zone of Yakutia 113 species were recorded only in the arctic tundra subzone.

In the hypoarctic tundra subzone 288 lichen species were recorded belonging to 4 orders, 31 families and 78 genera. Ten leading families comprise 77.7% of the subzone's lichen flora. The most dominating species are the representatives of the leading *Parmeliaceae* and *Cladoniaceae* families. The number of recorded lichens in the hypoarctic tundra reflects the role they play in the composition of the vegetation cover of tundra and mountainous associations. Thus, participation of species of the family *Pertusariaceae* and the genera *Pertusaria, Ochrolechia, Cetraria* is characteristic for the tundra zone. Because of the prominence of mountainous, rubbly landforms families *Lecideaceae, Parmeliaceae, Physciaceae, Umbilicariaceae* and the genera *Parmelia, Aspicilia, Lecidea, Umbilicaria* contribute many species. Analysis of the subzone's lichen flora revealed the absence of the families *Arthoniaceae, Placynthiaceae, Arctomiaceae, Ramalinaceae* and the genera *Arthonia, Thelidium, Placynthium, Arctomia, Mycobilimbia, Ropalospora, Schadonia, Tephromela, Haematomma, Lecania, Cetrelia, Masonhalea, Parmeliopsis, Ramalina, Icmadophyla, Lasallia, Thelocarpon, Phaeo-physcia, Xylographa* in the lichen flora of the arctic tundra. Of the 403 lichen species of the tundra zone of Yakutia 137 species were recorded only in the hypoarctic tundra subzone.

In the family spectra of both subzones the following families feature rather high positions: *Lecideaceae, Parmeliaceae, Cladoniaceae, Physciaceae.* The leading families are different for each subzone, though: *Aspiciliaceae, Verrucariaceae* in the arctic tundra and *Usneaceae, Umbilicariaceae* in the hypoarctic tundra. The role of the families *Lecanoraceae, Pertusariaceae* increase in southward direction with a simultaneous increase in number of boreal species of the families *Cladoniaceae, Parmeliaceae, Usneaceae, Peltigeraceae* and the genera *Cladonia, Parmelia, Lecanora.* The presence of *Parmeliaceae, Lecideaceae, Aspiciliaceae, Umbilicariaceae* and the genera *Lecidea, Parmelia, Aspicilia, Umbilicaria* is related to mountainous landscapes and rubbly substrates. The coenotic role of lichens increases from the southern tundra towards the polar deserts where lichens dominate in the zonal vegetation communities.

Within the Russian Arctic (Andreyev et al. 1996a) the following species were recorded only in the territory of Yakutia: *Aspicilia laevatoides, Bacidia rubella, B. xylophila, Biatora ocelliformis, Lecania cyrtella, Ramalina fraxinea* in *the*

*Anabar-Olenyok area*; *Buellia notabilis, Collema limosum, Lecidea karaënsis, L. lenensis, Porpidia thomsonii, P. zeoroides, Rhizocarpon formosum, Sarcogyne distinguenda, Thelocarpon tichomirovii, Umbilicaria caroliniana* in the *Kharaulakh area; Aspicilia cupreoatra, A. farinosa, A. reticulata., A. tromsoeënsis, Lecanora argopholis, Ramalina fastigiata, Sagiolechia protuberans, Thelidium olivaceum* in the *Yana-Kolyma area* and *Acarospora cervina, Aspicilia composita, A. contigua, A. subarctica, Caloplaca tenuis, Lecidea polycocca, Miriquidica atrofulva, Rhizocarpon glaucescens, Rh. pusillum, Rinodinella controversa, Xanthoria polycarpa* in *the Novosibirskie Islands.*

The Arctic region within the territory of Yakutia totals 480 lichen species (68.3% of the total species number) of 99 genera, 45 families, 10 orders. There are 201 lichen species of 76 genera recorded only in the Arctic region.

*The Olenyok floristic region (Ol).* The lichen flora of the region is studied very poorly. Incidental data are represented in the works of Ivanova (1961), Gorbach and Mashenkova (1966), Lukicheva (1963a, b). The monograph "Vegetation of the Viluy River basin" (Galaktionova et al. 1962) contains data on the lichens from the Olenyok region identified by V.P. Savich, Z.M. Smirnova, K.A. Rassadina, E.K. Stukenberg. At the SASY the lichen herbarium material is stored with samples collected and identified by V.I. Ivanova and L.A. Dobretsova (the Middle Olenyok River basin), and collected by I.M. Okhlopkov, identified by L.N. Poryadina (the Molodo River basin). The lichen flora of this region comprises 77 species (11.0%) belonging to 27 genera, 12 families, 4 orders.

*The Yana-Indigirka Floristic Region (Ya-I).* Data on lichens from the Region are mentioned in the geobotanical works of Yarovoy (1939), Sheludyakova (1938, 1948a, b), Dobretsova (1959), Kuvaev (1956), Prakhov (1957), Kildyushevsky (1960), Karavaev (1976). The lichen flora of Ya-I is also summarized in the works by Oksner (1939, 1940a, b, c), Kuvaev and Samarin (1961), Rykova (1976, 1980), Golubkova and Shapiro (1978). Afonina et al. (1979, 1980), Bredkina (1980), Fesko (1988a, b, 1990a), Poryadina (1998, 1999a, b, c, 2001b, c, d, 2003a, d), Zhurbenko (2003).

The lichens of the region are represented by 434 species (61.7%) of 124 genera, 44 families, 10 orders. There are records of 126 species of 72 genera rare for Yakutia. The species of the order *Lichinales (Lempholemma polyanthes, Psorotichia schaereri), Pezizales (Schaereria fuscocinerea),* the families *Coniocybaceae (Chaenotheca ferruginea, Ch. Furfuracea)* and *Mycocaliciaceae (Phaeocalicium praecedens)* were found only within the territory of the Yana-Indigirka FR. The core of the flora is composed of species with pluriregional, Holarctic and Eurasio-American distributions. The lichen flora of the region is arctic-alpine-montane-boreal representing species of various origins.

*The Kolyma Floristic Region (Kol).* Limited data on the lichens of the Region are given in the works by Elenkin (1904, 1909). The lichens were also studied by Rykova (1972, 2004). Presently, 80 species have been recorded (11.4%) of 24 genera, 12 families, 3 orders.

*The Central Yakutian Floristic Region (CYa).* The lichen data of the Region are summarized in the works by Oksner (1939, 1940a, b, c), Karavaev (1958, 1976),

Kildyushevsky (1964), Rykova (1987), Rykova and Chechurova (2001), Poryadina (2001a, 2003b, c), Zhurbenko (2003), Galaktionova et al. (1962). The lichen flora, characterized as boreal, totals 215 species (30.6%) of 62 genera, 25 families, 7 orders. The leading families are *Parmeliaceae* and *Cladoniaceae,* the leading genera *Cladonia, Lecanora, Melanelia, Peltigera, Caloplaca.* In the territory of Yakutia the following species were recorded solely within this region: *Endocarpon psorodeum, Lecanactis latebrarum, Megalaria laureri, Placidiopsis pseudocinerea, Toninia candida, T. physaroides, Thrombium epigaeum.*

*The Upper Lena Floristic Region (UL).* The scanty information on the lichens of this FR is given in the monograph by Galaktionova et al. (1962) and works by Karavaev (1958, 1976).

In the summer of 2004, during the comprehensive expedition of the Institute for Biological Problems of Cryolithozone SB RAS, for the first time a lichenological survey was conducted yielding collections from the Vitim and Lena Rivers basins, as well as from the territory of the Resource Reserve "Pilka". Presently, the lichen flora of the region comprises 114 species (16.2% of total lichen flora) of 41 genera, 21 family, 6 orders. The core of the flora is provided by the order *Lecanorales* (92 species): first of all by *Parmeliaceae* (38) and *Cladoniaceae* (28). The order *Peltigerales* is represented by the monogeneric families *Lobariaceae* (1), *Nephromataceae* (4) and *Peltigeraceae* (9). The order *Teloschistales* includes one family represented by 4 species of 2 genera. The orders *Arthoniales, Graphidales, Leothiales* are monogeneric and monospecific. The lichen flora of the region is characterized as boreal (Poryadina 2006). There are records of rare species in Yakutia such as *Collema furfuraceum, Diploicia canescens, Bryoria capillaris, Graphis scripta, Micarea cinerea, Ramalina calicaris, Usnea cavernosa, U. glabrescens, Nephroma bellum, N. resupinatum.*

*The Aldan Floristic Region (Ald).* The lichen flora of the Region was studied by Rabotnov (1936a), Fesko (1990b). The collection of T.A. Rabotnov from the Timpton (now Aldan) region was identified by Oksner (1939, 1940a, b, c). The lichen herbarium stored at the SASY contains collections made and identified by N.N. Fesko (Neryungri vicinity), collections of E.V. Sofronova and L.G. Mikhalyova identified by L.N. Poryadina (the Elga River basin). At present, the lichen flora of this FR totals 220 species (31.3%) of 60 genera, 28 families, 5 orders. The following species were recorded only within the Aldan FR: *Acarospora fuscata, Cyphelium inquinans, Collema minor, C. subnigrescens, Leptogium intermedium, L. palmatum, Cetrelia cetrarioides, Hypogymnia delavayi, Hypotrachyna sinuosa, Melanelia albertana, Nephromopsis laureri, Heterodermia leucomelos, Stereocaulon apocalypticum, S. grande, S. intermedium, S. nanodes, S. subcoralloides, Lobaria isidiosa, L. retigera, Nephroma isidiosum, Peltigera degenii, Lichenomphalina hudsoniana.* As a whole, the lichen flora is characterized as montane-boreal.

The Red Data book of Russia (Golovanov et al. 1988) contains the following rare species occurring in Yakutia. Category III "Rare species" (R): *Asahinea scholanderi* (Arct, Ya-I, CYa, Ald); *Lichenomphalina hudsoniana* (Ald), *Lobaria retigera* (Ald), *Nephromopsis laureri* (Ald), *Cetrelia alaskana* (Arct); Category II

"Vulnerable species" (V): *Cetraria steppae* (Ya-I), *Lobaria pulmonaria* (Ald), *Stereocaulon dactylophyllum* (Arct, Ya-I, UL). Seven species were listed into the Red Data book of Yakutia (Dolinin et al. 2000) being under protection of various Especially Protected Natural Territories: *Cetrelia alaskana, Sticta arctica, Lobaria linita, Asahinea scholanderi, Umbilicaria krascheninnikovii.*

## 2.5 Fungi

### L.G. Mikhalyova

Distributional and taxonomical data on Yakutian fungi are very uneven. Over the years the objects of study varied from parasitic fungi of agricultural plants, to edible fungi, pathogenic fungi causing forest diseases, and medicinally used fungi. Mycological investigations are most complete for Central and South Yakutia, while the northern and western parts of the republic were studied occasionally. Sometimes sampling of fungi was conducted by non-specialists which resulted in scanty and scrappy data from over the whole region. Results of Yakutian fungi investigations are found in Benua and Karpova-Benua (1973, 1979a, b); Nikadimova (1964, 1967); Guseva (1967, 1970); Parmasto (1967, 1975, 1976, 1977); Petrenko (1978); Kotiranta and Mukhin (2000). The mycological and phytopathological survey made by K.A. Benua in 1925–1926 is considered to be the first such study since no earlier record in this field is known. The expedition in which Benua participated covered the Lena-Amga and the Lena-Aldan Interfluves. This locality represents a vast plain with a slight altitudinal increase eastwards from the right bank of the Lena towards the Lena-Amga watershed. Thus, the samples were collected in Central and South–West Yakutia (the Central Yakutian Floristic Region). The records include 53 species of the aphyllophoroids including the interesting resupinate form of *Phellinus chrysoloma, f. laricis* (Benua and Karpova-Benua 1979a, b), 348 species and 108 varieties of parasitic fungi and about 10 agaricoid species.

Basidial xylotrophs represent an important fungal group playing a significant role in wood and wood litter decay and therefore in the biological cycling of forest biocoenoses as a whole. Expanded industrial exploitation of forest tracts necessitates the study of wood-rotters. Stem rot is a major defect decreasing commercial wood production and worsening forest quality. Nikadimova (1964, 1967) studied the wood-destroying fungi and the problem of wood faultiness in Yakutia. Her surveys in South Yakutia (the Upper and Middle Aldan River basin) yielded 17 species of aphyllophoroid fungi. Guseva (1967, 1970) studied the effect and distribution of root sponge in pine and larch forests of South–West Yakutia.

The comprehensive expedition of the Academies of the Estonian and Lithuanian Soviet Socialist Republics in the northern, eastern and central parts of the Republic in 1972 has made a great contribution to Yakutian mycobiota investigations. For the first time material from the Cold Poles of the Earth was collected. This survey resulted in a list of aphyllophoroid fungi (Parmasto 1975, 1976, 1977) including 87 species of 5 families (*Thelephoraceae, Ganodermataceae, Hymenochetaceae,*

*Poriaceae, Polyporaceae*) of the order. In the region of the Verkhoyansk Cold Pole 13 species were recorded, while in the Oymyakon Cold Pole 19 species were identified with fruitbodies at a height not exceeding the snow cover depth (Parmasto 1977).

The one of the most comprehensive reviews of micro- and macromycetes of Central Yakutia is presented in the monograph by Petrenko (1978) where the floristic composition of macromycetes, and their distribution as regards ecological groups in various forest types are given. According Petrenko, the records of fungi in the territory of the Republic comprise 249 species including 32% xylotrophs, as well as 41 species of aphyllophoroid fungi in 6 families, 199 species of agaricoid fungi, 4 species of *Pezizales,* 4 of gastromycetoids and 1 myxomycete species. For each fungus species the author gives a detailed description of their biology, phenology, distribution and confinement to various substrates in the forests of Central Yakutia.

In 1989 mycological collecting in Yakutia was carried out by specialists of the Yakut Institute of Biology (Siberian Branch of the Academy of the USSR) and resulted in numerous publications (Mikhalyova 1993, 1995, 1998, 1999, 2001a, b, c; Danilova 2005; Mukhin et al. 2004).

The *Agaricales* and *Boletales* were studied by I.F. Shurduk and N.S. Medvedeva (the Yakut Institute of Biology). The results of this investigation are presented in a manuscript report of 1967. The survey covered the Zhigansky and Lensky regions and the vicinity of Yakutsk. Zhigansk yielded 18 fungi species, while Nyuya (Lensky region) about 60 species.

In 1999 mycologists from Finland, Austria, Denmark, as well as from the Ural Institute of Plant and Animal Ecology conducted researches in the territory of the Republic (Tiksi and Yakutsk vicinities) collecting fungi of the *Aphyllophorales, Agaricales* and *Boletales*. Thus, the records from Tiksi comprise 7 species of aphyllophoroid macromycetes and about 10 species of agaricoid fungi. In 2003 a joint Russian-Finnish expedition investigated the Kolyma River basin. The results of this work are still subject to publish.

For unification of taxonomical data the system accepted in North-European countries was used (Hansen and Knudsen 1992, 1997): the kingdom division is given following Margelis (1983), divisions, classes and subclasses are given according to Müller and Löffler (1995).

The distribution of Yakutian fungi according to Floristic Regions is as follows:

*The Arctic Floristic Region (Arct)* is still subject to study. Mikhalyova conducted investigations in the vicinity of Tiksi as a member of the Russian-Scandinavian expedition. Seven species of polypores were recorded.

*The Olenyok Floristic Region (Ol).* Investigations were carried out in the vicinity of Zhigansk (collections made by Medvedeva in 1967 and Parmasto in 1972) and in the Molodo River basin. The records comprise 26 agaricoid and 6 aphyllophoroid species.

*The Yana-Indigirka Floristic Region (Ya-I)* is also characterized by lack of data. The survey covered the vicinities of Batagai, Oymyakon and Verkhoyansk. In the region of the Verkhoyansk Cold Pole 13 aphyllophoroid species were recorded, while in the Oymyakon Cold Pole region 19 species were found (collections made by Parmasto in1972).

*The Kolyma Floristic Region (Kol).* Mikhalyova investigated the Kolyma basin from the Shaitanikha River mouth through the settlement of Chersky (collections of 1991, 2003, 2006). About 30 species of aphyllophoroid fungi were recorded according to archive material and herbarium collections.

*The Central Yakutian Floristic Region (CYa).* The mycobiota of this region is best studied. The records include over 150 aphyllophoroid species, about 200 agaricoid species, 4 *Pezizales* species, 4 gastromycetes and 1 species of myxomycetes (collections made by Shurduk in 1967, Parmasto in 1972, Petrenko in 1975–1976, Mikhalyova in 1995–2007).

*The Upper Lena Floristic Region (UL).* The mycobiota of this region was studied as a forest component at the Research Stations Zakharovka (Lensky region), Olyokminsky and Kochegarovo (Olyokminsky region) (collections made by Nikadimova in 1964, Guseva in 1964, Shurduk in 1967, Medvedeva in 1967). Besides, the list was supplemented by herbarium data of L.G. Mikhalyova from the Tokko River basin and the vicinity of Torgo (1994). The records comprise about 200 agaricoid fungi and over 180 aphyllophoroids.

*The Aldan Floristic Region (Ald).* The mycobiota of this region includes about 190 aphyllophoroid species and about 150 agaricoids. The investigation was conducted in the Aldan, Amedichi, Algama, Uchur, and Timpton Rivers basins, Lake Bolshoye Toko, as well as in the vicinities of the Aldan, Neryungri, and Tommot between 1989 and 2007.

Thus, the records for the territory of Yakutia comprise 1 species of Protista and 451 fungi species in 32 orders and 62 families of macromycetes belonging to the divisions *ASCOMYCOTA* and *BASIDIOMYCOTA* and the classes *Ustilaginomycetes, Ascomycetes* and *Hymenomycetes*.

Many wood-attacking fungi of Yakutia grow at the northern or eastern boundary of their distribution areas. Investigations made clear that the distribution area of xylotrophic fungi is wider than that of tree species. Thus, for example, in Tiksi, characterized by the absence of tree vegetation, the xylotrophic fungi grow on timber that is brought in: saw-timber, wooden posts, basements, fences, etc. First of all, this demonstrates that the specific zonal climatic conditions are not a limiting factor for the growth of xylotrophic fungi. When there is suitable substrate even beyond the distribution area of a host-tree, fungi may successfully grow. This can be explained first of all by the fact that the main mass of a xylotrophic fungus (mycelium) is hidden in the substrate, and the chemical reaction of wood decomposition is accompanied by calorification. So, the most vulnerable moment of fungus survival is the process of spore formation. Under ultracontinental climatic conditions the fruitbodies are formed underneath the fallen trunks or at a height not exceeding the snow cover depth (Parmasto 1977). Besides, the fruitbodies often have a resupinate form which also protects the spore-bearing layer against unfavourable climatic conditions.

Data from the literature (Nikadimova 1970; Petrenko 1980; Andreyev et al. 1987) and our investigations (Mikhalyova and Protopopov 1999) document severe damages of near-tundra forests by fungal diseases. Practically 100% of the trees older than 300 years suffer from stem rot. The larch forests growing further south are affected by stem rot for 60%. This is caused by the most prominent phytopathogen of Yakutian forests *Fomitopsis pinicola* which often destroys the core of old living trees often without forming fruitbodies (Averensky and Mikhalyova 2000).

Other common larch pathogens are *Fomitopsis cajanderi* and *Trichaptum laricinum* affecting fallen trees and stubs. The rarest fungus is *Skeletocutis stellae* which has been recorded in Yakutia strictly north of the Polar Circle.

The quantitative distribution of fungi according to rot type is as follows: 78.2% of the fungi cause white rot and 21.8% are brown-rotting fungi. This is not typical for the coniferous forests of the nemoral zone, though normal for monospecific northern hard-wood *Larix* forests.

Most aphyllophoroid species are confined to deciduous trees having soft wood (*Dushekia, Betula)* or acid pH values of the bark (*Populus, Chosenia, Salix*), which considerably facilitates fungal spore penetration into the wood (Bondartseva and Solovyev 1992).

Geographically the aphyllophoroid mycobiota is mainly represented by plurizonal species (66.4%), while one third of the fungi (31.8%) are boreal. The nemoral element (1.8%) is also present and shows the low specificity of mycobiota of a boreal character. The majority of species are cosmopolites (21%) or multiregional or holarctic in distribution area. The Eurasian geoelement of Yakutia comprises several species: *Ceriporiopsis resinascens, Daedaleopsis septentrionalis, D. tricolor, Oligoporus rennyii. Antrodia mellita, Polyporus tubaeformis, Postia lateritia, Skeletocutis carneogrisea.*

The only species of the Manchurian group is *Perenniporia maackiae,* recorded in Central Yakutia. The distribution area of this species covers coniferous-broad-leaved forests of Primorsk Territory, Khabarovsk Territory and Sakhalin Island where it grows on the wood of *Maackia amurensis* (Lyubarsky and Vasilyeva 1975).

Besides the Manchurian group, the Yakutian mycobiota is also peculiar for the following species: *Datronia scutellata* is common in North America, the southern part of the Russian Far East and Yakutia though rare in Europe and the southern part of Central Siberia; and *Fomitopsis cajanderi* having two distribution areas – in Northeastern Asia and North America. The presence of these two species probably indicates that the Yakutian mycobiota entirely belongs to the Beringian mycobiotic group of North Eurasia (Lyubarsky and Vasilyeva 1975, Parmasto 1979, Mukhin et al. 2004). Besides, the following species were found in Central Yakutia which have not been recorded elswhere in Russia: *Antrodia mellita* and *Polyporus tubaeformis. Postia lateritia* and *Skeletocutis carneogrisea* are new species for Siberia (Mukhin 1993).

The aphyllophoroid mycobiota of Yakutia generally consists of the most adaptive fungi having di- and trimitic hyphal systems. They are characteristic for affected forests and habitats under extreme climatic conditions. It is clear that the distribution areas of some fungal species reaches beyond the northern boundary of the forest zone.

## 2.6 Algae

**I.I. Vasilyeva, P.A. Remigailo, V.A. Gabyshev, A.P. Ivanova, and L.I. Kopyrina**

The first records on algae of different geographical regions of Yakutia are given in publications of Skvortsov (1917), Kiselyov (1935), Alabyshev (1932), and Kosinskaya (1936). They represent taxonomical lists based on isolated collections made by the members of comprehensive expeditions studying the climate and natural and economical resources of the region.

Since 1947, when the Scientific Base of the Academy of Sciences of the USSR was established in Yakutsk, systematic investigation of algoflora of Yakutia has started. From 1948 through 1958 L.E. Komarenko, the founder of algological investigations in the Republic, collected, processed and published interesting material(for a regional flora study) on the algal composition of some stagnant and running water bodies in Central and North Yakutia. Those works (Komarenko 1955, 1956) provided the basis for further detailed study of algae in the north–eastern USSR.

From 1958 through 1975 L.E. Komarenko and I.I. Vasilyeva conducted a number of insightful investigations on algal composition, distribution and ecology of various types of water bodies in different botanical-geographical regions of Yakutia (Komarenko 1962, 1968; Komarenko and Vasilyeva 1967, 1972). Since the 1960s, along with routinal investigations, permanent observations of the developmental dynamics of phytoplankton in combination with hydrochemical, hydrobiological and botanical investigations, have been conducted. In 1962–1966, the photosynthetic activity, composition and dynamics of lake phytoplankton in the vicinity of Yakutsk were studied (Vasilyeva 1965, 1966, 1968).

Since 1970, the ecological-floristic study of algae in the Lena, Tatta, Viluy, Kolyma, Yana, Indigirka, and Anabar River basins, as well as in the Viluy Reservoir, in the Central Verkhoyansk region, in the Novosibirskie Islands, in the icings of the Momsky region, etc. have been conducted.

Also long-term permanent observations on phytoplankton dynamics have been carried out in connection with the creation of the Viluy Reservoir (Vasilyeva and Remigailo 1982). And since 1986 the Upper, Middle and Lower Lena River as well as some of its tributaries (the Aldan, Olyokma, etc.) have been studied (Vasilyeva-Kralina et al. 1997; Remigailo 1983; Remigailo and Gabyshev 2001).

The first record of soil algae of Yakutia was made by Rabotnov (1934) who described the alga *Nostoc commune Vauch.* on a ground layer of solonetzic soils. Dorogostaiskaya (1959) gives the results of an algal study near Tiksi, while several publications are dedicated to the algae of the steppes of North–East Yakutia (Pivovarova 1976, 1986a, b; Pivovarova et al. 1975; Berman et al. 1978). They represent descriptions of the soil microalgal flora of the mountainous isolated steppe landscapes in the Lena, Yana, and Indigirka River valleys. There is only one publication on soil algae of alases (Dubovik 1988) concerning the question of the distribution of the soil algoflora of the Isteehkh alas, situated in the Lena-Amga Interfluve, and based on samples of P.A. Gogoleva. Pshennikova (1992) provided a taxonomical study of the soil algoflora of alases of the Lena-Amga Interfluve, and the algae of alas lakes of the Lena-Amga Interfluve were also studied (Vasilyeva and Pshennikova 1992; Pshennikova 1995; Vasilyeva and Smirnov 1995). In late 1990s, the algae of tundra soils of natural and disturbed phytocoenoses of the Lower Yana were studied (Pshennikova 1995) and as from 2000, the study of soil algae of the Lena banks, as well as of forb meadows and larch forest of the National Park “Lena Pillars”, started.

In 1990, a program of monitoring 10 urban and suburban lakes near Yakutsk was launched (Ivanova 1999; Ivanova and Kopyrina 1999, 2002), while since 1994 phytoplankton dynamics near Yakutsk and its suburbs (the Middle Lena sector) have been observed (Gabyshev 1998; Remigailo and Gabyshev 1999).

Recently, the investigation of the diatoms from bottom sediments of Yakutian lakes has started along two basic lines: a palaeoecological study (Andreyev et al. 1995; Pestryakova 1999; Pestryakova and Ushnitskaya 2000; Pestryakova 2004) and a study of anthropogenic eutrophication processes (Pestryakova 1994; Pestryakova et al. 2004).

As from 1996, in Yakutia the first studies on the taxonomy of epiphytic algae have been initiated, with investigations of the epiphytic algae confined to 20 species of higher aquatic plants of the Middle Lena valley’s oxbow lakes (Kopyrina 1999a, b), of the phytoperifiton on various substrates of the Lena, Aldan, Kolyma, Indigirka, Vitim, Pilka, Muna, and Molodo River basins as well as of the lakes of Central Yakutia has started. These investigations were an integral part of geobotanical expeditions ((Kopyrina 2000a, b, c, d, 2003a, b, c; Zakharova et al. 2004b; Pshennikova et al. 2004).

During the whole period of investigations algologists of the Yakut Institute of Biology (Siberian Branch of the Academy of Sciences of the USSR and later Russian Academy of Sciences) and their predecessors, in collaboration with colleagues of other institutions have conducted over 130 field works of various types covering 169 running water systems and over 1,400 lakes.

The analysis of much material gathered during floristic and hydrobiological investigations has yielded eight summarizing monographs on composition and distribution of algae in the studied region (Komarenko and Vasilyeva 1975a, 1978; Vasilyeva 1987, 1989a, b) as well as 1 manual, 2 practical recommendations and other scientific publications. The systematic studies have proved the thriving of aquatic and soil algoflora under conditions of permafrost action underlying the surface of enormous territories in north–eastern Asia (cryolithozone) and the ultra continentality of the climate which determines a short vegetation period and a regional peculiarity in the algoflora. The following is list of main publications of algologists grouped by 14 large river basins of Yakutia, concerning algae species of floodplains, alas and tectonical lakes, as well as the icings of the Momsky region and the water bodies of the Novosibirskie Islands.

*Lena River basin.* Bening (1942); Vasilyeva (1968, 1987, 1989a, b); Vasilyeva and Remigailo (1986); Vasilyeva et al. (1984); Vasilyeva and Rizvanova (1976); Vasilyeva-Kralina et al. (1997); Gabyshev (1998, 2003, 2004); Gabyshev and Remigailo (2003); Egorova et al. (1991); Ivanova (1999); Ivanova and Kopyrina

(1999, 2002); Ivanova and Trofimova (2003); Kiselyov (1935); Komarenko (1955, 1956); Komarenko and Vasilyeva (1975a, 1978); Kopyrina (1999a, b, 2003a, b, c); Kosinskaya (1936); Pshennikova (1992, 1995); Remigailo (1983); Remigailo and Gabyshev (1999); Serkina (1969); Skvortsov (1917).

*Viluy River basin.* Kirillov et al. (1979); Vasilyeva and Remigailo (1982); Vasilyeva et al. (1981); Komarenko and Vasilyeva (1975a, 1978); Vasilyeva (1987).

*Yana River basin.* Vasilyeva-Kralina (1999); Vasilyeva-Kralina and Gabyshev (1996); Vasilyeva-Kralina et al. (2004); Komarenko (1968); Komarenko and Vasilyeva (1975a, 1978); Vasilyeva (1987).

*Aldan River basin.* Komarenko (1956); Komarenko and Vasilyeva (1975a, 1978); Vasilyeva (1987); Vasilyeva-Kralina and Pshennikova (1998); Remigailo and Gabyshev (2001).

*Kolyma River basin.* Vasilyeva and Pshennikova (1995, 1996); Vasilyeva and Remigailo (1980); Komarenko and Vasilyeva (1972, 1975a, 1978); Komarenko et al. (1974); Pshennikova (1986).

*Anabar River basin.* Komarenko and Vasilyeva (1975a, b, 1978); Vasilyeva (1987).

*Amga River basin.* Vasilyeva and Pshennikova (1992); Vasilyeva-Kralina and Pshennikova (2000); Pshennikova (1992, 1995).

*Indigirka River basin.* Komarenko (1955); Komarenko and Vasilyeva (1975a, 1978); Vasilyeva (1987)

*Olyokma River basin.* Vasilyeva (1987); Vasilyeva et al. (1984); Vasilyeva-Kralina et al. (1997); Komarenko and Vasilyeva (1975a, 1978).

*Khroma River basin.* Komarenko (1975); Komarenko and Vasilyeva (1975a, 1978); Vasilyeva (1987).

*Olenyok River basin.* Komarenko and Vasilyeva (1967); Komarenko and Vasilyeva (1975a, 1978); Vasilyeva (1987).

*Alazeya River basin.* Komarenko and Vasilyeva (1975a, 1978); Vasilyeva (1987).

*Omoloy River basin.* Komarenko and Vasilyeva (1975a, 1978); Vasilyeva (1987).

*Taxonomical analysis.* The algoflora of Yakutia totals 2,476 species (3,124 taxa of intraspecific level) of 508 genera, 160 families, 55 orders, 22 classes and 12 divisions (Table 2.6). The most numerous in species number are the green algae – 33.8% and diatoms – 23.2%. They are followed by the blue-green algae – 14.2%, golden algae – 9.9%, yellow-greens – 8.8%, and euglenoids – 5.7%. The dynophytes (3.1%), cryptophytes (0.6%), red algae (0.3%), raphidophytes (0.2%), brown algae and charophytes (0.1% each) are less diverse in species. The flora ratio is 1 : 3.2 : 15.5 : 19.5, the genera saturation with species is 4.9, or 6.1 with intraspecific taxa. Species variability is 1.3. A high genus saturation indirectly points to the autochtonous trends in the algoflora development: the allochthonous genera comprise less species than the indigenous ones (Malyshev and Peshkova 1984). The yellow-greens are largest in number of classes (6), while the green algae dominate in number of orders (13), families (49), and genera (208). Diatoms and euglenids typically have the highest species variability, i.e. their intraspecific-specific ratio is maximal (1.5 and 1.4 respectively), as well as genus saturation (11.5 and 5.6) and family saturation rates with species (26.1 and 28.2) and intraspecific taxa (39.9 and

**Table 2.6** Taxonomical composition of the algoflora of Yakutia

| Divisions, Classes, Orders, Families | Number of genera | Number of species and varieties |
|---|---|---|
| CYANOPHYTA, Chroococcophyceae, Chroococcales, Synechococcaceae | 10 | 40 |
| Holopediaceae | 5 | 20 |
| Beckiaceae | 5 | 20 |
| Merismopediaceae | 5 | 20 |
| Tetrapediaceae | 5 | 20 |
| Mycrocystidaceae | 5 | 20 |
| Gleocapsaceae | 5 | 20 |
| Coelospheriaceae | 5 | 20 |
| Gomphosphaeriaceae | 5 | 20 |
| Woronichiniaceae | 5 | 20 |
| Entophysalidales, Chlorogloeaceae | 5 | 20 |
| Chamaesiphonaceae, Pleurocapsales, Pleurocapsaceae | 5 | 20 |
| EUGLENOPHYTA, Euglenophyceae, Euglenales, Eutreptiaceae | 2 | 4 |
| Euglenaceae | 13 | 157 |
| Menoidiaceae | 2 | 6 |
| Peranematales, Peranemataceae | 5 | 13 |
| Petalomanadaceae | 4 | 6 |
| DINOPHYTA, Dinophyceae, Gymnodiniales, Gymnodoniaceae | 5 | 23 |
| Peridiniales, Peridiniaceae | 5 | 46 |
| Gloeodiniales, Gloeodiniaceae | 1 | 1 |
| Dinococcales, Phytodiniaceae | 6 | 12 |
| CRYPTOPHYTA, Cryptophyceae, Cryptomonadales, Cryptomonadaceae | 4 | 14 |
| Katablepharidaceae | 1 | 1 |
| RAPHYDOPHYTA, Raphydophyceae, Raphydales, Vacuolariaceae | 2 | 5 |
| Monomastigaceae | 1 | 1 |
| BACILLARIOPHYTA, Centrophyceae, Thalassiosirales, Thalassiosiraceae | 1 | 4 |

**Table 2.6** (continued)

| Divisions, Classes, Orders, Families | Number of genera | Number of species and varieties |
|---|---|---|
| Stephanodiscaceae | 3 | 29 |
| Pseudoposirales, Radialiplicataceae | 1 | 2 |
| Melosirales, Melosiraceae | 1 | 6 |
| Aulocosiraceae | 1 | 14 |
| Coscinodiscales, Coscinodiscaceae | 1 | 2 |
| Biddulphinales, Hemiaulaceae | 1 | 1 |
| Rhizosoleniales, Rhizosoleniaceae | 1 | 3 |
| Pennatophyceae, Araphales, Fragilariaceae | 4 | 88 |
| Diatomaceae | 2 | 13 |
| Tabellariaceae | 2 | 8 |
| Raphales, Naviculaceae | 11 | 311 |
| Achnanthaceae | 3 | 64 |
| Eunotiaceae | 1 | 54 |
| Rhoicospheniaceae | 1 | 1 |
| Cymbellaceae | 2 | 76 |
| Gomphonemataceae | 2 | 53 |
| Entomoneidaceae | 1 | 2 |
| Epithemiaceae | 2 | 14 |
| Rhopalodiaceae | 1 | 2 |
| Nitzschiaceae | 3 | 86 |
| Surirellaceae | 4 | 43 |
| Rhodophyta, Florideophyceae, Nemaliales, Acrochaetiaceae | 2 | 3 |
| Batrachospermaceae | 1 | 1 |
| Lemaniales | 1 | 2 |
| Thoreaceae | 1 | 1 |
| Hildenbrandtiaceae | 1 | 1 |
| PHAEOPHYTA, Phaeozoosporophyceae, Ectocarpales, Ectocarpaceae | 1 | 1 |
| Chordariales, Myrionemataceae | 1 | 1 |
| CHAROPHYTA, Charophyceae, Charales, Nitellaceae | 1 | 1 |
| Characeae | 1 | 1 |

40.6). This phenomenon probably indicates a good species adaptability to various environmental factors. The yellow-greens that in most cases do not have sexual propagation are characterized by low genus saturation (3.1) and species variability (1.0). This reduces speciation rates and points to their ancient origin and wide geographical distribution at high latitudes (Dogadina 1986).

Considering the classes, the *Chlorophyceae* and *Pennatophyceae* stand out for their species saturation (530 species or 21.4% each), but also the *Conjugatophyceae* (307 species or 12.4%) *Hormogoniophyceae* (242 species, 9.8%), and *Chrysophyceae* (238 species, 9.6%). The most important orders belong to the diatoms and green algae: *Raphales* (475 species, 19.2%), *Chlorococcales* (283, 11.4%), *Desmidiales* (267, 10.8%).

Geographical position affects the sequence of the families leading in species number and reflects the character of the algoflora. The first three places belong to *Desmidiaceae* (217 species, 8.8%), *Naviculaceae* (194, 7.8%) and *Oscillatoriaceae* (119, 4.8%).

The largest 10 families comprise 1,017 species (41.1% of total algae species number). Twenty two families are represented by 2 species and 34 families by one species. Thus, more than one third of the families are mono- or bispecific. The ten leading genera include 586 species (23.7% of the total algae species number). The first three places are taken by the representatives of the diatoms and green algae: *Cosmarium* (110 species, 4.4%), *Navicula* (93, 3.8%), and *Nitzschia* (59, 2.4%). There are 101 two-species and 207 monospecific genera.

The prevalence of monospecific genera and families is a characteristic feature of northern floras and, following some authors (Rebristaya 1977; Getsen 1987), this reflects the region's location at high latitudes.

The monospecific genera make up 40.8% in Yakutia. A high percentage of them are the green algae (41.5% of all monospecific algae genera), yellow-greens (14.5%) and golden algae (13.0%). The diatoms feature the largest number of genera with 10 and more species (15 genera or 30.0% of total diatom genera).

*Ecological analysis.* The algae of Yakutia inhabit terrestrial (*Trentepohlia, Stigonema, Gloeocapsa*, etc.), soil (*Nostoc, Oscillatoria, Phormidium*, etc.) and water biotopes.

The snow of the Verkhoyansk Range at the border of the tundra altitudinal belt and icings are dominated by the blue-green algae of the genera *Gloeocapsa* and *Nostoc*, along with the diatoms. The nival meadows of the Stanovoy Range's spurs (1,470 m a.s.l.) are inhabited mainly with *Cosmarium, Penium, Actinotaenium, Hantzschia, Oscillatoria* occurring among moss patches of *Amblystegiella*. The mountain swift streams are overgrown with species of *Prasiola, Hydrurus, Lemanea* in combination with *Didymosphenia geminata, Hannaea arcus, Meridion circulate*, and species of *Cymbella, Gomphonema, Achnanthes*. In the river valleys of the Verkhoyansk Range, in subalpine shrubby belt (700–800 m a.s.l.) rock debris often possesses reddish coatings of *Gloeocapsa magma, Stigonema minutum* and *Thorea sp*. The moss-covered slopes in the mountain tundra (the Lena River Delta) are also covered with yellow-green coatings of *Heterococcus chodatii*. Among patches of the moss *Andrea rupestris* and lichen *Siphula ceratites* the blue-green algae *Stigonema minutum* and *S. mamillosum* are the pioneer species on bare soil patches in the Arctic tundra of the delta's islands.

The largest amount of algae species is recorded from water bodies where they become a necessary component of hydro- and geobiocoenoses following the general hydrobiological laws, and structured by latitudinal and altitudinal zonation.

*River algae*. (For locations of the rivers see Fig. 2.1) *The Lena River*. The Lena basin (length 4,270 km, river net density 0.42 km/km$^2$, lake coefficient 0.7 (Nogovitsyn 1985)) (from now on these three measures will be given in brackets following each river name) is home for 77.8% of the total Yakutian algoflora, represented by 1,927 species (2,348 intraspecific taxa) of 11 divisions. The green algae and diatoms are largest in species number (35.5% and 21.7% respectively). They are followed by the blue-green algae – 15.5%, yellow-green algae – 9.8%, golden algae – 7.3%, and euglenids – 6.2%. The dynophytes (3.0%), cryptohpytes (0.6%), red algae (0.3%), raphidophytes and brown algae (0.1% each) possess less species.

The Lena flora is enriched via its tributary system. The two large tributaries – the Viluy (2,650; 0.27; 1.1) and Aldan (2,273; 0.51; 0.2) are inhabited by 421 and 308 algae species respectively. The dominating species there are the same as for the Lena. The records of 1,774, 520 and 590 species respectively were made in stagnant water bodies in those rivers' basins. The first four places belong to the diatoms, green, blue-green algae and euglenids. A characteristic feature of the flora of stagnant water bodies of the Lena basin is a high diversity of green algae (*Chlorococcales, Volvocales*). The only record of algae of the *Charophyta* division was made in the Aldan basin.

The basins of the rest of Yakutian rivers are situated at higher latitudes which is reflected in their floral composition. To be specific, the Omoloy river's flora makes up 1.7% of Yakutian algae, the Aldan 12.4%, the Khroma 4.0%, the Anabar 6.0%, the Olenyok 3.2%, the Indigirka 5.0%, the Kolyma 10.6%, and the Yana 30.0% of Yakutian algae.

*The Anabar River (939; 0.48; 0.8)*. There are records of 171 taxa of 8 divisions. Almost all of them occur in the river and its tributaries with predominance of the diatoms, green *Desmidiales* and yellow-greens.

*The Olenyok River (2,270; 0.49; 0.7)*. The flora of this river is not rich. There are 174 specific and intraspecific taxa of 6 divisions. It contains the flora of tributaries (75 species) and stagnant water bodies (84 species). The river itself contains 58 species and varieties. The rank sequence of the divisions is the same as for the Anabar River basin.

*The Omoloy River (593; 0.67; 1.76)*. The algae flora is also scanty – 71 taxa and 5 divisions. There are 32 species recorded for the river and stagnant water bodies each and 14 species for the tributaries. The second place is taken by the blue-green algae after the diatoms.

*The Yana River (872; 0.57; 1.4)*. The basin's flora is represented by 742 species and varieties of 11 divisions. The dominant taxa are the diatoms, green and blue-green algae both in the river and in the whole basin. There are 448 species occurring in stagnant water bodies with the following sequence of the first four places: 148

species of green algae, 138 species of diatoms, 61 of blue-green algae and 38 of euglenides.

*The Khroma River (711)* is poor in species, comprising 131 taxa. There are only 136 species of 5 divisions recorded in the river's basin. The flora of the river has predominantly green algae (*Desmidiales*), penetrating from the adjacent territories. The blue-green algae take the first place in the tributaries and stagnant water bodies. The diatoms are not numerous both in the river and in the basin, comprising 42 and 47 species respectively.

*The Indigirka River (1,726; 0.77; 1.6).* There are 263 species of 8 divisions occurring in the river's basin including 147 diatom species and varieties, 34 species of blue-green algae, 33 of green algae and 30 of yellow-green algae. The river's flora is enriched due to stagnant water bodies (188 species) and tributaries (65 species). There are 61 species in the river itself.

*The Alazeya River (1,520; 0.77; 14.4)* is still subject to study. At present, 99 algae species of 6 divisions are recorded in the basin. The flora is composed of representatives inhabiting stagnant water bodies: 46 species of diatoms, 26 of blue-green algae and 21 species of green algae.

*The Kolyma River (2,129)* There are records of 642 species and varieties of 7 divisions in the basin with a predominance of 230 species of diatoms, 148 of green algae, 120 of blue-greens, 63 of yellow-greens and 54 species of golden algae. Only 172 species were found in the river itself; 9 records were made for the tributaries, 533 taxa for stagnant water bodies. The river is dominated by the diatoms, green and blue-green algae, and the stagnant water bodies by green algae.

Thus, the further north a river bed is situated, the higher is the content of green *Desmidiales* algae, blue-green *Nostoc, Gloeocapsa spp.*, and yellow-green *Tribonema, Ophiocytium spp.* in the floral composition. Abundance of the latter depends on the number of stagnant water bodies in river basins, i.e. the higher the lake coefficient the higher the rank of those divisions.

Comparison of the algae flora of 11 rivers shows characteristic features of the flora of each basin, i.e. each river is peculiar. The highest similarity coefficient was assessed for the Lena River and its tributaries the Aldan and Viluy.

The flora of the rivers is characterized by predominance of the diatoms (66.2% of the species composition on average), green (15.9%) and blue-green algae (9.2%), whereas the algal flora of the basins is predominated by the green algae (37.9%), blue-green algae (20.2%) and the diatoms (15.7%). Prevalence of the green and blue-green algae is explained by effects of stagnant water bodies.

Due to the prolonged period of rivers breaking up in spring and early summer, the seasonal development and algae dynamics in the rivers of Yakutia are prolonged and differ in latitudinal direction. Increase in the number and biomass of cells depends on a water warming-up level and ranges from 10 to 30 thousand cells per litre in the Khroma, Lower Lena and Yana to 0.06–1.00 million cells per litre in the Aldan and Middle Lena. One peak in numbers is recorded in the first decade of August at water temperatures of 15–17°C concurring with a biomass peak (0.3–1.1 mg/l). The species leading in number and biomass belong to the diatoms (*Aulocosira, Melosira,*

*Paralia, Asterionella* genera), to the golden algae (*Dinobryon spp.)* and to the blue-green algae (*Dactylococcopsis spp).*

The data on biomass of the Yakutian river algae are comparable with those of the Taimyr tundra (up to 1.2 mg/l (Ermolaev et al., 1971)), some rivers of West Siberia (0.01–0.97 mg/l (Naumenko 1985)) and subarctic rivers of Canada (up to 0.2 mg/l (Moore 1977)).

Thus, the river algoflora composition has an obvious predominance of the diatomic complex, with small numbers (no more than 1 million cells per litre) and a low biomass (no more than 1.1 mg/l) during the short vegetation period. This allows assign the rivers of Yakutia to the northern oligotrophic water bodies enriched by their tributary nets and shallow water areas. The rather low diversity of the river flora of North Yakutia is the result of full or partial freeze-up of the rivers as well as of pollution caused by mining activity.

*Lake algae. Floodplain lakes* of the middle taiga are characterized by neutral or alkalescent reactions, while those of the tundra and northern taiga have acid or, rarely, neutral reactions. 43.4% of the whole Yakutian algal flora is recorded in lakes comprising 1,357 species and varieties of 7 divisions. The dominant taxa are the green algae (440 species), diatoms (328), blue-green algae (272), euglenids (157), and yellow-greens (86). A characteristic feature of the lake algal flora is the rather large number (average for a vegetation period) of 8,700 thousand cells per litre and a high biomass (11.5 mg/l). The indices vary depending on lakes location. Most abundant in numbers of algae of the tundra and taiga are the blue-green, green algae and diatoms. The highest biomass is characteristic for the green algae of *Desmidiales*, blue-green algae and diatoms in the tundra; and the blue-green, green algae of *Chlorococcales* and diatoms in the taiga. The biomass of low floodplain lakes does not exceed 10 mg/l in the taiga and 1–1.2 mg/l in the tundra. It may reach 15–17 mg/l in taiga lakes and 1.5 mg/l in tundra lakes that are replenished every three years. The biomass of the taiga lakes which are replenished every 10–15 years increases up to 46.6–50 mg/l. The relatively high biomass of these lakes is conditioned by shallow water (2.5–3 m on average), proper warming up (up to 18–26°C), increased water mineralization lasting from spring to the end of summer and with a winter maximum of up to 1.0–1.3 g/l. The lakes of the middle and upper floodplains possess a silt layer and sapropel (Vasilyeva 1983). Such lakes feature intensive water bloom caused by the blue-green algae: *Microcystis aeruginosa f. flos-aquae, M. pulverea, Aphanizomenon flos-aquae, Anabaena lemmermannii, A. flos-aquae*, etc. numbering up to 0.7–1.5 milliard cells per litre. The oxygen content in lakes reaches 100–154% of saturation, and carbon dioxide 0.0–17.6 mg/l. Fish kill is very common in floodplain lakes.

The photosynthetic activity is more intensive in oxbow lakes located in the former Lena River valley that are highly eutrophic and rich in biomass. Lakes with the depth of over 3 m and with a silt layer are characterized by a negative biological balance, even at photosynthetic rates of up to 5.4–6 g $O_2$ g/m$^2$ per day. A positive biological balance is observed only in lakes as deep as more than 3–5 m (Beloye Lake and the lakes of Tokorikan group) with a phototrophic layer of 0.5–2 m. Photosynthetic rates in such lakes are more intensive than the destruction process which results in organic matter accumulation. The photosynthetic index in the surface horizon in summer is two times higher than in autumn. At a depth of 1 m the amount of produced oxygen is equal both in summer and autumn. In summer respiration and other oxidative processes require 6 times more oxygen than in autumn. The average gross primary production is 2.08 g $O_2$ g/ $m^2$ daily, i.e. two times more than in water reservoirs. Destruction rates in lakes with a positive biotic balance are 1.28 g $O_2$ g/ $m^2$ daily. The photosynthesis – destruction ratio is 1.6 indicating a high net production which makes up 73.6 g $O^2$ g/ $m^2$ daily for the ice-free period (Vasilyeva 1966).

*Thermokarst lakes* are typical landscape elements of the cryolithozone. They are formed as a result of ground ice thawing when temperature conditions allow. Their algae species composition is 2–2.5 times poorer than that of floodplain lakes.

*Thermokarst subarctic and northern taiga lakes* of river valleys are very similar to their feeding rivers in feeding type, conditions and water chemical composition. The environmental reaction is neutral or subacid, oxygen content is 8.2–9.6 mg/l, $CO_2$ content is 12.2–79.2 mg/l, and biogenic element resources are low. The lakes are inhabited by 579 algal species and varieties of 8 divisions (18.5% of the flora) including dominant diatoms (167 taxa), green *Desmidiales* (140), blue-greens (136), and yellow-green algae (50). The insignificant depth (up to 4 m) and warming up level (not higher than 15°C) of the lakes affect the number of algae (average for a vegetation period), i.e.159 thousand cells per litre, and biomass (0.6 mg/l). That is 54 and 19 times lower than those of the floodplain lakes respectively. The most abundant in numbers and biomass are the diatoms and blue-green algae (Vasilyeva and Remigailo 1980; Vasilyeva and Pshennikova 1992).

*Thermokarst middle taiga (alas) lakes* of the Lena-Viluy and Lena-Amga interfluves are usually salinized and have no drainage. Mineralization rates range from 0.2 to 10 g/l (Anisimova 1959). The permanent drying out process alternates with water replenishment in 2–3-year periods with a positive water balance having a recurrence of 20–45 years. The complete desiccation-water replenishment cycle occurs within a 100–150 years period (Solovyov 1959). There are 559 species and varieties recorded in alas lakes (17.9%). The sequence of most important groups coincides with those of the subarctic and taiga thermokarst lakes with some prevalence of the diatoms (193) and green algae *Chlorococcales* (160). Considerable salinization favours inhabiting of the lakes with *Cyclotella spp*., like in the oxbow lakes of the Lena River valley. Compared to other thermokarst lakes, the alas lakes are characterized by a higher number of algae (3.4 million cells per litre) (average for a vegetation period) and biomass (4.1 mg/l). This is determined by the absence of a water cycle, the insignificant depth (up to 3 m), the strong shallowing, the proper warming up (up to 26°C), and the enrichment with organic matter. The following algae dominate both in number and biomass: the filamentous green algae of *Spirogyra, Ulothrix, Uronema*, and blue-green algae (*Aphanozomenon flos-aquae, Lyngbya limnetica, Microcystis aeruginosa f. flos-aquae)*.

*Tectonic lakes* (Greater and Lesser Toko Lakes) are situated at the northern foothills of the Stanovoy Range at an altitude of 903–919 m a.s.l. with the areas of 8,260 and 200 ha, and depths of 68 and 71 m respectively. The lakesides are rocky and steep. The water is fresh with mineralization rates of 10.1–15.8 mg/l. The lakes open up in the second half of June, the period of open water lasting 140–146 days (Konstantinov and Efimov 1973). Water transparency is high, and temperature rises up to 18°C. These lakes, as well as the lakes of the Timpton-Uchur mountainous region, are inhabited by 419 species and varieties of 8 divisions (13.4% of the flora). Quantitative prevalence is typical for the green algae of *Chlorococcales* and *Desmidiales (*134 species) penetrating into the lake via streams from waterlogged areas, diatoms (134), blue-green (102), and golden algae (35 species and varieties). The average number (18.2 thousand cells per litre) and biomass (0.3 mg/l) of algae of the G. and L. Toko during the vegetation period are low, almost approximating those of the tundra thermokarst lakes. The maximal numbers and biomass in plankton is provided by the species of the centric diatoms of *Melosira, Cyclotella, Cyclostephanos, Coscinodiscus* as well as by golden algae. Benthos and the littoral zone are inhabited by charophytes and diatoms. The Timpton-Uchur lakes have the same number (5.5 millions cells per litre on average) and biomass (5.5 mg/l) as the alas lakes of the middle taiga. They are predominated by *Anabaena lemmermannii, Mallomonas* and *Dictyosphaerium spp.* in numbers and biomass. Species of *Asterionella* dominate in autumn.

*Glacier lakes* of the Sordonakh Plateau and Verkhoyansk Range are situated at an altitude of 900–1,020 m a.s.l. having maximal depths of 52.6 m. The lakes are fresh. There are 334 species and varieties recorded there comprising 10.7% of the total algal flora. The dominant taxa are the diatoms represented by 214 species and varieties of *Cyclotella, Cymbella, Eunotia, Gomphonemaa* and other genera. There are also 48 species and varieties of the blue-green algae, 40 taxa of the green algae, 13 of golden algae, 12 of yellow-greens, 3 euglenides, 2 taxa of brown algae, and cryptophytes and red algae are represented by one species each. The glacier lakes are characterized by the lowest number (74 thousand cells per litre on average for a vegetation period) and biomass (0.2 mg/l), which is probably related to the late breaking up of the lakes in July and insufficient warming up of water (up to 9–11°C). The diatoms and golden algae dominate in numbers, biomass and species diversity (Vasilyeva-Kralina et al. 2004).

Comparison of the floras of 5 lake types proved flora specificity for each type.

The seasonal development of algae depends on both water body type and its location, open water period and warming up rates. Severe conditions of prolonged winters oppress the development of algae. The colonies of *Microcystis* and *Gloeocapsa* become looser, the mucous cover being thinner and the colouration paler, etc. Finds of considerable amounts of dead algae cells, empty valves and shells in water and lake silt prove that much algae die. Only species tolerant to unfavourable conditions remain in a vegetative state. The representatives of some alga groups hibernate in various forms of anoxybiosis: *Anabaena* and *Aphanizomenon* as spores; *Desmidiales* as zygospores; *Dinobryon, Ceratium, Peridinium* and other dynophytes as cysts; all diatoms and some green algae as vegetative cells; *Microcystis, Gloeocapsa, Scenedesmus*, etc. as colonies. The maximum concentration (4.4 million cells per litre; 1.1 mg/l) is observed in lakes with a depth of over 3 m, having favourable temperature and other conditions. Most of the mass of plankton algae descends to the bottom of these lakes to spend winter period in a complex with benthos algae. In shallow areas, where algae often freeze in the ice, their accumulation is minimal. Chromatophores and pigments preserve well in *Chlorococcales, Volvocales*, blue-green and diatom algae. Some species of *Dinobryon* and *Cyclotella,* followed later by the blue-green algae, start their intensive growth right after the ice disappears from a lake or its littoral side and proper light conditions are available (338 hours of sunshine in June; Gavrilova 1962).

The only peak in number and biomass of algae in the lakes of the taiga and tundra of Yakutia, in contrast to two peaks in the Kulunda steppe (Ermolaev 1964, 1965), is determined by climatic rigors. According to average long-term data, in the tundra it is observed in mid August with a number of 0.9 million cells per litre and a biomass of 1.8 mg/l; in the taiga in late July – early August with a number of 250 million cells per litre and a biomass of 27 mg/l. The tundra lakes are dominated in number by the diatoms (*Navicula, Pinnularia, Eunotia spp.*) and by the blue-green algae (*Gloeocapsa, Nostoc spp.*) and green *Desmidiales* in biomass. In the taiga lakes the most abundant are the blue-green algae (*Aphanizomenon, Microcystis, Anabaena spp.*), green *Volvocales* (*Pandorina, Chlamydomonas spp.*) and *Chlorococcales.*

The biomass level of the tundra lakes of Yakutia is similar to that of small lakes of North Skandinavia, 0.6 mg/l (Holmgren 1968); the Bolshaya zemlya tundra, 0.3 mg/l (Getsen 1976); subarctic lakes of Canada, up to 0.3 mg/l (Sheath et al. 1975); and shallow lakes of Alaska, up to 0.5 mg/l (Alexander et al. 1980). In the lakes of the taiga zone of Yakutia the biomass content is considerably higher than that of Canadian eutrophic ponds, up to 1.5 mg/l (Stanley 1976) and lakes, up to 1.8 mg/l (Moore 1979).

Thus, the algae species composition, their numbers and biomass allow to consider the floodplain lakes of Yakutia as eutrophic and highly eutrophic, the thermokarst alas lakes of the middle taiga as eutrophic, whereas the mountainous and thermokarst subarctic and northern taiga lakes are oligotrophic. A common feature for these lakes is one peak in number and biomass of phytoplankton growth.

*Bog algae.* The majority of bogs are situated in the Indigirka and Kolyma basins. The waterlogged landscapes occupy over half the territory of the tundra. They directly depend on thermokarst dimensions being characteristic for the cryolithozone. Following some authors (Shirshov 1935; Getsen 1987), we distinguish the algae of moss biotopes and bog hollows. Of the total of 343 species and varieties (11.0% of the flora) the 141 species of the green *Desmidiales* (*Closterium, Cosmarium, Xantidium spp.*) dominate. The second place is taken by the blue-green algae: *Nostoc, Gloeocapsa, Merismopedia spp.*, the species diversity of which is two times higher then that of the taiga mires. The yellow-greens take the third place (55 species). They include new species for the former USSR territory: *Gloeoskene turfosa, Chlorallantos attenuatus,* etc. The euglenids are also diverse, comprising up to 17 species, especially of *Trachelamonas* (9 species). The diatoms (53 species) are represented by *Eunotia, Pinnularia spp.* The cryptophytes and dynophytes have the same number of species as in rivers. The contribution of golden algae is very insignificant.

Taiga bogs are very common, occupying 5.1% of the Lena basin territory and 3.1% of the Kolyma basin territory (Andreyev et al. 1987). The records of algae in these landforms comprise 371 taxa (11.9% of the total algal flora). Like in the tundra zone, the first place is taken by *Desmidiales* (126 species), but at the same time the role of euglenids rises (*Trachelamonas, Phacus, Euglena* genera) as well as that of the diatoms (59 species of *Eunotia, Pinnularia, Cymbella, Navicula* genera). The diversity of yellow-green algae decreases to 46 species or less (*Ophiocytium, Characiopsis spp.*). This rank sequence is explained by confinement of the algae to the ponds with pH < 7, which is very characteristic for bogs (Kosinskaya 1960; Shtina et al. 1981).

Thus, the species diversity of Yakutian bog algae is rich (the third place after lakes and rivers), though the temperature conditions there are less favourable due to insufficient thawing of permafrost under the bogs (up to 40 cm).

*Water reservoir algae.* Construction of the Viluy Reservoir (for the Viluy Power Station) with the area of 2,170 km$^2$ and a water body volume of 36 km$^3$ has led to considerable transformation in the seasonal distribution of the Viluy's drainage. Of 330 species and varieties (10.6% of the Yakutian algal flora) recorded in the Reservoir the proportion of species of the leading divisions is the same as for running water ponds of Yakutia as a whole, i.e. 126 species of diatoms, 74 species of green algae and 65 of blue-green algae. Decrease of flowage in summer time, rising water temperature and a change in chemical composition due to decaying organics of the unprepared Reservoir's bottom result in increased contribution of the golden and yellow-green algae. Before the stream was regulatied 169 species of 26 genera of diatoms were recorded in the Viluy basin. Long-term studies of the Reservoir's dynamics revealed two peaks. The first peak was induced by the blue-green algae (up to 0.48 milliards cells per litre; 7.8 mg/l) and was observed during the Reservoir's flood. The second peak was recorded 10 years later (1977) with diatoms and golden algae dominating (the biomass ranged from 0.21 g/m$^3$ in 1976 to 0.26 g/m$^3$ in 1979). The average perennial number is 159 thousand cells per litre, the biomass being 2.6 mg/l. The seasonal development of the phytoplankton features one peak concurring with the peak of biomass in July–August. The species diversity is also high in this period. The vegetation period of the Reservoir's algae lasts 100–110 days which is 1.5–2 times shorter than that of the water bodies of the European part of the former USSR and of the moderate zone of Siberia (140–180 days). At a depth of up to 80 m vertical, daily and perennial changes in algal numbers and biomass are well measurable, but depend on water temperature conditions, transparency, turbulence and biological peculiarities of the algae, the photosynthetic activity of which has a primary role in the oxygen stratification and the development of hydrobionts, due to the complete absence of higher aquatic plants.

The photosynthetic activity in the Viluy Reservoir (Vasilyeva and Remigailo 1982) during its foundation varied depending on phytoplankton migration. Maximal daily gross production for 1 m$^2$ was 0.28–1.24 g. Organic matter decomposition was 0.95–38 g $O_2$/m$^2$ daily. This decomposition rate was, probably, the result of decaying sink wood. The average rate of photosynthetic gross production in 1973–1976 was 0.71 g $O_2/m^2$ daily, approximating that of the Bratsk Reservoir at 0.6 g $O_2/m^2$ daily (Kozhova 1969).

Thus, the Viluy Reservoir is an oligotrophic water body with mesotrophic features, i.e. having enough nutrition for zooplankton and young fish growth. The Reservior's algae have a permanent effect on the Viluy as far as 2–5 km away and episodic effects as far as 35–60 km away.

*Algae of ephemeral ponds.* The limited water containing capacity of frozen soils of the cryolithozone yields numerous ephemeral ponds (up to 1 ha in area) existing only for 0.5–1.5 months in Central Yakutia and up to 2–2.5 months in the rest of the territory. They feature low depths, warming up to 26–28°C and salinization. The algae of such ponds are represented by 208 species and varieties of 7 divisions (6.7% of the Yakutian algal flora). The dominating taxa are the green algae (83 species), the diatoms (50), the euglenids (40), and the blue-green algae (20) which easily withstand the alas soils salinization of the sodium type (pH 8.3), as well as the sulfate–chloride and chloride–sulfate salinizations of soils of the above-floodplain terraces (0.5%). These ponds surpass other water bodies of Yakutia in numbers (up to 10–15 milliard cells per litre) and biomass (up to 6 mg/l) and are mainly made up of *Enteromorpha, Chlamydomonas spp.* (11species), and *Hantzschia, Nitzschia, Oscillatoria, Spirulina, Phormidium spp.* Intensive anthropogenic pollution with domestic waste in populated areas provides favourable conditions for successful development of the euglenids of *Euglena spp.* (14 species), *Astasia* and *Phacus, Monomorphyna, Anisonema spp.* (3 species each) inducing a red or green "hyperbloom" of the water. This phenomenon is not observed in Yakutia under normal conditions.

The ephemeral ponds are the only hyper-eutrophic ponds in the region. This is the result of both the intensive summer sum radiation (up to 38–46 cKal/cm$^2$) and the excessive food capacity due to anthropogenic pressure. Unlike other ponds, the algae of puddles withstand freezing that lasts for as long as 8–9 months, and prolonged dryouts in Central Yakutia, being replenished with water only in rainy periods.

*Algae of icings.* Icings of the Moma River basin are inhabited by 83 species and varieties of algae (2.7% of Yakutian flora), including 73 specific and intraspecific taxa of diatoms and 10 of blue-green algae. The icing flora is 3 times scantier than that of the area surrounding the icing. The dominant genera are *Navicula, Pinnularia, Cymbella spp.*, comprising 10–16 species. Species of *Gloeocapsa* and *Nostoc* typically provide the largest biomass.

Thus, the algae of Yakutia feature a diversity of species and the biotopes they inhabit each have their peculiar species composition and growth rates. The diatoms of the *Raphales* take the first place in running water bodies, in reservoirs and icings. The green algae have the richest species diversity in lakes, where the species of the *Chlorococcales* are very numerous. The species of the *Desmidiales* are common for bogs, while the *Volvocales* for ephemeral ponds. The blue-green algae take the second place in icings and bog hollows at the expense of *Gloeocapsa* and *Nostoc spp.* In other water body types they take the third to fifth places. The euglenids are in a rather high rank position in bogs and ephemeral ponds, whereas the yellow-green algae are so in bog hollows.

## References

Abolin RI (1913) V taige Leno-Viluyskoy ravniny. In: Predvaritelny otchyot po organizatsii i issledovaniyu rabot po issledovaniyu pochv Aziatskoy Rossii v 1912 godu. Saint-Petersburg – *In the taiga of the Lena-Viluy Lowland* (in Russian)

Abolin RI (1929) Geobotanicheskoy i pochvennoye opisaniye Leno-Viluyskoy ravniny. In: Trudy komissii po izucheniyu Yakutskoy ASSR 10. Leningrad – *Geobotanical and soil description of the Lena-Viluy Lowland* (in Russian)

Afonina OM, Bredkina LI, Makarova II (1979) Mkhi i lishainiki lesostepnogo landshafta v srednem techenii reki Indigirki. Novosti sistematiki nizshikh rasteniy 16: 175–186 – *Mosses and lichens of the forest-steppe landscape of the Middle Indigirka River* (in Russian)

Afonina OM, Bredkina LI, Makarova II (1980) Raspredeleniye lishainikov i mkhov v lesostepnykh landshaftakh v srednem techenii reki Indigirki. Botanicheskiy zhurnal 65 (1): 66–82 – *Distribution of lichens and mosses of the forest-steppe landscapes of the Middle Indigirka River* (in Russian)

Afonina OM, Duda J (1993) Pechenochnye mkhi Chukotki. Botanicheskiy Zhurnal 78 (3): 77–93 – *The liverworts of Chukotka* (in Russian)

Alabyshev VV (1932) Zonalnost ozernykh otlozheniy. Izvestiya sapropelevogo komiteta 6: 1–45 – *Zoning of lake sediments* (in Russian)

Alexander V, Stanley DW, Daley RJ, McRoy CR (1980) Primary producers. In: Hobbie JE (ed) Limnology of Tundra Ponds, Barrow, Alaska. US/IB Synth Ser 13. Dowden, Hutchinson & Ross, Stroudsburg, Pennsylvania.

Alexandrova VD (1960) Flora sosudistykh rasteniy Bolshogo Lyakhovskogo (Novosibirskiye ostrova). Botanichesky zhurnal 45 (11): 1687–1693 – *The flora of vascular plants of the Bolshoy Lyakhovsky Island (Novosibirskie Isalnds)* (in Russian)

Alexandrova VD (1961) Vliyaniye snezhnogo pokrova na rastitelnost v arkticheskoy tundre. In: Alexandrova VD, Kuvaev VB, Tikhomirov BA et al. (eds) Materialy po rastitelnosti Yakutii. Izd-vo LTA, Leningrad – *Influence of snow cover on vegetation in the Arctic tundra* (in Russian)

Alexandrova VD (1962) O podzemnoy structure nekotorykh rastitelnykh soobschestv arkticheskoy tundry na ostrove Bolshom Lyakhovskom. In: Problemy botaniki. Izd-vo AN SSSR, Moscow–Leningrad 6: 148–160 – *On underground structure of some plant communities of the Arctic tundra on the Bolshoy Lyakhovsky Isalnd* (in Russian)

Alexandrova VD (1963) Ocherk flory i rastitelnosti ostrova Bolshogo Lyakhovskogo. In: Novosibirskiye ostrova. Morskoy transport, Leningrad – *Essay on the flora and vegetation of the Bolshoy Lyakhovsky Isalnd* (in Russian)

Alexandrova VD (1970) Dinamika mozaichnosti rastitelnykh soobschestv pyatnistykh tundr v arkticheskoy Yakutii. In: Mozaichnost rastitelnykh soobschestv i eyo dinamika. Vladimir – *Dynamics of mosaicity of plant communities of spotted tundra in the Arctic Yakutia* (in Russian)

Almquist E (1883) Lichenologische Beobachtungen an der Nordkuste Sibiriens. In: Die wissenschaftliche Ergebnisse der Vega-Expedition, herausgegeben von AE Nordenskiold. I Band. Leipzig: 50–74.

Andreyev AA, Zhirkov II, Ivanov KP, Pestryakova LA (1995) Tsentralnaya Yakutia. Ozero Ulakhan Chabyda. In: Istoriya ozyor Severa Azii. Seria Istoriya ozyor. Saint-Petersburg: 70–81 – *Central Yakutia. Lake Ulakhan Chabyda* (in Russian)

Andreyev MP (1983) Lishainiki ostrova Chetyryokhstolbovogo (Medvezhyi ostrova, Vostochno-Sibirskoye more). Novosti sistematiki nizshikh rasteniy 20: 133–139 – *The lichens of the Chetyryokhstolbovoy Island (Medvezhyi Islands, East-Siberian sea)* (in Russian)

Andreyev MP (1984) Sistematicheskiy sostav likhenoflory Anuyskogo nagorya. Novosti sistematiki nizshikh rasteniy 21: 137–140 –*Taxonomical composition of the lichen flora of the Anuy Plateau* (in Russian)

Andreyev MP, Kotlov YuV, Makarova II (1996a) Biologicheskoye raznoobrazie lishainikov Russkoy Arktiki (taksonomicheskiy sostav i predvaritelny analiz). Novosti sistematiki nizshikh

rasteniy 31: 82–94 – *Biological diversity of the Russian Arctic lichens (taxonomical composition and preliminary analysis)* (in Russian)
Andreyev MP, Kotlov YuV, Makarova II (1996b) Checklist of lichens of the Russian Arctic (including lichenicolous fungi). Bryologist 99 (2): 137–169.
Andreyev VN (ed) (1974) Tebenyovochnye pastbischa Severo-Vostoka Yakutii. Knizhnoye izd-vo, Yakutsk – *Winter pastures of North–East Yakutia* (in Russian)
Andreyev VN, Nakhabtseva SF (1974) Podzony yakutskoy tundry. In: Biologicheskiye problemy Severa. 4th Symposium. Issue 3. Izd-vo YaF SO AN SSSR: 40–45 – *Subzones of the Yakut tundra* (in Russian)
Andreyev VN, Galaktionova TF, Mikhalyova VM et al. (1975) Luga Yakutii. Nauka, Moscow – *Meadows of Yakutia* (in Russian)
Andreyev VN, Perfilyeva VI, Nakhabtseva SF (1980) Flora okrestnostey posyolka Saskylakh na reke Anabar (Severo-Zapadnaya Yakutia). Botanichesky zhurnal 65 (11): 1560–1568 – *Flora of the Saskylakh settlement vicinity on the Anabar River (North–West Yakutia)* (in Russian)
Andreyev VN, Galaktionova TF, Perfilyeva VI, Scherbakov IP (1987) Osnovnye osobennosti rastitelnogo pokrova Yakutskoy ASSR. Izd-vo YaF SO AN SSSR, Yakutsk – *Basic features of vegetation cover of the Yakutian ASSR* (in Russian)
Anisimova NP (1959) Khimichesky sostav podzemnykh i poverkhnostnykh vod i nekotorye zakonomernosti ego izmeneniya v raione srednego techeniya reki Leny. Yakutskoye knizhnoye izdatelstvo, Yakutsk – *Chemical composition of underground and surface waters and some regularities of its changes in the Middle Lena River region* (in Russian)
Arnell WH (1913) Zur Moosflora des Lena-Tales. Arkiv for Botanik Band 2: 1–92.
Averensky AI, Mikhalyova LG (2000) Vrediteli i bolezni lesov Yakutii: Uchebnoye posobie. Izd-vo YaGU, Yakutsk – *Pests and diseases of the forests of Yakutia: Textbook* (in Russian)
Baikov KS (ed) (2005) Konspekt flory Sibiri: sosudistye rasteniya. Nauka, Novosibirsk – *Conspectus of the flora of Siberia: Vascular plants* (in Russian)
Bakalin VA (1999) Pechenochniki Karelii. Arctoa 8: 17–26. – *The liverworts of Karelia* (in Russian)
Bakalin VA (2003) Notes on Lophozia IV. Some new taxa of *Lophozia* sensu stricto. Ann Bot Fenn 40 (1): 47–52
Bakalin VA (2005) Monographicheskaya obrabotka roda *Lophozia (Dumort.) Dumort. s. str.* Nauka, Moskow. – *Monographical processing of the Lophozia (Dumort.) Dumort. s. str.* (in Russian)
Bakalin VA (2007) Pechenochniki Kamchatki: itogi izucheniya. In: Sokhranenie bioraznoobrazia Kamchatki i prilegayuschikh morey. Proceedings of the VII International Conference. Petropavlovsk-Kamchatsky: 6–14 – *The liverworts of Kamchatka: Investigation outcomes* (in Russian)
Bardunov LV (1974) Listostebelnye mkhi Altaya i Sayan. Nauka, Novosibirsk – *Leafy mosses of Altai and Sayan* (in Russian)
Bardunov LV (1992) Ocherk brioflory Sibiri. Nauka, Novosibirsk – *Review of bryoflora of Siberia* (in Russian)
Bening AL (1942) O planktone reki Leny. Izvestia Biologo-geograficheskogo NII pri Vostochno-Sibirskom Universitete imeni Zhdanova 9 (3–4): 217–230 – *On the Lena River plankton* (in Russian)
Benua KA, Karpova-Benua EI (1973) Parazitnye griby Yakutii. Nauka, Novosibirsk – *Parasitic fungi of Yakutia* (in Russian)
Benua KA, Karpova-Benua EI (1979a) Derevorazrushayuschiye i parazitnye bazidialnye griby Yakutskoy ASSR. Novosti sistematiki nizshikh rasteniy 9: 139–151 – *Wood-attacking and parasitic basidial fungi of the Yakutian ASSR* (in Russian)
Benua KA, Karpova-Benua EI (1979b) Derevorazrushayuschiye i parazitnye trutovye griby Yakutskoy ASSR. Mikologiya i fitopatologiya 13 (6): 509 – *Wood-attacking and parasitic polypore fungi of the Yakutian ASSR* (in Russian)

Berman DI, Pivovarova ZhF, Gelman VB (1978) O neravnomernosti raspredeleniya pochvennykh vodorosley pod *Artemisia santolinifolia* Turcz v stepyakh severo-vostochnoy Yakutii. Botanichesky zhurnal 63 (8): 1196–1198 – *On uneven distribution of soil algae under Artemisia santolinifolia Turcz. in the steppes of North–East Yakutia* (in Russian)

Bezdelev AB, Bezdeleva TA (2006) Zhiznennye formy semennykh rasteniy Rossiyskogo Dalnego Vostoka. Dalnauka, Vladivostok – *Life forms of seed-bearing plants of the Russian Far East* (in Russian)

Birkengof AL (1932) Lesnoy pokrov i lesnye resursy Severo-Vostochnogo kraya Yakutskoy ASSR. In: Trudy SOPS AN SSSR. Issue 3. Izd-vo AN SSSR, Leningrad: 7–49 – *The forest cover and forest resources of the North–East of the Yakutian ASSR* (in Russian)

Bondartseva MA, Solovyev VA (1992) Nauchnye osnovy ustoychivosti lesov k derevorazrushayuschim gribam. Nauka, Moscow – *Scientific principles of forest tolerance to wood-attacking fungi* (in Russian)

Borodin AM et al. (eds) (1984) Krasnaya kniga SSSR: Redkie i nakhodyaschiesya pod ugrozoy ischeznoveniya vidy zhivotnykh i rasteniy Tom 2 Chast 10. Lesnaya promyshlennost, Moscow – *The Red Data Book of the USSR: Rare and threatened species of animals and plants Vol 2 Part 10* (in Russian)

Bredkina LI (1980) *Toninia groenlandica* Lynge em. Bredk. v lesostepnykh landshaftakh Yakutskoy ASSR. Novosti sistematiki nizshikh rasteniy 17: 109–111 – *Toninia groenlandica Lynge em. Bredk. in forest-steppe landscapes of the Yakut ASSR* (in Russian)

Bykov BA (1973) Geobotanicheskiy slovar. Nauka, Alma-ata – *Geobotanical glossary* (in Russian)

Buks II (1963) Nekotorye voprosy dinamiki rastitelnogo pokrova Leno-Olenyokskogo mezhdurechya. In: Vestnik MGU Series Geography 6: 45–49 – *Some questions in vegetation cover dynamics of the Lena-Olenyok Interfluve* (in Russian)

Buks II (1966) Listvennichnye redkolesya na severe Leno-Olenekskogo mezhdurechya. Byulleten MOIP. Otdel Biologiya 71 (4): 71–82 – *The sparse larch forests in the North of the Lena-Olenyok Interfluve* (in Russian)

Cajander AK (1901) Lenan rannoilta. Luonnon Ystävä 5: 162–164 – *From the Lena banks* (in Finnish)

Cajander AK (1903) Beiträge zur Kenntniss der Vegetation der Alluvionen des nordlichen Eurasiens. I. Die Alluvionen des unteren Lena-Thales. Act Soc Sc Fenn 32 (1): 1–182.

Cherepanov SK (1995) Sosudistye rasteniya Rossii i sopredelnykh gosudarstv (v predelakh byvshego SSSR). Mir i semya, Saint-Petersburg – *Vascular plants of Russia and adjacent countries (within limits of the former USSR)* (in Russian)

Cherosov MM (2005) Sinantropnaya rastitelnost Yakutii. Izd-vo YaNTs SO RAN, Yakutsk – *Synanthropic vegetation of Yakutia* (in Russian)

Damsholt K (2002) Illustrated flora of Nordic Liverworts and Hornworts. Lund.

Danilova NS (ed) (2005) Raznoobrazie rastitelnogo mira Yakutii. Izd-vo SO RAN, Novosibirsk – *Diversity of the vegetation world of Yakutia* (in Russian)

Dobretsova LA (1959) Rastitelnost verkhoviy reki Moly (Charky). In: Nauchnye soobscheniya YaF SO AN SSSR. Issue 2. Izd-vo YaF SO AN SSSR, Yakutsk – *Vegetation of the upper reaches of the Mola (Charky) River* (in Russian)

Dobretsova LA (1961) Yernikovye formatsii v basseyne reki Viluya. In: Materialy po rastitelnosti Yakutii. Leningrad – *The yernik formations in the Viluy River basin* (in Russian)

Dobretsova LA (1962) Rastitelnost Zhiganskogo raiona. In: Nauchnye soobscheniya. Issue 8: Biologia. Knizhnoye izd-vo, Yakutsk – *Vegetation of the Zhigansk Region* (in Russian)

Dogadina TV (1986) Zheltozelyonye vodorosli SSSR: Flora, sistematika, evolyutsiya, filogeniya. Abstract of Thesis for a Doctor Degree. Kiev – *The yellow-green algae of the USSR: Flora, taxonomy, evolution, phylogeny* (in Russian)

Dolenko GI (1913) Dolina reki Leny bliz Yakutska. In: Predvaritelny otchyot po organizatsii i issledovaniyu rabot po issledovaniyu pochv Aziatskoy Rossii v 1912 godu. Saint-Petersburg – *The Lena River valley in Yakutsk vicinity* (in Russian)

Dolenko GI (1916) Chasti Leno-Viluyskogo vodorazdela Yakutskoy oblasti. In: Predvaritelny otchyot po organizatsii i issledovaniyu rabot po issledovaniyu pochv Aziatskoy Rossii v 1914 godu. Saint-Petersburg – *The parts of the Lena-Viluy watershed of the Yakut Region* (in Russian)

Dolinin IN et al. (eds) (2000) Krasnaya kniga Respubliki Sakha (Yakutia) Tom 1: Redkie i nakhodyaschiesya pod ugrozoy ischeznoveniya vidy rasteniy i gribov. NIPK 'Sakhapoligrafizdat', Yakutsk – *The Red Data Book of the Republic of Sakha (Yakutia) Vol 1: Rare and endangered species of plants and fungi* (in Russian)

Dorogostaiskaya EV (1959) K voprosu o pochvennoy algoflore pyatnistykh tundr Krainego Severa. Botanichesky zhurnal 44 (3): 312–321 – *On the question of the soil algoflora of the spotted tundra of the Far North* (in Russian)

Drobov VP (1910) Rastitelnye formatsii Lensko-Kirenskogo kraya Verkholenskogo uezda. In: Trudy pochvenno-botanicheskoy ekspeditsii po issledovaniyu kolonizatsionnykh raionov Aziatskoy Rossii. Part 2: Botanicheskiye issledovaniya. Saint-Petersburg – *Vegetation formations of the Lena-Kirensk Territory of the Verkhnelensk District* (in Russian)

Drobov VP (1914) Rastitelnost v raione Yakutsk-Ust-Maiskogo trakta Yakutskoy oblasti i okruga. In: Trudy pochvenno-botanicheskoy ekspeditsii po issledovaniyu kolonizatsionnykh raionov Aziatskoy Rossii. Part 2: Botanicheskiye issledovaniya. Saint-Petersburg – *Vegetation adjacent to the Yakutsk-Ust-Maya post road of the Yakut Oblast and District in 1912* (in Russian)

Drobov VP (1927) Kratky otchyot o rastitelnosti Leno-Aldanskogo plato. In: Materialy komissii po izucheniyu Yakutskoy ASSR. Issue 8. Izd-vo AN SSSR, Leningrad – *Brief report on the vegetation of the Lena-Aldan Plateau* (in Russian)

Dubovik IE (1988) O nekotorykh zakonomernostyakh raspredeleniya pochvennoy algoflory v alasakh Tsentralnoy Yakutii. Biologicheskiye nauki 2: 52–55 – *On some principles of soil algoflora distribution in alases of Central Yakutia* (in Russian)

Efimova AP, Shurduk IF (2002) Klassifikatsiya ivnyakov nizkoy poimy Yakutskogo i Namskogo uchastkov sredney Leny. In: Rol selskokhozyaistvennoy nauki v stabilizatsii i razvitii agropromyshlennogo proizvodstva Krainego Severa. Proceedings of scientific-practical conference. SO RAFKhN, Novosibirsk: 77–81 – *Classification of willow communities of a lower floodplain of the Yakutsk and Namsky sections of the Middle Lena River* (in Russian)

Efimova AP, Shurduk IF, Sivtseva TA, Nikitin IA (2003) K klassifikatsii lesnykh soobschestv nizkoy poimy Yakutskogo i Namskogo uchastkov sredney Leny. In: Isaev AP (ed) Problemy botanicheskikh i lesovodstvennykh issledovaniy v RS (Ya) i Finlyandii. Proceedings of Sakha-Finnish Conference dedicated to 100th Anniversary of Cajander's expedition along the Lena. Izd-vo YaF SO RAN, Yakutsk: 96–101 – *On the classification of the forest communities of a lower floodplain of the Yakutsk and Nam sections of the Middle Lena River* (in Russian)

Egorova AA (2001) Konspekt flory sosudistykh rasteniy Lenskikh Stolbov. In: Solomonov NG, Isaev AI, Ivanova EI (eds) Natsionalny prirodny park "Lenskiye stolby": geologia, pochvy, rastitelnost, zhivotny mir, okhrana i ispolzovanie. Collection of articles. Izd-vo YaGU, Yakutsk – *Conspectus of the flora of vascular plants of the Lena Pillars* (in Russian)

Egorova AA (2004) Flora sosudistykh rasteniy resursnogo rezervata "Chonsky" (Zapadnaya Yakutia). In: Problemy izucheniya rastitelnogo pokrova Yakutii. NIPK "Sakhapoligrafizdat", Yakutsk – *Flora of vascular plants of the Resource Reserve "Chonsky" (West Yakutia)* (in Russian)

Egorova AA (2006) Redkiye sosudistye rasteniya i novye floristicheskiye nakhodki na territorii Yugo-Zapadnoy Yakutii. In: Volpert YaL (ed) Pochvy, rastitelny i zhivotny mir Yugo-Zapadnoy Yakutii. Nauka, Novosibirsk – *Rare vascular plants and new floristic finds in the territory of South–West Yakutia* (in Russian)

Egorova AA, Ivanova EI (2001) Flora resursnogo rezervata "Pilka" (Lensky Ulus). In: Timofeev PA, Shmatkova LE (eds) Flora i fauna osobo okhranyaemykh prirodnykh territoriy Respublikanskoy sistemy Ytyk Kere Sirder. Kuduk, Yakutsk –*The flora of the Resource Reserve "Pilka" (Lensky Ulus)* (in Russian)

Egorova AA, Vasilyeva II, Stepanova NA, Fesko NN (1991) Flora tundrovoy zony Yakutii. Izd-vo YaNTs SO RAN, Yakutsk – *The flora of the tundra zone of Yakutia* (in Russian)

Elenevsky RA (1933) Priroda Aldanskogo kraya. In: Prirodnye resursy Yuzhnoy Yakutii v svyazi s sotsialisticheskoy rekonstruktsiey selskogo khozyaistva. Selkhozgiz, Moscow–Leningrad – *The nature of the Aldan territory* (in Russian)

Elenkin AA (1904) Lishainiki russkoy flory i pogranichnykh vostochnykh oblastey. Proceedings of the Saint-Petersburg Botanical garden. 24: 118 – *The lichens of the Russian flora and adjacent eastern areas* (in Russian)

Elenkin AA (1909) Lishainiki polyarnogo poberezhya Sibiri (Scientific results of the Russian polar expedition of baron E.V. Toll) Issue 1. Notes of the Academy of Sciences Series 8 26 (1): 1–53 – *The lichens of the polar coastline of Siberia* (in Russian)

Ermolaev VI (1964) Fitoplankton ozera Krivogo sistemy reki Karasuk. In: Vodorosli i griby Zapadnoy Sibiri 1 (8). Izd-vo SO AN SSSR, Novosibirsk: 82–96 – *The phytoplankton of Krivoye Lake of the Karasuk River system* (in Russian)

Ermolaev VI (1965) Fitoplankton ozera Studyonogo sistemy reki Karasuk. In: Vodorosli i griby Zapadnoy Sibiri 2 (10). Izd-vo SO AN SSSR, Novosibirsk: 50–56 – *The phytoplankton of Studyonoye Lake of the Karasuk River system* (in Russian)

Ermolaev VI, Levadnaya GD, Safonova TA (1971) Algoflora vodoyomov okrestnostey Taimyrskogo statsionara. In: Biogeotsenozy Taimyrskoy tundry i ikh produktivnost. Nauka, Leningrad: 116–129 – *The algofora of water bodies of the Taimyr Research Station vicinity* (in Russian)

Fedosov VE (2006) Novye nakhodki mkhov v Taimyrskom avtonomnom okryge. Arctoa 15: 258–260 – *New finds of mosses in the Taimyr Autonomous District* (in Russian)

Fesko NN (1988a) Flora makrolishainikov verkhovya reki Kele. In: Abstratcs of 7th scientifical-practical conference of young scientists and specialists Part 3: Problemy selkhozproizvodstva, okhrana okruzhayuschey sredy, meditsina. Yakutsk: 50 – *The flora of macrolichens of the Upper Kele River* (in Russian)

Fesko NN (1988b) Flora makrolishainikov Vostochnogo Verkhoyanya (Yakutia). In: Abstracts of 2nd Symposium of mycologists and lichenologists of Baltic republics and Belarus. Tallin: 179–182 – *The flora of macrolichens of East Verkhoyanye (Yakutia)* (in Russian)

Fesko NN (1990a) K flore makrolishainikov Tsentralnogo Verkhoyanya. Yakutsk (Manuscript deposited in All-Russian Institute of Scientific and Technical Information No2045-B90) – *On the flora of macrolichens of Central Verkhoyanye* (in Russian)

Fesko NN (1990b) K flore lishainikov Tokinskogo Stanovika. Yakutsk (Manuscript deposited in All-Russian Institute of Scientific and Technical Information No2044-B90) – *On the lichen flora of Tokinsky Stanovik Range* (in Russian)

Gabyshev VA (1998) Vidovoy sostav fitoplanktona reki Leny v rayone Yakutska (Rossiya). Algologiya 8 (3): 260–267– *Species composition of the phytoplankton of the Lena River in the vicinity of Yakutsk (Russia)* (in Russian)

Gabyshev VA (2003) O sostave i raspredelenii fitoplanktona na uchastke reki Leny v rayone Yakutska i ego prigoroda. In: Isaev AP (ed) Problemy botanicheskikh i lesovodstvennykh issledovaniy v RS (Ya) i Finlyandii. Proceedings of Sakha-Finnish Conference dedicated to 100th Anniversary of Cajander's expedition along the Lena. Izd-vo YaF SO RAN, Yakutsk: 31–36 – *On composition and distribution of the phytoplankton in the Lena River section in the vicinity of Yakutsk and its suburbs* (in Russian)

Gabyshev VA (2004) Vodorosli planktona vodoyomov basseyna reki Molodo v rayone almazodobychi. In: Solomonov NG, Okhlopkov IM (eds) Ekologicheskaya bezopasnost pri razrabotke rossypnykh mestorozhdeniy almazov. Collection of articles. Institute of Biological Problems of Cryolythozone, Yakutsk: 100–106 – *The algae of the plankton in water bodies of the Molodo River basin in the area of diamond mining activity* (in Russian)

Gabyshev VA, Remigailo PA (2003) Fitoplankton vodoyomov basseyna reki Molodo (Yakutia) v zone vliyaniya almazodobyvayuschego proizvodstva. Sibirsky ekologichesky zhurnal 10 (4): 423–426 – *The phytoplankton of water bodies of the Molodo River basin (Yakutia) in the area of diamond mining activity* (in Russian)

Gakkel YaYa, Korotkevich ES (eds) (1962) Severnaya Yakutia: fiziko-geograficheskaya kharakteristika. Collection of articles. Morskoy transport, Leningrad – *North Yakutia: Physiogeographic characteristics* (in Russian)

Galaktionova TF, Permyakova VA (1964) Luga srednego techeniya reki Aldana. In: Issledovaniya pochvennykh, rastitelnykh i faunisticheskikh resursov. Knizhnoye izd-vo, Yakutsk – *The meadows of the Middle Aldan River* (in Russian)

Galaktionova TF, Petrov AM (1965) Praktika ispolzovaniya lugov doliny reki Leny v Namskom i Kobyayskom raionakh. In: Priroda Yakutii i eyo okhrana. Knizhnoye izd-vo, Yakutsk – *Practical usage of meadows of the Lena River valley in the Nam and Kobyay Regions* (in Russian)

Galaktionova TF, Petrov AM (1967) Luga doliny reki Leny v Namskom i Kobyayskom raionakh. In: Uchyonye zapiski YaGU. Seria biologicheskikh i khimicheskikh nauk. Issue 17: 31–34 – *The meadows of the Lena River valley in the Nam and Kobyay Regions* (in Russian)

Galaktionova TF, Dobretsova LA, Permyakova AA, Usanova VM (1962) Rastitelnost basseyna reki Viluy. Proceedings of the Institute of Biology YaF SO AN SSSR Issue 8. Moscow–Leningrad – *Vegetation of the Viluy River basin* (in Russian)

Gambaryan SK (1992) Antotserotovye i pechenochniki Yuzhnogo Primorya. Izd-vo DVO RAN, Vladivostok. – *Hornworts and liverworts of the South Primorye* (in Russian)

Gavrilova MK (1962) Klimat Tsentralnoy Yakutii. Yakutsk – *The climate of Central Yakutia* (in Russian)

Getsen MV (1976) Fitoplankton tundrovykh ozyor Kharbeiskoy sistemy. In: Produktivnost ozyor vostochnoy chasti Bolshezemelskoy tundry. Leningrad – *The phytoplankton of tundra lakes of the Kharbey system* (in Russian)

Getsen MV (1987) Problema mesta i znacheniya vodorosley v slozhenii ekosistem Krainego Severa (na primere Bolshezemelskoy tundry). Thesis for a Doctor degree. Leningrad – *The problem of a place and significance of the algae in the Far North ecosystem composition (case study of the Bolshaya Zemlya tundra)* (in Russian)

Gogoleva PA (1999) Stepi Tsentralnoy Yakutii. In: Timofeyev PA, Ivanova EI, Isaev AP (eds) Flora i rastitelnost Yakutii. FIPS, Moscow – *The steppes of Central Yakutia* (in Russian)

Gogoleva PA, Cherosov MM (1987) Ruderalnaya rastitelnost Verkhney Kolymy. Moscow. Manuscript deposited in All-Russian Institute of Scientific and Technical Information No 6561-B87 – *Ruderal vegetation of the Upper Kolyma River* (in Russian)

Golovanov VD et al. (eds) (1988). Krasnaya kniga RSFSR. Gosagropromizdat, Moscow –*The Red Data Book of the RSFSR* (in Russian)

Golubkova NS (1977) Otnoshenie lishainikov k substratu i drugim factoram vneshney sredy. Zhizn rasteniy. Vodorosli, lishainiki 3: 426 – *Relation of lichens to substrate and other environmental factors* (in Russian)

Golubkova NS (1983) Analiz flory lishainikov Mongolii. Nauka, Leningrad – *Analysis of the lichen flora of Mongolia* (in Russian)

Golubkova NS, Shapiro IA (1978) Vidy roda *Acarospora* Massal. podroda *Xanthothallia* H.Magn. v SSSR. Novosti sistematiki nizshikh rasteniy 15: 156–175 – *The species of genus Acarospora Massal. subgenus Xanthothallia H.Magn. in the USSR* (in Russian)

Golyakov PV (1994) Sosudistye rasteniya Olyokminskogo zapovednika. In: Flora i fauna zapovednikov. Issue 54. Moscow – *The vascular plants of the Olyokminsk Sanctuary* (in Russian)

Golyakov PV (1996) Materialy k flore sredney Olyokmy (Yuzhnaya Yakutia). In: Flora i rastitelonst Altaya. Proceedings of the South Siberian Botanical Garden. Barnaul: 103–115 – *Materials on the flora of the Middle Olyokma River (South Yakutia)* (in Russian)

Golyakov PV (1997) Flora Olyokminskogo zapovednika. Abstract of Thesis for a Candidate Degree. Saint-Petersburg – *The flora of the Olyokminsk Sanctuary* (in Russian)

Gorbach NV, Mashenkova AI (1966) Lishainiki Tuvinskoy, Yakutskoy ASSR i Yamalo-Nenetskogo natsionalnogo okruga. In: Botanika. Issledovaniya 8. Minsk: 82–83 – *The lichens of Tuvinian and Yakutian ASSRs and Yamal-Nenets National Okrug* (in Russian)

Gorodkov BN (1956) Rastitelnost i pochvy ostrova Kotelnogo (Novosibirsky Arkhipelag). In: Rastitelnost Krainego Severa SSSR i eyo osvoyeniye. Izd-vo AN SSSR, Moscow–Leningrad 2: 7–132 – *Vegetation and soils of the Kotelny Isalnd (Novosibirsky Archipelago)* (in Russian)

Grolle R (1983) Nomina generica Hepaticarum; references, types and synonymies. Acta Bot Fenn 121: 1–62.

Guseva AN (1967) *Fomitopsis annosa* (Fr.) Karst. (kornevaya gubka) v sosnyakakh Yugo-Zapadnoy Yakutii. Mikologiya i fitopatologiya 1 (3): 232 – *Fomitopsis annosa (Fr.) Karst. (root sponge) in Pinus sylvestris forests of South–West Yakutia* (in Russian)

Guseva AN (1970) Kornevaya gubka v sosnyakakh Yugo-Zapadnoy Yakutii. In: Vodorosli i griby Sibiri i Dalnego Vostoka Part 1 (3). Nauka, Novosibirsk – *Root sponge in Pinus sylvestris forests of South–West Yakutia* (in Russian)

Hansen L, Knudsen H (eds) (1992) Nordic macromycetes Vol 2: Polyporales, boletales, agaricales, russulales. Copenhagen.

Hansen L, Knudsen H (eds) (1997) Nordic macromycetes Vol 3: Heterobasidioid, aphyllophoroid and gasteromycetoid Basidiomycetes. Copenhagen

Hawksworth DL, Kirk PM, Sutton BC, Pegler DN (1995) Ainsworth and Bisbis Dictionary of the Fungi. CAB International, Egham.

Holmgren SK (1968) Phytoplankton production in a lake north of the Arctic Circle: Fil. Lie. Thesis Uppsala University.

Ignatov MS (1996) Brioflora Altaya i briogeografiya Severnoy Palearktiki. Thesis for a Doctor Degree. Moscow – *Bryoflora of Altai and bryogeography of the North Palaeoarctic* (in Russian)

Ignatov MS, Afonina OM, Igantova EA et al. (2006) Spisok mkhov Vostochnoy Evropy I Severnoy Azii. Arctoa 15: 1–130 – *List of mosses of Eastern Europe and Northern Asia* (in Russian)

Ignatov MS, Ivanova EI, Igantova EA, Krivoshapkin KK (2001) K flore mkhov Ust-Mayskogo ulusa (Respublika Sakha (Yakutia), Vostochnaya Sibir). Arctoa 10: 165–184 – *On the moss flora of Ust-Maysky ulus (the Republic of Sakha (Yakutia), Eastern Siberia)* (in Russian)

Igantova EA (2001) Novy vid *Barbula (Pottiaceae, Musci)* iz Sibiri. Arctoa 10: 161–164 – *New species Barbula (Pottiaceae, Musci) from Siberia* (in Russian)

Ivanov BI (ed) (2003) Atlas lekarstvennykh rasteniy Yakutii 1: Lekarstvennye rasteniya ispolzuyemye v nauchnoy meditsine. YaF Izd-va SO RAN, Yakutsk – *Atlas of medicinal herbs 1: Medicinal herbs used in official medicine* (in Russian)

Ivanov BI (ed) (2005) Atlas lekarstvennykh rasteniy Yakutii 2: Lekarstvennye rasteniya ispolzuyemye v narodnoy meditsine. YaF Izd-va SO RAN, Yakutsk – *Atlas of medicinal herbs 2: Medicinal herbs used in folk medicine* (in Russian)

Ivanova AP (1999) Sovremennoye sostoyanie vodorosley ozyor doliny Sredney Leny. In: Mikheyeva TM (ed) Ozyernye ekosistemy: biologicheskie protsessy, antropogennaya tranformatsiya, kachestvo vody. Abstracts of International Conference. Izd-vo BGU, Minsk: 45 – *Present condition of the algae of the Middle Lena valley lakes* (in Russian)

Ivanova AP, Kopyrina LI (1999) Bioraznoobrazie vodorosley ozyor Sredney Leny. Algologiya 9 (2): 52 – *Biodiversity of algae of the Middle Lena lakes* (in Russian) ****

Ivanova AP, Kopyrina LI (2002) Vodorosli ozera Belogo (Yakutiya). In: Biologiya vnutrennikh vod: problemy ekologii i bioraznoobraziya. IBVV RAN, Borok: 77–88 – *The algae of Beloye Lake (Yakutia)* (in Russian)

Ivanova AP, Trofimova TP (2003) Vosstanovleniye fitoplanktona posle reconstruktsii ozera Tyoplogo (Yakutsk). In: Mikheyeva TM (ed) Ozyornye ekosistemy: biologicheskie protsessy, antropogennaya transformatsiya, kachestvo vody. Izd-vo BGU, Minsk: 268–270 – *The phytoplankton recovery after reconstruction of Tyoploye Lake (Yakutsk)* (in Russian)

Ivanova EI (2001) Listostebelnye mkhi Yuzhnoy Yakutii. Izd-vo SO RAN, Novosibirsk – *Leafy mosses of South Yakutia* (in Russian)

Ivanova EI (2005) Predvaritelnaya otsenka vidobogo raznoobraziya listostebelnykh mkhov Yakutii. In: Afonina OM (ed) Aktualnye problemy briologii. Saint-Petersburg: 79–89 – *Preliminary estimation of species diversity of leafy mosses of Yakutia* (in Russian)

Ivanova EI, Krivoshapkin KK (1998) K flore listostebelnykh mkhov Yugo-Zapadnoy Yakutii (Vostochnaya Sibir). Novosti sistematiki nizshikh rasteniy 32: 135–146 – *On the flora of leafy mosses of South–West Yakutia (East Siberia)* (in Russian)

Ivanova EI, Pisarenko OYu (2007) Eksikaty mokhoobraznykh Sibiri Part 1. Novosibirsk – *Exicates of bryophytes of Siberia* (in Russian)

Ivanova VP (1961) Kratkiy ocherk rastitelnosti srednego techeniya reki Olenyok. In: Alexandrova VD, Kuvaev VB, Tikhomirov BA et al. (eds) Materialy po rastitelnosti Yakutii. Izd-vo LTA, Leningrad – *Brief essay on the vegetation of the Middle Olenyok River* (in Russian)
Ivanova VP (1967) O stepnoy rastitelnosti v doline sredney Leny. In: Uchyonye zapiski YaGU. Seria biologicheskikh i khimicheskikh nauk. Issue 7: 11–19 – *On the steppe vegetation in the Middle Lena River valley* (in Russian)
Ivanova VP (1971a) Stepnaya rastitelnost v doline sredney Leny (geobotanicheskaya kharakteristika). Abstract of Thesis of Candidate Degree. Tomsk – *The steppe vegetation in the Middle Lena River valley* (in Russian)
Ivanova VP (1971b) Stepnye fitotsenozy s *Cleistogenes squarrosa (Trin.) Keng*. v doline reki Leny. In: Uchyonye zapiski YaGU. Seria biologicheskikh i khimicheskikh nauk. Issue 20: 70–75 – *The steppe phytocoenoses with Cleistogenes squarrosa (Trin.) Keng in the Lena River valley* (in Russian)
Ivanova VP (1981) Tipchakovye stepi – odin iz etapov pastbischnoy digressii rastitelnosti v doline reki Leny. In: Andreyev VN (ed) Rastitelnost Yakutii i eyo okhrana. Izd-vo YaF SO AN SSSR, Yakutsk – *Fescue steppes as one of the stages of vegetation degradation of pastures in the Lena River valley* (in Russian)
Ivanova VP (1986) Vysshiye rasteniya okrestnostey Yakutska (opredelitel): uchebnoye pospbiye. Izd-vo YaGU, Yakutsk – *Higher plants of Yakutsk vicinity (Identifier): Tutorial* (in Russian)
Ivanova VP (1990) Dvudolnye rasteniya okrestnostey Yakutska (opredelitel): uchebnoye pospbiye. Izd-vo YaGU, Yakutsk – *Dicotyledons of Yakutsk vicinity (Identifier): Tutorial* (in Russian)
Ivanova VP, Perfilyeva VI (1972) Sokhranit kovylnye stepi Yakutii. In: Priroda Yakutii i eyo okhrana. Knizhnoye izd-vo, Yakutsk –*Preserving the Stipa steppes of Yakutia* (in Russian)
Karavaev MN (1945) Kratky analiz flory stepey Tsentralnoy Yakutii (predvaritelnoye soobscheniye). Botanichesky zhurnal 30 (2): 62–76 – *Brief analysis of the steppe flora of Central Yakutia (preliminary report)* (in Russian)
Karavaev MN (1955) K voprosu o geobotanicheskom raionirovanii tayozhnoy zony Yakutii (kratkiye soobscheniya). Vestnik MGU: Seria fiziko-matematicheskikh i estestvennykh nauk 5 (8): 109–115 – *On the problem of geobotanical regionalization of the taiga zone of Yakutia (brief reports)* (in Russian)
Karavaev MN (1958) Konspekt flory Yakutii. Izd-vo AN SSSR, Moscow–Leningrad – *Conspectus of the flora of Yakutia* (in Russian)
Karavaev MN (1965) Rastitelny pokrov Yakutii. In: Gerasimov IP (ed) Yakutia. Nauka, Moscow – *Vegetation cover of Yakutia* (in Russian)
Karavaev MN (1968) *Agropyron karavaevii P. Smirn.* kak eschyo odin primer amerikano-aziatskikh stepnykh svyazey. Botanichesky zhurnal 53 (10): 1457–1461 – *Agropyron karavaevii P. Smirn. as another exmaple of the American-Asian steppe relations* (in Russian)
Karavaev MN (1976) O sostave sinuziy napochvennykh lishainikov v reliktovykh stepnykh soobschestvakh. Botanicheskiy zhurnal 61(7): 956–960 – *On composition of ground lichen sinusiae in the relic steppe communities of Yakutia* (in Russian)
Karavaev MN, Dobretsova LA (1964) Kratky ocherk rastitelnosty doliny reki Nery v eyo nizhnem techenii (basseyn verkhney Indigirki). Botanichesky zhurnal 49 (11): 1544–1559 – *The brief review of vegetation of the Nera River valley in its lower reaches (the Upper Indigirka River basin)* (in Russian)
Karavaev MN, Skryabin SZ (1971) Rastitelny mir Yakutii. Yakutknigoizdat, Yakutsk – V*egetation world of Yakutia* (in Russian)
Kharkevich SS (ed) (1985) Sosudistye rasteniya Dalnego Vostoka 1. Nauka, Leningrad – *Vascular plants of the Far East 1* (in Russian)
Kharkevich SS (ed) (1987) Sosudistye rasteniya Dalnego Vostoka 2. Nauka, Leningrad – *Vascular plants of the Far East 2* (in Russian)
Kharkevich SS (ed) (1988) Sosudistye rasteniya Dalnego Vostoka 3. Nauka, Leningrad – *Vascular plants of the Far East 3* (in Russian)
Kharkevich SS (ed) (1989) Sosudistye rasteniya Dalnego Vostoka 4. Nauka, Leningrad – *Vascular plants of the Far East 4* (in Russian)

Kharkevich SS (ed) (1991) Sosudistye rasteniya Dalnego Vostoka 5. Nauka, Saint-Petersburg – *Vascular plants of the Far East 5* (in Russian)
Kharkevich SS (ed) (1992) Sosudistye rasteniya Dalnego Vostoka 6. Nauka, Saint-Petersburg – *Vascular plants of the Far East 6* (in Russian)
Kharkevich SS (ed) (1995) Sosudistye rasteniya Dalnego Vostoka 7. Nauka, Saint-Petersburg – *Vascular plants of the Far East* 7 (in Russian)
Khokhryakov AP (1985) Flora Magadanskoy oblasti. Nauka, Moscow – *Flora of the Magadan Region* (in Russian)
Kildyushevsky ID (1960) Formirovanie rastitelnosti yugo-vostochnoy chasti Verkhoyanskogo khrebta v raznykh usloviyakh sredy. Problemy botaniki 5: Materialy po izucheniyu flory i rastitelnosti vysokogoriy. Moscow–Leningrad: 62–71 – *Formation of vegetation in southeastern part of the Verkhoyansk Range under various environmental conditions* (in Russian)
Kildyushevsky ID (1964) K flore verkhovyev Vilyuya. In: Scherbakov IP (ed) Lesa Yuzhnoy Yakutii. Nauka, Moscow – *On the flora of the upper reaches of the Viluy River* (in Russian)
Kildyushevsky ID (1966) K flore Verkhoyanskogo khrebta. In: Rasteniya Severa Sibiri i Dalnego Vostoka. Nauka, Moscow–Leningrad – *On the flora of the Verkhoyansk Range* (in Russian)
Kirillov FN, Kirillov AF, Labutina TM et al. (1979) Biologiya Viluyskogo vodokhranilischa. Nauka, Novosibirsk – *Biology of the Viluy Reservoir* (in Russian)
Kiselyov IA (1935) Fitoplankton ozyor Tsentralnoy Yakutii po materialam limnosyemki 1932 goda. In: Issledovaniya ozyor SSSR. Proceedings of State Hydrologial Institute Vol 8. Izd-vo GGI, Leningrad: 51–84 – *The phytoplankton of lakes of Central Yakutia based on limnological survey of 1932* (in Russian)
Komarenko LE (1955) Flora vodorosley basseyna srednego techeniya reki Leny i reki Indigirki. Thesis for a Candidate Degree. Leningrad – *Algae flora of the Middle Lena River and Indigirka River* (in Russian)
Komarenko LE (1956) Kharakteristika flory vodorosley i zooplanktona vodoyomov basseyna srednego techeniya reki Leny. In: Proceedings of Institute of Biology of YaB SB AS USSR Vol 2. Obl gos izd-vo, Irkutsk: 145–212 – *Description of algae flora and zooplankton of water bodies of the Middle Lena River basin* (in Russian)
Komarenko LE (1962) Plankton levykh pritokov Viluya – rek Markha i Tyung. Trudy Institute biologii YaF SO AN SSSR 8: 162–163 – *The plankton of the Markha and Tyung Rivers, the left tributaries of the Viluy River* (in Russian)
Komarenko LE (1968) Plankton basseyna reki Yany. Nauka, Moscow – *The plankton of the Yana River basin* (in Russian)
Komarenko LE (1975) Vodorosli basseyna reki Khromy. In: Botanicheskiye materialy po Yakutii. Yakutsk: 87–99 – *The algae of the Khroma River basin* (in Russian)
Komarenko LE, Vasilyeva II (1967) K issledovaniyu vodorosley (mikroflory) basseyna reki Olenyok. In: Lyubite i okhranyaite prirodu Yakutii. Yakutsk – *On the study of the algae (microflora) of the Olenyok River basin* (in Russian)
Komarenko LE, Vasilyeva II (1972) Sostav i dinamika algoflory Kolymo-Indigirskoy nizmennosti v otkryty period 1966–1967. In: Rybokhozyaistvennoye osvoenie ozyor basseyna sredney Kolymy. Yakutsk – *Composition and dynamics of the algoflora of the Kolyma-Indigirka lowland in the ice-free period of 1966–1967* (in Russian)
Komarenko LE, Vasilyeva II (1975a) Presnovodnye diatomovye i sinezelyonye vodorosli vodoyomov Yakutii. Nauka, Moscow – *The freshwater diatoms and blue-green algae of the water bodies of Yakutia* (in Russian)
Komarenko LE, Vasilyeva II (1975b) Vodorosli basseyna reki Anabar. In: Botanicheskiye materially po Yakutii. Yakutsk: 78–86 – The algae of the Anabar River (in Russian)
Komarenko LE, Vasilyeva II (1978) Presnovodnye zelyonye vodorosli vodoyomov Yakutii. Nauka, Moscow – *The fresh water green algae of the water bodies of Yakutia* (in Russian)
Komarenko LE, Vasilyeva II, Remigailo PA (1974) Sine-zelyonye vodorosli vodoyomov subarkticheskoy tundry. In: Biologicheskiye problemy Severa: sporovye rasteniya, introduktsiya

rasteniy i polevoye kormoproizvodstvo. Proceedigns of the 6th Symposium. Issue 4. Yakutsk: 19–24 – *Blue-green algae of water bodies of the Subarctic tundra* (in Russian)

Komarov VL (1926) Vvedeniye v izucheniye rastitelnosti Yakutii. Trudy komissi po izucheniyu Yakutskoy ASSR. Vol 1. Leningrad – *Introduction to vegetation study of Yakutia* (in Russian)

Komarov VL (1927) Ocherk rastitelnosti Yakutii. In: Gerasimov II (ed)Yakutia. Izd-vo AN SSSR, Leningrad – *Review of vegetation of Yakutia* (in Russian)

Kononov KE (1971) Poymennye luga sredney Leny. Knizhnoye izd-vo, Yakutsk – *Floodplain meadows of the Middle Lena River* (in Russian)

Konstantinov AF, Efimov AS (1973) Predvaritelnye resultaty obsledovaniya ozera Bolshoye Toko. In: Voprosy energetiki Yakutskoy ASSR. Yakutsk: 189–203 – *Preliminary results of Bolshoye Toko Lake investigation* (in Russian)

Konstantinova NA (2000) Analiz arealov pechenochnikov severa Golarktiki. Arctoa 9: 29–94. – *Analyzis of distribution areas of the liverworts of northern Holarctic* (in Russian)

Konstantinova NA, Potemkin AD, Schlyakov RN (1992) Chesk-list of the Hepaticae and Anthocerotae of the former USSR. Arctoa 1: 87–127.

Kopyrina LI (1999a) Epifitnye vodorosli starichnykh ozyor srednego techeniya Leny. In: Timofeyev PA, Ivanova EI, Isaev AP (eds) Flora i rastitelnost Yakutii. Collection of articles. FIPS, Moscow: 43–50 – *The epiphytic algae of oxbow lakes of the Middle Lena* (in Russian)

Kopyrina LI (1999b) Obrastaniya na rastitelnom substrate v ozere Saisary (Yakutia). In: Ozyornye ekosistemy: biologicheskiye protsessy, antropogennaya transformatsiya, kachestvo vody. Proceedings of International Scientific Conference. Minsk: 120 – *Accreting on plant substrate in Saisary Lake (Yakutia)* (in Russian)

Kopyrina LI (2000a) Taksonomicheskoye raznoobraziye epifitnykh vodorosley ozera Belogo (sredneye techeniye reki Leny). In: Proceedings of the International Conference dedicated to 100th Anniversary of the organization and cryptogamous botany in Komarov Botanical Institute RAS.Izd-vo SPb gosudarstvennoy khim-farmakademii, Saint-Petersburg: 439–441 – *Taxonomical diversity of epiphytic algae of Beloye Lake (the Middle Lena River)* (in Russian)

Kopyrina LI (2000b) Taksonomicheskaya kharakteristika epifitnykh vodorosley ozera Saisary. In: Timofeev PA, Shmatkova LE (eds) Flora i fauna osobo okhranyaemykh prirodnykh territoriy respublikanskoy sistemy "Ytyk Kere Sirder". Kuduk, Yakutsk – *Taxonomical diversity of the epiphytic algae of Saisary Lake* (in Russian)

Kopyrina LI (2000c) Taksonomichesky analiz epifitnykh vodorosley ozera Sergelyakh. In: Ozyora kholodnykh regionov. Proceedings of International Conference Part 2. Izd-vo YaGU, Yakutsk: 82–93 – *Taxonomical analysis of the epiphytic algae of Sergelyakh Lake* (in Russian)

Kopyrina LI (2000d) Algae-epiphytes of higher aquatic plants of oxbow lakes of the middle river Lena. In: Kolchanov NA (ed) Biodiversity and Dinamics of Ecosystems in North Eurasia 5 (1). Izd-vo SO RAN, Novosibirsk: 39–40.

Kopyrina LI (2003a) Vodorosli perifitona vodoyomov NPP "Lenskiye Stolby" (Srednyaya Lena). In: Protasov AA, Tolstikov AV (eds) Perifiton kontinentalnykh vod: sovremennoye sostoyaniye izuchennosti i perspektivy dalneyshikh issledovaniy. Proceedings of International Symposium. OOO "Optsion TM-Kholdinh", Tyumen: 36–37 – *The algae of periphyton of the water bodies of the National Natural Park "Lena Pillars"* (in Russian)

Kopyrina LI (2003b) Fitoperifiton na raznykh substratakh vodoyomov Sredney Leny. In: Mikheyeva TM (ed) Ozyornye ekosistemy: biologicheskie protsessy, antropogennaya transformatsiya, kachestvo vody. Izd-vo BGU, Minsk: 281–284 – *The phytoperiphyton on various substrates of the Middle Lena valley's water bodies* (in Russian)

Kopyrina LI (2003c) Phytoperiphyton in Some Reservoirs of the Middle Lena river. In: Isaev AP (ed) Problemy botanicheskikh i lesovodstvennykh issledovaniy v RS (Ya) i Finlyandii. Proceedings of Sakha-Finnish Conference dedicated to 100th Anniversary of Cajander's expedition along the Lena. YaF SO RAN, Yakutsk: 33–36.

Korobkov AA (1981) Polyni Severo-Vostoka SSSR. Nauka, Leningrad – *Wormwoods of the North–East USSR* (in Russian)

Koropachinsky IYu (ed) (1996) Zelyonaya kniga Sibiri: redkiye i nuzhdayuschiesya v okrane rastitelnye soobschestva. Nauka, Novosibirsk – *Green Book of Siberia: Rare and requiring protection plant communities* (in Russian)

Korzhuev SS (1965) Relyef i geologicheskoye stroeniye. In: Gerasimov IP (ed) Yakutia. Nauka, Moscow – *Relief and geological structure* (in Russian)

Kosinskaya EK (1936) Desmidievye vodorosli iz Arktiki. In: Proceedings of Botanical Institute AS USSR Series 2 Issue 3. Izd-vo AN SSSR, Moscow–Leningrad – *The Desmids from the Arctic* (in Russian)

Kosinskaya EK (1960) Flora sporovykh rasteny SSSR Tom 5: Konyugaty ili stseplyanki (2) Desmidiyevye vodorosli 1. Nauka, Moscow–Leningrad – *The flora of spore-bearing plants of the USSR Volume 5: Conjugates (2) The desmids Issue 1* (in Russian)

Kotiranta H, Mukhin VA (2000) Aphyllophorales of Tiksi. Karstenia 40: 65–69.

Kozhevnikov YuP (1981) Botaniko-ekologicheskiye nablyudeniya na reke Kolyme v raionakh srednego techeniya reki Beryozovka i posyolka Chersky. In: Biologiya i ekologiya rasteniy basseyna Kolymy. DVNTs AN SSSR, Vladivostok – *Botanical-ecological observations on the Kolyma River within the Middle Beryozovka River and Chersky settlement* (in Russian)

Kozhova OM (1969) Rastitelnaya produktsiya Bratskogo vodokhranilischa. In: Biologicheskaya produktivnost vodoyomov Sibiri. Nauka, Moscow: 71–76 – *Plant production of the Bratsk Reservoir* (in Russian)

Krasnoborov IM (ed) (1988) Flora Sibiri: *Lycopodiaceae-Hydrocharitaceae*. Nauka, Novosibirsk – *The flora of Siberia: Lycopodiaceae-Hydrocharitaceae* (in Russian)

Krasnoborov IM (ed) (1997) Flora Sibiri: *Asteraceae*. Nauka, Novosibirsk – *The flora of Siberia: Asteraceae* (in Russian)

Krasnoborov IM, Malyshev LI (eds) (1992) Flora Sibiri: *Salicaceae-Amaranthaceae*. Nauka, Novosibirsk – *The flora of Siberia: Salicaceae-Amaranthaceae* (in Russian)

Krivoshapkin KK (1998a) Listostebelnye mkhi nalednykh polyan yugo-zapadnoy Yakutii. Botanicheskiy zhurnal 83 (8): 58–65 – *Leafy mosses of the icings in Southwestern Yakutia* (in Russian)

Krivoshapkin KK (1998b) Listostebelnye mkhi Yugo-Zapadnoy Yakutii. Abstract of Thesis for a Candidate Degree. Saint-Petersburg – *Leafy mosses of South–West Yakutia* (in Russian)

Krivoshapkin KK (2001) Listostebelnye mkhi pamyatnikov prirody "Ulakhan Taryn" i "Bulus" (Tsentralnaya Yakutia). In: Timofeev PA, Shmatkova LE (eds) Flora i fauna osobo okhranyaemykh prirodnykh territoriy Respublikanskoy sistemy Ytyk Kere Sirder. Kuduk, Yakutsk – *Leafy mosses of the nature monuments "Ulakhan Taryn" and "Bulus"* (Central Yakutia) (in Russian)

Krivoshapkin KK, Ivanova EI (2000) Bryoflora Yakutii: Uchebnoe posobie Chapter 1. Izd-vo YaGU, Yakutsk – *Bryoflora of Yakutia: Textbook Part 1* (in Russian)

Kuvaev VB (1955) Luga i pastbischa Olyokminskogo Raiona. In: Trudy Instituta Biologii. Izd-vo AN SSSR, Moscow–Leningrad 1: 103–135 – *Meadows and pastures of the Olyokminsk Region* (in Russian)

Kuvaev VB (1956) Rastitelnost Vostochnogo Verkhoyanya. Rastitelnost Krainego Severa i eyo osvoyeniye 2: 133–186 – *Vegetation of the East Verkhoyansk Region* (in Russian)

Kuvaev VB (1957) Rastitelny pokrov yugo-zapadnoy Yakutii i ego kormovye resursy. In: Materialy o prirodnykh usloviyakh i selskom khozyaistve yugo-zapada Yakutskoy ASSR. Issue 2. Izd-vo AN SSSR, Moscow – *Vegetation cover of South–western Yakutia* (in Russian)

Kuvaev VB, Samarin VP (1961) Kormovye rasteniya severnogo olenya v Tomponskom rayone Yakutskoy ASSR (Verkhoyanskiy khrebet). In: Alexandrova VD, Kuvaev VB, Tikhomirov BA et al. (eds) Materialy po rastitelnosti Yakutii. Izd-vo LTA, Leningrad – *Fodder plants of a reindeer in Tomponsky Region of Yakutian ASSR (the Verkhoyansk Range)* (in Russian)

Kuznetsova LV (1997) Perspektivy izucheniya bioraznoobraziya Olyokmo-Charskogo nagorya (Yugo-Zapad Yakutii). In: Strategiya sokhraneniya i ustoychivogo ispolzovaniya borealnykh lesov. Proceedings of International seminar. AO NPP "Sapfir", Moscow: 145–146 – *Outlook of biodiversity investigation of the Olyokma-Chara upland region* (in Russian)

Kuznetsova LV (1998) O nekotorykh redkikh rasteniyakh Yuzhnoy Yakutii. In: Problemy botaniki na rubezhe 20–21 vekov. Vol 2. Proceedigns of 2nd (10th) Congress of the Russian Botanical Society. Botanical Institute RAS, Saint-Petersburg: 252 – *On some rare plants of South Yakutia* (in Russian)

Kuznetsova LV (1999) Ekologo-geographicheskaya kharakteristika lekarstvennykh rasteniy Yakutii. In: Timofeyev PA, Ivanova EI, Isaev AP (eds) Flora i rastitelnost Yakutii. Collection of articles. FIPS, Moscow: 129–139 – *Ecological-geographical charactertistics of medicinal herbs of Yakutia* (in Russian)

Kuznetsova LV (2001) K flore redkikh, ischezayuschikh i novykh vidov vysshikh sosudistykh rasteniy basseyna reki Tokko i okrestnostey Olyokminska. In: Timofeev PA, Shmatkova LE (eds) Flora i fauna osobo okhranyaemykh prirodnykh territoriy Respublikanskoy sistemy Ytyk Kere Sirder. Kuduk, Yakutsk – *On the flora of rare, endangered and new species of higher vascular plants of the Tokko River basin and Olyokminsk vicinity* (in Russian)

Kuznetsova LV (2003a) Floristicheskiye nakhodki na khrebte Udokan. In: Cherosov MM, Simonova SV (eds) Flora i rastitelnost kriolitozony Yakutii. Proceedings of the Republican Scientific Conference. Izd-vo YaNTs SO RAN, Yakutsk: 69–75 – *Floristic finds in the Udokan Range* (in Russian)

Kuznetsova LV (2003b) Floristicheskiye nakhodki na Olyokmo-Charskom nagorye i Severo-Vostochnoy chasti khrebta Udokan. In: Botanicheskiye issledovaniya v Aziatskoy Rossii. Proceedings of 11th Congress of the Russian Botanical Society. Vol 1. Izd-vo "Azbuka", Barnaul: 360–362 – *Floristic finds on the Olyokma-Chara upland and in the Northeastern part of the Udokan Range* (in Russian)

Kuznetsova LV, Ivanova EI (2001) Elovo-pikhtovye lesa Olyokmo-Charskogo nagorya. In: Timofeev PA, Shmatkova LE (eds) Flora i fauna osobo okhranyaemykh prirodnykh territoriy Respublikanskoy sistemy Ytyk Kere Sirder. Kuduk, Yakutsk – *Picea-Abies forests of the Olyokma-Chara upland region* (in Russian)

Kuznetsova LV, Isaev AP, Lytkina LP (2006a) Kratkoye opisaniye rastitelnogo pokrova i redkiye rastitelnye soobschestva verkhnego techeniya reki Amgi. In: Lesnye issledovaniya v Yakutii: itogi, sostoyaniye i perspectivy. Izd-vo YaGU, Yakutsk – *Brief description of vegetation cover and rare plant communities of the upper reaches of the Amga River* (in Russian)

Kuznetsova LV, Isaev AP, Fyodorov IA (2006b) Redkiye i endemichnye rasteniya verkhnego techeniya reki Amgi. In: Lesnye issledovaniya v Yakutii: itogi, sostoyaniye i perspectivy. Izd-vo YaGU, Yakutsk –*Rare and endemic plants of the upper reaches of the Amga River* (in Russian)

Lazarenko AS (1956) Osnovni zasadi klassifikatsii arealiv listyanykh mokhiv Radyanskogo Dalekogo Skhodu. Ukrainsky Botanicheskiy Zhurnal 13 (1) – *Main principles of classification of distribution areas of leafy mosses of the Soviet Far East* (in Ukranian)

Lukicheva AN (1960) Rastitelny pokrov kak indikator kimberlitovykh trubok. Geologiya i geofizika 11: 35–48 – *Vegetation cover as an indicator of a kimberlite pipe* (in Russian)

Lukicheva AN (1963a) Rastitelnost severo-zapada Yakutii i eyo svyaz s geologicheskim stroeniem mestnosti. Izd-vo AN SSSR, Moscow–Leningrad – *Vegetation of North–West Yakutia and its relationship with the local geological structure* (in Russian)

Lukicheva AN (1963b) Rastitelnost verkhovyev reki Muny (Severo-Zapadnaya Yakutia) I vliyaniye na neyo noveyshey tektoniki. Botanicheskiy zhurnal 48 (3): 328–340 – *Vegetation of the Upper Muna River (North–West Yakutia) and the effect on it caused by the modern tectonics* (in Russian)

Lyubarsky LV, Vasilyeva LN (1975) Derevorazrushayuschiye griby Dalnego Vostoka. Nauka, Novosibirsk – *Wood-attacking fungi of the Far East* (in Russian)

Makarova II (1983) Lishainiki yugo-vostoka Chukotskogo poluostrova (bukhta Penkigney). Novosti sistematiki nizshikh rasteniy 20: 146–150 – *Lichens of the south–east of Chukotka Peninsula (Penkigney Bay)* (in Russian)

Makarova II (1985) Novye vidy dlya likhenoflory Yakutii. Novosti sistematiki nizshikh rasteniy 22: 178–180 – *New species for the lichen flora of Yakutia* (in Russian)

Makarova II (1989) K flore lishainikov nizovyev reki Leny. Novosti sistematiki nizshikh rasteniy 26: 118–124 – *On the lichen flora of the Lower Lena River* (in Russian)

Makarova II (1996) Lishainiki karbonatnykh mestoobitaniy Ust-Lenskogo Zapovednika. Novosti sistematiki nizshikh rasteniy 31: 126–130 – *The lichens of calcareous habitats of the Ust-Lena Reserve* (in Russian)

Makarova II (1998) Dopolnenie k likhenoflore Ust-Lenskogo zapovednika. Novosti sistematiki nizshikh rasteniy 32: 52–55 – *Supplement to the lichen flora of the Ust-Lena Reserve* (in Russian)

Makarova II, Perfilyeva VI (1984) K flore lishainikov severo-zapada Yakutii. Novosti sistematiki nizshikh rasteniy 21: 150–160 – *On the lichen flora of the north–west of Yakutia* (in Russian)

Makarova II, Perfilyeva VI (1985) Dopolnenie k likhenoflore severo-zapada Yakutii. Novosti sistematiki nizshikh rasteniy 22: 180–184 – *Addition to the lichen flora of the north–west of Yakutia* (in Russian)

Makarova II, Perfilyeva VI (1989) K flore lishainikov Ust-Lenskogo zapovednika. In: Vodorosli, lishainiki, griby i mokhoobraznye v zapovednikakh RSFSR. Collection of articles. Moscow: 53–59 – *On the lichen flora of the Ust-Lena Reserve* (in Russian)

Makarova II, Perfilyeva VI. Nikolin EG (1988) K flore lishainikov Novosibirskikh ostrovov. Novosti sistematiki nizshikh rasteniy 25: 127–134 – *On the lichen flora of the Novosibirskie Islands* (in Russian)

Malme G (1932) Lichenes orae Sibiriae Borealis inde ab insula Minin usque ad promantorium Ryrkajpia in Expeditione Vegae lecti. Arkiv för Botanik 25A (2): 1–42

Malyshev LI (1972) Ploschad vyyavleniya flory v sravnitelno-floristicheskikh issledovaniyakh. Botanichesky zhurnal 57 (2): 182–197 – *An area size for flora evaluation in comparative floristic studies* (in Russian)

Malyshev LI (ed) (1997) Flora Sibiri: *Pyrolaceae-Lamiaceae*. Nauka, Novosibirsk – *The flora of Siberia: Pyrolaceae-Lamiaceae* (in Russian)

Malyshev LI, Peshkova GA (1984) Osobennosti i genezis flory Sibiri (Predbaikalye i Zabaikalye). Nauka, Novosibirsk – *Characteristic features and genesis of Siberian flora* (Prebaikalia and Transbaikalia) (in Russian)

Malyshev LI, Peshkova GA (eds) (1987) Flora Sibiri: *Araceae-Orchidaceae*. Nauka, Novosibirsk – *The flora of Siberia: Araceae-Orchidaceae* (in Russian)

Malyshev LI, Peshkova GA (eds) (1990a) Flora Sibiri: *Poaceae*. Nauka, Novosibirsk – *The flora of Siberia: Poaceae* (in Russian)

Malyshev LI, Peshkova GA (eds) (1990b) Flora Sibiri: *Cyperaceae*. Nauka, Novosibirsk – *The flora of Siberia: Cyperaceae* (in Russian)

Malyshev LI, Peshkova GA (eds) (1993) Flora Sibiri: *Portulacaceae-Ranunculaceae*. Nauka, Novosibirsk – *The flora of Siberia: Portulacaceae-Ranunculaceae* (in Russian)

Malyshev LI, Peshkova GA (eds) (1994) Flora Sibiri: *Berberidaceae-Grossulariaceae*. Nauka, Novosibirsk – *The flora of Siberia: Berberidaceae-Grossulariaceae* (in Russian)

Malyshev LI, Peshkova GA, Baikov KS (eds) (2003) Flora Sibiri: dopolnitelnye ispravleniya i alfavitnye ukazateli. Nauka, Novosibirsk – *The flora of Siberia: Supplementary revision and alphabetical index* (in Russian)

Margelis L (1983) Rol simbioza v evolyutsii kletki. Mir, Moscow – *The role of symbiosis in cell evolution* (in Russian)

Mattick F (1953) Lichenologische Notizen 1. Der Flechten-Koefficient und seine Bedeutungfür Pflanzengeographie. Ber Deutsch Bot Ges. Bd 66 H 7

Mikhalyova LG (1993) Predvaritelnye materialy k izucheniyu derevorazrushayuschikh afilloforovykh gribov Yakutii. Mikologiya i fitopatologiya 27 (5): 15–33 – *Preliminary data on the study of wood-attacking aphyllophorous fungi of Yakutia* (in Russian)

Mikhalyova LG (1995) Afilloforovye griby okrestnostey Tommota (Yuzhnaya Yakutia). In: Proceedings of the 5th Youth Conference of botanists. Saint-Petersburg: 51–54 – *Aphyllophorous fungi in the vicinity of Tommot (South Yakutia)* (in Russian)

Mikhalyova LG (1998) Dopolneniye k spisku afilloforovykh gribov Yakutii. In: Problemy botaniki na rubezhe 20–21 vekov. Proceedings of the II(X) Congress of the Russian botanical society Vol 2. Botanichesky Institut RAN, Saint-Petersburg: – *Supplementation to the list of aphyllophorous fungi of Yakutia* (in Russian)

Mikhalyova LG (1999) Afilloforovye griby Yakutii (sistematichesky spisok). In: Timofeyev PA, Ivanova EI, Isaev AP (eds) Flora i rastitelnost Yakutii. FIPS, Moscow: 51–54 – *Aphyllophorous fungi of Yakutia (taxonomical list)* (in Russian)

Mikhalyova LG (2001a) Afilloforovye griby Natsionalnogo prirodnogo parka "Lenskiye stolby". In: Solomonov NG, Isaev AI, Ivanova EI (eds) Natsionalny prirodny park "Lenskie stolby": geologia, pochvy, rastitelnost, zhivotny mir, okhrana i ispolzovanie. Izd-vo YaGU, Yakutsk – *Aphyllophorous fungi of the National natural park "Lena Pillars"* (in Russian)

Mikhalyova LG (2001b) Derevorazrushayuschiye griby Natsionalnogo prirodnogo parka "Lenskiye stolby". In: Timofeev PA, Shmatkova LE (eds) Flora i fauna osobo okhranyaemykh prirodnykh territoriy Respublikanskoy sistemy Ytyk Kere Sirder. Kuduk, Yakutsk – *Wood-attacking fungi of the National natural park "Lena Pillars"* (in Russian)

Mikhalyova LG (2001c) The expansion of areas of rare fungi species in connection with the global change. In: The role of permafrost ecosystem in global climate change. Yakutsk Scientific Center Publishing House, Yakutsk: 129–132.

Mikhalyova LG, Protopopov AV (1999) Otsenka sovremennogo sostoyaniya lesov basseyna reki Siney. In: Timofeyev PA, Ivanova EI, Isaev AP (eds) Flora i rastitelnost Yakutii. Collection of articles. FIPS, Moscow: 96–100 – *Evaluation of present forest condition in the Sinyaya River basin* – (in Russian)

Moore JW (1977) Ecology of algae in a subarctic stream. Can J Bot 55 (13): 1838–1847.

Moore JW (1979) Distribution and abundance of attached littoral algae in 21 lakes and streams in the Northwest Territories. Can I Bot 57 (6): 568–577.

Muchnik EE (1991) Likhenoflora okhranyaemykh territoriy Voronezhskoy oblasti. Botanicheskiy zhurnal 76 11: 1629–1637 – *The lichen flora of the protected areas of Voronezh Region* (in Russian)

Muchnik EE (1997) Epilitnye lishainiki Tsentralnogo Chernozemya. Botanicheskiy zhurnal 82 4: 46–53 – *The epilithic lichens of Central Chernozemye* (in Russian)

Mukhin VA (1993) Biota ksilotrofnykh bazidiomitsetov Zapadno-Sibirskoy ravniny. UIF Nauka. Ekaterinburg – *The biota of xylotrophous basidiomycetes of the West Siberian Plain* (in Russian)

Mukhin VA, Mikhalyova LG, Ushakova NV (2004) Rasprostraneniye *Fomitopsis cajanderi* (P.Karst) Kotlaba et Pouzar na Urale i v Aziatskoy chasti Rossii. In: Proceedings of the Institute of Bioresources and Applied Ecology Issue 4. Izd-vo OGPU, Orenburg: 19–25 – *Distribution of Fomitopsis cajanderi (P.Karst) Kotlaba et Pouzar in the Ural and Asian part of Russia* (in Russian)

Müller E, Löffler V (1995) Mikologiya. Mir, Moscow – *Mycology* (in Russian)

Naumenko YuV (1985) Fitoplankton Obi, Nizhnego Irtysha i ego izmeneniya pod vozdeystviyem antropogennykh faktorov. Thesis for a Candidate degree. Novosibirsk – *The phytoplankton of the Ob, Lower Irtysh Rivers and its transformation caused by anthropogenic impact* (in Russian)

Nikadimova NF (1964) Kratky obzor bolezney i gribnoy flory drevostoev Verkhnego i Srednego Aldana. In: Pozdnaykov (ed) Lesa Yuzhnoy Yakutii. Izd-vo AN SSSR. Nauka, Moscow – *Brief review of diseases and fungi flora of forests of the Upper and Middle Aldan* (in Russian)

Nikadimova NF (1967) O gribnykh zabolevaniyakh listvennitsy i sosny v Yugo-Zapadnoy Yakutii. In: Izvestiya SO AN SSSR: Series Biologiya i meditsina 10 (130) Issue 2: 98–103 – *On fungi caused diseases of Larix and Pinus syvestris in South–West Yakutia* (in Russian)

Nikadimova NF (1970) Fitopatologicheskoye sostoyaniye listvyagov zelyonoy zony Yakutska. In: Vodorsoli i griby Sibiri i Dalengo Vostoka 1 (3): 119–120 – *Phytopathological condition of larch forests of the green zone of Yakutsk* (in Russian)

Nikitina NV (2001) Okhrana i ratsionalnoye ispolzovaniye kedrovykh lesov Olyokminskogo ulusa. In: Timofeev PA, Shmatkova LE (eds) Flora i fauna osobo okhranyaemykh prirodnykh territoriy Respublikanskoy sistemy Ytyk Kere Sirder. Kuduk, Yakutsk – *Protection and rational use of Pinus sibirica forests of Olyokminsk Ulus* (in Russian)

Nikitina NV (2004) Formirovaniye temnokhvoino-kedrovykh lesov v Yakutii. In: Problemy izucheniya rastitelnogo pokrova Yakutii. NIPK "Sakhapoligrafizdat", Yakutsk – *Formation of dark coniferous Pinus sibirica forests in Yakutia* (in Russian)

Nikitina NV, Isaev AP (1999) Sostav i struktura kedrovykh lesov Olyokminskogo ulusa. In: Timofeyev PA, Ivanova EI, Isaev AP (eds) Flora i rastitelnost Yakutii. Collection of articles. FIPS, Moscow – *Composition and structure of Pinus sibirica forests of the Olyokminsk Ulus* (in Russian)

Nikolin EG (1991) Flora i rastitelny pokrov Tsentralnogo Verkhoyanya. Abstract of Thesis of Candidate Degree. Novosibirsk – *The flora and vegetation cover of the Central part of the Verkhoyansk region* (in Russian)

Nikolin EG (1992) Obschiye cherty structurnoy organizatsii rastitelnosti Tsentralnogo Verkhoyanya. In: 10 Vsesoyuznoye soveschaniye po izucheniyu flory i rastitelnosti vysokogoriy. Proceedings. TsSBS SO RAN, Novosibirsk: 106 – *Common features in structural organization of vegetation of the Central part of Verkhoyansk region* (in Russian)

Nikolin EG, Baryshev EV (1992) Redkiye rasteniya Tsentralnogo Verkhoyanya. In: Botanicheskiye issledovaniya v kriolitozone. Collection of articles. Izd-vo YaNTs SO RAN, Yakutsk: 54–56 – *The rare plants of the Central part of Verkhoyansk region* (in Russian)

Nikolin EG, Petrovsky VV (1988) *Saxifraga lyallii (Saxifragaceae)* – novy vid dlya flory SSSR. Botanichesky zhurnal 73 (7): 1026–1027 – *Saxifraga lyallii (Saxifragaceae) is a new species for the flora of the USSR* (in Russian)

Nogovitsyn DD (1985) Vodnye resursy Yakutskoy ASSR i ikh ispolzovaniye. Izd-vo YaF SO AN SSSR, Yakutsk – *Water resources of Yakutian ASSR and their utilization* (in Russian)

Oksner AM (1939) Lishainiki baseynu rik Indigirki, Yani, Leni i Pivdennogo Pribaikallya I. Zhurnal Institutu botaniki AN URSR 23 (31): 117–139 – *The lichens of the Indigirka, Yana and Lena River basins and South Prebaikalia* I (in Ukranian)

Oksner AM (1940a) Lishainiki baseynu rik Indigirki, Yani, Leni i Pivdennogo Pribaikallya II. Botanicheskiy zhurnal AN URSR 1 (1): 77–100 – *The lichens of the Indigirka, Yana and Lena River basins and South Prebaikalia* II (in Ukranian)

Oksner AM (1940b) Lishainiki baseynu rik Indigirki, Yani, Leni i Pivdennogo Pribaikallya III. Botanicheskiy zhurnal AN URSR 1 (2): 313–324 – *The lichens of the Indigirka, Yana and Lena River basins and South Prebaikalia* III (in Ukranian)

Oksner AM (1940c) Lishainiki baseynu rik Indigirki, Yani, Leni i Pivdennogo Pribaikallya IV. Botanicheskiy zhurnal AN URSR 1 (3–4): 31–57 – *The lichens of the Indigirka, Yana and Lena River basins and South Prebaikalia* IV (in Ukranian)

Oksner AM (1974) Opredelitel lishianikov SSSR Vypusk 2: Morfologiya, sistematika i geograficheskoe rasprostraneniye. Nauka, Leningrad – *Handbook of the lichens of the USSR Issue 2: Morphology, taxonomy and geographical distribution* (in Russian)

Otnyukova TN, Ignatova EI, Ignatov MS, Fedosov VE (2004) Novye nakhodki *Tortella alpicola* Dix. v Evrazii. Arctoa 13: 197–202 – *New finds of Tortella alpicola Dix. in Eurasia* (in Russian)

Parmasto EH (1967). Trutovye griby Severa Sovetskogo Soyuza. Mikologiya i fitopatologiya 1: 280–286 – *The polypores of the North of the Soviet Union* (in Russian)

Parmasto EH (1975) Studies on Jakutian fungi. I Corticiaceae. Thelephoraceae. Izvestiya AN ESSR Biologiya 24 (1): 3–15.

Parmasto EH (1976) Studies on Jakutian fungi. II Ganodermataceae. Hymenochaetaceae. Polyporaceae. Izvestiya AN ESSR Biologiya 25 (4): 316–321.

Parmasto EH (1977) Studies on Jakutian fungi. III Polyporales. Poriaceae sl. Izvestiya AN ESSR Biologiya 26 (1): 15–28.

Parmasto EH (1979) Genezis flory derevorazrushayuschikh afilloforovykh gribov Sovetskogo Dalnego Vostoka. In: Proceedings of the 14th Pacific scientific congress (Botany). Moscow:

39–40 – *Genesis of the flora of wood-attacking aphyllophorous fungi of the Soviet Far East* (in Russian)

Paton JA (1999) The Liverwort Flora of the British Isles. Harley Books, Essex

Pavlov PD (1981) Geograficheskoye rasprostraneniye eolovykh peskov v Tsentralnoy Yakutii. In: Eolovye obrazovaniya Tsentralnoy Yakutii. Izd-vo IMZ SO RAN, Yakutsk – *Geographical distribution of eolian sands in Central Yakutia* (in Russian)

Perfilyeva VI (1977) Osnovnye cherty rastitelnosti gornoy chasti basseyna reki Yany. In: Teoreticheskiye i prikladnye problemy biologii na Severo-Vostoke SSSR. Collection of articles. Izd-vo YaF SO AN SSSR, Yakutsk: 85–94 – *Basic features of vegetation of the mountainous part of the Yana River basin* (in Russian)

Perfilyeva VI, Rykova YuV (1975a) Arkticheskaya tundra v ustye reki Chukochyey. In: Botanicheskie issledovaniya v Yakutii. Izd-vo YaF SO AN SSSR, Yakutsk – *The Arctic tundra in the Chukochya River mouth* (in Russian)

Perfilyeva VI, Rykova YuV (1975b) K kharakteristike pritundrovykh severo-tayozhnykh lesov Kolymskoy nizmennosti. In: Biologicheskiye resursy Yakutii. Collection of articles: 30–32 – *On description of the near-tundra northern taiga forests of the Kolyma Lowland* (in Russian)

Perfilyeva VI, Shurduk IF (1999) Rastitelnost ostrovov Ust-Viluyskogo zakaznika. In: Timofeyev PA, Ivanova EI, Isaev AP (eds) Flora i rastitelnost Yakutii. Collection of articles. FIPS, Moscow: 78–82 – *Vegetation of islands of the Ust-Viluysky Reserve* (in Russian)

Perfilyeva VI, Egorova AA, Stepanova NA (1981) K flore Kondakovskogo ploskogorya (Severo-Vostochnaya Yakutia). In: Andreyev VN (ed) Rastitelnost Yakutii i eyo okhrana. Izd-vo YaF SO AS SSR, Yakutsk – *On the flora of the Kondakovskoye Plateau (north–eastern Yakutia)* (in Russian)

Perfilyeva VI, Teterina LV, Karpov NS (1991) Rastitelny pokrov tundrovoy zony Yakutii. Izd-vo YaNTs SO RAN, Yakutsk – *Vegetation cover of the tundra zone of Yakutia* (in Russian)

Permyakova AA (1973) Fragmenty stepnoy rastitelnosti v Srednekolymskom raione. In: Pochvy i rastitelnost merzlotnykh raionov SSSR. Proceedings of 5th All-Union Symposium. Magadan: 141–147 – *Fragments of steppe vegetation in the Srednekolymsk Region* (in Russian)

Peshkova GA (ed) (1996) Flora Sibiri: *Geraniaceae-Cornaceae*. Nauka, Novosibirsk – *The flora of Siberia: Geraniaceae-Cornaceae* (in Russian)

Pestryakova LA (1994) Evolyutsiya ozyornykh ekosistem Tsentralnoy Yakutii po materialam diatomovogo analiza donnykh otlozheniy. Thesis for a Candidate Degree. Yakutsk – *Evolution of lake ecosystems of Central Yakutia based on the data of the bottom sediments diatom analysis* (in Russian)

Pestryakova LA (1999) *Baccilariophyta* ozyornykh otlozheniy Tsentralnoy Yakutii. Algologiya 9 (2): 112–113 – *Baccilariophyta of lake sediments of Central Yakutia* (in Russian)

Pestryakova LA (2004) Diatomovye kompleksy poverkhnostnykh osadkov ozyor Yakutii. Nauka i obrazovaniye 3 (35): 19–25 – *Diatom complexes of surface sediments of the lakes of Yakutia* (in Russian)

Pestryakova LA, Ushnitskaya LA (2000) Diatomei donnykh otlozheniy ozyor Tiksi (Respublika Sakha). In: Ozyora kholodnykh regionov Part 4: Voprosy paleoklimatologii i paleoekologii. Izd-vo YaGU, Yakutsk: 151–160 – *The diatoms of bottom sediments of Tiksi lakes (Sakha Republic)* (in Russian)

Pestryakova LA, Pshennikova EV, Sobakina IG, Ushnitskaya LA (2004) Otsenka sostoyaniya ozyor naselyonnykh punktov verkhovya reki Tatta. Dalnevostochny meditsinsky zhurnal Appendix 1: 42–43 – *Evaluation of lakes' condition in the settlements of the Upper Tatta River basin* (in Russian)

Petrenko IA (1978) Makro- i mikromitsety lesov Yakutii. Nauka, Novosibirsk – *The macro- and micromycetes of the forests of Yakutia* (in Russian)

Petrenko IA (1980) Sinuzii derevorazrushayuschikh gribov v lesnykh fitotsenozakh Tsentralnoy Yakutii. In: Problemy lesnoy biogeotsenologii. Nauka, Novosibirsk: 248–264 – *Synusiae of the wood-attacking fungi in forest phytocenoses of Central Yakutia* (in Russian)

Petrov AM (1957) Kormovaya baza zhivotnovodstva v Lenskom i Olyokminskom raionakh. In: Materialy o prirodnykh usloviyakh i selskom khozyaistve yugo-zapada Yakutskoy ASSR. Issue 2. Izd-vo AN SSSR, Moscow – *Fodder base for cattle-breeding in the Lensk and Olyokminsk Regions* (in Russian)

Petrov VA (1930) Flora Yakutii Tom 1: Paporotniki-zlaki. Leningrad – *Flora of Yakutia Volume1: Ferns-grasses* (in Russian)

Petrovsky VV, Plieva TV (1994) Flora Leno-Olenekskogo vodorazdela (severnaya Yakutia). Botanichesky zhurnal 79 (9): 1–11 – *The flora of the Lena-Olenyok watershed* (North Yakutia)

Petrovsky VV, Zaslavskaya TM (1981) K flore pravoberezhya reki Kolymy bliz eyo ustya. Botanichesky zhurnal 65 (5): 1520–1528 – *On the flora of the Kolyma River's right bank near its mouth* (in Russian)

Pisarenko OY u (2006) Ob izmenchivosty i ekologii *Pteregoneurum subsessile* i *P. kozlovii (Pottiaceae, Bryophyta)*. Arctoa 15: 169–182 – *On variability and ecology of Pteregoneurum subsessile and P. kozlovii (Pottiaceae, Bryophyta)* (in Russian)

Pivnik SA (1958) Rastitelnost prilenskikh otrogov Verkhoyanskogo khrebta v raione ustya Viluya. In: Rastitelnost Krainego Severa SSSR i eyo osvoyeniye. Izd-vo AN SSSR, Moscow–Leningrad 3: 128–153 – *Vegetation of the near-Lena spurs of the Verkhoyansk Range in the territory of the Viluy River mouth (Novosibirsky Archipelago)* (in Russian)

Pivovarova ZhF (1976) Nekotorye osobennosti gruppirovok vodorosley, inditsiruyuschikh step-noy tip pochvoobrazovaniya. In: Biologicheskaya diagnostika pochv. Nauka, Moscow – *Some peculiarities of the algae groups indicating the steppe type of soil formation* (in Russian)

Pivovarova ZhF (1986a) Vodoroslevye gruppirovki kashtanovykh pochv Kolymskikh stepey. Botanichesky zhurnal 71 (7): 888–891 – *The algae groups of chestnut soils of the Kolyma steppes* (in Russian)

Pivovarova ZhF (1986b) Osobennosti floristicheskogo sostava i fitotsenoticheskoy organizatsii vodoroslevykh gruppirovok gornykh stepey Severo-Vostochnoy Azii. Botanichesky zhurnal 71 (4): 521–527 – *Characteristic features of floristic composition and phytocoenotic organization of the algae groups of mountainous steppes of North–East Asia* (in Russian)

Pivovarova ZhF, Berman DI, Volkovintser VM (1975) O flore mikrovodorosley stepey Yakutii. In: Botanicheskiye issledovaniya v Yakutii. Collection of articles. Izd-vo YaFSO AN SSSR, Yakutsk: 109–121– *On the microalgae flora of the steppes of Yakutia* (in Russian)

Piyn TKh (1984) Flora i rasprostraneniye napochvennykh lishainikov yuzhnykh tundr Taimyra. In: Flora i gruppirovki nizshikh rasteniy v prirodnykh i antropogennykh ekstremalnykh usloviyakh sredy. Tallin – *The flora and distribution of ground lichens of the southern tundra of Taimyr* (in Russian)

Poelt J (1973) Classification. In: The lichens. New York, London

Polozhy AV, Malyshev LI (eds) (1988) Flora Sibiri: *Rosaceae*. Nauka, Novosibirsk – *The flora of Siberia: Rosaceae* (in Russian)

Polozhy AV, Malyshev LI (eds) (1994) Flora Sibiri: *Fabaceae*. Nauka, Novosibirsk – *The flora of Siberia: Fabaceae* (in Russian)

Polozhy AV, Peshkova GA (eds) (1996) Flora Sibiri: *Solanaceae-Lobeliaceae*. Nauka, Novosibirsk – *The flora of Siberia: Solanaceae-Lobeliaceae* (in Russian)

Poryadina LN (1998) Novye i redkiye vidy lishainikov dlya flory Azii. Novosti sistematiki nizshikh rasteniy 32: 76–81 – *New and rare lichen species for the flora of Asia* (in Russian)

Poryadina LN (1999a) Novye i redkiye vidy dlya likhenoflory Yakutii. Novosti sistematiki nizshikh rasteniy 33: 153–158 – *New and rare species for the lichen flora of Yakutia* (in Russian)

Poryadina LN (1999b) Lishainiki Tomponskogo ulusa Yakutii. Botanicheskiy zhurnal 84 (4): 66–72 – *The lichens of the Tomponsky ulus of Yakutia* (in Russian)

Poryadina LN (1999c) Lishainiki doliny reki Momy i khrebta Ilin-Tas. In: Timofeyev PA, Ivanova EI, Isaev AP (eds) Flora i rastitelnost Yakutii. FIPS, Moscow: 54–58 – *The lichens of the Moma River delta and Ilin-Tas Range* (in Russian)

Poryadina LN (2001a) Lishainiki basseynov rek Sinyaya i Buotama. In: Solomonov NG, Isaev AI, Ivanova EI (eds) Natsionalny prirodny park "Lenskie stolby": geologia, pochvy, rastitelnost, zhivotny mir, okhrana i ispolzovanie. Izd-vo YaGU, Yakutsk – *The lichens of the Sinayay and Buotama River basins* (in Russian)

Poryadina LN (2001b) Lishainiki zakaznika 'Suntar-Khayata'. Novosti sistematiki nizshikh rasteniy 34: 167–176 –*The lichens of the reserve 'Suntar-Khayata'* (in Russian)

Poryadina LN (2001c) Redkiye vidy lishainikov gor Vostochnoy Yakutii. In: Timofeev PA, Shmatkova LE (eds) Flora i fauna osobo okhranyaemykh prirodnykh territoriy Respublikanskoy sistemy Ytyk Kere Sirder. Kuduk, Yakutsk – *The rare lichen species of the mountains of East Yakutia* (in Russian)

Poryadina LN (2001d) Relict lichens (eastern Yakutia) Proceedings of international conference The role of permafrost ecosystems in global climate change. Yakutsk: 139–140.

Poryadina LN (2003a) Comparative analysis of lichen flora of Suntar-Khayata mountain ridge and Novosibirsk Islands. In: Proceedings of 2nd International Conference "The role of permafrost ecosystems in global climate change". Yakutsk: 259–264.

Poryadina LN (2003b) THE Lichens of the Lyapiske River Basin (Central Yakutia). In: Isaev AP (ed) Problemy botanicheskikh i lesovodstvennykh issledovaniy v RS (Ya) i Finlyandii. Proceedings of Sakha-Finnish Conference dedicated to 100th Anniversary of Cajander's expedition along the Lena. Izd-vo YaF SO RAN, Yakutsk: 45–48.

Poryadina LN (2003c) Lishainiki NPP "Ust-Viluyskiy" (Tsentralnaya Yakutia). Sibirskiy ekologichesky zhurnal 4: 513–516 – *The lichens of the National Natural Park "Ust-Viluyskiy" (Central Yakutia)* (in Russian)

Poryadina LN (2003d) The Nemoral Relicts of Lichen Flora of the Aldan-Indigirka Interfluve (Yakutia). In: Proceedigns of the International Conference "Phytogeography of North Asia". Vladivostok: 72.

Poryadina LN (2006) Spisok lishainikov reki Pilka i nizhnego techeniya reki Vitim. In: Volpert YaL (ed) Pochvy, rastitelny i zhivotny mir Yugo-Zapadnoy Yakutii. Nauka, Novosibirsk – *The list of lichens of the Pilka River and the Lower Vitim River* (in Russian)

Potemkin AD (2001) Evolyutsia, filogenia i klassifikatsiya semeystva *Scapaniaceae (Hepaticae)*. Thesis for a Doctor degree. Saint-Petersburg. – *Evolution, phylogenesis and classification of the Scapaniaceae (Hepaticae) family* (in Russian)

Povarnitsyn SA (1932a) Lesa doliny reki Aldan ot Tommota do ustya reki Notory. In: Lesnye resursy Yakutii. Izd-vo AN SSSR, Leningrad – *The forests of the Aldan River from Tommot town through the Notora River mouth* (in Russian)

Povarnitsyn SA (1932b) Lesa Leno-Aldanskogo vodorazdela po Sanyyakhtakhskoy trope. In: Lesnye resursy Yakutii. Izd-vo AN SSSR, Leningrad – *The forests of the Lena-Aldan watershed along the Sanyyakhtakh pathway* (in Russian)

Povarnitsyn SA (1933) Lesa doliny reki Aldan ot Tommota do ustya reki Uchura. In: Trudy Instituta po izucheniyu lesa. Izd-vo AN SSSR, Leningrad 1: 155–231 – *The forests of the Aldan River from Tommot town through the Uchur River mouth* (in Russian)

Poyarkova AI (1953) *Sorbocotoneaster pozdnjakovii Pojark.* – novy estestvenny mezhrodovoy gibrid. In: Botanicheskiye materialy gerbariya BIN AN SSSR. Leningrad 15: 92–108 – *Sorbocotoneaster pozdnjakovii Pojark. is a new natural bigener* (in Russian)

Pozdnyakov LK (1952) Drevovidnaya forma kedrovogo stlanika. Botanichesky zhurnal 37 (5): 688–691 – *Arborescent form of Pinus pumila* (in Russian)

Pozdnyakov LK (1955) Lesa basseina Olyokmy. In: Priroda. Izd-vo AN SSSR, Moscow – *The forests of the Olyokma River basin* (in Russian)

Pozdnyakov LK (1961a) Lesa verkhnego techeniya Yany. In: Trudy Instituta biologii YaF SO AN SSSR. Izd-vo AN SSSR, Moscow 7: 162–242 – *The forests of the Upper Yana River* (in Russian)

Pozdnyakov LK (1961b) Listvennichnye i sosnovye lesa verkhnego Aldana. Izd-vo AN SSSR, Moscow – *The Larix and Pinus sylvestris forests of the Upper Aldan River* (in Russian)

Pozdnyakov LK (ed) (1964) Lesa Yuzhnoy Yakutii. Nauka, Moscow – *The forests of South Yakutia* (in Russian)

Pozdnyakov LK (1969) Lesa Yakutskoy ASSR. In: Lesa SSSR: Lesa Urala, Sibiri i Dalnego Vostoka. Nauka, Moscow – *The forests of the Yakut ASSR* (in Russian)
Pozdnyakov LK, Gortinsky VI (1960) Lesa i lesnye resursy Yuzhnoy Yakutii. Izd-vo AN SSSR, Moscow – *The forests and forest resources of South Yakutia* (in Russian)
Prakhov NI (1957) Osnovnye elementy rastitelnosti Verkhoyanskogo khrebta. Trudy Instituta biologii YaF SO AN SSSR. Izd-vo AN SSSR, Moscow 3: 39–67 – *Basic elements of vegetation of the Verkhoyansk Range* (in Russian)
Pristyazhnyuk SA (1996) Zhiznennye formy lishainikov subarkticheskikh tundr poluostrova Yamal I: Sistema zhiznennykh form. Botanicheskiy zhurnal 81 (3): 34–42 – *The life forms of the lichens of the subarctic tundra of the Yamal Peninsula I: The system of life forms* (in Russian)
Pshennikova EV (1986) Vodorosli nekotorykh ozyor delty reki Kolymy. In: Biologicheskiye problemy Severa. Proceedings of the 11th Symposium. Issue 2. Yakutsk: 50–51 – *The algae of some lakes of the Kolyma River delta* (in Russian)
Pshennikova EV (1992) Pochvennye vodorosli alasa Allakh Leno-Amginskogo mezhdurechya. In: Botanicheskiye issledovaniya v kriolitozone. Izd-vo YaNTs SO RAN, Yakutsk: 12–22 – *The soil algae of the Allakh alas in the Lena-Amga Interfluve* (in Russian)
Pshennikova EV (1995) Pochvennye vodorosli alasov Leno-Amginskogo mezhdurechya (Yakutia, Rossiya). Algologiya 5 (3): 269–275 – *The soil algae of alases of the Lena-Amga Interfluve (Yakutia, Russia)* (in Russian)
Pshennikova EV, Ivanova AP, Salova TA, Kopyrina LI (2004) Sovremennoye sostoyaniye ozyor Tsentralnoy Yakutii i otsenka kachestva vody po biologicheskim pokazatelyam. Dalnevostochny meditsinsky zhurnal Appendix 1: 43–45 – *Present condition of the lakes of Central Yakutia and water quality estimation by biological indices* (in Russian)
Rabotnov TA (1933a) Prirodnye usloviya selskogo khozyaistva doliny reki Amgi (po materialam Yakutskoy ekspeditsii Narkomzema SSSR 1931 goda). In: Prirodnye resursy Yuzhnoy Yakutii v svyazi s sotsialisticheskoy rekonstruktsiey selskogo khozyaystva. Selkhozgiz, Moscow–Leningrad – *Natural conditions for agriculture in the Amga River valley (based on the data of the Yakut expedition of the People's Land Committee of the USSR of 1931)* (in Russian)
Rabotnov TA (1933b) Veynik Langsdorfa v Tsentralnoy Yakutii i ego kormovoe znacheniye. Botanichesky zhurnal 18 (5): 407–411 – *Rough blue-joint (Calamagrostis langsdorfii) in Central Yakutia and its fodder value* (in Russian)
Rabotnov TA (1934) *Nostoc commune* L. v napochvennom pokrove solontsevatykh pochv Yakutii. Sovetskaya botanika 2 – *Nostoc commune* L. in the ground cover of the solonetzic soils of Yakutia (in Russian)
Rabotnov TA (1935a) Landshafty peschanykh obrazovaniy v nizovyakh Vilyuya. Zemlevedeniye 37 (4): 321–338 – *Landscapes of sand formations in the lower reaches of the Viluy River* (in Russian)
Rabotnov TA (1935b) Za uluchsheniye lugov Tsentralnoy Yakutii. Sovetskoye kraevedeniye 8: 42–45 – *For improvement of meadows of Central Yakutia* (in Russian)
Rabotnov TA (1936a) Ekologocheskie nablyudeniya nad lishainikami v Yuzhnoy Yakutii. Sovetskaya botanika 6: 149–153 – *Ecological observations of the lichens in South Yakutia* (in Russian)
Rabotnov TA (1936b) Lesa iz *Betula ermanii* v Yuzhnoy Yakutii. Priroda 4: 96–97 – *Forests of Betula ermanii in South Yakutia* (in Russian)
Rabotnov TA (1939) O bolotakh Tsentralnoy Yakutii. Izvestiya geograficheskogo obschestva 71 (4): 5447–550 – *On marshlands of Central Yakutia* (in Russian)
Rabotnov TA (1945) O stepyakh Tsentralnoy Yakutii. Priroda 2: 65–66 – *On the steppes of Central Yakutia* (in Russian)
Rebristaya OV (1977) Flora vostoka Bolshezemelskoy tundry. Nauka, Leningrad – *The flora of the east of the Bolshaya Zemlya tundra* (in Russian)
Remigailo PA (1983) Algoflora vodoyomov proektiruemogo Ust-Lenskogo zapovednika. In: "Biologo-ekonomicheskiye issledovaniya mestnoy rastitelnosti i introdutsiruemoy flory Yakutii". Bulletin of scientific-technical information. Izd-vo YaF SO AN SSSR, Yakutsk: 5–6 – *The algoflora of the water bodies of the projected "Ust-Lensky" Sanctuary* (in Russian)

Remigailo PA, Gabyshev VA (1999) Osobennosti razvitiya fitoplanktona sredney Leny (Rossiya). Algologiya 9 (2): 122–123 – *Characteristic features of phytoplankton development in the Middle Lena River* (in Russian)

Remigailo PA, Gabyshev VA (2001) Taksonomicheskaya struktura i vidovoye raznoobraziye fitoplanktona verkhovyev reki Aldan. Sibirskiy ekologicheskiy zhurnal 8 (4): 385–387 – *The taxonomical structure and species diversity of phytoplankton of the Upper Aldan River* (in Russian)

Rykova YuV (1972) Ekologiya, geographicheskoye rasprostraneniye i fitomassa osnovnykh vidov kustistykh lishainikov Nizhnekolymskogo raiona. In: Proceedings of the scientific session dedicated to 50th anniversary of education of the USSR and YaASSR. Yakutsk: 39–40 – *Ecology, geographical distribution and phytomass of basic species of fruticose lichens of the Nizhnekoplymsky Region* (in Russian)

Rykova YuV (1976) Raspredeleniye kustistykh lishainikov v Vostochnom Verkhoyanye. In: Abstracts of 7th Symposium “Biologicheskiye problemy Severa” Series Botanika. Petrozavodsk: 188–190 – *Distribution of fruticose lichens in the East Verkhoyanye* (in Russian)

Rykova YuV (1978) Opredelitel listovatykh i kustistykh lishainikov Yakutii. Izd-vo YaGU, Yakutsk – *Identifier of foliose and fruticose lichens of Yakutia* (in Russian)

Rykova YuV (1980) Rasprostraneniye i zapasy kustistykh lishainikov na severo-vostoke Yakutii. In: Rastitelnost i pochvy subarkticheskoy tundry. Novosibirsk – *Distribution and reserves of fruticose lichens in the north–east of Yakutia* (in Russian)

Rykova YuV (1987) Likhenoflora khvoynykh lesov Tsentralnoy Yakutii. In: Ekologo-biologicheskiye osnovy lesovodstvennykh mer v Yakutii. Collection of articles: 67–74 – *The lichen flora of coniferous forests of Central Yakutia* (in Russian)

Rykova YuV (2004) K flore lishainikov listvennichnykh lesov Kolymskoy nizmennosti. In: Problemy izucheniya rastitelnogo pokrova Yakutii. Yakutsk: 33–35 – *On the lichen flora of larch forests of the Kolyma lowland* (in Russian)

Rykova YuV, Chechurova NV (2001) Flora epifitnykh lishainikov pamyatnika prirody ‘Ulakhan-Taryn’. In: Timofeev PA, Shmatkova LE (eds) Flora i fauna osobo okhranyaemykh prirodnykh territoriy Respublikanskoy sistemy Ytyk Kere Sirder. Kuduk, Yakutsk – *The flora of epiphytic lichens of the natural monument “Ulakhan-Taryn”* (in Russian)

Samarsky MA, Sokolova MV, Zhurbenko MP, Afonina OM (1997) O flore i rastitelnosti ostrova Zhokhova (Novosibirskie ostrova). Botanicheskiy zhurnal 82 (4): 62–70 – *On the flora and vegetation of the Zhokhov Island (Novosibirskie Isls)* (in Russian)

Scherbakov IP (1964) Tipy lesa Yuzhnoy Yakutii. In: Pozdnyakov LK (ed) Lesa Yuzhnoy Yakutii. Nauka, Moscow – *The forest types of South Yakutia* (in Russian)

Scherbakov IP (1975) Lesnoy pokrov Severo-Vostoka SSSR. Nauka, Novosibirsk – *The forest cover of the North–East USSR* (in Russian)

Scherbakov IP (1992) Lesa verkhnego i srednego techeniya reki Viluy. In: Botanicheskiye issledovaniya v kriolotozone. Collection of articles. Izd-vo YaNTs SO RAN, Yakutsk – *The forests of the Upper and Middle Viluy River* (in Russian)

Schuster RM (1966) The Hepaticae and Anthocerotae of North America east of the hundredth meridian 1. Columbia University Press, New York-London: 643–691

Schuster RM (1969) The Hepaticae and Anthocerotae of North America east of the hundredth meridian 2. Columbia University Press, New York-London

Schuster RM (1974) The Hepaticae and Anthocerotae of North America east of the hundredth meridian 3. Columbia University Press, New York-London

Schuster RM (1992a) The Hepaticae and Anthocerotae of North America east of the hundredth meridian 5. Field Museum of Natural History, Chicago

Schuster RM (1992b) The Hepaticae and Anthocerotae of North America east of the hundredth meridian 6. Field Museum of Natural History, Chicago

Sedelnikova NV (1985) Likhenoflora nagorya Sangilen. Nauka, Novosibirsk – *The lichen flora of the Sangilen table-land* (in Russian)

Sedelnikova NV (1990) Lishainiki Altaya i Kuznetskogo nagorya: Konspekt flory. Nauka, Novosibirsk – *The lichens of Altai and Kuznetsk table-land* (in Russian)

Sedelnikova NV (1993) Lishainiki Salaira. In: Flora Salairskogo kryazha. Izd-vo TsSBS SO RAN, Novosibirsk: 33–78 – *The lichens of Salair* (in Russian)

Sedelnikova NV, Laschinskiy NN (1990) Lishainiki karbonatnykh mestoobitaniy Salaira. Izvestiya SO AN SSSR Series Biologicheskiye nauki 2: 51–58 – *The lichens of calcareous habitats of Salair* (in Russian)

Sekretaryova NA (2004) Sosudistye rasteniya Rossiyskoy Arktiki i sopredelnykh territoriy. Tovarischestvo nauchnykh izdaniy KMK, Moscow – *Vascular plants of the Russian Arctic and neighbouring territories* (in Russian)

Serkina PA (1969) Plankton i bentos reki Leny i eyo primorskikh uchastkov. In: Trudy Yakutskogo otdeleniya SibNII rybnogo khozyaistva Issue 3: 118–196 – *Plankton and benthos of the Lena River and its maritime sectors* (in Russian)

Sheath RG, Munawar M, Hellebust JA (1975) Fluctuations of phytoplankton biomass and its summer composition in a subarctic lake during summer. Can J Bot 53 (19): 2240–2246

Sheludyakova VA (1938) Rastitelnost basseyna reki Indigirki. Sovetskaya botanika 4–5: 43–79 – *Vegetation of the Indigirka River basin* (in Russian)

Sheludyakova VA (1943) Chosenia v Yakutskoy ASSR. Botanichesky zhurnal 28 (1): 30–34 – *Chosenia in the Yakutian ASSR* (in Russian)

Sheludyakova VA (1948a) Rastitelnost severo-vostoka Yakutii. In: Doklady na pervoy nauchnoy sessii Yakutskoy bazy AN SSSR. Gosizdat, Yakutsk – *Vegetation of the North–eastern Yakutia* (in Russian)

Sheludyakova VA (1948b) Rastitelnost Verkhoyanskogo raiona Yakutskoy ASSR (botaniko-geografichesky ocherk). Gosizdat, Yakutsk – *Vegetation of the Verkhoyansk Region of the Yakutian ASSR (botanical-geographical essay)* (in Russian)

Sheludyakova VA (1957a) Kratky ocherk lugov doliny reki Leny. In: Trudy Instituta Biologii YaF SO AN SSSR. Izd-vo AN SSSR, Moscow 3: 127–138 – *Brief review of meadows of the Lena River valleys* (in Russian)

Sheludyakova VA (1957b) Stepnaya rastitelnost Yakutskogo Zapolyarya. In: Trudy Instituta Biologii YaF SO AN SSSR. Izd-vo AN SSSR, Moscow 3: 139–156 – *The steppe vegetation of the Yakut Arctic* (in Russian)

Sheludyakova VA (1958) Izucheniye prirodnykh rastitelnykh resursov Yakutskoy ASSR. In: Sostoyaniye i perspektivy izucheniya rastitelnykh resursov SSSR. Moscow-Lennigrad – *Study of natural plant resources of the Yakut ASSR* (in Russian)

Sheludyakova VA (1959) Luga i pastbischa Tsentralnoy i Yugo-Zapadnoy Yakutii. Knizhnoye izd-vo, Yakutsk – *Meadows and pastures of Central and South–West Yakutia* (in Russian)

Sheludyakova VA, Skryabin SZ (1969) Charany – landshafty lesostepnogo tipa v Tsentralnoy Yakutii. Botanichesky zhurnal 54 (4): 539–556 – *Charans, the forest-steppe landscapes in Central Yakutia* (in Russian)

Sheludyakova VA, Karavaev MN, Petrov AM (1954) Luga i pastbischa Tsentralnoy Yakutii. In: Materialy o prirodnykh usloviyakh i selskom khozyaistve Tsentralnoy Yakutii. Izd-vo AN SSSR, Moscow – *Meadows and pastures of Central Yakutia* (in Russian)

Shirshov PP (1935) Ekologo-geographicheskiy ocherk presnovodnykh vodorosley Novoy Zemli i Zemli Frantsa-Iosifa. Proceedings of the Arctic Institute14: 73–162 – *Ecological-geographical essay on the freshwater algae of the Novaya Zemlya and Franz Joseph Land* (in Russian)

Shlyakov RN (1961) Flora listostebelnykh mkhov Khibinskikh gor. Murmansk – *Leafy moss flora of the Khibin Mnts.* (in Russian)

Shlyakov RN (1979a) Pechenochnye mkhi Severa SSSR. Pechenochniki: Gerbertovye-Geokaliksovye Issue 2. Nauka, Leningrad – *The liverworts of the northern USSR: Herbertaceae – Geocalicaceae* (in Russian)

Shlyakov RN (1979b) Pechenochnye mkhi Severa SSSR. Pechenochniki: Lofozievye-Mezoptikhievye Issue 3. Nauka, Leningrad – *The liverworts of the northern USSR: Lophoziaceae – Mesoptychiaceae* (in Russian)

Shlyakov RN (1981) Pechenochnye mkhi Severa SSSR. Pechenochniki: Yungermannievye – Skapanievye Issue 4. Nauka, Leningrad – *The liverworts of the northern USSR: Jungermanniaceae – Scapaniaceae* (in Russian)

Shlyakov RN (1982) Pechenochnye mkhi Severa SSSR. Pechenochniki: Lofokoleyevye – Richchievye Issue 5. Nauka, Leningrad – *The liverworts of the northern USSR: Lophocoleaceae – Ricciaceae* (in Russian)

Shtina EA, Antipina GS, Kozlovskaya LS (1981) Algoflora bolot Karelii i eyo dinamika. Nauka, Leningrad – *The algoflora of the bogs of Karelia and its dynamics* (in Russian)

Skryabin SZ (1968) Stepnaya rastitelnost v srednem techenii Indigirki. Abstract of Thesis for a Candidate degree. Yakutsk – *The steppe vegetation in the Middle Indigirka River* (in Russian)

Skryabin SZ (1976) Fragmenty gornoy petrofitnoy stepi v nizovye reki Tokko (Yuzhnaya Yakutia). Botanichesky zhurnal 61 (12): 1748–1751 – *The fragments of a mountainous petrophytic steppe in the lower reaches of the Tokko River (South Yakutia)* (in Russian)

Skryabin SZ, Karavaev MN (1991) Zelyony pokrov Yakutii. Knizhnoye izd-vo, Yakutsk – *The green cover of Yakutia* (in Russian)

Skvortsov BV (1917) Materialy po flore vodorosley Aziatskoy Rossii. Zhurnal Russkogo Botanicheskogo obschestva 11: 10–12 – *Materials on the algae flora of Asian Russia* (in Russian)

Skye E (1968) Lichens and Air Pollution. Almqvist & Wiksells Boktryckeri AB, Uppsala

Sochava VB (1933) Tundry basseyna reki Anabary. Izvestiya gosudarstvennogo geographicheskogo obschestva 65 (4) – *The tundra of the Anabar River basin* (in Russian)

Sochava VB (1934) Rastitelnye assotsiatsii Anabarskoy tundry. Botanichesky zhurnal 19 (3) – *Plant communities of the Anabar tundra* (in Russian)

Sochava VB (1957) Taiga na severo-vostoke Sredne-Sibirskogo ploskogorya. Botanicheskiy zhurnal 42 (9): 1408–1415 – *Taiga in the northeastern Middle Siberian Plateau* (in Russian)

Sofronov RR (2003) Floristicheskiye nakhodki redkikh i novykh vidov dlya nizhnego techeniya reki Vitim. In: Isaev AP (ed) Problemy botanicheskikh i lesovodstvennykh issledovaniy v RS (Ya) i Finlyandii. Proceedings of Sakha-Finnish Conference dedicated to 100th Anniversary of Cajander's expedition along the Lena. Izd-vo YaF SO RAN, Yakutsk: 75–77 – *Floristic finds of rare and new speices in the lower reaches of the Vitim River* (in Russian)

Sofronova EV (2003) Pechenochnye mkhi yakutskoi chasti Vostochnogo Verkhoyanya. Thesis for a Candidate Degree. Saint-Petersburg. – *The liverworts of the yakutian part of the Eastern Verkhoyansk region* (in Russian)

Sofronova EV (2005) Pechenochnye mkhi. In: Danilova NS (ed) Raznoobrazie rastitelnogo mira Yakutii. Izd-vo SO RAN, Novosibirsk. – *The liverworts* (in Russian)

Sofronova EV (2008a) Pechenochniki prirodnogo parka 'Lenskie Stolby' (sredneye techenie reki Leny, Tsentralnaya Yakutia). Novosti sistematiki nizshikh rasteniy 41: (in press) – *The liverworts of the Natural Park 'Lena Pillars' (the Middle Lena River, Central Yakutia)* – (in Russian)

Sofronova EV (2008b) Redkiye vidy pechenochnikov vo flore Yakutii. In: Geobotanicheskie i resursovedcheskie issledovaniya v Arktike. Proceedings of Conference dedicated to 100th anniversary of professor VN Andreev's birthday. Izd-vo YaF SO RAN, Yakutsk (in press) – *Rare liverwort species in the flora of Yakutia* (in Russian)

Sokolov FV (1923) Ocherk yugo-vostochnoy chasti Yakutiskoy oblasti. Izvestiya RGO. 1919–1923. Saint-Petersburg 55 (1): 181–236 – *Essay on the south–eastern part of Yakutian Region* (in Russian)

Solomonov NG, Isaev AI, Ivanova EI (eds) (2001) Natsionalny prirodny park "Lenskie stolby": geologia, pochvy, rastitelnost, zhivotny mir, okhrana i ispolzovanie.Collection of articles. Izd-vo YaGU, Yakutsk – N*ational Natural Park "Lena Pillars": Geology, soils, vegetation, fauna, conservation and utilization* (in Russian)

Solovyov PA (1959) Kriolotozona severnoy chasti Leno-Amginskogo mezhdurechya. Izd-vo AN SSSR, Moscow – *The cryolithozone of the northern part of the Lena-Amga Interfluve* (in Russian)

Sosina NK, Isaev AP (2003) Novye nakhodki Redowskia sophiifolia (Brassicaceae) v Yakutii. In: Isaev AP (ed) Problemy botanicheskikh i lesovodstvennykh issledovaniy v RS (Ya) i Finlyandii. Proceedings of Sakha-Finnish Conference dedicated to 100th Anniversary of Cajander's expedition along the Lena. Izd-vo YaF SO RAN, Yakutsk: 81–83 – *New finds of Redowskia sophiifolia (Brassicaceae) in Yakutia* (in Russian)

Sosina NK, Sofronov RR (2004) K izucheniyu flory vysshikh sosudistykh rasteniy basseyna reki Muny (nizhnee techeniye reki Leny). In: Ekologicheskaya bezopasnost pri razrabotke rossypnykh mestorozhdeniy almazov. Yakutsk – *On the study of the higher vascular plants flora of the Muna River basin (lower reaches of the Lena River)* (in Russian)

Sosina NK, Zakharova VI (2004) K izucheniyu rastitelnosti basseyna reki Muny (nizhnee techeniye reki Leny). In: Ekologicheskaya bezopasnost pri razrabotke rossypnykh mestorozhdeniy almazov. Yakutsk – *On the study of vegetation of the Muna River basin (lower reaches of the Lena River)* (in Russian)

Stanley DW (1976) Production of epipelic algae in tundra ponds and a lake near Barrow, Alaska. Ecology 57: 1015–1024.

Stepanova NA (1986) Konspekt flory mkhov tundry Yakutii. Izd-vo YaF SO AN SSSR, Yakutsk – *Synopsis of the moss flora of Yakutian tundra* (in Russian)

Sumina OI (1986) Dopolneniye k flore i rastitelnosti ostrova Kotelnogo i Zemli Bunge (Novosibirskiye ostrova). Botanichesky zhurnal 71 (7): 903–911 – *Supplement to the flora and vegetation of the Kotelny Island and Zemlya Bunge (Novosibirskie Islands)* (in Russian)

Tikhomirov BA, Shtepa VS (1956) K kharakteristike lesnykh forpostov v nizovyakh reki Leny. Botanichesky zhurnal 41 (8): 1107–1122 – *On characteristics of the forest outposts in the lower reaches of the Lena River* (in Russian)

Tikhomirov BA, Petrovsky VV, Yurtsev BA (1966) Flora okrestnostey bukhty Tiksi (Arkticheskaya Yakutia). In: Rasteniya Severa Sibiri i Dalnego Vostoka. Nauka, Moscow–Leningrad – *Flora of the Tiksi Bay vicinity (the Arctic Yakuia)* (in Russian)

Timofeyev PA (1992) Bereznyaki doliny sredney Leny. In: Botanicheskiye issledovaniya v kriolotozone. Collection of articles. Izd-vo YaNTs SO RAN, Yakutsk – *The birch forests of the Middle Lena River valley* (in Russian)

Timofeyev PA (2003) Lesa Yakutii: sostav, resursy, ispolzovaniye i okhrana. Izd-vo SO RAN, Novosibirsk – *The forests of Yakutia: Composition, resources, utilization and protection* (in Russian)

Timofeyev PA, Isaev AP, Scherbakov IP et al. (1994) Lesa srednetayozhnoy podzony Yakutii. Izd-vo YaNTs SO RAN, Yakutsk – *The forests of the middle taiga subzone of Yakutia* (in Russian)

Tolmachev AI (1960) Arkticheskaya flora SSSR. Vypusk 1:*Polypodiaceae-Butomaceae*. Nauka, Leningrad – *The Arctic flora of the USSR. Issue 1: Polypodiaceae-Butomaceae* (in Russian)

Tolmachev AI (1963) Arkticheskaya flora SSSR. Vypusk 4: *Lemnaceae-Orchidaceae*. Nauka, Leningrad – *The Arctic flora of the USSR. Issue 4: Lemnaceae-Orchidaceae* (in Russian)

Tolmachev AI (ed) (1964) Arkticheskaya flora SSSR. Vypusk 2: *Graminae*. Nauka, Leningrad – *The Arctic flora of the USSR. Issue 2: Graminae* (in Russian)

Tolmachev AI (ed) (1966a) Arkticheskaya flora SSSR. Vypusk 3: *Cyperaceae*. Nauka, Leningrad – *The Arctic flora of the USSR. Issue 3: Cyperaceae* (in Russian)

Tolmachev AI (ed) (1966b) Arkticheskaya flora SSSR. Vypusk 5: *Salicaceae-Portulacaceae*. Nauka, Leningrad – *The Arctic flora of the USSR. Issue 5: Salicaceae-Portulacaceae* (in Russian)

Tolmachev AI (ed) (1971) Arkticheskaya flora SSSR. Vypusk 6: *Caryophyllaceae-Ranunculaceae*. Nauka, Leningrad – *The Arctic flora of the USSR. Issue 6: Caryophyllaceae-Ranunculaceae* (in Russian)

Tolmachev AI (ed) (1974) Opredelitel vysshikh rasteniy Yakutii. Nauka, Novosibirsk – *Identifier of higher plants of Yakutia* (in Russian)

Tolmachev AI (ed) (1975) Arkticheskaya flora SSSR. Vypusk 7: *Papaveraceae-Cruciferae*. Nauka, Leningrad – *The Arctic flora of the USSR. Issue 7: Papaveraceae-Cruciferae* (in Russian)

Tolmachev AI, Yurtsev BA (eds) (1980) Arkticheskaya flora SSSR. Vypusk 8 (1): *Geraniaceae-Scrophullariaceae*. Nauka, Leningrad – *The Arctic flora of the USSR. Issue 8 (1): Geraniaceae-Scrophullariaceae* (in Russian)

Tolmachev AI, Yurtsev BA (eds) (1983) Arkticheskaya flora SSSR. Vypusk 8 (2): *Orobanchaceae-Plantaginaceae*. Nauka, Leningrad – *The Arctic flora of the USSR. Issue 8 (2): Orobanchaceae-Plantaginaceae* (in Russian)
Trufanova ER (1972) Rastitelnost ozyor doliny srednego techeniya reki Kolymy. In: Pochvennye i botanicheskiye issledovaniya v Yakutii. Knizhnoye izd-vo, Yakutsk – *Vegetation of lakes of the Kolyma River valley* (in Russian)
Trutnev YuP et al. (eds) (2008) Krasnaya kniga Rossiyskoy Federatsii (rasteniya i griby). Tovarischestvo nauchnykh izdaniy KMK, Moscow – *The Red Data Book of the Russian Federation (Plants and fungi)* (in Russian)
Tsvelev NN, Yurtsev BA (1984) Novye vidy oduvanchikov *(Taraxacum Wigg., Asteraceae)* flory SSSR. Novosti sistematiki nizshikh rasteniy 21: 187–202 – *New species of Taraxacum Wigg. (Asteraceae) in the flora of the USSR* (in Russian)
Tyrtikov AP (1955) Rastitelnost nizovyev reki Yany. Byulleten MOIP (Biologiya) 60 (5): 135–146 – *Vegetation of the lower reaches of the Yana River* (in Russian)
Tyrtikov AP (1958) Nekotorye svedeniya o rastitelnosti nizovyev reki Indigirki. Byulleten MOIP (Biologiya) 63 (1) – *Some data on vegetation of the lower reaches of the Indigirka River* (in Russian)
Tyulina LN (1956) Na ozere Toko i severnom sklone Stanovogo Khrebta (kratky geobotanichesky ocherk). In: Akademiku VN Sukachevu k 75-letiyu so dnya rozhdeniya. Izd-vo AN SSSR, Moscow–Leningrad – *On Toko lake and north-facing slope of the Stanovoy Range (brief geobotanical review)* (in Russian)
Tyulina LN (1957) Ocherk lesnoy rastitelnosty verkhnego techeniya reki Aldan. In: Trudy Instituta Biologii YaF SO AN SSSR. Izd-vo AN SSSR, Moscow 3: 83–138 – *Essay on the forest vegetation of the Upper Aldan River* (in Russian)
Tyulina LN (1959) Lesnaya rastitelnost srednego i nizhnego techeniya reki Yudomy i nizovyev Mai. Izd-vo AN SSSR, Moscow – *The forest vegetation of the Middle and Lower Yudoma River and lower reaches of the Maya River* (in Russian)
Tyulina LN (1962) Lesnaya rastitelnost sredney i nizhney chasti basseina reki Uchur. Izd-vo AN SSSR, Moscow–Leningrad – *The forest vegetation of the middle and lower parts of the Uchur River basin* (in Russian)
Usanova VM, Perfilyeva VI (1966) Opredelitel kormovykh rasteniy Yakutii (zlakovye, osokovye, bobovye). Nauka, Moscow-Lennigrad – *Identifier of fodder plants of Yakutia (grasses, sedges, legumes)* (in Russian)
Utkin AI (1959) Kratkaya kharakteristika listvennichnykh lesov Tsentralnoy Yakutii. Izvestiya Sibirskogo otdeleniya AN SSSR 3: 59–97 – *Brief description of the larch forests of Central Yakutia* (in Russian)
Utkin AI (1961) Kedrovy stlanik na severo-zapadnoy okraine areala i istoriya ego rasprostraneniya. In: Trudy Instituta lesa i drevesiny: Voprosy lesovodstva i lesovedeniya 50: 104–119 – *Pinus pumila at the northwestern border of its distribution area and the history of its distribution* (in Russian)
Utkin AI (1965) Lesa Tsentralnoy Yakutii. Nauka, Moscow – *The forests of Central Yakutia* (in Russian)
Vasilyeva II (1965) Sezonnye izmeneniya sostava sine-zelyonykh vodorosley v prigorodnykh ozyorakh Yakutska. Ivestiya SO AN SSSR. Seria biologo-meditsinskikh nauk 1 (4): 139–141 – *Seasonal dynamics of the blue-green algae composition in the suburban lakes of Yakutsk* (in Russian)
Vasilyeva II (1966) Fotosintez i vertikalnoye raspredelenie fitoplanktona nekotorykh ozer okrestnostey Yakutska. Izvestia SO AN SSSR. Seria biologo-meditsinskikh nauk 4: 28–32 – *Photosynthesis and vertical distribution of phytoplankton of some lakes in the vicinity of Yakutsk* (in Russian)
Vasilyeva II (1968) Sostav i sezonnaya dinamika fitoplanktona ozyor okrestnostey Yakutska. Abstract of Thesis for a Candidate Degree. Novosibirsk – *Composition and seasonal dynamics of phytoplankton of the lakes in the vicinity of Yakutsk* (in Russian)

Vasilyeva II (1983) Vodorosli nekotorykh ozer Yuzhnoy Yakutii. In: Biologo-ekologicheskiye issledovaniya mestnoy rastitelnosti i introdutsiruemoy flory Yakutii. Bulletin of scientific-technical information. Izd-vo YaF SO AN SSSR, Yakutsk: 30–35 – *Algae of some lakes of South Yakutia* (in Russian)

Vasilyeva II (1987) Presnovodnye evglenovye i zheltozelyonye vodorosli vodoyemov Yakutii. Nauka, Leningrad – *Freshwater euglenide and yellow-green algae of Yakutian water bodies* (in Russian)

Vasilyeva II (1989a) Analiz vidovogo sostava i dinamiki razvitiya vodorosley vodoyemov Yakutii. Izd-vo YaF SO AN SSSR, Yakutsk – *Analysis of species composition and growth dynamics of algae of Yakutian water bodies* (in Russian)

Vasilyeva II (1989b) Vodorosli vodoyomov krioltiozony SSSR: Sistematichesky sostav, ekologiya, rasprostraneniye (na primere Yakutii). Thesis for a Doctor Degree. Kishinev. – *Algae of the water bodies of the cryolithozone of the USSR: Taxonomical composition, ecology, distribution (case study of Yakutia)* (in Russian)

Vasilyeva II, Pshennikova EV (1992) Sezonnye izmeneniya chislennosti i biomassy vodorosley nekotorykh alasnykh ozyor Leno-Amginskogo mezhdurechya. In: Ekologiya rastitelnogo mira Yakutii. Conference proceedings. Yakutsk: 9–10 – *Seasonal changes in algae numbers and biomass of some alas lakes of the Lena-Amga Interfluve* (in Russian)

Vasilyeva II, Pshennikova EV (1995) Sovremennoye sostoyaniye diatomovykh vodorosley basseina reki Kolymy v svyazi so stroitelstvom Ust-Srednekanskoy GES. Proceedings of 6th International school on diatoms. Minsk: 14 – *Modern condition of diatoms of the Kolyma River basin in relation to construction of the Ust-Srednekansk Power Station* (in Russian)

Vasilyeva II, Pshennikova EV (1996) Vodorosli reki Kolymy i vodoyemov eyo basseyna (Rossiya). Algologia 6 (1): 35–41 – *Algae of the Kolyma River and the water bodies of its basin* (in Russian)

Vasilyeva II, Remigailo PA (1980) Flora vodoyomov subarkticheskoy tundry v raione statsionara "Pokhodsk". In: Rastitelnost i pochvy subarkticheskoy tundry. Nauka, Novosibirsk – *Flora of water bodies of the subarctic tundra near the Research station "Pokhodsk"* (in Russian)

Vasilyeva II, Remigailo PA (1982) Vodorosli Viluyskogo vodokhranilischa. Izd-vo YaF SO AN SSSR, Yakutsk – *The algae of the Viluy Reservoir* (in Russian)

Vasilyeva II, Remigailo PA (1986) Flora presnovodnykh vodorosley Lenskoy delty. Yakutsk. Manuscript deposited in All-Russian Institute of Scientific and Technical Information No2414-B86 – *Flora of freshwater algae of the Lena Delta* – (in Russian)

Vasilyeva II, Rizvanova RG (1976) Vodorosli delty reki Leny In: "Biologicheskie problemy Severa". Bulletin of scientific-technical information. Izd-vo YaF SO AN SSSR, Yakutsk: 8–9 – *The algae of the Lena River Delta* (in Russian)

Vasilyeva II, Remigailo PA, Kuzina TS (1981) Sanitarno-biologicheskaya kharakteristika vody Viluyskogo vodokhranilischa. In: Biologoya gidrobiontov v vodoyomakh Yakutii s razlichnym gidrobiologicheskim rezhimom. Collection of articles. Yakutsk: 14–22 – *Sanitary-biological characteristics of water of the Viluy Reservoir* (in Russian)

Vasilyeva II, Remigailo PA, Sokolova LS (1984) Sanitarno-biologicheskaya kharakteristika otdelnykh uchastkov rek Leny, Olyokmy, Charuodokana. In: "Biologicheskie problemy Severa". Bulletin of scientific-technical information. Izd-vo YaF SO AN SSSR, Yakutsk: 10–12 – *Sanitary-biological characteristics of some sections of the Lena, Olyokma, and Charuodokan Rivers* (in Russian)

Vasilyeva II, Smirnov NN (1995) The biota of Lake Abalakh (Yakutia, eastern Siberia, Russian Federation). Int J Salt Res 4: 57–63.

Vasilyeva-Kralina II (1999) Burye i krasnye vodorosli Yakutii. In: Timofeyev PA, Ivanova EI, Isaev AP (eds) Flora i rastitelnost Yakutii. Collection of articles. FIPS, Moscow: 35–36 – *Brown and red algae of Yakutia* (in Russian)

Vasilyeva-Kralina II, Gabyshev VA (1996) K izucheniyu *Bacillariophyta* gornykh vodoyomov Verkhoyanya (Yakutiya). Algologiya 6 (4): 394–400 – *On the study of Bacillariophyta of mountainous water bodies of the Verkhoyansk region (Yakutia)* (in Russian)

Vasilyeva-Kralina II, Pshennikova EV (1998) Vodorosli vodoyemov basseyna nizhnego techeniya reki Uchur (Yakutiya). Sibirsky ekologichesky zhurnal 5 (2): 151–160 – *The algae of the Lower Uchur River basin* (in Russian)

Vasilyeva-Kralina II, Pshennikova EV (2000) Vodorosli. In: Savvinov DD, Savvinov GN, Prokopyev NP et al. (eds) Prikladnaya ekologiya Amgi. Izd-voYaF SO RAN, Yakutsk: 80–84 – *The algae* (in Russian)

Vasilyeva-Kralina II, Ivanova AP, Pshennikova EV (1997) Sostav i dinamika razvitiya vodorosley ozyyor Yakutska i ego okrestnostey (sredneye techeniye Leny). Algologiya 7 (1): 30–34 – *Composition and development dynamics of lake algae in Yakutsk and its vicinity (the Middle Lena)* (in Russian)

Vasilyeva-Kralina II, Gabyshev VA, Pshennikova EV, Ivanova AP (2004) Vodorosli gornykh vodoyemov Verkhoyanya. Biologiya vnutrennikh vod 3: 3–15 – *The algae of mountainous water bodies of Verkhoyansk region* (in Russian)

Vodopyanova NS (1980) Novye dannye o flore Olenyokskogo raiona Yakutii. Botanichesky zhurnal 65 (8): 1185–1189 – *New data on the flora of the Olenyok Region of Yakutia* (in Russian)

Volotovsky KA (1987) Redkiye vidy sosudistykh rasteniy Ust-Viluyskogo zakaznika i ikh okhrana. In: Izucheniye, okhrana i ratsionlanoye ispolzovaniye prirodnykh resursov. Collection of articles. Ufa: 63 – *The rare vascular plant species of the Ust-Viluysky Reserve and their protection* (in Russian)

Volotovsky KA (1992) Tipy poyasnosti rastitelnogo pokrova na Aldanskom nagorye. In: Proceedings of the All-Union meeting on the study of flora and vegetation. Novosibirsk: 105 – *Altitudinal zonation of vegetation cover of the Aldan upland region* (in Russian)

Volotovsky KA (1994) Novy vid roda *Dendranthema (DC.) Des Moul. (Asteraceae)* iz Yuzhnoy Yakutii. Botanichesky zhurnal 79 (8): 102–105 – *New species of genus Dendranthema (DC.) Des Moul. (Asteraceae) from South Yakutia* (in Russian)

Volotovsky KA (1996) Novy vid roda *Anoplocaryum* Ledeb. (*Boraginaceae*) iz Yuzhnoy Yakutii. Novosti sistematiki vysshikh rasteniy 79(8): 102–105 – *New species of the Anoplocaryum Ledeb. genus (Boraginaceae) from South Yakutia* (in Russian)

Volotovsky KA, Chevychelov AP (1991) Kamennoberyozovye lesa Yakutii. Botanichesky zhurnal 76 (6): 39–47 – *The forests of Betula ermanii in Yakutia* (in Russian)

Volotovsky KA, Kuznetsova LV (1998) Novye dannye o estestvennom gibride *Sorbocotoneaster pozdnjakovii (*Rosaceae*)*. Botanichesky zurnal 83 (1): 94–103 – *New data on the natural hybrid Sorbocotoneaster pozdnjakovii (Rosaceae)* (in Russian)

Volpert YaL (ed) (2006) Pochvy, rastitelny i zhivotny mir Yugo-Zapadnoy Yakutii. Nauka, Novosibirsk – *Soils, plant and animal world of South–West Yakutia* (in Russian)

Yarovoy MI (1939) Rastitelnost basseyna reki Yany i Verkhoyanskogo khrebta. Sovetskaya botanika 1: 21–40 – *Vegetation of the Yana River basin and the Verkhoyansk Range* (in Russian)

Yurtsev BA (1959) Vysokogornaya flora gory Sokuydakh i eyo mesto v ryadu gornykh flor arkticheskoy Yakutii. Botanichesky zhurnal 44 (8): 1171–1177 – *Alpine flora of the Sokuydakh Mountian and its position in the series of montane floras of the Arctic Yakutia* (in Russian)

Yurtsev BA (1961) K kharakteristike podzony severo-tayozhnykh listvennichnikov v zapadnoy chasti basseina reki Yany. In: Alexandrova VD, Kuvaev VB, Tikhomirov BA et al. (eds) Materialy po rastitelnosti Yakutii. Izd-vo LTA, Leningrad – On *characteristics of subzone of the northern taiga larch forests in the western part of the Yana River basin* (in Russian)

Yurtsev BA (1962) Botaniko-geograficheskiye nablyudeniya u severnogo predela rasprostraneniya listvennitsy na reke Olenyok. In: Problema botaniki. Izd-vo AN SSSR, Moscow-Leninrad – *Botanical-geographical observations at the northern limits of larch distribution area on the Olenyok River* (in Russian)

Yurtsev BA (1964) Botaniko-geograficheskiy ocherk Indigirskogo sklona gornogo uzla Suntar-Khayata. In: Rastitelnost zarubezhnykh stran. Nauka, Moscow–Leningrad – *Botanical-geographical essay on the Indigirka slope of the Suntar-Khayata Ridge* (in Russian)

Yurtsev BA (1968) Flora Suntar-Khayata: Problemy istorii vysokogornykh landshaftov Severo-Vostoka Sibiri. Nauka, Leningrad – *The flora of Suntar-Khayata: The issues of history of alpine landscapes of North–East Siberia* (in Russian)

Yurtsev BA (1981) Reliktovye stepnye kompleksy Severo-Vostochnoy Azii: problemy rekonstruktsii kriokseroticheskikh landshaftov Beringii. Nauka, Novosibirsk – *The relic steppe complexes of the North–East Asia: Problems of reconstruction of the cryoxeric landscapes of Beringia* (in Russian)

Yurtsev BA (ed) (1984) Arkticheskaya flora SSSR. Vypusk 9 (2): *Droseraceae-Rosaceae*. Nauka, Leningrad – *The Arctic flora of the USSR. Issue 9–1: Droseraceae-Rosaceae* (in Russian)

Yurtsev BA (ed) (1986) Arkticheskaya flora SSSR. Vypusk 9 (2): *Leguminosae*. Nauka, Leningrad – *The Arctic flora of the USSR. Issue 9 (2): Leguminosae* (in Russian)

Yurtsev BA (ed) (1987) Arkticheskaya flora SSSR. Vypusk 10: *Rubiaceae-Compositae Leguminosae*. Nauka, Leningrad – *The Arctic flora of the USSR. Issue 10: Rubiaceae-Compositae* (in Russian)

Zagrebina NL (1960) O svyazi rastitelnosti s geomorfologicheskim i geologicheskim stroeniem v basseyne srednego techeniya reki Daldyn. Trudy laboratorii aerometodov 9: 13–21 – *On relationship between vegetation and geomorphologic and geological features in the basin of the Middle Daldyn River* (in Russian)

Zakharova VI (2001) Sosudistye rasteniya reki Buotamy. In: Solomonov NG, Isaev AI, Ivanova EI (eds) Natsionalny prirodny park "Lenskiye stolby": geologia, pochvy, rastitelnost, zhivotny mir, okhrana i ispolzovanie. Collection of articles. Izd-vo YaGU, Yakutsk –*The vascular plants of the Buotama River* (in Russian)

Zakharova VI (2004a) Flora sosudistykh rasteniy natsionalnogo prirodnogo parka "Ust-Viluysky" (Zapadnoye Verkhoyanye). In: Problemy sokhraneniya raznoobraziya rastitelnogo pokrova Vnutrenney Azii. Proceedings of All-Russian scientific conference. Ulan-Udeh: 137–138 – *The flora of vascular plants of the National Natural Park "Ust-Viluysky" (West Verkhoyanye)* (in Russian)

Zakharova VI (2004b) Sosudistye rasteniya natsionalnogo prirodnogo parka "Ust-Viluysky". In: Problemy izucheniya rastitelnogo pokrova Yakutii. NIPK "Sakhapoligrafizdat", Yakutsk – *The vascular plants of the National Natural Park "Ust-Viluysky"* (in Russian)

Zakharova VI (2006) Flora vysshikh sosudistykh rasteniy prirodnogo parka "Kolyma". In: Lesnye issledovaniya v Yakutii: itogi, sostoyaniye i perspektivy. Issue 2: Lesnye resursy. Flora i rastitelnost lesnykh territory. Izd-vo YaGU, Yakutsk – *The flora of higher vascular plants of the Natural Park "Kolyma"* (in Russian)

Zakharova VI, Isaev AP, Ivanova EI, Sosina NK, Mikhalyova LG, Chikidov II (2004a) Rastitelny pokrov, flora i mikobiota srednego techeniya reki Molodo. In: Ekologicheskaya bezopasnost pri razrabotke rossypnykh mestorozhdeny almazov. Collection of articles. Sakhapoligrafizdat, Yakutsk: 133–142 – *Vegetation cover, flora and mycobiota of the Middle Molodo River* (in Russian)

Zakharova VI, Kopyrina LI, Poryadina LN, Ivanova EI (2004b) Bioraznoobraziye flory antropogennykh soobschestv Vostochnoy Yakutii (bassein verkhney Indigirki) In: Problemy sokhraneniya raznoobraziya rastitelnogo pokrova Vnutrenney Azii. Proceedings of All-Russian scientific conference. Ulan-Udeh: 138–140 – *Biodiversity of the flora of anthropogenic communities of East Yakutia (the Upper Indigirka basin)* (in Russian)

Zakharova VI, Nikiforova EN, Timofeyev AP (2007a) Pozdnepleistotsenovye stepi na territorii prirodnogo parka "Lenskiye Stolby". In: Prirodny park "Lenskiye Stolby": proshloye, nastoyascheye i buduscheye. Izd-vo YaNTs SO RAN, Yakutsk – *The Late Pleistocene steppes in the territory of the Natural Park "Lena Pillars"* (in Russian)

Zakharova VI, Sosina NK, Nikiforova EN, Ivanova NS (2007b) Redkiye rasteniya prirodnogo parka "Lenskiye Stolby". In: Prirodny park "Lenskiye Stolby": proshloye, nastoyascheye i buduscheye. Izd-vo YaNTs SO RAN, Yakutsk – *The rare plants of the Natural Park "Lena Pillars"* (in Russian)

Zhurbenko MP (2003) Novye i redkiye vidy lishainikov (*LICHENES)* iz Respubliki Sakha-Yakutia i Magadanskoy oblasti. Botanicheskiy zhurnal 88 (1): 111–118 – *New and rare lichen species (LICHENES) from the Republic of Sakha-Yakutia and Magadan Region* (in Russian)
Zhurbenko MP, Hansen ES (1992) New, rare or otherwise interesting lichen species from the Siberian Arctic. Mycotaxon 45: 275–284.
Zhurbenko MP, Chernyadyeva IV, Kozhevnikov YuP (2002) Lishainiki, likhenofilnye griby, mkhi i sosudistye rasteniya ostrova Samoilovsky (Ust-Lensky zapovednik, arkticheskaya Yakutia). Novosti sistematiki nizshikh rasteniy 36: 100–113 – *Liches, lichenophilous fungi, mosses and vascular plants of the Samoylovsky Island (Ust-Lena Reserve, the Arctic Yakutia)* (in Russian)

# Chapter 3
# Vegetation of Yakutia: Elements of Ecology and Plant Sociology

**A.P. Isaev, A.V. Protopopov, V.V. Protopopova, A.A. Egorova, P.A. Timofeyev, A.N. Nikolaev, I.F. Shurduk, L.P. Lytkina, N.B. Ermakov, N.V. Nikitina, A.P. Efimova, V.I. Zakharova, M.M. Cherosov, E.G. Nikolin, N.K. Sosina, E.I. Troeva, P.A. Gogoleva, L.V. Kuznetsova, B.N. Pestryakov, S.I. Mironova, and N.P. Sleptsova**

## 3.1 Principles of Geobotanical Regionalization

**A.P. Isaev**

The vegetation cover of Yakutia is part of the Holarctic kingdom of the vegetation of the Earth. The Holarctic kingdom covers the vast tracts of tundra, taiga, steppe and forest of both Eurasia and North America. The largest units of the botanical-geographical regionalization are the regions that reflect the latitudinal zonation of Eurasia. Yakutia is covered by the three regions: the Arctic tundra, the European-Siberian forest-tundra and the Eurasian coniferous taiga. Most part of the Republic, occupied by taiga or mountainous vegetation, belongs to the taiga region, or the Yakut province of the East-Siberian light coniferous forests.

Diversity of the vegetation cover of Yakutia, main principles of its distribution, as well as the characteristic features for each region have been considered in numerous publications: Cajander (1903, 1904); Komarov (1926); Abolin (1929); Birkengof (1932); Karavaev (1955, 1965); Scherbakov (1962, 1975); Galaktionova et al. (1962); Utkin (1965); Pozdnyakov (1969); Karavaev and Skryabin (1971); Timofeyev (1980, 2003); Perfilyeva et al. (1991); Skryabin and Karavaev (1991); Timofeyev et al. (1994); Danilova (2005); etc. However, the most comprehensive review of Yakutian vegetation is given in the book "Basic features of the vegetation cover of the Yakut ASSR" by Andreyev et al. (1987). This monograph provides the basis for this chapter.

The vegetation cover of Yakutia is very uneven. The territory of the Republic, stretching for 2,000 km from the north southwards and for 2,500 km from the west eastwards, features a clear latitudinal and longitudinal zonation of vegetation. 40% of the territory is occupied by mountains and, thus, is characterized by altitudinal belts. The peculiarities of the distribution of the vegetation are determined by

---

A.P. Isaev (✉)
Institute for biological problems of the cryolithozone, 41 Lenin Ave., Yakutsk, 677980, Russia
e-mail: forest_forest@ibpc.ysn.ru

E.I. Troeva et al. (eds.), *The Far North: Plant Biodiversity and Ecology of Yakutia*, Plant and Vegetation 3, DOI 10.1007/978-90-481-3774-9_3,

the combined influence of both modern physiographic conditions and complex processes of the Quaternary period, as well as by differences in geological age of the bedrocks of various regions. Various types of the anthropogenic activity also have an effect on the vegetation cover inducing the occurrence of pyrogenic, pasture, technogenic, and other secondary communities.

The territory of Yakutia is characterized by two vegetation types: arctic (26%) and boreal (74%). The arctic group comprises the arctic deserts and semi-deserts, the arctic and subarctic tundra, tundra bogs, the fell-fields and mountain tundra, as well as the riparian and maritime vegetation within the Arctic.

The arctic vegetation is confined to the Novosibirskie Islands, the coastline washed by the Laptev and East-Siberian Seas, and highlands, i.e. the regions situated northwards of the latitudinal tree-line and above the altitudinal tree-line. The region is characterized by most rigorous climatic conditions.

The mountainous and lowland tundra are characterized by some differences in ecological conditions. However, their floristic similarity and common ecosystem peculiarities allow to combine them into one group.

The boreal group is prevailed by the taiga type with common light coniferous larch forests (mainly of *Larix gmelinii* and *L. cajanderi*). The forested area occupies 67% of the territory of Yakutia, while the forests themselves cover 54% of the territory. We distinguish the near-tundra, northern taiga, middle taiga, and mountain taiga forest types. The taiga zone also bears shrubberies, bogs, and riparian vegetation which belong to the intrazonal vegetation type. The boreal vegetation occupies the lowlands south of the Arctic, and in mountains it is limited by an alpine belt. The climatic conditions are more favourable. Most of the boreal region is characterized by arid climatic conditions, while in the tundra and in the mountains the amount of precipitation exceeds the evaporation rates.

The distribution patterns of the arctic and boreal vegetation determine the basic contours of the vegetation cover of Yakutia. Their border lies along 72° N in the West, gradually descending eastwards, and reaching 69° N in the Kolyma River basin. In highlands the arctic vegetation penetrates the boreal zone even in the South of the Republic. It occurs widespread in the North–eastern part of Yakutia (the Verkhoyansk, Chersky, and Momsky Ranges), and less in the North–West (the Anabar Plateau) and in the South (the Stanovoy Range). Southwards, the areas of the arctic vegetation are reduced due to increasing altitudinal limits of the tree-line and decreasing heights of the mountains. So, in South and South–West Yakutia the arctic vegetation represents small "islands" among the dominating taiga.

The latitudinal zonation of the lowland territories is expressed by vegetation changes from the North southwards with a slight deviation to the South–East. In the arctic region the borders of zones and subzones are more definite, while in the boreal region they are less pronounced.

The zonal structure is as follows:

The zone of the arctic stone deserts and semi-deserts;

The tundra zone:

- the arctic subzone (the northern and southern arctic tundra);
- the subarctic subzone (the northern and southern subarctic tundra);

The taiga zone:

- The subzone of near-tundra taiga;
- The subzone of the northern taiga;
- The subzone of the middle taiga.

The altitudinal zonation includes the following belts:

Alpine:

- The nival belt;
- The belt of fell-fields;
- The belt of mountain tundra;

Boreal:

- The subalpine belt of mountain shrubberies;
- The belt of mountain sparse forests;
- The belt of mountain forests.

To a certain extent, the altitudinal belts are similar to the latitudinal subzones, though there are significant differences too. The nival belt in Yakutia has no analogues in lowlands. The fell-fields more or less correspond with the zone of stone deserts and semi-deserts. The belt of the mountain tundra can sometimes be divided into two sub-belts, corresponding with the lowland tundra subzones. However, the mountain tundra is not as expanded as the lowland tundra. Besides, the latter is clearly divided into smaller units (see above). The belt of mountain shrubberies is not always clearly defined due to the peculiarities of snow cover conditions, and has no analogues in lowlands. The belt of mountain forests is closely associated with the surrounding lowland forests. Besides, a distinct feature of altitudinal belt formation should be noted, the so called "belt inversion" due to accumulation of cooled air masses at foothills. As a result, the tundra and shrubbery vegetation appear under the belt of mountain forests. Such "lower" tundra communities differ from those of the mountain tundra belts. They are more common in the near-tundra forest subzone and less common in the northern taiga subzone. The peculiarities of mountain vegetation (especially in lower belts) are determined by latitudinal position: they markedly change along the gradient from North to South. This is noticeable both by an increasing number of belts, and some changes in floristic composition and phytocoenotic structure.

Besides the zonal principles of distribution, the vegetation of Yakutia also has the provincial distinct features determined by longitudinal and local climate characteristics, as well as by topography and geological history. The complex influence of zonal and provincial factors produce the ultimate picture of Yakutian vegetation, dividing it into geobotanical regions. Figure 3.1 shows the geobotanical regionalization. The estimates of the areas occupied by geobotanical regions (Table 3.1) is based on the "Map of vegetation" (Matveyev et al. 1989).

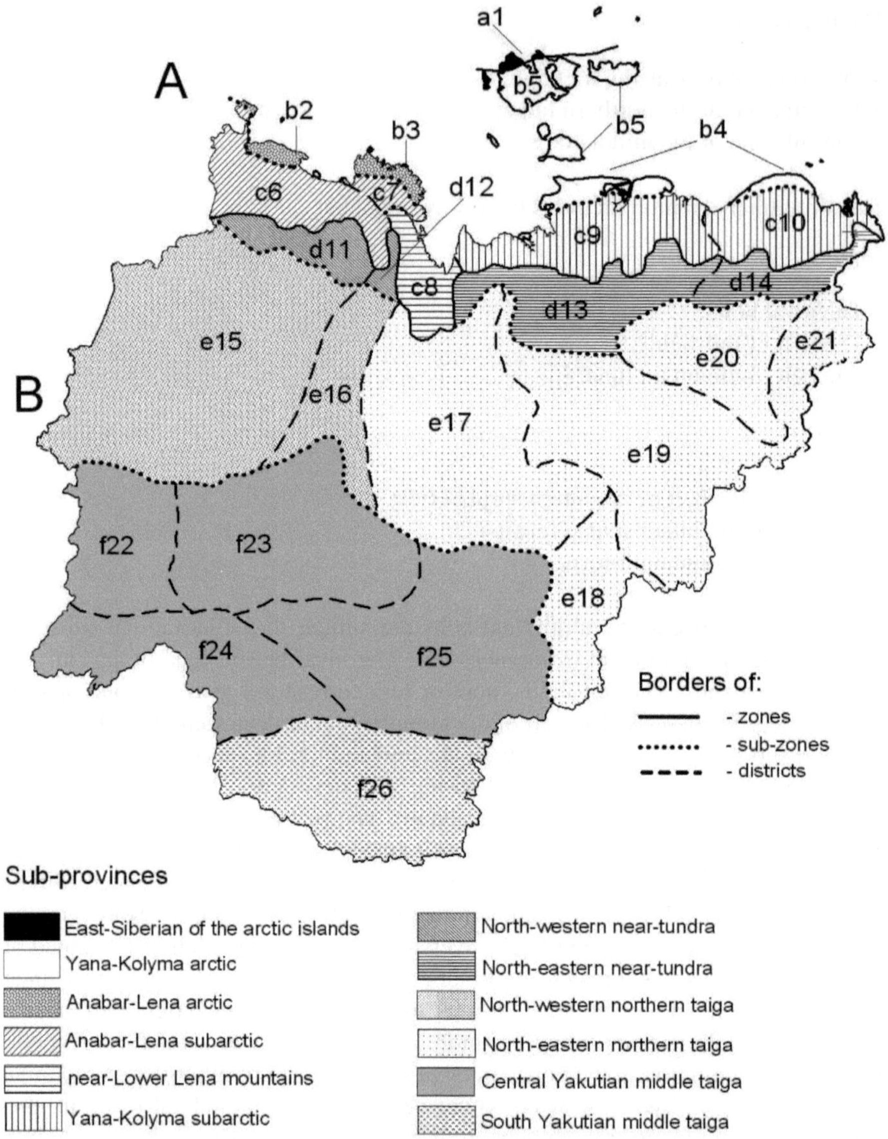


**Fig. 3.1** Geobotanical regionalization of Yakutia (Andreyev et al. 1987): **a**– Arctic region, **b** – boreal region; small letters – sub-zones; numbers – districts

In the zone of the arctic deserts and semi-deserts its southernmost part within Yakutia can be distinguished as a special subzone of polar semi-deserts (a). It covers the *district of De Long Islands* (District 1).

Further to the South the tundra zone is situated. In the territory of Yakutia only the East Siberian Province is situated in this zone. In Russia, this Province covers also the eastern part of the Taimyr tundra.

**Table 3.1** Basic units of geobotanical regionalization of Yakutia

| Region | Area | |
|---|---|---|
| | Thousand km$^2$ | % |
| **A. Arctic region** | **394.3** | **12.6** |
| ***Zone of the arctic deserts and semi-deserts*** | ***3.1*** | ***0.1*** |
| a. Subzone of polar semi-deserts | 3.1 | 0.1 |
| Sub-province, East Siberian of the arctic islands | 3.1 | 0.1 |
| *1. District of the De Long islands* | *3.1* | *0.1* |
| ***Tundra zone*** | ***391.2*** | ***12.5*** |
| Province, East Siberian arctic | ***391.2*** | ***12.5*** |
| b. Subzone of the arctic tundra | 74.7 | 2.3 |
| Sub-province, Anabar-Lena arctic | 13.7 | 0.4 |
| *2. Khatanga-Olenyok District* | *6.2* | *0.2* |
| *3. District of the Lena River Delta shoreline* | *7.5* | *0.2* |
| Sub-province, Yana-Kolyma arctic | 61.0 | 1.9 |
| *4. Indigirka-Kolyma District* | *26.8* | *0.8* |
| *5. District of Novosibirskie Islands* | *34.2* | *1.1* |
| c. Subzone of the subarctic tundra | 316.5 | 10.2 |
| Sub-province, Anabar-Lena subarctic | 105.5 | 3.4 |
| *6. Anabar-Lower Olenyok District* | *86.9* | *2.8* |
| *7. District of the Lena River Delta* | *18.6* | *0.6* |
| Sub-Province of near-Lower Lena mountains | 40.3 | 1.3 |
| *8. Chekanovsky-Kharaulakhsky District* | *40.3* | *1.3* |
| Sub-province, Yana-Kolyma subarctic | 170.7 | 5.5 |
| *9. Yana-Indigirka District* | *102.4* | *3.3* |
| *10. Alazeya-Lower Kolyma District* | *68.3* | 2.2 |
| **B. Boreal region** | **2708.9** | **87.4** |
| ***Taiga Zone*** | ***2708.9*** | ***87.4*** |
| Yakutian Province of East Siberian light coniferous forests | ***2708.9*** | ***87.4*** |
| d. Subzone of near-tundra forests | 245.2 | 7.9 |
| Sub-province, North–western near-tundra | 55.9 | 1.8 |
| *11. Anabar-Olenyok District* | *49.7* | *1.6* |
| *12. Lower Lena District* | *6.2* | *0.2* |
| Sub-province North–eastern near-tundra | 189.3 | 6.1 |
| *13. Omoloy-Indigirka District* | *136.5* | *4.4* |
| *14. Kolyma District* | *52.8* | *1.7* |
| e. Subzone of the northern taiga forests | 1374.5 | 44.4 |
| Sub-province, North–western northern taiga | 508.8 | 16.4 |
| *15. Olenyok District* | *418.9* | *13.5* |
| *16. Zhigansk District* | *89.9* | 2.9 |
| Sub-province North–eastern northern taiga | 865.7 | 28.0 |

**Table 3.1** (continued)

| Region | Area | |
|---|---|---|
| | Thousand km$^2$ | % |
| *17. Verkhoyansky (Upper Yana) District* | *322.7* | *10.4* |
| *18. Suntar-Khayata District* | *89.9* | *3.0* |
| *19. Indigirka District* | *279.3* | *9.0* |
| *20. Abiy-Kolyma District* | *124.1* | *4.0* |
| *21. Yukagirsky District* | *49.7* | *1.6* |
| f. Subzone of the middle taiga forests | 1089.2 | 35.1 |
| Sub-province, Central Yakutian middle taiga | 788.2 | 25.4 |
| *22. Upper Viluy District* | *121.0* | *3.9* |
| *23. Viluy District* | *248.3* | *8.0* |
| *24.Upper Lena District* | *136.5* | *4.4* |
| *25. Aldan-Lena District* | *282.4* | *9.1* |
| Sub-province, South Yakutian middle taiga | 301.0 | 9.7 |
| *26. Uchur-Olyokma District* | *301.0* | *9.7* |
| **Total** | **3103.2** | **100.0** |

The subzone of the arctic tundra (b) includes two sub-provinces: the Anabar-Lena and the Yana-Kolyma. The Anabar-Lena sub-province is characterized by widespread communities of *Carex stans* with participation of *Dryas* and *Cassiope* tundra, as well as by the presence of dwarf *Salix* species. It is divided into the following districts:

District 2 *The Khatanga-Olenyok*: Abundant green mosses and prostrate *Salix* species (*Salix polaris, S. reticulata*).

District 3 *The Lena River Delta shoreline*: Fissured tundra with *Carex stans, Siphula ceratites*; unique spotted tundra with *Andreaea rupestris* in spots; and maritime meadows with *Carex subspathacea*.

The Yana-Kolyma sub-province bears a more arctic character representing the habitats for dwarf shrub communities growing in fissured and spotted microrelief. The districts of this sub-province are:

District 4 *The Indigirka-Kolyma*: Widespread hummocky tundra.

District 5 *The Novosibirskie Islands* (Anju and Lyakhov Islands.): The northern arctic tundra with a discontinuous vegetation cover. The *baidjarakhs* (earth mounds) at different stages of their evolution are also abundant.

In the subzone of the subarctic tundra (c) there are three sub-provinces. The Anabar-Lena subarctic sub-province is characterized by more pronounced shrub vegetation in its southern part.

District 6 *The Anabar-Lower Olenyok*: Widespread shrubberies, thick moss layer, and indistinct polar tree-line. Lichen-rich reindeer pastures.

District 7 *The Lena River Delta*: Waterlogged territory with widespread polygonal-ridged tundra-bog complexes and tundra with prevalence of the moss species *Andreaea rupestris*. The subarctic tundra here borders on the subzone of the near-tundra forests.

The sub-province of the near-Lower Lena mountains in District 8 *Chekanovsky-Kharaulakhsky* is characterized by prevalence of mountain tundra landscapes with a rich flora.

The Yana-Kolyma sub-province features extensive tussock tundra with *Eriophorum vaginatum* and hummocky tundra with dwarf shrubs.

District 9 *The Yana-Indigirka*: most pronounced ultra-continental climate, widespread tundra and bog complexes on permafrost induced microrelief, appearance of *Carex lugens* in the tundra of *Eriophorum vaginatum*, and relatively weak shrubberies and moss layer.

District 10 *The Alazeya-Lower Kolyma*: less continental climate, increased participation of shrubs and mosses, widespread hummocky tundra and tundra-bog complexes.

As has been mentioned before, the boreal region of Yakutia is represented by the taiga zone covered by the Yakutian Province of East Siberian light coniferous forests.

The relatively narrow (60–220 km wide) subzone of the near-tundra forests (d) consists of two sub-provinces. The North–western sub-province represents lowland landscapes covered by sparse *Larix gmelinii* forests.

District 11 *The Anabar-Olenyok*: *Larix gmelinii* woodlands (sparse forest) of various ages, shrub tundra, and polygonal-ridged tundra bogs with forested ridges.

District 12 *The Lower Lena*: Prevalent mountain tundra, near-tundra and riparian larch forests along the mountain rivers, and forests of *Chosenia arbutifolia*. Habitat for numerous rare species (*Caragana jubata, Papaver leucotrichum*, etc.).

The North–eastern near-tundra sub-province is characterized by the mountain and lowland forests of *Larix cajanderi*.

District 13 *The Omoloy-Indigirka*: Alternation of forested and forest-free areas due to various altitudinal characteristics of the landscape. The altitudes of 50–100 m a.s.l. are characterized by forests retreat giving way to tundra. The ground layer of the *Larix cajanderi* forests is composed of fruticose lichens, *Sphagnum spp*., and *Eriophorum vaginatum* in waterlogged foothills. The thin shrub layer includes *Pinus pumila, Betula divaricata* (lichen types of the forests), *Betula exilis, Salix pulchra* (*Sphagnum* and *Eriophorum vaginatum* types). The near-icing vegetation with *Equisetum variegatum* and fragments of the cryophile steppes are also typical.

District 14 *The Kolyma*: abundant lakes and extensive wetlands. The forested area is smaller compared to District 13. *Larix* woodlands alternate with polygonal-ridged tundra bogs with forested ridges and hillocky shrub tundra.

The subzone of the northern taiga forests (e) also consists of two sub-provinces. The North–western northern taiga sub-province is characterized by *Larix gmelinii* forests with an admixture of *Picea obovata*.

District 15 *The Olenyok*: Widespread *Larix gmelinii* woodlands with an admixture of *Picea obovata* on calcareous bedrocks of earth stripes.

District 16 *The Zhigansk*: *Larix gmelinii* woodlands with common dwarf shrub-*Carex* bogs among them. Small patches of the *Pinus sylvestris* communities are observed reaching 68° N. *Pinus pumila* shrubberies grow on sandy ridges reaching the left bank of the Lena River. Steppe fragments with *Festuca lenensis, Phlox sibirica,* and *Dianthus repens* occur as far as the northern limits of the sub-province. The valley vegetation is properly developed represented by *Larix* forests with significant participation of *Picea obovata* and by communities of *Salix viminalis* and *S. dasyclados*.

The North–eastern northern taiga sub-province consists of 5 districts and is represented by dominating mountain forests of *Larix cajanderi*.

District 17 *The Verkhoyansky* (i.e. Upper Yana): covers highlands with mountain tundra and fell-fields and tablelands with mountain *Larix* forests. The altitudinal tree-line lies as high as 1,600 m a.s.l. The belt of mountain forests features also cryophilic steppes. The Verkhoyansk mountains represent a natural barrier for penetration of many plant species (including *Pinus sylvestris*) to North–East Yakutia though they could find favourable growing conditions there.

District 18 *The Suntar-Khayata*: mountain tundra, stony deserts, icing remnants. Compared to District 17, the climate is less continental and more humid. Owing to this, the upper limit of *Larix cajanderi* lies at a lower altitude.

District 19 *The Indigirka*: most part is represented by ranges and tablelands with icing remnants. The district is characterized by the process of establishment of a vegetation cover on vast areas after the last glaciation period. The riparian vegetation is represented by the forests of *Chosenia arbutifolia* and *Populus suaveolens*. Cryophilic steppes are also common.

District 20 *The Abiy-Kolyma*: vast waterlogged and shrubby lowlands covered by *Larix cajanderi*. The climate is less continental. Reduction of the areas of micro- and mesocomplexes induced by permafrost action. Vast communities of dominating *Calamagrostis langsdorffii* around lakes. The relic steppes are of minor importance.

District 21 *The Yukagirsky*: Prevalence of mountain forests. In some places mountain forest-tundra occurs.

The middle taiga subzone (f) consists of two sub-provinces. The Central Yakutian middle taiga sub-province is dominated by *Larix* forests. *Pinus sylvestris* forests occur in watershed areas.The district is characterized by distinct alas landscapes. The Lena, Aldan and Viluy Rivers feature vast valleys with rich meadows. Steppe and forest-steppe landscapes are also common. The sub-province comprises four districts.

District 22 *The Upper Viluy*: Prevalence (80–90% of the District's area) of *Larix gmelinii* forests of green moss and dwarf shrub types with significant participation of lichens (*Cladonia arbuscula, Cetraria cucullata*). Small areas are occupied by *Pinus sylvestris* forests mainly of the *Arctostaphylos uva-ursi* type; stripes of *Betula exilis* and *B. fruticosa* yerniks, *Sphagnum* bogs; and small river valley meadows.

District 23 *The Viluy*: Widespread larch forests mainly of *Vaccinium vitis-idaea* and *Ledum palustre* types. *Pinus sylvestris* forests are more common. The watershed areas feature alas meadows of *Alopecurus arundinaceus*, *Scolochloa festucacea*, and *Puccinellia tenuiflora* communities. Dry belts of alas vegetation are represented by the *Carex duriuscula* steppes. In the floodplains of the Lower Viluy River *Calamagrostis* communities occur. The watershed areas feature vast unique sandy landscapes (*tukulan*) with sparse vegetation of *Pinus sylvestris* and *P. pumila*. *Betula fruticosa* shrubberies grow along the rivers.

District 24 *The Upper Lena*: Most favourable climatic conditions have yielded a rich flora here. The dominating light coniferous forests (83.4% of the territory of the District) include patches of dark conifers. The tree species include *Pinus sibirica, Abies sibirica,* and *Larix sibirica*, and they grow here at the northern border of their distribution areas. The Upper Lena River valley and its tributaries bear fragments of cryophile steppes. The floodplain supports polydominant forb-grass meadows with *Alopecurus pratensis, Festuca pratense, Trifolium pratense*, etc.

District 25 *The Aldan-Lena*: Prevalence (82.7%) of the middle taiga *Larix cajanderi* forests (*Vaccinium vitis-idaea, Arctostaphylos uva-ursi*, and forb types) with inclusions of the *Pinus sylvestris* communities. The watershed areas support the alas meadows with *Puccinellia tenuiflora, Hordeum brevisubulatum, Alopecurus arundinaceus,* and *Calamagrostis langsdorffii*. The communities with *Calamagrostis langsdorffii* cover large areas in the lower reaches and in the mouth of the Aldan River. The floodplains of the Middle Lena and Middle Aldan Rivers also feature *Hordeum brevisubulatum* and *Alopecurus arundinaceus* meadows mostly with participation of *Salix viminalis*. Steppe and forest-steppe landscapes are also characteristic, occupying the terraces above floodplains and valley slopes. The small river valleys represent the habitats for the *Calamagrostis* + *Carex* tussock meadows and yerniks of *Betula fruticosa*.

The South Yakutian middle taiga sub-province comprises only one District 26 *The Uchur-Olyokma*. It is situated within the Aldan and Timpton-Uchur uplands with altitudes of 500 m a.s.l. and more. Mountain *Larix* forests prevail (76.5%). Mountain tundra appears from 1,200 to 1,300 m a.s.l., occupying less then 5% of the territory. Compared to the Central Yakutian sub-province, the flora is more diverse due to penetration of species from the South (*Botrychium virginatum, Listera sovatieri*, etc.). The *Larix* taiga is also characterized by an admixture of dark coniferous forests of *Picea ajanensis* on the lower parts of the slopes. Communities of *Betula ermannii* are common along the border with the mountain tundra. The dwarf shrub layer of the mountain forests is abundant in *Dryas crenulata* and *Rhododendron aureum*. Shrubberies of *Pinus pumila* may occur with lichen, green moss or *Sphagnum* covers.

## 3.2 History of Vegetation Development

**A.V. Protopopov and V.V. Protopopova**

The history of the development of the Yakutian vegetation is closely associated with all climatic changes that have taken place throughout the last seventy million years.

Paleobotanical investigations in Yakutia date back to 1950s, and have put emphasis on the palynology of the Tertiary and Quaternary periods. Detailed spore-pollen spectra of the region have resulted of the research works conducted by V.I. Gromov, S.F. Biske, A.A. Arkhangelov, Yu.P. Baranova, A.P. Vaskovsky, R.E. Giterman, M.P. Grichuk, I.A. Kulkova, O.A. Ivanova, V.I. Kalayainen, A.I. Tomskaya, G.M. Savinova, G.G. Kartashova, T.N. Kaplina, A.V. Lozhkin, etc. The reconstruction of vegetation development was made by Tolmachev (1954), Skryabin and Karavaev (1991), Giterman (1985), Kozhevnikov and Ukraintseva (1996), Tomskaya (2000), and Ukraintseva (2002).

The Palaeogene (67... 25.5–26 Ma BP (million years before present)) is characterized by wide distribution of broad-leaved forests all over Yakutia. They featured a rich tree species composition and participation of subtropical coniferous and evergreen flowering plants, palms and arboreous ferns. The topography consisted of plain landforms with predominating marshlands. Intensive processes of carbon formation took place in this period. Paleobotanical investigations show presence in the flora of such plants as *Sequoia, Glyptostrobus, Dammara, Taxodium, Araucaria, Trochodendron, Platanus*, and *Nyssa*. This allows the assumption that the vegetation of that period resembled the modern subtropical vegetation of southern North America and south–eastern China (Skryabin and Karavaev 1991).

The Early and Middle Neogene (26–25.5... 10 Ma BP) is characterized by active mountain formation processes. The Verkhoyansk and Chersky Ranges acquired their modern outlines. The polar ocean started to cool down. Decreasing winter temperatures along with stable snow cover resulted in alteration in the vegetation cover. The Neogene in the territory of modern Yakutia featured the stage of prevalent coniferous-broad-leaved forests with some participation of evergreen plants. The trees and shrubs were represented by various species of *Betula, Quercus, Acer, Carpinus, Tilia, Ulmus, Fraxinus, Fagus, Castanea, Corylus, Viburnum*, etc. The mountains were the habitats for coniferous forests, with presumed alpine vegetation in modern fell-field belt. Gradual cooling of the climate favoured a significant expansion of meadows and an increase of the species diversity of herbs. As a whole, the climate of that period was rather warm and humid.

The coniferous flora consisted mainly of various species of *Pinus, Picea, Larch, Abies, Tsuga, Pseudotsuga, Taxus*, etc. The vegetation resembled that of the modern Manchurian-Chinese and North American forests. During that period a vigorous exchange of plant species took place between Asia and North America via the Beringian land bridge. Later, Beringia would periodically appear (resuming thus the flora-fauna exchange) during the periods of the Arctic Ocean drying up (Hopkins 1967; Yurtsev 1981).

The Late Neogene is characterized by significant changes in the vegetation cover of Yakutia. The zoning now resembles the southern parts of modern Eurasia (Tolmachev 1954). Light coniferous *Pinus + Larix* forests took their leading position in the northern parts of modern Yakutia, beyond the Polar Circle. This was the first development of the light coniferous taiga in Yakutia. The tundra zone did not exist at that time.

The Early Anthropogene (1.5–2 Ma – 600 Ka BP) was characterized by further chilling and the formation of icings in Siberia that greatly influenced the characteristics of the modern vegetation. The Arctic Ocean had become isolated and covered with continuous ice. In the southern and eastern parts of Siberia high ranges isolated the vast territory from the warming effects of the Pacific Ocean. The process of the development of a cold-resistant flora in the near-Polar and highland regions that had started in the Late Neogene became more vigorous. As a result, the tundra appeared during that period (Giterman 1963).

Climate cooling in the territory of modern Central Yakutia has led to the replacement of rich warmth-requiring coniferous-broad-leaved forests by dark coniferous forests with a scantier species diversity. They resembled the modern taiga of the Far East. The tree layer was composed of various *Picea spp.*, *Larix sibirica, Abies spp., Pinus sylvestris, P. sibirica, P. monticola, Tsuga,* etc. The small-leaved species were represented by *Betula lanata* and *B. davurica* (Vaskovsky 1959).

The light coniferous forests with a pronounced herb cover continued to expand in the North. In the mountains the shrubberies of *Pinus pumila* developed. The stripe of tundra was also characterized by further expansion, slowly but steadily (Giterman 1985).

In the Middle Anthropogene (500–100 Ka BP) a series of glaciations took place: the Riss glaciation in Europe and Samar glaciation in Russia divided into the Oka, Moscow and Dnieper periods. The climatic conditions in the territory of modern Yakutia became especially rigorous. The Arctic and its neighboring regions were covered with a continuous ice sheet. The glaciers in the mountains were at the peak of their development. Sliding down, they changed river courses and induced ice blockages resulting in flooding of vast lowland areas. Winters were cold though milder than modern ones. Summers, on the contrary, were colder. All these factors resulted in an expansion of the tundra southwards. Its area exceeded the modern territory significantly. During this period tundra species penetrated the flora of modern Central Yakutia. Some of them (*Dryas, Juncus, Armeria, Loiseleuria,* etc.) are still the constituent part of the Central Yakutian flora (Skryabin and Karavaev 1991).

The most widespread vegetation type were sparse larch forests of *Larix gmelinii* with an admixture of *Pinus* and *Betula*. They alternated with swamped meadows, peat bogs and yerniks. One of the basic limiting factors was permafrost that acquired a continuous character (Tomskaya 1981). During the Dnieper glaciation the climate of Yakutia became more continental. The glaciated areas in the mountains decreased significantly. Winters were colder, while summers were warmer and more arid. The waterlogged plains in lowlands started drying out. The steppe vegetation appeared with participation of *Artemisia spp.* and *Chenopodiaceae* species. At the same time, the *Larix-Pinus-Betula* sparse forests developed south of the Viluy River in complex with xerophilous grass-forb and *Artemisia* associations. In the mountains the retreating glaciers were replaced by the mountain tundra formations. Tundra and forest-tundra were the dominating landscapes.

About 100,000 years ago the Kazan (Riss-Würm) interglacial period started that was later followed by the Zyryan (or Würm I) glaciation and the Kargin interglacial (or Würm II). The Kazan interglacial period was warm and humid enough for restoration of the dark coniferous forests in Yakutia along with the light coniferous *Pinus* and *Larix* forests. The South–eastern regions of Yakutia featured *Ulmus* and *Corylus*. During that period the forest tracts forwarded far to the North. In the Middle Yana and Indigirka Rivers basins the complex of grassland and forest-steppe formations was widespread. The zonal vegetation type of North Yakutia during that period was light coniferous forests predominated by *Larix gmelinii*. The territory of the modern tundra was covered by sparse larch forests with *Alnus* and arboreous forms of *Betula*. In the Lower Yana River basin the *Larix* communities included *Pinus* and *Picea*, presently being absent east of the Verkhoyansk Range. Tundra vegetation was represented along the shoreline of the Arctic Ocean and on arctic islands. In the mountains mountain tundra was common, and in the Upper Indigirka mountain steppe communities (Giterman et al. 1968).

The zonal vegetation type of Central Yakutia was dark coniferous forest with participation of *Picea, Abies, Pinus sibirica* and *Larix sibirica*. Meadows were also abundant and highly productive. They provided enough forage to maintain the populations of large herbivorous mammals of the mammoth fauna: mammoths, wooly rhinoceroses, bison, and horses. Some specialists assume that forest-steppe and xerophilous meadow steppe formations prevailed in Central Yakutia (Skryabin and Karavaev 1991; Yurtsev 1981; Ukraintseva 2002). Large areas were also occupied by the *Pinus* + *Larix* and *Betula* forests.

In the South–eastern parts of Yakutia the dark coniferous forests were enriched by nemoral elements penetrating from the Far East. The mountain river valleys in the southern spurs of the Verkhoyansk Range represented the habitats for *Populus, Ulmus* and *Corylus*. This was possible under much more humid and warmer climate conditions as compared to those of modern Yakutia. This was the last record of the representatives of broad-leaved forests in Yakutia (Lazarev and Tirskaya 1975).

During the Sartan glaciation (or Würm III) (30–10 KaBP) the climate became significantly colder and more arid. Summer temperatures were much higher than modern ones, while humidity was lower. The forest tracts were reduced drastically giving way to the forest-steppe landscapes that bordered on the tundra in the North (Fig. 3.2).

In the Yana-Kolyma lowland the only forest-forming tree species was *Larix cajanderi*. In Central Yakutia islands of the *Pinus* + *Larix* and *Betula* forests were observed. A number of steppe species penetrated the region from the Trans-Baikalian and Mongolian steppes. The steppes east of the Verkhoyansk Range had genetic relations with the North American prairies via the Beringia bridge (Yurtsev 1974, 1981). Meadow-steppe associations along with *Salix* shrubberies represented rich fodder lands promoting the flourishing of large herbivorous mammals (bison, mammoth, wooly rhino). Steppe became the zonal vegetation type in Yakutia. In the North the tundra and steppe zones jointed to form a distinct vegetation type, the tundra-steppe, or the arctic steppe (Cwynar and Ritche 1980; Lavrenko 1981). This combination was possible only under the unique climatic conditions, when dry air, withering the soil surface, facilitated the growth of xerophytes, while the

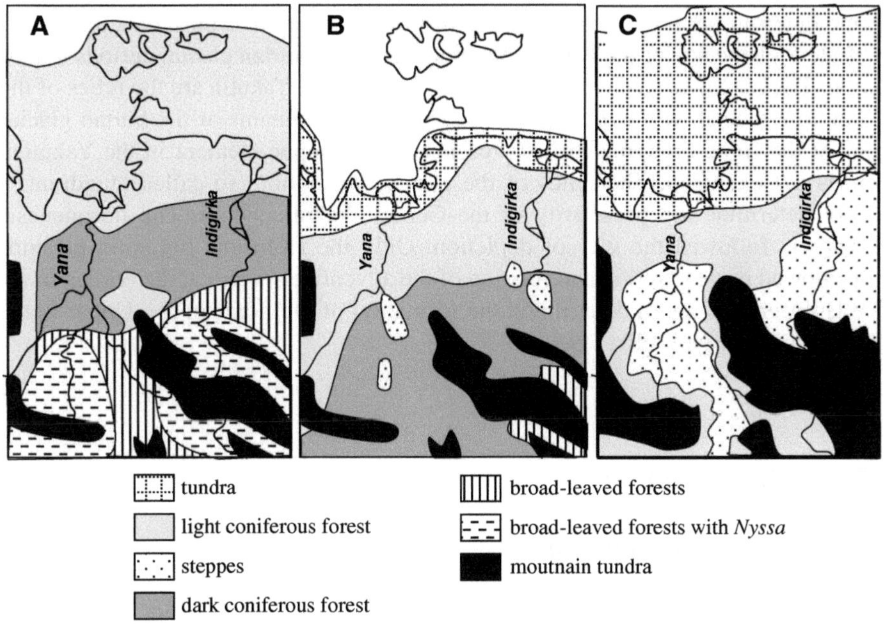


**Fig. 3.2** Dynamics of forest vegetation in the Pleistocene (the Yana-Indigirka Interfluve). **a** – Early Pleistocene, **b** – Late Pleistocene (Kargin Interglacial), **c** – Late Pleistocene (Sartan Glacial)

subsoil frozen grounds saturated the upper soil horizons with moisture. This, in turn, favoured the growth of tundra-bog species. There is a reasonable hypothesis that under conditions of a very hot summer and extremely cold winter the thermokarst processes were especially vigorous and actively changed the microrelief. As a result, the depressions were covered by bog species, while the hillocks (even small ones) supported steppe plants. Relic steppe communities are presently observed in Central Yakutia (the Middle Lena River basin), and in the Yana-Indigirka Interfluve (middle and upper reaches of the rivers) (Yurtsev 1981). Those landscapes provided the basis of distinguishing the Central Yakutian and Yana-Indigirka Floristic Regions. During the Sartan glacial, *Larix cajanderi* positioned its major forest-forming role in Yakutia (Protopopov 2005).

In the Holocene, when the glaciation was over, the Yakutian vegetation acquired its present appearance. Even the Holocene optimum, that took place 6–7 thousand years ago, did not bring radical changes into the vegetation patterns. Significant warming has led to a shift of the forest zone northwards. During the Holocene optimum the territory of the modern tundra was covered with forest-tundra with participation of *Larix*, arboreous *Betula* and *Picea*. Consequently, the subzone of the northern taiga *Larix* forest also proceeded further North (Giterman et al. 1968).

Thus, the vegetation of Yakutia had passed the long and complex history before it achieved its modern appearance. The boundaries of vegetation zones shifted northwards during the interglacial periods and southwards during the glaciations. The period determining the modern patterns of the vegetation and flora of Yakutia was the Late Pleistocene, or the Kazan interglacial and the Sartan glacial periods in particular. The boreal-taiga elements of the present flora of Yakutia are the relics of the Kazan interglacial, while the steppe species are the remnants of the Sartan glaciation. At the same time, the earlier presence of the steppe element in the Yakutian flora is not excluded. The relics of the Samara glaciation, so called "tundrants", do not determine the peculiarity of the Central Yakutian flora. The florogenesis of Yakutia followed the way of depletion. Only the Holocene optimum presumably allowed re-penetration and fixation of the adventive species at the border areas. The Holocene optimum determined the formation of the floras of the Upper-Lena, Kolyma and Olenyok FRs.

## 3.3 Arctic Vegetation

### *3.3.1 Maritime Vegetation*

**A.A. Egorova**

The seashore vegetation is composed of lowland grasses, sometimes hummocky-hollow swamped areas and small fragments of maritime meadows on marsh soils. The maritime salt meadows and swamps are characteristic for the lower floodplain. They are found within the arctic tundra sub-zone, penetrating the Hypoarctic: along the coast-line, in the mouths of channels within the coast-line, as well as along the shore-line of islands. The largest areas of maritime vegetation are recorded on the Nordwick Bay, the Buor-Khaya inlet, the Novosibirskie Islands, the Yana-Indigirka and Alazeya-Chukochya Interfluves. The meadows are formed on saline soils and are regularly irrigated by the salt waters of rising tides of the sea.

The species composition of the maritime meadows is determined by the colonization stage of the saline alluvial marine silty sediments. The initial colonization stage on strongly salinized dark-grey viscid silty sediments within the rising tide area is characterized by the rooting sprouts of *Puccinellia phryganodes.* They have a reddish colour (Andreyev and Perfilyeva 1980; Labutin et al. 1985; Andreyev et al. 1987; Skryabin and Karavaev 1991; Perfilyeva et al. 1991). They form continuous homogenous or spotty grass stands as high as 5–7 cm. Sometimes the community may include *Stellaria humifusa* or *Pleuropogon sabinii, Ranunculus hyperboreus,* and *Arctophila fulva* on desalinated silt of river mouths of the Novosibirskie Islands (Gorodkov 1956). With rising elevation of the seacoast, the *Puccinellia* community is replaced by *Carex subspathacea* coenoses with constant and abundant *Calamagrostis deschampsioides*, as well as *Pleuropogon sabinii, Rhodiola borealis, Stellaria humifusa, Potentilla egedii, Tephroseris palustris, Dupontia fisheri* (sometimes abundant), *Dendranthema arcticum ssp. polare,* etc. Very seldom *Wilhelmsia physodes* and *Honckenya peploides* are observed. Marsh soddy soils are formed under those communities. The grass stands are not high. Further increase in altitude (within the northern hypoarctic tundra) yields less moist conditions. When drier, the maritime meadows may support *Salix reptans*. Very often such meadows are characterized by cryogenic hummocky nanorelief. In the autumn the meadows acquire bright red colours sharply standing out against the background of the blue sea and the monotonous brownish-green tundra vegetation. At lake-sides the maritime meadows are replaced by communities of *Arctophila fulva* and *Carex concolor.*

## *3.3.2 Arctic Tundra Vegetation*

The flora and vegetation of the Arctic tundra of Yakutia is covered by the numerous publications (Sochava 1933a, 1933b, 1934a, 1934b; Gorodkov 1956; Kartushin 1963; Mikhailov 1963; Alexandrova 1960, 1961, 1962, 1963, 1970, 1977; Safronova 1980, 1982; Sumina 1975, 1986; Andreyev and Nakhabtseva 1974; Perfilyeva and Rykova 1975; Andreyev et al. 1976; Andreyev and Perfilyeva 1980; Matveyeva 1980; Labutin et al. 1985; Nikolin 1986; Andreyev et al. 1987; Perfilyeva et al. 1991; Egorova et al. 1991).

The vegetation cover of the Yakutian Arctic tundra is characterized by a polydominant structure and spatial heterogeneity. This mosaic structure is determined by the peculiarities of the microrelief, mainly of permafrost origin (pingo (*bulgunnyakh*), earth mounds (*baidjarakh*), hummocks, microdepressions, patches of bare ground). They specify the set of ecological factors, such as moisture, temperature and light conditions, snow cover characteristics, exposure to winds, etc. The complex of hummocky spotted and polygonal small-grass tundra types is characteristic for the Arctic tundra. The polygonal tundra occupies more than 20% of the territory (Fig. 3.3), while the polygonal bogs and riparian vegetation cover 70% (Skryabin and Karavaev 1991).

**Fig. 3.3** Polygonal arctic tundra (Tit-Aryy Island, the Lena River Delta)

The watershed areas with loamy soils are covered by herb – dwarf shrub – moss spotty tundra dominated by *Alopecurus alpinus* and *Salix polaris* (Gorodkov 1956; Mikhailov 1963; Alexandrova 1963; Andreyev and Perfilyeva 1980; Perfilyeva et al. 1991). The small prostrate sprouts of dwarf *Salix* species do not emerge above the surrounding vegetation, displaying only their aments out of the moss cover. *Salix polaris* is a small arctic-alpine species as high as 1 cm. This is the only dwarf shrub that is able to propagate vigorously forming a large phytomass. Its underground organs are thickly entangled in the moss mat. *Alopecurus alpinus* is a low arctic-alpine grass species. The tundra with *Dryas punctata* is widespread both in lowlands and mountains on soils of various mechanical composition, chemical reaction, hydrothermal conditions and rates of cryogenic processes. It loses vigour only in the northern stripe of the Arctic tundra. Besides the dominant species, the following plants are characterized by a high constancy: *Papaver polare, Cerastium bialynickii, Oxyria digyna, Saxifraga cespitosa, S. nivalis, S. oppositifolia,* and *Luzula confusa. Salix reticulata, S. reptans, S. nummularia* are observed more rarely. The moss cover is represented by constant *Dicranum, Aulacomnium turgidum, Hylocomium splendens var. alaskanum, Ditrichum flexicaule, Racomitrium lanuginosum, Distichium capillaceum, Orthothecium chryseum,* and *Timmia austriaca.* Lichens are not numerous.

The tundra of the interior Arctic borders on the northern subarctic tundra. Their watershed communities are very identical both in structure and dominant composition. The diagnostic feature of the Arctic tundra of that area is a lack of dwarf shrubs, especially of *Betula exilis.* The participation of dwarf *Salix* species (almost prostrate forms of *Salix glauca, S. reptans),* as well as of *Vaccinium vitis-idaea* and *V. uliginosum* is sporadic. Many species feature modifications of their habitus (formation of cushions, thick tussocks). The participation of subarctic species in associations is not high, except for *Eriophorum polystachion.* The occurrence of *Ranunculus sabinii, Puccinellia angustata, Draba subcapitata, D. oblongata, Saxifraga platysepala* is common in these plant communities.

The growth activity of the Arctic tundra plants after the long winter dormancy starts under conditions of below-zero temperatures. The major growth season at snow free sites takes place in late May-early June. By early August most plants finish blossoming, and the Arctic tundra acquires a monotonous greenish-brown colour.

*The Anabar-Lena arctic sub-province.* The narrow strip along the shoreline of the Laptev Sea is dominated by the herb – green moss tundra in the watershed area. That includes communities with *Carex concolor, Eriophorum polystachion* and *Salix polaris*. Places with loamy soils are characterized by a continuous cover dominated by *Carex arctisibirica* and mosses: *Aulacomnium turgidum, Hylocomium splendens var. alaskanum,* etc. There are also spotty *Dryas punctata*, *Cassiope tetragona* and *Dryas* + *Cassiope* types of the tundra. Closer to the Sea, the complicated relief composed of ridges and earth mounds determines complex combinations of associations with *Salix polaris* and *Dryas punctata.* According to Matveyeva (1980), the Bolshoy Begichev Island is a habitat for the dwarf shrub hummocky and spotted psammophytic tundra, resembling the tundra landscapes of the Lena River delta.

**Fig. 3.4** Psammophytic tundra in the Lena River Delta

In the district of the Lena River delta the psammophytic tundra on aeolian sands is observed (Fig. 3.4). On Erge-Muora-Siseh Island *Carex concolor* grows on sandy landscapes at lower altitudinal levels. The higher elevations of the island support the dwarf shrub spotted tundra with *Salix nummularia, Cassiope tetragona.* At the shoreline of the eastern part of the Lena River delta waterlogged polygonal bogs and herb – green moss tundra are widespread. On the ridges of polygons *Carex concolor* or *Salix polaris* grows. The polygons are partly covered by water and partly by *Carex concolor, Eriophorum scheuchzeri,* or sometimes *Arctophila fulva.* A moss layer of *Drepanocladus spp.* is also common.

The presence of loose sands guarantees proper drainage of the soils. This determines a high specificity of the vegetation in this region: psammophytes are dominant and sub-dominant, and lichens become important (Labutin et al. 1985). The wet soils of the Lena River delta support the endemic lichen – green moss tundra with participation of *Siphula ceratites* and *Andraea rupestris.*

*The Yana-Kolyma arctic sub-province.* The Indigirka-Kolyma district covers the East-Siberian Sea shoreline within the southern strip of the Arctic tundra. It is characterized by a combination of the hummocky spotted dwarf shrub tundra dominated by *Salix polaris* and *Dryas punctata, Carex arctisibirica* and polygonal small-grass tundra. Eastwards of the Indigirka River *Carex lugens* appears in the Kolyma River basin. The tussock tundra with *Eriophorum vaginatum* is also common there.

From the Lower Kolyma River the Arctic hummocky spotted dwarf shrub and polygonal small-grass tundra associations were described. The microrelief is composed of hummocks of 10–15 cm high and 10–35 cm in diameter. In some places small hollows are observed as deep as 5–7 cm. Patches of bare ground on the top of hummocks may occupy up to 10%. *Dryas punctata* and *Salix reticulata* are the dominant species. *Arctagrostis latifolia* is also abundant. The communities also contain *Calamagrostis holmii, Poa arctica, Tephroseris atropurpurea, Potentilla hyparctica, Luzula confuza, Saxifraga cernua,* etc. Lichens occur scattered: *Thamnolia vermicularis, Cetraria cucullata.*

The Arctic polygonal small-grass tundra covers large areas in the District. The polygons are low; the fractures are narrow and rather dry. The forb + *Eriophorum* association is dominated by *Eriophorum scheuchzeri* or *E. polystachion,* and *Dupontia fischeri* with aboveground cover values of up to 70%. *Arctagrostis latifolia, Calamagrostis holmii, Poa arctica, Tephroseris atropurpuruea, Saxifraga cernua, Luzula confuza* are also common. The moss layer is represented by *Polytrichum* and *Sphagnum* species. Common lichens are *Peltigera aphthosa, Cladonia amaurocraea.*

The North-facing slopes of ridges in the Arctic tundra may represent the *baidjarakh* complexes covered by a rich vegetation. Between the earth mounds narrow stripes of *Eriophorum polystachion, Carex concolor* and *Dupontia fischeri* grow.

The Novosibirskie Islands, situated within the northern strip of the Arctic tundra, feature widespread earth mound microcomplexes with a vegetation cover interspersed by polar deserts (Perfilyeva et al. 1991). The central part of Kotelny Island is occupied by the ancient Lower Paleozoic plateau with an altitude of 170–180 m, reaching in some places 374 m. It is covered by polygonal spotted calciphilous tundra communities. The edges of the island are plain. Earth stripes are also common there. On the Malakatyn-Tas mountain the belt of Arctic desert is distinct. Bunge Land represents a sandy level area. Only the northern- and southernmost extremities, as well as the Evsekyu-Bulgunnyakh hill support pioneer vegetation. The islands Bolshoy and Maly Lyakhovsky, Faddeeyvsky and Novaya Sibir are characterized by lowlands and ridges and earth mounds are common. The watershed areas are covered by hummocky and partly spotted tundra. Bennett Island is characterized by a mountainous landscape with a glacier in its central part. The ice-free basalt detritus supports synusiae of crustose lichens, which are characteristic for polar deserts. The melkozem soils of the watershed areas are dominated by the herb – *Salix* – green moss tundra with abundant *Alopecurus alpinus* and *Salix polaris*. In the valley of the Lagernaya River *Dryas punctata* tundra was observed with a cover of up to 50%. The Tollevskaya River's banks are covered by thick mats of *Alopecurus alpinus* as high as 20 cm.

### 3.3.3 Subarctic Tundra Vegetation

The vegetation cover of the subarctic tundra develops under more favourable climatic conditions compared to the Arctic tundra. They are characterized by a longer and warmer vegetation period, a more regular distribution of the snow cover, and some increase in summer precipitation rates. The permafrost induced nanorelief is very common there, represented by polygons and hummocks. The vegetation is formed of prostrate shrubs, dwarf shrubs, perennial herbs, mosses and lichens that grow in places where the cool summer and other factors impede tree growth. The subarctic tundra is characterized by a richer flora, a more complicated structure of thick sods, a significant participation of shrubs, and the appearance of isolated trees in the southern part.

Compared to Russia, the species composition of the Yakutian subarctic tundra is not so diverse, reaching 200–400 species (216 in the Pokhodsk settlement vicinity, 268 in Ambarchik vicinity, 204 in Sukharnaya, 241 in Krutaya Dresva, and 392 in Tixi vicinity) (Petrovsky and Korolyova 1979, 1980; Petrovsky and Zaslavskaya 1981; Egorova et al. 1991).

The northern strip of the subarctic tundra covers only about 1.3% of the territory of Yakutia (Andreyev et al. 1987). The climatic conditions there are still severe for a proper development of shrubberies. So, only in the river floodplains they may form thickets (*Salix glauca)*, while in the watershed areas they are represented by low plants that do not rise over the herb-dwarf shrub layer (20–30 cm). The most common species there is *Salix pulchra*. The dominating vast landscapes of the northern subarctic tundra are the dwarf shrub-green moss tundra westwards of the Lena River and the *Eriophorum vaginatum* tundra eastwards. The *Carex* – green moss associations are also common on the peaty gleyey soils. The moss layer forms the basis of the vegetation cover there, and is dominated by *Aulcomnium turgidum, Hylocomium splendens var. alaskanum, Dicranum spp.*, and *Polytrichum spp.* The thick moss mat includes scattered lichens: species of *Cladonia, Cetraria, Thamnolia, Peltigera,* etc. The herb-dwarf shrub layer is predominated by *Carex arctisibirica* (in the West). The dwarf shrubs are the prostrate willow species: *Salix pulchra, S. reptans, S. reticulata,* etc. *Dryas punctata*, *Pedicularis lapponica, Claytonia arctica, Lagotis minor, Saxifraga spp.* also occur with low abundance. Sometimes *Ledum decumbens, Vaccinium vitis-idaea, V. uliginosum,* and *Casiope tetragona* are observed.

The northern subarctic tundra with *Eriophorum vaginatum* in the eastern part of Yakutia is characterized by relatively dry soils and weakly developed tussocks. They are as high as 20–40 cm and occupy about 50% of a community's area. Green mosses grow between the tussocks (*Aulacomnium turgidum, Hylocomium splendens var. alaskanum, Dicranum spp.,* and *Polytrichum spp.)* with an admixture of lichens and sometimes *Sphagnum* mosses. The tussocks may serve as a substrate for isolated dwarf shrubs (*Vaccinium vitis-idaea, V. uliginosum, Ledum decumbens*, *Cassiope tetragona, Dryas punctata,* and *Diapensia lapponica ssp. obovata),* as well as for some grass and *Carex* species.

In the Khroma-Indigirka and Indigirka-Kolyma Interfluves, on prominent elements (25–30 m a.s.l.) of the relief, the *Eriophorum vaginatum* tundra is observed in a combination with hummocky tundra. The latter resembles the Arctic hummocky tundra by a similar species composition of mosses and lichens. However, they feature a different species composition of flowering plants and diverse microcomplexes of the vegetation due to a complicated microrelief (Andreyev et al. 1987; Perfilyeva et al. 1991). *Salix pulchra, S. reptans, S. sphenophylla, Vaccinium vitis-idaea, Arctous alpina* and *Diapensia lapponica ssp. obovata* are common there.

The dwarf shrub-herb tundra takes an intermediate position in the ecological series between the hummocky and *Eriophorum vaginatum* tundra stages. They are more widespread to the West of the Lena River. The constant shrub and dwarf shrub species are *Salix pulchra, S. glauca, S. polaris, Vacciniun vitis-idaea, Cassiope tetragona*; the herbs are represented by *Carex concolor, Eriophorum polystachion,* sometimes *Eriophorum vaginatum.* The ground layer is formed of *Aulacomnium turgidum* and *Hylocomium splendens var. alaskanum.*

The southernmost part of the tundra zone is represented by the southern strip of the subarctic tundra. It occupies 3.5% of the total territory of Yakutia (Andreyev et al. 1987). This subzone is best pronounced in the western part of the Asian Arctic, while in Yakutia it is less represented (Skryabin and Karavaev 1991; Andreyev et al. 1987). The most widespread communities include *Betula exilis* and *Salix pulchra* with aerial covers of 10–20% and a height of up to 50 cm. The *Eriophorum vaginatum* and hummocky tundra are common eastwards of the Lena River and the hummocky and lichen tundra westwards (Andreyev et al. 1987). The high diversity of tundra types is determined by the topological peculiarities and soil charactertistics.

The dense *yernik* tundra of *Betula exilis* is not common in Yakutia due to ultra-continentality of the climate and the low amount of snow. The only large thicket of *Betula exilis* was recorded in North–West Yakutia in the lower reaches of the Anabar, Buolkalakh and Olenyok Rivers. *Betula exilis* forms there thick (cover 40–60%) though not high (up to 60 cm, very rarely reaching 1.5 m) shrubberies with a dense moss layer of *Aulacomnium turgidum, Hylocomium splendens var. alaskanum, Dicranum spp.,* and *Polytrichum spp.* and sparse lichens (*Cetraria cucullata)* depending on moisture conditions.

The southern subarctic tundra with *Eriophorum vaginatum* forms a series from the lichen – green moss to the green moss – *Sphagnum* types (Andreyev et al. 1987; Perfilyeva et al. 1991). The shrub layer is composed of *Betula exilis* and *Salix pulchra* as high as 50 cm and a cover of 10–20%. The moss layer is represented by *Aulacomnium turgidum, Hylocomium splendens var. alaskanum, Ptilidium ciliare, Tomenthypnum nitens, Sphagnum spp.* The common dwarf shrubs are *Ledum decumbens* and *Vaccinium vitis-idaea.*

The hummocky subarctic tundra is best developed west of the Lena River occupying the prominent parts of the relief. This spotted pattern (15%) is most pronounced in the Anabar River basin, while towards the East (the Indigirka and Kolyma Rivers' basins) this feature is reduced, not exceeding 5–10%. The prevailing association is the sparse low shrub – dwarf shrub – lichen – green moss tundra (Andreyev et al. 1987). The shrub layer is composed of *Betula exilis, Salix pulchra,* and *S. glauca* (height up to 50 cm, cover 10–20%), dwarf shrubs are represented by *Vaccinium vitis-idaea, V. uliginosum, Ledum decumbens, Arctous alpina.* The ground cover is formed by *Aulacomnium turgidum, Hylocomium splendens var. alaskanum, Dicranum spp., Polytrichum spp., Cetraria spp.*

In the basins of the Indigirka and Kolyma Rivers vast areas are occupied by the hummocky-tussocky *Eriophorum vaginatum* tundra. It represents an intermediate stage of succession between the hummocky and *Eriophorum vaginatum* tundra stages (Andreyev and Perfilyeva 1980). The determining factor here is increase in elevation. However, the effects of grazing reindeer should also be taken into consideration. The Kondakovskoye tableland lacks such intermediate associations, since this stressing factor does not occur there (Perfilyeva et al. 1981). In this sub-zone *Duschekia fruticosa* and *Betula divaricata* appear in the valleys, as well as isolated specimens of *Larix.*

The southern subarctic tundra also commonly features the green moss – lichen associations, especially in the Anabar River basin, where they occupy significant areas (Sochava 1933b). This can be explained by a general increase of the altitudinal level southwards. They inhabit the well-drained tops of hills and ridges composed of sandy alluvial or eluvial deposits. Depending on topology and local ecological factors, the lichens may differ in structure and species composition. Basically *Alectoria* and *Cetraria* formations can be distinguished. The most widespread communities are represented by *Alectoria ochroleuca*, especially to the West of the Anabar and Olenyok Rivers. This fruticose lichen species forms a rather thick (up to 10 cm) cover with an admixture of other lichens. *Cetraria cucullata* communities are also common. As opposed to *Cladonia* species, *Cetraria* is less prefered by reindeer as a food. In the Yana River basin the *Cetraria* tundra is represented as a combination with waterlogged areas with *Sphagnum* and *Eriophorim vaginatum*, while in the tundra-bog complexes they are interspersed by patches of green and *Sphagnum* mosses. Higher plants form a very scanty layer. It is composed of the arctic-alpine prostrate and adpressed dwarf shrubs (*Dryas punctata, Cassiope tetragona).* Patches of bare ground are also common.

The subarctic tundra consists of three sub-provinces (see Table 3.1).

*The Anabar-Lena subarctic (hypoarctic) sub-province* covers the Anabar and Olenyok Rivers' basins, as well as the southern hypoarctic part of the Lena River Delta. In the North of the sub-province the herbaceous tundra with *Carex concolor* and *Eriophorum polystachion* is a prevailing landscape, while in the South the lichen and dwarf shrub hummocky tundra predominates. Besides, the Pronchischev Ridge supports the green moss and lichen stony-spotted, as well as other types of the mountain tundra with participation of *Salix polaris* and *Dryas punctata.* Southwards, the lichen tundra alternates with the widespread hummocky tundra with hypoarctic dwarf shrubs. Compared to other regions, the *yernik* tundra is also common in this sub-province.

In the Lena River Delta the herbaceous tundra associations with *Carex concolor* predominate. On the Erge-Muora-Siseh and Jeppiries-Siseh Islands they occupy the lower elevations along with the desiccating polygonal-ridged tundra-bog complexes. The upper levels are inhabited by the psammophytic tundra with *Salix nummularia* and *Cassiope tetragona.* The residual islands (Khardang-Siseh, Kurugnakh-Siseh) are dominated by the polygonal-ridged tundra-bog complexes being at early stages of desiccation. The dried out tundra-bogs are replaced by the flat hummock-hollow complexes and sometimes communities of the herbaceous tundra. The elevated sites are covered by the *Eriophorum vaginatum* tundra.

*The Sub-Province of near-Lower Lena mountains* covers the Chekanovsky Ridge and the Kharaulakh Range. It is characterized by high diversity of the lichen and dwarf shrub mountain tundra types. The Kharaulakh Range, up to the latitude of the Tiksi settlement, features wide-spread earth strip complexes.

*The Yana-Kolyma subarctic Sub-province* stretches as a strip along the coastline of the Yana Bay eastward of the Kharaulakh Range as far as the right bank of the Alazeya River. Then it goes to an open sea from the Krestovsky Cape to the Kolyma River. The lowlands of the Indigirka-Alazeya Interfluve are dominated by the polygonal-ridged and flat-hillocky tundra-bog complexes of various composition. The dominating landscapes in the North of the sub-province are represented by the *Eriophorum vaginatum* tundra, while in the South it occurs in complex with the lichen or hypoarctic dwarf shrub hummock tundra occupying elevated parts of the relief. The basins of the Khroma and Berelekh Rivers feature flat-hillock – hollow complexes with the *Eriophorum vaginatum* and *yernik* tundra on hummocks.

The Alazeya-Kolyma Interfluve is characterized by the *Eriophorum vaginatum* and dwarf shrub hummock tundra on ridges, and polygonal-ridged tundra-bogs in depressions and level areas.

## 3.4 Boreal Vegetation

### *3.4.1 General Characteristic of Boreal Forests*

#### A.P. Isaev and P.A. Timofeyev

The boreal vegetation of Yakutia is represented by the near-tundra, northern taiga, middle taiga, and mountain taiga forests, as well as by various types of extra- and intrazonal vegetation.

According to the data of the Forest Department of the Ministry of nature conservation of the Republic of Sakha (Yakutia) as of 01.01.2008, the general area occupied by the forest zone is 256.1 million hectares, or 82.5% of its territory. The forests themselves cover 158.0 ha, or 61.7% of the forest zone and 51.3% of the territory of Yakutia. The forest cover is very uneven, from 11.5% in the Verkhoyansk mountainous region to 91.7% in South Yakutia.

The major forest forming trees species are *Larix cajanderi, L. gmelinii, L. sibirica, L. czekanowskii* (*L. sibirica* x *L. gmelinii*), that occupy 122.6 million hectares or 77.5% of the forest zone. Less common species are *Pinus sylvestris* (10.2 million ha or 6.5%), and *Picea obovata* (0.36 million ha or 0.24%). South–West Yakutia features *Pinus sibirica* (0.41 million ha or 0.25%) and *Abies sibirica* (0.02 million ha or 0.01%). In the South mountainous region *Picea ajanensis* occurs. The deciduous tree species (*Betula pendula, B. pubescens, B. ermannii, Populus suaveolens, P. tremula, Chosenia arbutifolia,* and arborescent *Salix spp.*) occupy about 2 million ha or 1.24% of the forest zone. *Pinus pumila* covers large areas (7.2 million ha or 4.6%), basically in the mountains. Other shrubs (mainly *Betula nana ssp. exilis, B. divaricata, B. fruticosa*, *Duschekia fruticosa*, frutescent *Salix spp.*) occupy 15.2 million ha or 9.6% of the forest zone. The distribution areas of the basic forest forming tree species are depicted in Figs. 3.5 and 3.6.

The general timber reserve of the forests of Yakutia is assessed as 9.2 milliard $m^3$, of which 95.6% is coniferous. One hectare contains: 58 $m^3$ of forest reserves on average, 83 $m^3$ of ripe and over-ripe tree stands, 62 $m^3$ of *Larix spp.*, 104 $m^3$ of *Pinus sylvestris*, 130 $m^3$ of *Picea spp.*, 188 $m^3$ of *Pinus sibirica*, or 41 $m^3$ of *Betula spp*.

Due to continuous and thick permafrost and vigorous cryogenic processes, the forests of Yakutia differ from the boreal forests of both Russia and the Northern Hemisphere as a whole. They are peculiar for growing conditions, ecological

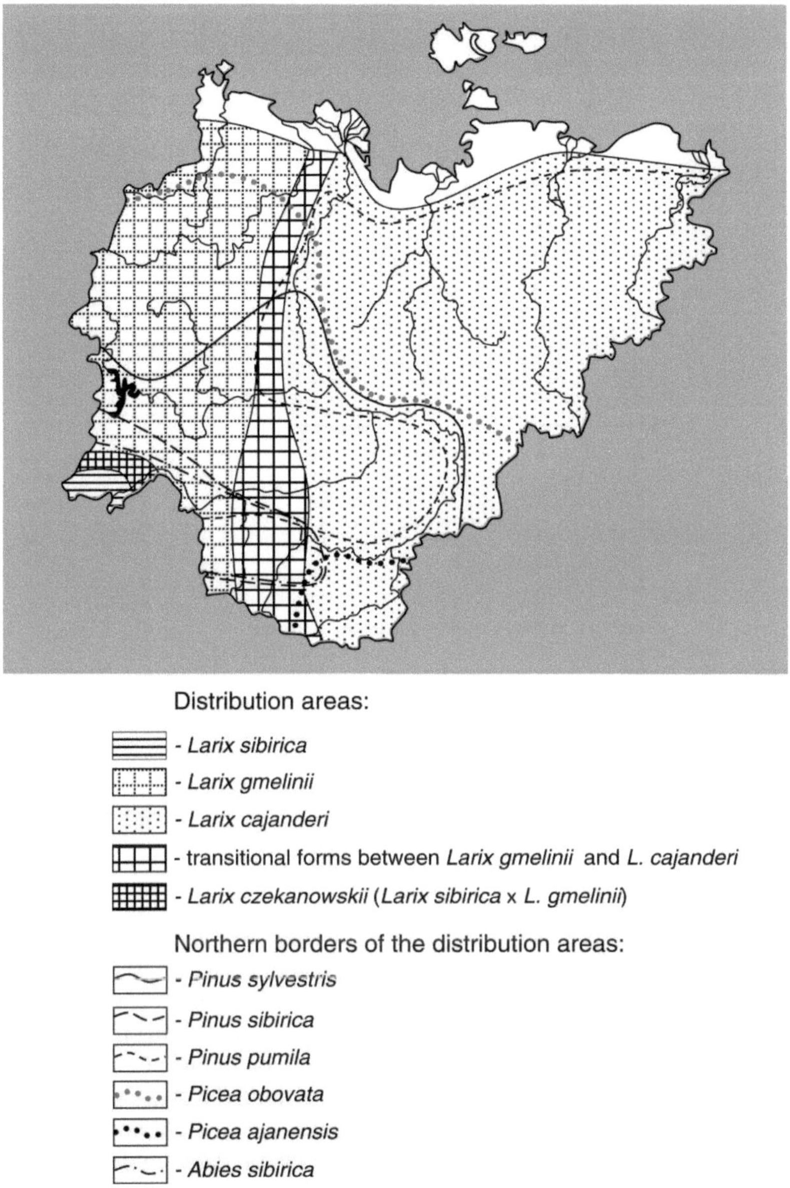


**Fig. 3.5** Distribution areas of the basic forest-forming coniferous species

composition of the flora and fauna, typological diversity, spatial and functional structure of the vegetation, tree stand productivity and recovery capability. The following regional peculiarities of Yakutian forests are considered for utilization, restoration and protection activities.

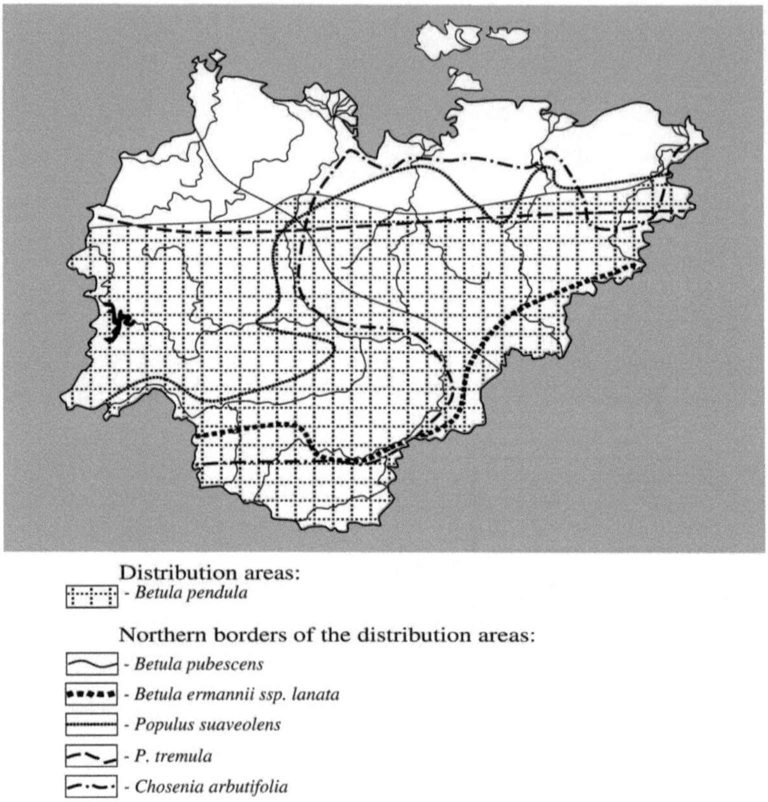


**Fig. 3.6** Distribution areas of the basic forest-forming deciduous species

The forest cover is predominantly light coniferous forests of *Larix spp.*, as well as *Pinus sylvestris* to a lesser extent. The dark coniferous forests of *Picea spp., Abies sibirica* and *Pinus sibirica* occur either in the river valleys with rather moderate microclimatic and growing conditions, or in the mountainous regions of South Yakutia characterized by more favourable moisture conditions of air and soils. The original belt communities of deciduous forests are limited to the large river valleys and alases. Thus, the zonal vegetation is characterized by light coniferous forests and secondary deciduous forest communities, while the dark conifers and original deciduous forests represent the extra- and intrazonal vegetation.

The forests are characterized by a distinct coenomorphic and ecological composition comprising meadow, steppe and bog species; by a prevalence of light-demanding and mesotrophic woods and herbs; a predomination of vegetative propagation both for trees and perennial herbs. The herb vegetation lacks annual plants. The wide ecological variety of the higher vascular plants of the forests of the cryolithozone is conditioned by low values of canopy density (crown closure) and significant fluctuations in the hydrothermal regimes of the soils during the growing season. The moss and lichen floras are rather rich, though the forests of the arid regions of Central and South–East Yakutia are often characterized by a scarce moss-lichen layer. Perennial plants of the forests growing on permafrost annually pass a full development cycle and successfully propagate asexually. This provides a timely vegetation recovery when disturbed by outer factors.

The air and soil temperatures vary significantly in the vertical profile of the forest biogeocoenoses. This is especially pronounced in mid summer, when all life processes of the plants are most intensive. In summer time, the above-ground parts of plants, especially trees and shrubs, function under high temperatures (25° – 28° or higher), and the underground organs under low positive temperatures (4– 5° or lower) (Timofeyev 1980; Pozdnyakov 1983, 1986; Isaev and Mikhalyova 2000). This puts the forest plants under extremely harsh conditions at the peak of vegetative growth and regeneration, when water and mineral supplies are not sufficient, while transpiration is very intensive. Owing to this, the perennial plants are not able to use their growing and developmental capacity at full and acquire a dwarf shape. This is characteristic for all tree and perennial herb species. The major forest-forming species (*Larix spp., Pinus sylvestris*) reach 5–7 m in height in the near-tundra zone, up to 12–20 m in the northern taiga and 25–35 m in the middle taiga zone.

The forests of the cryolithozone possess a rather plain vertical coenotic structure: clean and one-layered, uneven-aged tree stands with low values of tree cover; a weakly developed undergrowth; a more or less thick dwarf shrub-herb layer composed of few species; a scarce moss or lichen layer. They are characterized by low productivity of both trees and the lower layers of vegetation. This is also true for the forests of Central Yakutia, which thanks to the natural-climatic conditions and permafrost activity, is considered the geographical centre of the cryolithozone. Owing to this, the forests growing on permafrost have weaker edificatory and protective properties. The frozen soils under forests are less favourable for phytomass accumulation. These forests of a plain spatial structure are inhabited by small numbers of animal species despite the fact that there are sufficient and various reserves of forage resources. It is the weak protective function of the forest biogeocoenoses that forces the animals to migrate to other sub-zones or forest formations during the unfavourable seasons of a year.

There are three moisture sources in the forests of the cryolithozone. They are vertical precipitations (rain, snow, etc.), horizontal precipitations (dew, hoarfrost, etc.), and thaw water of frozen grounds that significantly contribute to soil moisture content. Each moisture source has a noticeable effect on the physiology of forest plants, their life rhythms, growth and development (Utkin 1976; Timofeyev 1980). Thaw water compensates the moisture deficiency in soils of the forest biogeocoenoses that is often observed during the most of the growing season in the forests of dry and mesic types, and in the middle of the growing season in the moderately wet forests.

The cycles of energy and matter in the forest ecosystems are characterized by rather low rates due to the cooling effect of permafrost, this unique ecological factor that is absent in other regions of the Earth. The frozen soils lack a loss of products of organic matter decomposition. This results in a potential soil salinization and an intensity of allelopathic interrelations of plants. The latter determines the biological productivity of plants in forest biogeocoenoses (Pozdnyakov 1962; Timofeyev 1976, 1979, 1990; Borisova and Popova 1987).

The natural seed reproduction of trees under a forest canopy has a shifting character due to periodicity of fruit bearing and frequent forest fires. Tree reproduction under the canopy of *Larix* forests in North Yakutia is promoted by quick surface fires that induce heat melioration by partial destruction of the dwarf shrub-herb and moss-lichen layers (Pozdnyakov 1963, 1975; Stepanov 1981, 1987; Tarabukina and Savvinov 1990; Timofeyev and Protopopov 1996; Isaev and Mikhalyova 2000).

The forests on permafrost are very vulnerable. Any disturbance may yield a strong transformation of the vegetation, up to desertification. The thawing of ground ice lens leads to thermokarst slumps and caldrons that transform into thermokarst lakes and, eventually, into alases (Utkin 1976; Pozdnyakov 1983, 1986; Desyatkin 1984; Bosikov 1991; Isaev and Mikhalyova 2000). The anthropogenic impact on forest ecosystems (forest fires, timber felling, forest tracts development for industrial and agricultural purposes) intensifies the thermokarst modification of landscapes. For instance, in the Tabaga settlement vicinity (the Lena-Amga Interfluve) the elimination of the *Larix cajanderi-Ledum palustre* forests by stable fires has led to the development of 104 small thermokarst lakes in a territory of 300 km$^2$.

## *3.4.2 Dendrochronology: Response of Trees to Various Nature Conditions*

**A.N. Nikolaev**

The systematic dendrochronological study in Yakutia started in 1990s. In 1991–1992 and 1994 a number of joint expeditions were conducted in collaboration with Russian and foreign specialists (Vaganov et al. 1996). The study has yielded tree-ring chronologies for the whole territory of North Yakutia based on core samples of *Larix cajanderi*. Some samples showed the life-span of over 600 years (Vaganov et al. 1996; Hughes et al. 1999). The results of the study have proved the significant potential of dendrochronological studies in Yakutia.

In 1997–1999 the scientists from the V.N. Sukachev Forest Institute SB RAS (FI) (Krasnoyarsk) conducted an expedition in the valley of the Kusagan-Mastakh River (North Yakutia). They recorded living trees as old as over 850 years (Vaganov et al. 1996). The core samples of the trees of North–East Yakutia were used for building the long chronology covering 2,358 years from 359 B.C. through 1998 A.D. (Sidorova and Naurzbaev 2002). Based on this tree-ring chronology, a reconstruction of the June-July and average annual air temperatures was made at the FI. The results of the analysis show that the warmest period for North–East Yakutia took place in the beginning of first millennium (Naurzbaev et al. 2003; Sidorova and Naurzbaev 2005). Later on, in 2004, the joint expedition of the FI, Institute of Permafrost SB RAS (IP) and Laboratory of dendrochronology of Arizona University (Tucson, USA) was conducted. It revealed interesting specimens, such as a 945 year old living tree and remains of a dead tree as old as 1,216 years.

Starting from 1996, the dendrochronological material was collected in many regions of Yakutia, especially in its central part (Nikolaev 2000). Previous work stated that *Larix* in Central Yakutia grows as old as 250–400 years. However, in 1999 the joint expedition of FI and the Institute for Biological Problems of Cryolithozone SB RAS recorded the nearly 600-year-old trees in the Kyrbakan River basin (northern part of the Aldan Region) (Nikolaev et al. 2001). All of the abovementioned proves the uniqueness of the natural conditions of Yakutia due to the rigorous climate and the presence of perennially frozen grounds.

The radial growth of *Pinus sylvestris* and *Larix cajanderi* stems in Yakutian forests depends on external factors, especially weather conditions (air temperature, amount of precipitation, hydrothermal conditions of soils). Temperature conditions have the most significant effect on annual growth in the beginning of growing season, i.e. in May for Central Yakutia and in June for North Yakutia. The amount of precipitations has little effect on *Larix*, though remarkable on *Pinus*, especially in June-July. High temperatures in August, causing the soil to dry up, have a slight negative effect on *Pinus* growth and a positively influence on *Larix* growth.

Comparative analysis of radial growth of *Larix cajanderi* and *Pinus sylvestris* with air temperature and amount of precipitation given by five days intervals (from April through late September) has revealed sharp differences in the dates of the start of the growth season for both tree species (Figs. 3.7 and 3.8) as regards geographic location. The most pronounced differences were revealed for *Pinus sylvestris*. The radial growth of this species at the northern border of its distribution area is almost one month late compared to *Pinus* growing in Central Yakutia.

This phenomemnon is also true for *Larix cajanderi*, though the time lag is not so pronounced. This shows a better adaptability of *Larix* to heat perception under conditions of permafrost.

The analysis also determined the threshold values of air temperature that activate the radial growth of the trees. The positive correlation of radial growth starts from temperatures a little higher than 4° C. Most vigorous growth is observed at air temperatures close to 10° C. Further increase in temperatures (up to 12° C for *Pinus* and 14° C for *Larix*) showed insignificant correlations between air temperatures and radial growth. The duration of the influence of air temperature on tree growth in the North is longer than that in Central Yakutia, as clearly shown in the diagrams.

The influence of precipitation on *Larix* growth is not strong. As regards *Pinus*, precipitations have a significant effect in Central Yakutia and an insignificant at the northern border of the distribution area of *Pinus sylvestris*. Most likely, *Pinus sylvestris* is characterized by a higher competitive ability than *Larix cajanderi* on the dry sandy soils of Central Yakutia and experiences a lack of moisture during the period of vigorous growth. Thus, the temperature regime has an obvious limiting role in the beginning of growth season in North Yakutia, while for Central Yakutia

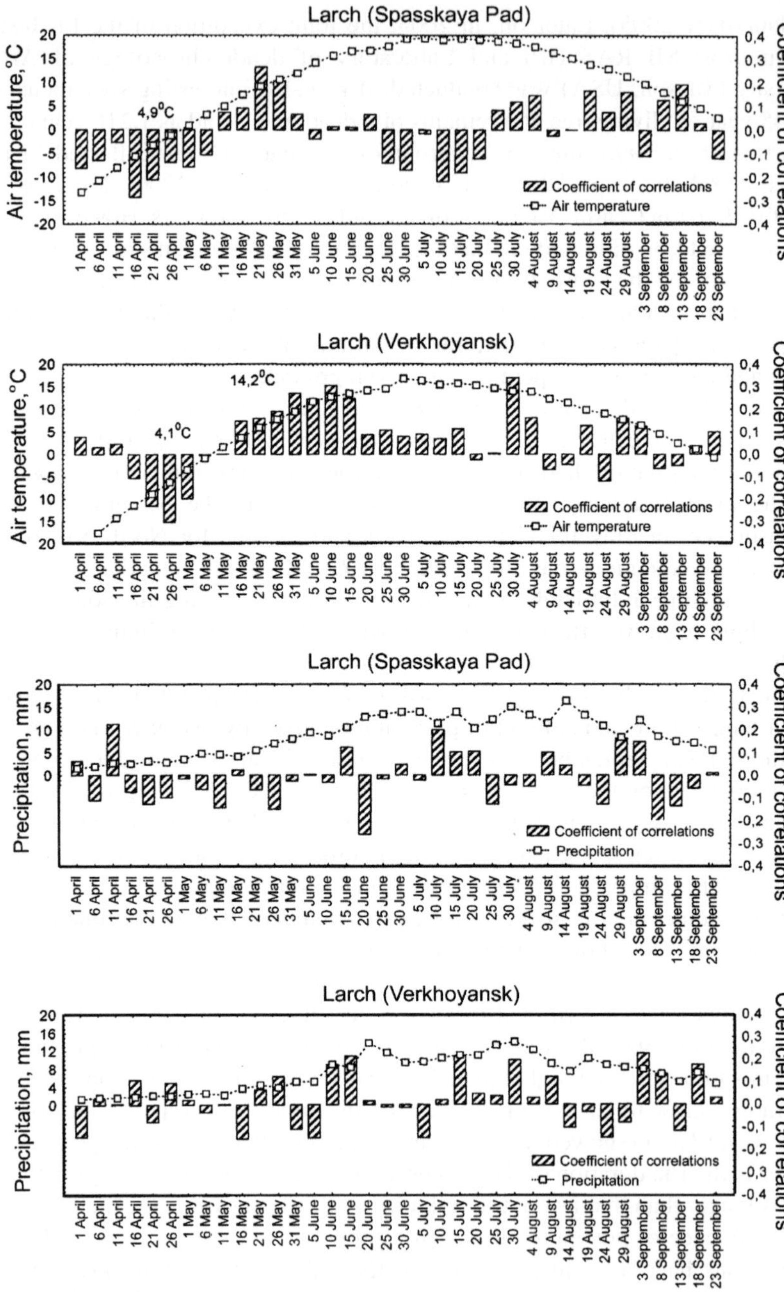


**Fig. 3.7** Correlation of the tree-ring chronologies of *Larix cajanderi* from Central Yakutia (the Reasearch Station "Spasskaya Pad") and North–East Yakutia (town of Verkhoyansk) with air temperature and amount of precipitation

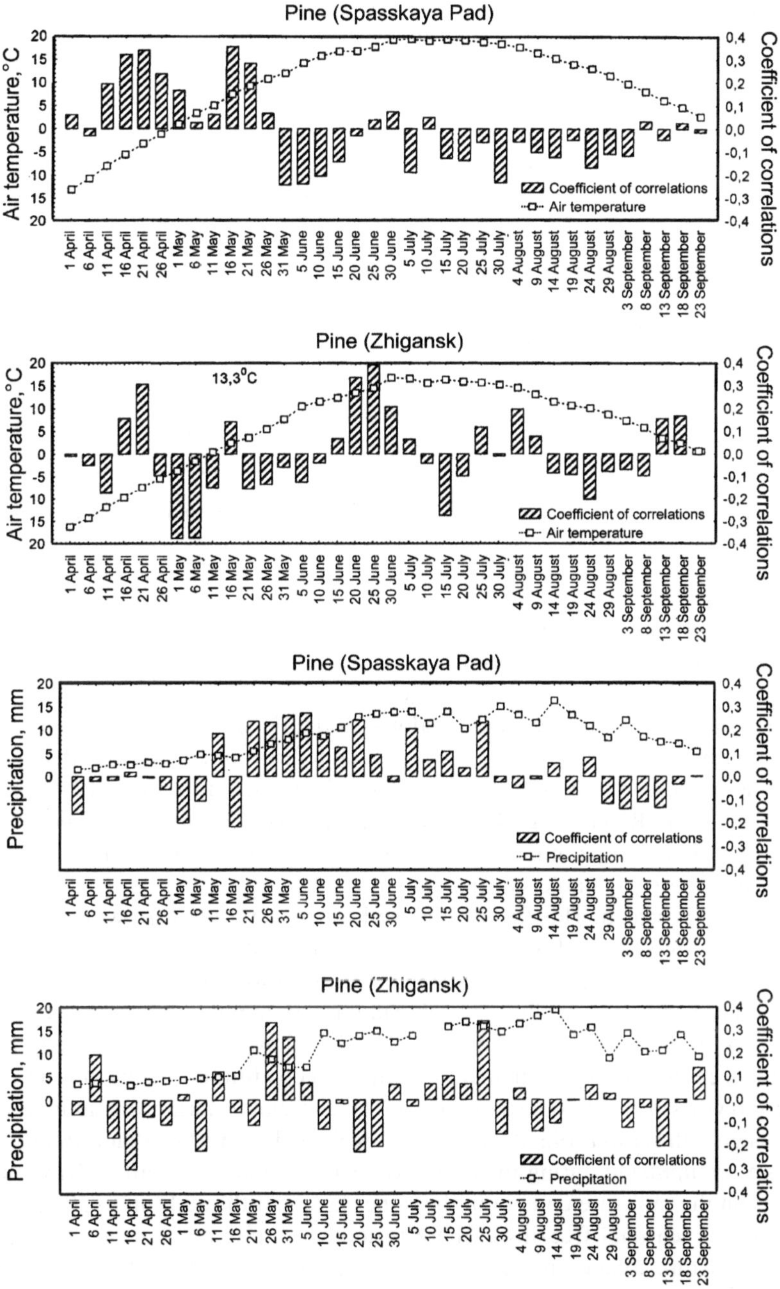


**Fig. 3.8** Correlation of the tree-ring chronologies of *Pinus sylvestris* from Central Yakutia (the Reasearch Station "Spasskaya Pad") and North Yakutia (Zhigansk settlement) with air temperature and amount of precipitation

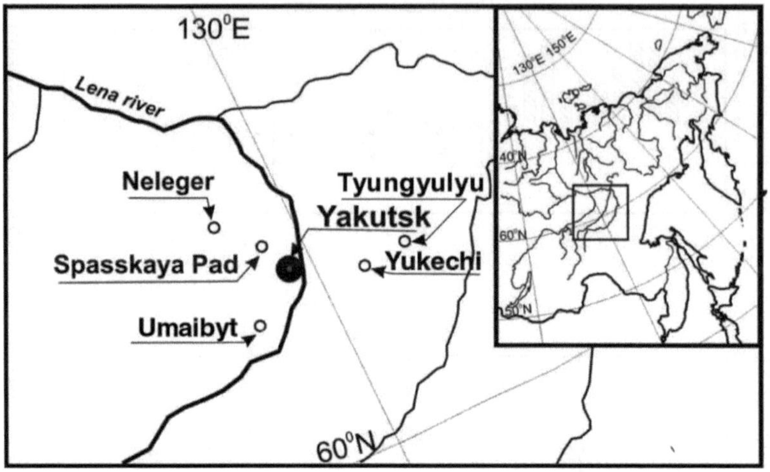


**Fig. 3.9** The sites in Central Yakutia where the dendrochronological study was conducted

the most important factors are the air temperature in early summer and precipitations in the first half of the growth season.

One of most important factors influencing the growth of trees is the hydrothermal condition of the soil (Nikolaev and Fyodorov 2004; Fyodorov et al. 2007). Relations of radial growth of *Larix cajanderi* and ground temperature at various depths were studied at the Research Station "Spasskaya Pad" (20 km North–west of Yakutsk, left bank of the Lena River) (Fig. 3.9). The analysis revealed a most significant influence of the winter temperatures of the soils on the radial growth of *Larix*. This must be related to more profitable starting conditions: the higher the soil temperature, the faster the soil is heated. This favours the well-timed start of tree growth at the beginning of growing season (Fig. 3.10). Summer temperatures do not limit the radial growth of trees since the accumulated heat is enough for proper growth. The results of the study conducted at the research Station "Tyungyulyu" (45 km North–east of Yakutsk, right bank of the Lena River) have shown well-pronounced correlations between the radial growth and hydrothermal conditions of soils throughout the whole growth season (Fyodorov et al. 2007), especially during the autumn of the previous growth season (Fig. 3.11). This is explained by the fact that *Larix* uses soil moisture accumulated during the previous year. The radial growth of *Larix* mainly depends on the moisture content in the upper layers of the soil (up to 50 cm during the period of seasonal freezing). There is another strong relation of radial growth and soil moisture content in September-October at a depth of 80–100 cm, i.e. at the end of the growth season, when the root system of *Larix* uses moisture from lower horizons due to the drying up of the upper horizons in summer.

Thus, the radial growth of *Larix* on frozen soils proves to be closely related with temperature and moisture conditions of the active soil layer during the growth season.

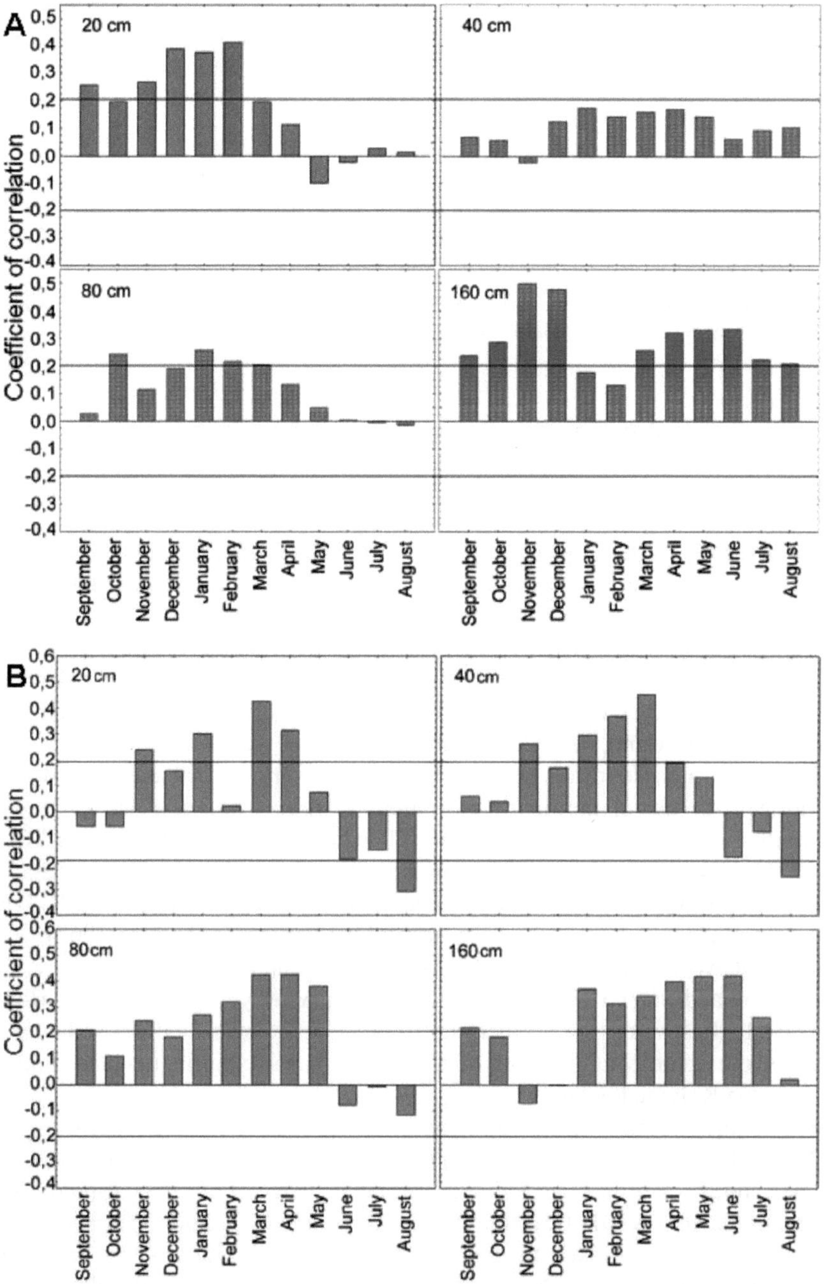


**Fig. 3.10** Results of correlation analysis of tree-ring chronology of *Larix cajanderi* at Spasskaya Pad (**a**) and Tyungyulyu (**b**) with soil temperature at various depths

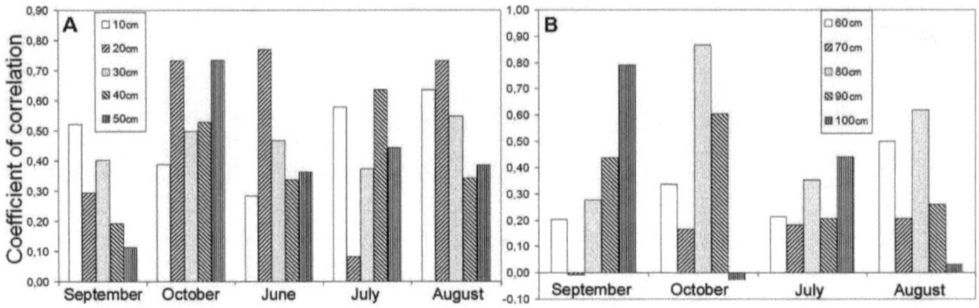


**Fig. 3.11** Results of correlation analysis of tree-ring chronologies of *Larix cajanderi* with soil moisture content at depths of 10 to 50 cm (**a**) and 60 to 100 cm (**b**)

## *3.4.3 The Near-Tundra Forests and Woodlands*

### A.P. Isaev and A.V. Protopopov

The near-tundra forests and woodlands are formed by *Larix gmelinii* in the West and *L. cajanderi* in the East. These forests are mainly old-aged (See Section 3.4.2). The old *Larix* trees are affected by rot, have curved trunks with dry tops, are curled and have cross-fibered wood. Their crowns are irregular, often flag-shaped.

The near-tundra forests of Yakutia form a stripe of *Larix* woodlands as wide as 60–220 km (Andreyev et al. 1987), being a peculiar northern outpost of the boreal zone. They are lowland, foothill and mountainous forests and woodlands.

The northern border of the near-tundra forests is very distinct and coincides with the northern border of the distribution areas of *Larix cajanderi* and *L. gmelinii*. The northernmost occurrence of larch woodlands is recorded in the West of Yakutia at 72° 30′ N at the Srednyaya River mouth, the Anabar River tributary (Sochava 1933b; Karpov 1980; Trufanova et al. 1981; Andreyev et al. 1987); at 72° 37′ N at the Olenyok River, 7 km south of the Taymylyyr settlement (Yurtsev 1962); at 72° N at the Tit-Aryy Island in the Lower Lena River (Cajander 1903; Komarov 1926; Tikhomirov and Shtepa 1956; Scherbakov 1965; Andreyev et al. 1987; Perfilyeva et al. 1991). Eastwards, the northernmost border of the near-tundra woodlands lies further to the South: within 70°30′ – 71° N in the Yana-Indigirka lowland and 69° – 70° N in the Kolyma lowland.

The southern border of the sub-zone is less pronounced, since the near-tundra larch woodlands gradually change into the northern taiga sparse forests.

The natural conditions determine the variable character of the near-tundra landscapes. In the very North–East (the Lower Kolyma River) region, characterized by numerous thermokarst lakes, the woodlands alternate with tundra-bogs and tundra. Westwards, in the lower reaches of the Indigirka and Yana Rivers, the alternation of the woodlands and the tundra is specified by the mountainous and foothill relief. The lower reaches of the Lena River, where mountainous landforms predominate, are characterized by combination of the mountainous woodlands and mountain tundra.

In North–West Yakutia, the Lower Olenyok and Anabar Rivers, the sparse forests of *Larix gmelinii* alternate with the typical tundra and tundra-bog landscapes, that are characteristic for the Subarctic tundra.

The near-tundra forests feature uneven-aged, thin tree stands of low productivity (Pozdnyakov 1948). Tree cover values range between 20 and 30%, the reproductivity is weak (Medvedeva 1971, 1973) and the edificatory role of *Larix* is also low compared to the southern regions. Distinct features of the near-tundra forests and woodlands are the more vigorously developed dwarf shrub – herb and ground layers compared to the true taiga; a similarity with the Subarctic tundra associations; and a pronounced complexity of coenotic components, especially of the lower layers due to presence of the tundra microrelief (hummocks, hollows, etc.). The convex parts are inhabited by xeric and mesic species (*Vaccinium vitis-idaea, Arctous alpina, Dryas punctata, Hierochloe alpina, Cetraria cucullata, Alectoria ochroleuca, Cetraria nivalis, Cladonia rangiferina,* etc.) characteristic for the dwarf shrub – lichen tundra. The concave elements of the microrelief are the habitats for hygrophilous species (*Ledum decumbens, Eriophorum vaginatum, Rubus chamaemorus, Chamaedaphne calyculata, Andromeda polifolia, Calamagrostis lapponica, Arctagrostis latifolia, Sphagnum spp.*).

The growth conditions determine the typology of the near-tundra forests. The *Larix* forests of the hilly landforms resemble the northern taiga forest communities (Andreyev et al. 1987) and are represented by the dwarf shrub – lichen – green moss, dwarf shrub – green moss, *Vaccinium vitis-idaea* – green moss, and *yernik* – green moss types (Cajander, 1903, 1904; Perfilyeva and Rykova 1975; Andreyev et al. 1987; Boychenko and Isaev 1992). They develop on shallow mountain loamy podburs with detritus. The tree stands are thin (crown closure 10–30%), represented by isolated curved *Larix* trees as high as 4–6 m (up to 15 m in the South). The undergrowth is properly developed and composed of *Salix pulchra, S. glauca, S. reptans, Duschekia fruticosa* and *Betula exilis.* The dwarf shrub-herb layer is dominated by *Ledum decumbens, Vaccinium vitis-idaea, Arctous alpina, Vaccinium uliginosum, Empetrum sibiricum,* and *Cassiope tetragona. Dryas punctata, Oxytropis middendorffii, Saxifraga spinulosa, Hierochloe alpina* are also common there. The moss layer (cover 30–60%) is dominated by *Aulacomnium turgidum,* and the lichens (40–50%) by *Cetraria cucullata,* and often *Alectoria ochroleuca.*

The water-logged areas on lowlands or foothills of ridges in highlands support sparse forests of the *Eriophorum vaginatum* – green moss type (Perfilyeva and Rykova 1975; Andreyev et al. 1987; Boychenko and Isaev 1992) that combine the *Larix* thin tree stand and *Eriophorum vaginatum* – green moss tundra as a ground layer. This forest type is the most swampy type. The trees are undersized (5–6 (8) m), the canopy density is 0.2–0.3. Dead wood is abundant. The trees are strongly covered with *Bryoria simplicior.* The undergrowth is composed of *Betula exilis* with a cover up to 60%. Among the dwarf shrubs *Ledum decumbens* is common. *Eriophorum vaginatum* covers 50–60%. Isolated specimens of *Calamagrostis lapponica, Arctagrostis arundinacea,* and *Valeriana capitata* occur as well. The moss layer is predominated by *Sphagnum warnstorfii, S. girgesohnii, S. lenense,* etc. Green mosses also are well represented with *Aulacomnium turgidum* the most abundant. Lichens are sparse.

The forests developing on the hillocky-hollow nanorelief are very peculiar. They occur in the near-tundra – forest-tundra transitional zone, particularly in the Kolyma River basin (Perfilyeva and Rykova 1975; Andreyev et al. 1987), the lower reaches of the Yana River (Tyrtikov 1955; Boychenko and Isaev 1992), and the Anabar River basins (Andreyev et al. 1987) in the small river valleys. The characteristic feature is a cryogenic nanorelief: hill diameter is up to 2 m, their heights range between 0.5 and 1 m. The uneven-aged tree stand is characterized by crown closure values of 10–20% and heights of 3 to 5 (9) meters. It is often given the epithet the “drunk” forest. The trees are covered with abundant *Bryoria simplicior.* The undergrowth is weakly (Boychenko and Isaev 1992) to properly developed (Tyrtikov 1955) and is made up of *Duschekia fruticosa, Betula exilis, Salix myrtilloides, S. fuscescens.* The properly developed dwarf shrub-herb layer (40 to 80%) is characterized by complexity. On hill tops the dwarf shrubs are predominantly *Vaccinium vitis-idaea* and *Ledum decumbens*, while at their base *Eriophorum vaginatum* strengthens its position. *Rubus chamaemorus, Chamaedaphne calyculata, Andromeda polifolia, Calamagrostis lapponica* and *Arctagrostis latifolia* are abundant. The synusia of mosses and lichens also significantly differ: *Cetraria cucullata, Alectoria ochroleuca, Cetraria nivalis,* and *Cladonia rangiferina* are characteristic for hill tops, patches of *Aulacomnium turgidum, Polytrichum hyperboreum, P. juniperinum, P. alpestre, Dicranum sp.* and *Ptilium crista-castrensis* grow on microslopes, while *Sphagnum warnstorfii, S. girgesohnii, S. lenense, S. fallax, S. balticum,* and *S. flexuosum* are confined to the inter-hill spaces.

The two abovementioned forest types are often represented by the *Rubus chamaemorus* and *Chamaedaphne* (*Chamaedaphne calyculata, Andromeda polyfolia, Oxycoccus microcarpus*) variants (Perfilyeva and Rykova 1975; Andreyev et al. 1987; Boychenko and Isaev 1992).

In the mountainous regions the well-heated south- and west-facing slopes support small patches of the dwarf shrub – lichen forests (Boychenko and Isaev 1992). They are also characterized by a sparse (cover density 20%) and undersized (4 to 7 m) tree stand. The shrubs are represented by isolated specimens of *Betula exilis,* or *Pinus pumila* in North–East Yakutia. Cover of the dwarf shrub-herb layer is 30–40%. It is dominated by *Vaccinium vitis-idaea* with participation of *Vaccinium uliginosum, Empetrum sibiricum,* and *Arctous alpina*. Isolated plants of *Hierochloe alpina, Claytonia acutifolia* and *Saxifraga spinulosa* are also present. The continuous lichen cover (80–90%) is composed of *Cetraria cucullata, Alectoria ochroleuca, Cetraria nivalis*, and here and there *Stereocaulon alpinum* and *Cladonia stellaris*. Mosses are practically absent.

The peculiar *Cetraria – Ptilidium* association of the dwarf shrub – lichen forest formation is very limited in the area and confined to relatively steep (up to 10–15°) and properly heated slopes. The multiple-aged tree stand is thin (30–40%) and low (4–5, sometimes up to 12 m). The undergrowth is represented by isolated specimens of *Betula exilis*. The dwarf shrub-herb layer (50%) consists basically of *Ledum decumbens,* while *Vaccinium vitis-idaea, Vaccinium uliginosum,* and *Arctous alpina* are less common. *Poa arctica* is prominent in the herb layer. Isolated *Claytonia acutifolia* occurs. Mosses and lichens form a distinct cover, predominated by *Cetraria cucullata, C. nivalis, C. laevigata,* and *Nephroma arcticum*, while the "cracks" in the lichen carpet are abound with *Ptilidium ciliare*.

Specialists have long been concerned with the question of *Larix* reproduction in the North and its significance for the forest – tundra relationship: whether forest advances and replaces tundra or vice versa. Some authors report a relatively good *Larix* reproduction at the northern border of its distribution area (Andreyev 1954, 1956, 1975; Andreyev et al. 1978; Tikhomirov and Shtepa 1956; Kryuchkov 1968; Andreyev et al. 1987; etc.) and consider it as one of most important proofs of present-day forest progression toward the North. Other experts contradict this (Tolmachev 1932; Gorodkov 1946; Isaev and Mikhalyova 2000; etc.). Indeed, the retreat of the forest vegetation in favour of the tundra landscapes is more often observed in nature. This phenomenon is usually induced by disastrous or anthropogenic effect. The "forest – tundra" border is characterized by the pyrogenic *Eriophorum vaginatum* – green moss tundra communities that appeared 100–150 years ago (Boychenko and Isaev 1992). This time span was not enough for forest vegetation development, since the very thick moss layer impedes successful recovery of *Larix*.

As has been mentioned before, the present northernmost woodlands represent the old-aged communities lacking young tree undergrowth. Here the scarce trees of the lower sub-layer are undersized old specimens (Boychenko and Isaev 1992; Isaev and Mikhalyova 2000) resembling young trees. Even the bark of such trees looks like that of young plants, smooth and without cracks. A very good example of such a tree stand is the *Larix* forest of the Tit-Aryy Island, first described by the Finnish scientist Cajander (1903, 1904) (Fig. 3.12). It was thought that during the World War

**Fig. 3.12** Old-aged *Larix* forest on Tit-Aryy Island

II (1941–1945) the whole forest was cut completely, but this did not prove correct. Presently, only thin "young" growth can de seen there with a height up to 2–3 m. An analysis of the age structure revealed that most of these "young" trees are older then 100 years, i.e. they were there long before the "complete" cutting of the forest. This growth was most likely favoured by the following ecological coincidence: long warm summers that are sufficient for *Larix* seeds ripening; elimination of the thick moss layer that impedes seed germination and sprout fixation; and drying up of the wet mineral substrate; etc.

In conclusion it seems that the system "forest – tundra" presently is in a stable equilibrium, but this can be disturbed by a combination of extraordinary circumstances.

### *3.4.4 Northern Taiga*

**A.P. Isaev and A.V. Protopopov**

One of the first explorers of the vegetation cover of the northern taiga subzone was the Finnish botanist, public and political figure Aimo Kaarlo Cajander. Basic results of his expedition were published in a number of works (Cajander, 1901, 1902, 1903, 1904; Cajander and Poppius 1903).

Further investigations of the northern taiga was continued by numerous specialists who described forest and non-forest communities of the region in details (Stark 1933; Rabotnov 1935; Sheludyakova 1938, 1948, 1957c; Yarovoy 1939; Pozdnyakov 1946, 1956, 1961a, 1969; Kuvaev 1956; Sochava 1957; Prakhov 1957; Pivnik 1958; Lukicheva 1960, 1963a, Yurtsev 1961, 1968, 1981; Dobretsova 1962; Karavaev and Dobretsova 1964; Scherbakov 1965; Buks 1966; Skryabin 1968; Medvedeva 1971, 1973; Perfilyeva 1977; Karpel and Medvedeva 1977; Stepanov 1981, 1987, 1988; Boychenko 1987; Nikolin 1992; Boychenko and Isaev 1996; Sosina and Zakharova 2004; Isaev et al. 2004; Zakharova et al. 2004, etc.).

The vegetation of the northern taiga subzone differs from the near-tundra forests by having a more proper forest character, with higher trees (up to 15–18 m), higher values of tree canopy coverage (40–60%), absence of tundra communities as a significant element of the vegetation cover, as well as a reduction of the "tundra" component in the understory. Most of the subzone are northern taiga forests of *Larix cajanderi* and *L. gmelinii*. Sparse forests prevail with canopy covers of 30–50%. Other species, such as *Pinus sylvestris, Picea obovata, Betula spp., Populus tremula, P. suaveolens, Chosenia arbutifolia, Salix* (*S. viminalis, S. udensis, S. schwerinii, S. rorida,* etc.) occur sporadically and in practice do not play any role in the forest cover.

The northern taiga forests west of the Verkhoyansk Mountains are characterized by *Larix* forests and woodlands of the *Vaccinium uliginosum* – green moss – lichen type (*Vaccinium uliginosum – Aulacomnium palustre + Cetraria cucullata + Cladonia rangiferina*) and *Vaccinium uliginosum* – lichen type (*Vaccinium uliginosum – Cetraria cucullata + Cladonia rangiferina*). Communities on calcareous bedrocks sometimes include *Picea obovata* as an admixture. These forests of low productivity are usually attributed to growth class Va or Vb. The western part of the northern taiga subzone features *Pinus sylvestris* and *Picea obovata*. Small patches of *Pinus sylvestris* forests reach the latitude of Zhigansk settlement (66° 46′ N and up to 68° N), practically the latitude of the North Polar Circle. Northwards the edaphic distribution area of *Pinus sylvestris* becomes limited to sandy landscapes (*tukulan*), where it occurs in isolated stands of several trees (Boychenko 1987). The northernmost occurrence of *Pinus sylvestris* was recorded 40–50 km north of the Zhigansk settlement, in the valleys of the Khoruonka, Ukhunku and Appanay Rivers. The northernmost records of *Picea obovata* were made in the Anabar and Olenyok Rivers, a little north of 70° N, while in the Lena River valley it is found up to 67°. However, it rarely forms independent communities within the northern taiga subzone, but usually occurs in narrow strips along ridges on river banks. More often *Picea obovata* participates in *Larix* forests or grows as single trees in river valleys.

The eastern part of the northern taiga subzone includes the mountain systems of the Verkhoyansk, Chersky and Momsky Ranges, the Oymyakon Upland, the Alazeya and Yukagir Tablelands, as well as of the Abyy and Kolyma Lowlands. The inter-ridge depressions and hollows feature rigorous and very specific conditions. There the coldest spots of the populated areas of the Earth are located, the Oymyakon settlement and the town of Verkhoyansk. The vegetation of mountain systems is described in Section 3.4.6, so here we just characterize the lowland northern taiga. The forest lowland vegetation is concentrated mostly in the Yana, Indigirka and Kolyma River valleys.

The Yana River basin features relatively highly productive forests in the valleys of large rivers and less productive forests towards the highlands. The river valleys are occupied by *Larix* forests of the *Vaccinium vitis-idaea* type. They are characterized by growth class IV to V, a tree coverage of 30–70%, and wood reserves up to 80–120 $m^3$/ha. Large areas of ancient terraces and gentle slopes of low ridges are covered by *Larix* sparse forests of the *yernik* type: *Ledum palustre* – lichen (*Betula divaricata – Ledum palustre – Cetraria cucullata + Cladonia rangiferina*). They alternate with *Larix* forests of the *Vaccinium vitis-idaea – Aulacomnium palustre* and *Vaccinium uliginosum – Aulacomnium palustre* types or with swamped *Larix* sparse forests (*Betula divaricata – Ledum palustre – Aulacomnium palustre + Sphagnum s.l.*) in depressions.

The vegetation of the Indigirka River basin consists of low-productive *Larix* forests and woodlands of three basic types: riparian *Equisetum pratense* and *Equisetum pratense – Calamagrostis langsdorffii* forming narrow belt communities along large rivers; *Larix* – lichen – green moss forests of the *yernik* type (*Betula nana ssp. exilis – Ledum palustre – Cetraria cucullata + Cladonia rangiferina + Aulacomnium palustre*) of Va–Vb growth classes and a canopy cover of 30–60% on slopes and ridge summits; *Larix – Carex* – moss and *Larix – Ledum palustre* – moss sparse forests of the Vb growth class and a canopy cover of 10–30% in lake and bog depressions.

In the Kolyma River basin most productive forests grow in river valleys. Narrow bank stripes bear the *Populus suaveolens, Chosenia arbutifolia* and *Salix* communities alternating with *Larix – Equisetum pratense* and *Larix – Equisetum pratense + Calamagrostis langsdorffii* forests of II–III growth classes, tree heights of 30–35 m and wood reserves over 300 m$^3$/ha. On terraces, with deteriorated drainage, these forests are replaced by the *Larix* woodlands of the *Vaccinium uliginosum – Aulacomnium palustre* and *Sphagnum* (*Betula nana ssp. exilis – Sphagnum s.l.*) types of Va or Vb growth classes. The slopes, where drainage is proper, are covered by *Larix* forests of the *Vaccinium vitis-idaea*, *Vaccinium vitis-idaea – Cetraria cucullata + Cladonia rangiferina*, and *Vaccinium vitis-idaea – Aulacomnium palustre* types.

### *3.4.5 Middle Taiga*

**A.P. Isaev, P.A. Timofeyev, I.F. Shurduk, L.P. Lytkina, N.B. Ermakov, and N.V. Nikitina**

The basic vegetation type of the middle taiga sub-zone of Yakutia is coniferous forest vegetation. The Braun-Blanquet approach attributes it to two large plant-geographical and phytosociological types which can be included in two classes: the hemiboreal *RHYTIDIO-LARICETEA sibiricae* Korotkov et Ermakov 1999 and the *VACCINIO-PICEETEA* Br.-BL. in Br.-Bl. et al. 1939 (Ermakov et al. 2000, 2002). The first class includes xeromesophilous, light coniferous, herb-rich forests of ultracontinental regions in Mongolia and southern and eastern Siberia. The true taiga forests of the class *VACCINIO-PICEETEA* absolutely prevail in the zonal vegetation of the vast territory of Central Yakutia. The special local climate and permafrost result in some phytosociological peculiarities in comparison with other types of the widespread taiga of northern Eurasia: (1) the predominance of light coniferous trees of *Larix cajanderi* and *L. gmelinii*, forming a light, open canopy; (2) the group of typical boreal herbs and shrubs (diagnostic species of the class *VACCINIO-PICEETEA*) is not numerous and does not play a distinct phytosociological role in the light coniferous forests in contrast to the dark coniferous taiga of regions with a weaker continentality; (3) the lack of mesophilous and moderately thermophilous boreal ferns, shrubs and herbs which are typical plants for regions with a lower degree of climate continentality; (4) the participation of xerophilous and meso-xerophilous species in boreal forests; (5) the insignificant role of mosses which are diagnostic species of the *VACCINIO-PICEETEA* forests in the main part of their range, e.g. *Pleurozium schreberi*, *Hylocomium splendens*, *Ptilium crista-castrensis*, *Dicranum polysetum*; and (6) a frequent presence of lichens of the genus *Cladonia* (*C. stellaris*, *C. rangiferina*, *C. arbuscula*, *C. amaurocraea*, *C. mitis*) in the lichen-moss layer of most associations (Ermakov and Cherosov 2005). These phytosociological peculiarities have led to the formation of a special name "light coniferous dry continental taiga" in the Russian geobotanical literature. In Central Yakutia, these forests are at the extreme northern limit of the class range as components of azonal isolated patches of forest-steppe.

The forests of the middle taiga subzone differ from the northern taiga by a high diversity of tree species. Besides *Larix gmelinii*, *L. cajanderi*, and *Pinus pumila,* the conifers are represented by *Larix sibirica, Pinus sibirica, P. ajanensis,* and *Abies sibirica.* The role of *Pinus sylvestris* and *Picea obovata* is increased. The tree crown canopy is more closed (40–80 sometimes up to 100% cover), the tree stand is more productive (of III–IV, rarely Va growth classes), the timber reserves in mature forests usually exceeds 120–200 $m^3$/ha, and the understory and herb-dwarf shrub layer are floristically more diverse.

The light coniferous forests are predominated by *Larix gmelinii* (westwards of 120–122° of eastern longitude (E)), *L. cajanderi* (eastwards of 120–122° E) and *L. sibirica* (south–westernmost Yakutia). They make up 79% of the forest-covered area. *Larix* forms a zonal type of vegetation. The ecological characteristics of *Larix* are very wide. It occupies various ecotopes at almost all forms of topology. This yields a very diverse typology of *Larix* forests. Since moisture is a leading ecological factor in the arid climate of the lowland middle taiga, the groups of forest types are distinguished with respect to moisture content in a certain ecotope (Timofeyev et al. 1994). They are the *Arctostaphylos uva-ursi* group of dry and fresh habitats, the *Vaccinium vitis-idaea* group of mesic habitats, the green moss group of wet habitats with stagnant moisture conditions, the herb group of wet habitats with running moisture conditions, and the *Sphagnum* group of waterlogged habitats. Specific sets of ecological conditions determine the predomination of the abovementioned types of *Larix* forests in different regions. For instance, Central Yakutia, characterized by a most arid climate, has predominantly *Larix* forests of fresh and mesic habitats (*Arctostaphylos uva-ursi,* xerophytic herb and *Vaccinium vitis-idaea* types). Ermakov et al. (2002) distinguishes 3 associations of *Larix* forests using the Braun-Blanquet approach. They are *Limnado stelleri-Laricetum cajanderi, Aquilegio parviflorae-Laricetum cajanderi* and *Ledo palustris-Laricetum cajanderi,* enumerated along a gradient of increasing moisture availability.

The association *Limnado stelleri-Laricetum cajanderi* belongs to the class *RHYTIDIO-LARICETEA sibiricae* and includes moderately xerophilous, open, herb-rich *Larix cajanderi* forests with participation of *Betula pendula.* It is found in the alases in combination with azonal patches of eastern Asian steppes of the class *Cleistogenetea squarrosae*, which occur on watersheds in the middle taiga zone with a summer-dry ultracontinental climate, thus taking the intermediate position between steppes and zonal taiga. Communities of this association occupy the periphery of alases on dark grey loamy soils and occurrence of permafrost at a depth of 1 m. The herb layer includes mesophilous and moderately xerophilous species which are typical for steppes, meadows and hemiboreal forests. It is well-developed with a herb layer cover of 60–80%, a height of 40–60 cm and a species richness of 43 to 60 species per 200 $m^2$. The moss layer is weakly developed (1–5%) and is represented by *Rhytidium rugosum, Abietinella abietina* and *Ptilidium ciliare*.

The association *Aquilegio parviflorae-Laricetum cajanderi* belongs to the class of true coniferous taiga forests of northern Eurasia, the *VACCINIO-PICEETEA*. It represents the typical communities of zonal light coniferous taiga of central Yakutia. It predominates in vast areas of flat and convex parts of watersheds with well-drained loamy soils and permafrost at a depth of 0.4–0.7 m. The characteristic feature of these sites is their uneven micro-relief. It is represented by numerous small knolls up to 5–10 cm which are related to local thermokarst phenomena.

*Larix cajanderi* forms a tree layer with low cover values of 20–40% and a height of 16–18 m. In some places *Betula pendula* and *Pinus sylvestris* occur in these communities. The shrub layer is poorly developed with a cover of 1 to 5%. *Vaccinium vitis-idaea* is the dominant of the herb layer and shows cover values of 15–80%. The herb layer includes mesophilous and xeromesophilous species which are typical of southern hemiboreal forests, widespread in the forest-steppe zone of Dauria and southern Siberia. The xeric features in the floristic composition, combined with the occurrence of the *VACCINIO-PICEETEA* species, are related to the ultracontinental climate of central Yakutia, which is characterized by large differences in seasonal indices of warmth and precipitation. The latter causes a poorly developed moss layer. The typical boreal mosses, *Pleurozium schreberi*, *Hylocomium splendens*, *Ptilium crista-castrensis* and *Dicranum polysetum,* have low cover and constancy values. The moss layer is dominated by *Ptilidium ciliare*, *Aulacomnium turgidum*, *A. acuminatum* and *Rhytidium rugosum*. Common boreal lichens (*Cladonia stellaris*, *C. rangiferina*, *C. amaurocraea*, *C. arbuscula*) occur in this association but they are never dominants.

The *Ledo palustris-Laricetum cajanderi* also represents an association of the *VACCINIO-PICEETEA*. It unites mesophilous larch forests of cool, often waterlogged soils. They are confined to wide depressions on watersheds and higher river terraces, with permafrost occurrence at a depth of 10–30 cm. Like in the previous association, the characteristic feature of the sites is an uneven micro-relief due to local thermokarst phenomena. Numerous higher knolls are as high as 10–35 cm. *Larix cajanderi* is the only dominant of the tree layer, showing low cover values of 25 to 35%. The low shrub layer has a cover of 5–10% being formed by *Duschekia fruticosa*, *Rosa acicularis*, *Salix myrtilloides*, and the dwarf shrubs *Vaccinium uliginosum* and *Ledum palustre*. The herb layer has a cover of 30–50% and a low species richness of 10–18 species per 200 $m^2$. *Vaccinium vitis-idaea* plays a leading role there. The uneven micro-relief supports some xero-mesophilous (and xerophilous in some places) plants occupying the tops of small knolls. The moss layer is well developed (cover values of 70 to 90%) and contains *Aulacomnium palustre*, *A. turgidum*, *A. acuminatum*, *Polytrichum juniperinum*, *Sphagnum girgensohnii*, *S. warnstorfii*, *S. fuscum* and *Tomenthypnum nitens*. Lichens include *Cladonia stellaris*, *C. rangiferina*, *C. arbuscula*, *Cetraria islandica*, *Peltigera aphthosa* and *P. canina* and occur predominantly at the tops of small knolls.

In South and South–West Yakutia, with more moderate climate features, the *Larix* forests of mesic and wet habitats occur (the *Vaccinium vitis-idaea, Ledum palustre, Vaccinium vitis-idaea – Hylocomium splendens,* and *Vaccinium myrtillus – Hylocomium splendens* types). In the western regions the *Larix* forests of wet habitats predominate (*Vaccinium vitis-idaea – Aulacomnium palustre*, *Ledum palustre -Aulacomnium palustre,* etc.).

The *Pinus sylvestris* forests cover significantly less area, occupying 10.7% of the forest-covered area. The ecological range of *Pinus sylvestris* is very narrow. In the northern part of the middle taiga subzone it occurs as monospecific tree stands, only on the most heated sandy and loamy sandy ecotopes: summits, terrace edges, south-facing slopes, etc. The typology of the *Pinus sylvestris* forests shows very little variation, and is represented mainly by *Arctostaphylos uva-ursi,* lichen (*Cladonia stellaris, Cl. rangiferina, Cetraria cucullata*) or related types. In South–West Yakutia *Pinus sylvestris* grows on more or less heated elevated landscape elements with dry or rarely mesic moisture conditions: watershed areas among the *Larix* taiga, terraces above floodplains, small ridge summits with loamy, sandy loamy, sandy or stony soils with proper drainage. *Pinus sylvestris* often participates in the *Larix* or even in the dark coniferous forests.

In Central Yakutia *Pinus sylvestris* forests are represented by two associations of the abovementioned classes of boreal forests (Ermakov et al. 2002): the *Festuco lenensis-Pinetum sylvestris* of the hemiboreal *RHYTIDIO-LARICETEA sibiricae* and the *Saxifrago bronchialis-Pinetum sylvestris* of the *VACCINIO-PICEETEA*. The first association represents most xerophilous coniferous forests in central Yakutia. Small areas of *Pinus sylvestris* forests in combination with steppes occur on steep south-facing slopes of watersheds of the Lena River valley. These moderately warm, eroded slopes are covered by shallow sandy-loamy soils, which are favourable for *Pinus sylvestris*. During summer the permafrost is found at a depth of 1.5 m. Communities are characterized by the low cover of the tree layer (20–40%) and sometimes are represented by sparse groves. The main feature of the herb layer is the prevalence of meso-xerophilous and xerophilous plants. Mesophilous species of the taiga class *VACCINIO-PICEETEA* occur rarely. The moss layer is represented by scattered patches of the xerophilous moss *Rhytidium rugosum*.

The *Saxifrago bronchialis-Pinetum sylvestris* includes *Pinus sylvestris* forests with participation of meso-xerophilous plants. They are typical of extensive areas of sandy deposits widespread on watersheds and higher terraces of the Lena River valley. Communities of the association occupy dry, moderately warm tops of small hills and convex parts of well-drained summits of watersheds with poor sandy soils and permafrost at a depth of 1.4 m. The tree layer has a low cover (30–40%). The shrub layer is poorly developed or absent. The herb layer is dominated by *Arctostaphylos uva-ursi* forming dense patches in combination with areas of open sandy soil and sometimes with patches of lichens. Other higher plants are scattered over the surface and have low cover values. Among them, facultative psammophytes predominate, such as *Thymus serpyllum, Potentilla bifurca, Carex vanheurckii, Artemisia commutata, Saxifraga bronchialis* and *Dianthus versicolor*. The moss-lichen layer varies in cover between 5 and 60% and depends on periodical fires.

The southern *Pinus sylvestris* forests are more diverse and are represented by the following types: *Vaccinium vitis-idaea, Rhododendron dauricum + Vaccinium vitis-idaea, Dryas punctata, Vaccinium vitis-idaea – Hylocomium splendens, Ledum palustre – Aulacomnium palustre*, etc.

The *Picea obovata* forests are observed all over the middle taiga subzone of Yakutia. However, their growth and productivity characteristics significantly change in latitudinal and longitudinal directions. Southern and south–western regions feature the *Picea-Larix* forests on watersheds and pure tree stands of *Picea obovata* on river banks and islands, i.e. it forms communities both under zonal and intrazonal conditions. In central regions, more continental and arid, *Picea obovata* acts as an edificatory or co-dominant species of an intra-zonal, narrow belt of communities along rivers and streams, and on islands. The only exclusion is the *Picea obovata* forests forming narrow strips around alases (Scherbakov 1992). Owing to this, the *Picea* forests are divided into two groups: the herb riparian and green moss watershed groups (Timofeyev et al. 1994). The riparian *Picea* forests are represented by the following types forming a spatial-chronological series: *Equisetum* or *Pyrola* type, *Equisetum – Sanionia uncinata, Equisetum – Rhytidiadelphus triquetrus, Vaccinium vitis-idaea – Rhytidiadelphus triquetrus, Vaccinium vitis-idaea – Hylocomium splendens, Vaccinium vitis-idaea – Pleurozium schreberi* and related types. The watershed *Picea obovata* forest diversity includes the *Vaccinium vitis-idaea – Hylocomium splendens + Pleurozium schreberi,* green moss (*Hylocomium splendens* and *Pleurozium schreberi*), fern – green moss and dwarf shrub – *Sphagnum* types.

Under the conditions of the humid climate (annual precipitations up to 1,000 mm) of the highlands of South Yakutia east of the Timpton River pure stands of *Picea ajanensis* occur. There they grow on acid bedrocks near the upper border of the forest belt, in intermountain depressions and gullies on North-facing slopes. The most favourable conditions for *Picea ajanensis* forest growth are observed in the upper part of forest belt from 600 m a.s.l. to its upper limit (1,600 m a.s.l.) (Timofeyev et al. 1994). The tree stands of *Picea ajanensis* often include an admixture of *Betula ermannii* that increases its role with altitude. It also can be part of the *Picea obovata* and *Picea obovata + Larix* forests, though always avoids calcareous bedrock. There are 4 types of *Picea ajanensis* forests distinguished for Yakutia: shrub (*Pinus pumila – Vaccinium vitis-idaea*), green moss (*Hylocomium splendens, Vaccinium vitis-idaea – Hylocomium splendens, Pleurozium schreberi, Gymnocarpium jesoense – Hylocomium splendens*), herb (*Calamagrostis langsdorffii* and *Pyrola incarnata*), and *Sphagnum* (*Ledum palustre – Sphagnum, Lycopodium annotinum – Sphagnum*) (Timofeyev et al. 1994). Small areas in South–west and partially South Yakutia are covered by *Pinus sibirica* forests, confined to the regions with abundant precipitation, humid air, and a less outspoken continentality. In Yakutia they grow at the North–eastern border of their distribution area (Fig. 3.5). Under the conditions of South–west Yakutia the *Pinus sibirica* forests develop as a result of a long-term successional transformation of light coniferous forests. There are four active stages distinguished as regards the edificatory-dominant criterion: *Larix* forests with an understory of *Pinus sibirica* → *Larix* forests with and an admixture of *Pinus sibirica* up to 3 units → *Pinus sibirica* forests with a participation of the species over 3 units → dark coniferous forests of *Pinus sibirica, Picea obovata* and *Abies sibirica* (Nikitina 2004). The typology of the *Pinus sibirica* forests is not diverse and comprises the following types: *Vaccinium vitis-idaea, Vaccinium vitis-idaea – Hylocomium splendens + Pleurozium schreberi, Vaccinium myrtillus – Hylocomium splendens + Pleurozium schreberi, Duschekia fruticosa – Vaccinium vitis-idaea – Hylocomium splendens + Pleurozium schreberi.*

The distribution area of *Abies sibirica* almost coincides with that of *Pinus sibirica*, but *Abies sibirica* is a relatively rare species. The dark coniferous forests of pure *Abies sibirica* or forest in which it predominates occur only in the very south–east of Yakutia (the North–western outskirts of the Patom upland), while the isolated trees or small groups can be observed along the Lena River as far as the Daban settlement (119°10′ E), and in the Aldan upland a little west of the town of Tommot

(126°30′ E). *Abies sibirica* also often forms elfin wood (long and rooted lower branches of adult or sometimes young trees). The *Abies sibirica* communities are represented only by the green moss type (*Hylocomium splendens* + *Pleurozium schreberi*). *Abies* often grows as a constituent part of multi-specific tree stands (*Larix sibirica, L. gmelinii, Pinus sibirica, Picea obovata, Abies sibirica, Pinus silvestris, Betula pendula, Populus tremula*) identified as *Pinus sibirica+Larix* or *Pinus sibirica* forests.

The *Pinus pumila* coenoses form a very peculiar element of the forest cover of Yakutia. They cover large areas in mountainous regions on stony soils and often form the subalpine belt. Under the conditions of the Yakutian middle taiga *Pinus pumila* participates in the undergrowth of lowland *Larix* forests (the Aldan River basin) or, like in the northern taiga subzone, forms independent communities in the sandy landscapes of the Lower Viluy River basin. The Aldan upland features the largest tracts of *Pinus pumila* shrubberies in the middle taiga subzone, due to widespread plateau-like forms of relief. There they occur at the altitudes ranging from 800 to 1,600 m a.s.l. The phytocoenotic diversity is represented by 3 main types (Boychenko 1992; Timofeyev et al. 1994): the lichen type (*Cladonia stellaris, Cl. rangiferina, Cetraria cucullata,* and *Alectoria ochroleuca*) in a complex with the lichen tundra, the *Cassiope*-lichen type (*Cassiope tetragona* and *Cladonia stellaris, Cl. rangiferina, Cetraria cucullata* and *Alectoria ochroleuca*) and the *Vaccinium uliginosum*-lichen type (*Vaccinium uliginosum* and *Cladonia stellaris, Cl. rangiferina, Cetraria cucullata,* and *Alectoria ochroleuca*).

In the *Larix* forests of South–West Yakutia proper conditions (light and soil moisture) occur under the tree canopy for the growth and development of young trees of both *Larix* and dark conifers, such as *Abies sibirica* and *Picea obovata*.

Based on typological diversity and growth conditions, four long-term successional series of the dark coniferous forests (*Pinus sibirica*) are distinguished: *Vaccinium vitis-idaea, Vaccinium vitis-idaea* – green moss, *Vaccinium myrtilllus* – green moss, *Duschekia* – *Vaccinium vitis-idaea* – green moss (Fig. 3.13). Since the *Larix* species are characterized by their light requirement, their relatively rapid growth and other pioneer properties, they participate in the early stages of the successions. Only in special cases, when the recovery series are disrupted by periodical fires, the *Larix* communities may be considered as quasi-climax forests supported by a permanent pyrogenic background at the stage of *Larix* forest.

The role of deciduous species (*Betula pendula, B. pubescens, B. ermannii, Populus tremula, Populus suaveolens, Chosenia arbutifolia, Salix spp.*) in the forest cover formation is not significant. Small patches of original birch forests (*Betula pendula, B. pubescens*), so called "charan", occur in Central Yakutia in mesodepressions, often surrounding alases (the Lena-Amga Interfluve) and in stepped valleys of large rivers. Most typical birch forests are represented by the stepped shrubby forb (*Rosa acicularis, Spiraea salicifolia, Ribes glabellum,* and xerophytic herbs) and *Calamagrostis epigeios* variants. However, the white-barked birches more often occur in the second layer or as an admixture in *Larix* and *Pinus sylvestris* forests. They also inhabit cuttings, post-fire areas, neglected agricultural lands, where they form secondary *Betula* forests of the *Calamagrostis langsdorffii* and

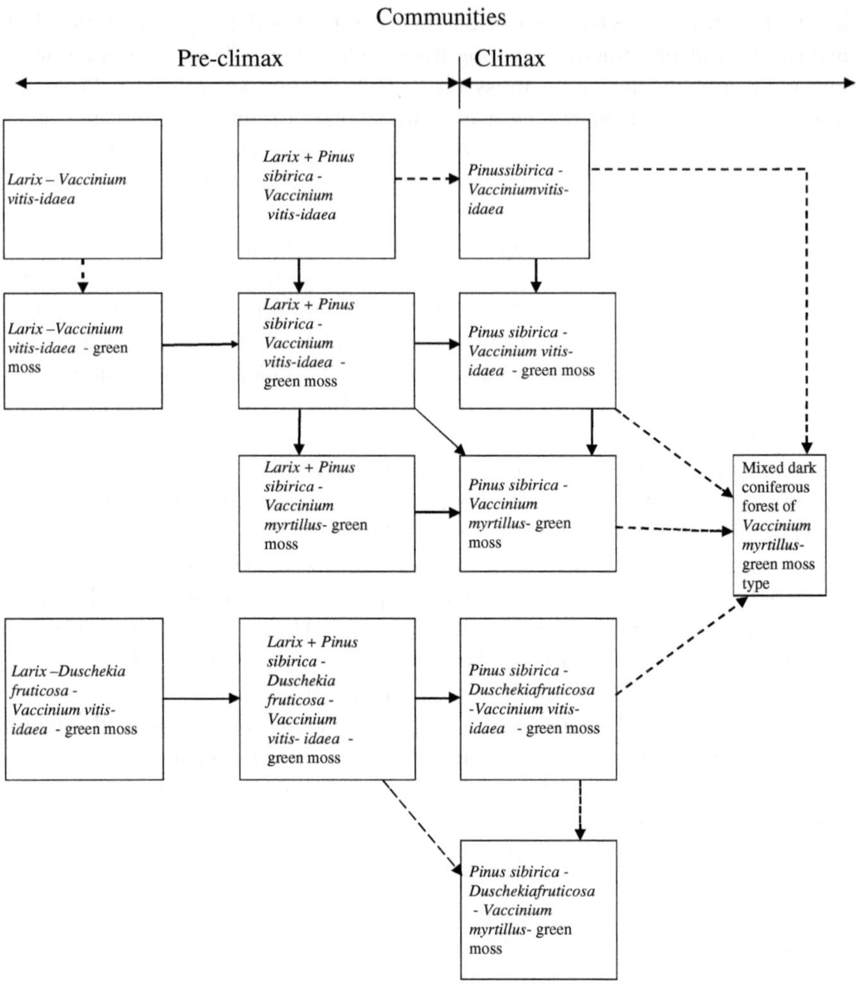


**Fig. 3.13** Scheme of succession series of dark coniferous forest formation in South–West Yakutia

forb (xerophytic herbs) types. A similar strategy is characteristic for *Populus tremula*. In South–West Yakutia it often becomes a constituent part of the dark coniferous forests. Rarely *Populus tremula* is observed on river terraces where it may form pure tree stands.

Original forests of *Betula ermannii ssp. lanata* grow in the mountains of South Yakutia on stony soils at the upper limits of the forest vegetation, seldom descending to a valley complex of mountain rivers. Like *Picea ajanensis,* this species of *Betula* prefers acid bedrock and often contributes to the formation of the upper tree-line in the mountains of South Yakutia. Most favourable conditions for development of the *Betula ermannii ssp. lanata* forests are found on steep (30–40°)

slopes of various aspects that are leeward from winter winds, more often under cliffs in intermountain depressions, and on trough valleys bottoms near river heads (Timofeyev et al. 1994). The following types of the *Betula ermannii* forests are distinguished: *Calamagrostis langsdorffii – Diplazium sibiricum*, herb – *Calamagrostis langsdorffii*, *Pinus pumila – Vaccinium vitis-idaea*, *Pinus pumila – Rhododendron aureum*, *Duschekia fruticosa – Vaccinium myrtillus*, *Hylocomium splendens*, and *Lycopodium annotinum – Hylocomium splendens*.

Pure forests of *Populus suaveolens* are very rare, being mainly an element of intrazonal plant communities. *Populus suaveolens* occurs in the valleys of mountainous or, rarely, lowland rivers on properly drained and bare pebble or sandy alluvium. It usually avoids silty substrates. Such conditions promote fast heating of the grounds in summer time, allowing a proper development of a strong tap root of a tree. The most characteristic types of *Populus suaveolens* forests are as follows: *Equisetum pratense*, herb – *Equisetum* (mesophilous large herbs + *Equisetum pratense*), shrubby – herb (*Swida alba* – mesophilous large herbs).

*Chosenia arbutifolia* communities play an insignificant role in the Yakutian forest cover. They are confined to floodplains representing the intrazonal type of vegetation. *Chosenia arbutifolia* occurs in the lower (pioneer vegetation), middle and rarely upper levels of floodplains of mountain rivers, with time (after one generation) giving way to conifers. This can be explained by the fact that the mother canopy changes the forest growth conditions, and *Chosenia arbutifolia* is no longer able to produce its own young growth. As regards the age stages of *Chosenia* forests, the following types are distinguished: young forest lacking a ground layer, middle-aged and ripening *Chosenia arbutifolia – Equisetum pratense*, and mature and over-ripen *Chosenia arbutifolia – Calamagrostis langsdorffii* + mesophilous large herbs.

The communities of arboreous *Salix spp.* are observed only as an element of intrazonal vegetation, so they are described in Section 3.5.8.

Among non-forest vegetation that occurs in the middle taiga subzone, most widespread are the shrubberies of fruticose *Betula* species (*yernik*), meadows, steppes, grass and moss bogs, mountain tundra, etc. They are reviewed in detail in the following works: Karavaev and Skryabin (1971), Andreyev et al. 1987; Skryabin and Karavaev 1991; Timofeyev et al. 1994, as well as Rabotnov 1935; Kuminova 1936, etc.; Sheludyakova 1957, etc.; Kuvaev 1955, etc.; Karavaev 1958, etc.; Dobretsova 1961; Galaktionova et al. 1962; Ivanova 1967; etc. These vegetation types are described in other parts of this book.

### 3.4.6 Mountain Taiga

**A.P. Isaev, L.V. Kuznetsova, A.P. Efimova, E.G. Nikolin, and N.B. Ermakov**

The vegetation cover of the mountainous territories of Yakutia forms a landscape-botanical complex that does not fit fully into the zonation principles of the lowland vegetation. The Transasian mountainous region, that includes the highlands of

Yakutia, featured various specific vegetation covers during the different historical periods: broad-leaved forests, steppe or tundra communities. The centuries-old interaction of different phytocoenotic species pools against the background of severe climatic conditions has resulted in development of the specific arctic-alpine floral element with widespread *Dryas punctata, Cassiope ericoides, Rhododendron parviflorum, Pinus pumila, Betula lanata,* etc. They have had a key influence on the peculiar mountain vegetation (Safronov and Safronov 2007).

The vegetation cover of the highlands of Yakutia was studied by Rabotnov (1936, 1937, 1940), Sheludyakova (1938, 1948, 1961), Yarovoy (1939), Kuvaev (1956, 1961, 1964), Prakhov (1957), Pivnik (1958), Dobretsova (1962), Skryabin (1968, 1976), Kildyushevsky (1960), Pozdnyakov (1961a, b), Tyulina (1956, 1957, 1959, 1962), Utikn (1961), Scherbakov (1964, 1965), Yurtsev (1959, 1961, 1964, 1968), Perfilyeva (1977), Belyaeva (1975), Nikolin (1990), Timofeyev et al. (1994), etc. General information on the mountain vegetation is given in the works by Komarov (1926), Karavaev and Skryabin (1971), Scherbakov (1975), Andreyev et al. (1987), Safronov and Safronov (2007), etc.

As regards their latitudinal zonation, the highland regions are situated both in the tundra (the Pronchischev and Chekanovsky Ridges, the Kharaulakh Range, the Kondakovskoye tableland, etc.) and taiga zones (in the near-tundra forests and northern taiga: the Anabar Plateau, the Verkhoyansk Range, the Chersky Range, the Alazeya and Yukagir tablelands, etc.; in the middle taiga: the Patom and Olyokma-Chara uplands, the Amga Ridge, the Aldan upland, the Stanovoy Range, etc.).

The vegetation structure in the mountainous regions is determined by the edaphic conditions, chemical composition of the bedrock, and orographical peculiarities yielded by cryogenic processes (earth stripes, medallions, hummocks, fissures and various polygons). The highland vegetation shows a clear altitudinal (belt) zonation in the dominating vegetation types covering the slopes (Fig. 3.14).

Three types of belt zonation are distinguished in the Yakutian mountains. The North East Siberian (Yakutian) type is characteristic for the highlands of North–East Yakutia (the Chersky, Momsky, Verkhoyansk, Orulgan Ranges, etc.). The Okhotsk type is typical for the southern spurs of the Verkhoyansk Range (the Suntar-Khayata and Sette-Daban Ranges). The Near-Baikalian altitudinal zonation is characteristic for the South Yakutian highlands (the Aldan upland and Stanovoy Range). They differ in the prevailing types of life forms of plants, the predominant types of plant communities and the degree of belt development. Under the conditions of Yakutia the following belts are clearly expressed: the forest belt, the belt of subalpine shrubberies, the tundra belt, the belt of epilithic lichens, and the nival belt. The latter three belts represent the alpine zone.

In the highlands with narrow and deep river valleys a separate complex of valley vegetation should be distinguished apart from the forest belt. The valley complex is represented by aestisilvae (summer green) conifers (*Larix*), by winter green coniferous forests (*Picea obovata, P. ajanensis*) in the mountains of South Yakutia, by deciduous small-leaved communities of *Populus suaveolens, Chosenia arbutifolia, Betula pendula, B. pubescens,* and *B. ermannii,* by deciduous shrubberies (*Betula*

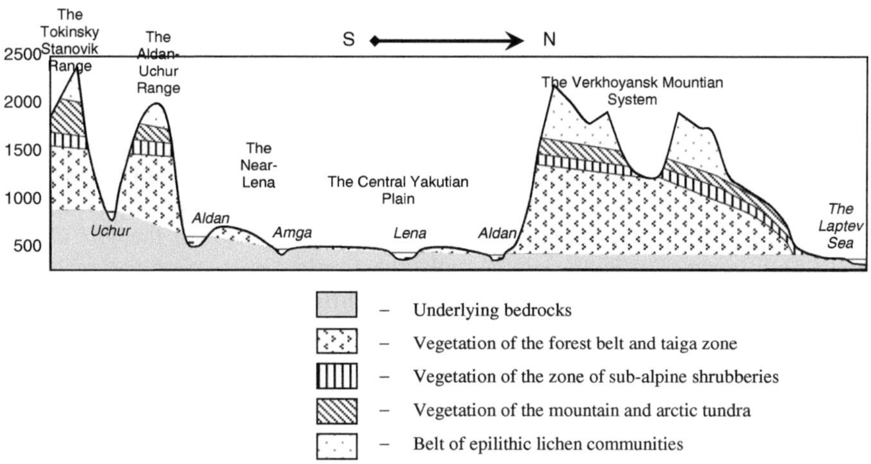


**Fig. 3.14** Altitudinal distribution of vegetation along the macroprofile 128° – 132° E (from the highlands of South Yakutia to the Laptev Sea shoreline)

*nana ssp. exilis, B. divaricata, B. fruticosa*), by riparian deciduous shrubberies mainly of *Salix spp.*, as well as by mixed forests, meadows, steppes, bogs and aquatic vegetation. In North–East Yakutia the valley steppe landscapes are more pronounced than in South Yakutia, while in the valleys of South Yakutia *Picea* and *Betula* forests occur. The vegetation of the valley complex is described in Section 3.5.8.

*The forest belt* is predominated by the aestisilvae coniferous forests and woodlands of *Larix cajanderi*. The dominant species of the shrub layer are *Pinus pumila, Betula divaricata, B. exilis, Juniperus sibirica, Duschekia fruticosa*, and the dominant dwarf shrubs are *Ledum palustre, Vaccinium uliginosum* and *V. vitis-idaea*. The moss-lichen layer is represented by sphagnum species (*Sphagnum warnstorfii, S. fuscum*, etc.), green mosses and fruticose lichens such as *Cladonia stellaris, C. arbuscula, C. rangiferina, Cetraria islandica, C. laevigata, C. cucullata,* and *C. nivalis* in mesic habitats and by the epilithic lichens *Umbilicaria, Parmelia, Hypogimnia* on dry stony slopes. In some places winter green forests of *Pinus sylvestris* (west-facing slopes of the Ust-Viluysk Range (the western Verkhoyansk mountain system)) and *Picea ajanensis* (the Stanovoy Range) occur as well as deciduous forests of *Populus tremula* (widespread), *Betula pendula* (the Verkhoyansk, Sette-Dabaan, Suntar-Khayata Ranges and the mountains of South Yakutia) and *Betula pubescens* (the Yana tableland). Mixed forests of *Populus suaveolens, Larix cajanderi, Picea obovata* and *Betula pubescens* with an undergrowth of *Sorbus sibirica, Spiraea dahurica, Rosa acicularis, Atragene sibirica,* etc. also occur. In the Yana, Indigirka and Kolyma River basins vast areas within the forest belt are covered with steppe vegetation. The position of the upper tree-line greatly varies depending on latitude and heights of the mountains: from 600 m in the North to 1,200–1,300 m in the Sette-Daban, Suntar-Khayata and Stanovoy Ranges.

In South Yakutia the forests include *Pinus sibirica, Picea ajanensis, Abies sibirica,* and *Betula ermannii*. The forests of the North-facing slopes of the Stanovoy Range were thoroughly studied and classified using the Braun-Blanquet approach (Ermakov and Cherosov 2005). The *Larix* forests are attributed to the class of true coniferous taiga forests *VACCINIO-PICEETEA*. At the altitude of 550–850 m the area is characterized by a more humid and cold climate during the summer period in comparison with the Central Yakutian lowland. As a result, xeric elements disappear in the boreal forests and the amount of characteristic species of the *VACCINIO-PICEETEA* increases. Phytosociologically the forests are a combination of two associations of *Picea-Larix* forests occurring at different parts of the gentle slopes of the mountains.

The association *Aconito-Laricetum cajanderi* predominates in foot-hills and on mountain slopes. It occupies convex and flat parts of gentle slopes with drained soils and permafrost at a depth of 60–80 cm. *Larix cajanderi* and *Picea obovata* form an open tree layer with low cover values of 20–40% and a height of 16–18 m. In some places *Betula pendula, Pinus sibirica* and *Pinus sylvestris* occur in these communities. The shrub layer (covering 1 to 5%) is dominated by *Duschekia fruticosa, Lonicera caerulea* and *Rosa acicularis*. *Vaccinium vitis-idaea* is the dominant of the herb layer and shows cover values of 15–80%. A well-developed herb layer is characterized by a cover of 50–60% and a species richness of 10–18 species per 200 $m^2$. The dwarf-shrubs *Vaccinium vitis-idaea, V. uliginosum* and *Ledum palustre* play a leading role there. The moss layer is dominated by *Pleurozium schreberi, Hylocomium splendens, Aulacomnium turgidum, A. palustre*. The fruticose lichens *Cladonia stellaris, C. rangiferina, C. amaurocrea, C. arbuscula* and *Cetraria islandica* are constant species in the association but they are never dominants.

The association *Ledo palustris-Laricetum cajanderi* var. *Arctous erytrocarpa* is confined to shallow depressions, cool, water-logged soils and permafrost at a depth of 40–70 cm on gentle northern slopes. Unlike the typical association described from the Central Yakutian lowland (Section 3.4.5), this variant occurs in a more humid climate as indicated by the presence of *Picea obovata* in the tree layer, higher values of species diversity of the herb layer and the predominance of the boreal bryophytes *Pleurozium schreberi, Hylocomium splendens, Dicranum polysetum* in the moss layer.

The association *Betulo divaricatae-Laricetum cajanderi* forms the higher part of the forest altitudinal belt and is replaced by subalpine communities of *Pinus pumila* at higher altitudes. The community predominates on gentle mountain slopes of different aspects, at sites with permafrost at a depth of 40–60 cm. The main peculiarity of this community is a well-developed shrub layer (cover values of 30–45%) formed by a combination of boreal (*Duschekia fruticosa, Lonicera caerulea, Rosa acicularis, Sorbus sibirica*) and subalpine (*Pinus pumila, Rhododendron aureum, Betula divaricata*) species. The closeness of the association to the more humid subalpine belt is indicated by a group of mesic herbs (*Aconitum ranunculoides, Mitella nuda, Equisetum arvense, Corydalis paeoniifolia, Geranium albiflorum, Saussurea parviflora, Bistorta major, Luzula parviflora, Streptopus streptopoides, Tofieldia coccifera, Parnassia palustris, Thalictrum alpinum, Hieracium korshinskyi*) and by the occurrence of *Vaccinium myrtillus* and *Abies sibirica*.

*The belt of subalpine shrubberies* in North–East Yakutia is fragmentary and is characterized by a colder climate, a moderate precipitation and widespread permafrost at a depth of 20–40 cm. The amount of typical boreal elements, including characteristic species of the *VACCINIO-PICEETEA*, decreases rapidly there. A characteristic feature of the subalpine communities is a well-developed ground layer of various lichens. The vegetation is composed mainly of the aestisilvae (summergreen) shrubs of *Betula divaricata, B. nana ssp. exilis, Duschekia fruticosa*, and to a lesser extent of *Pinus pumila* shrubberies. In the central and southern parts of the Verkhoyansk mountain system, as well as in the Stanovoy Range and Aldan upland its role is significantly more prominent. In the very North the belt of subalpine shrubberies is represented by fruticose and elfin forms of *Larix cajanderi*, the basic forest-forming species of the region. The belt borders on the forest and tundra belts at its lower and upper parts, respectively. The lower border ranges between 600 and 700 (900) and 1,100–1,400 m a.s.l. The width of the subalpine belt does not exceed 150–200 m.

In South Yakutia the phytocoenotic diversity of the belt of subalpine shrubberies contains four associations at different altitudes. All syntaxa described from the subalpine belt of the Stanovoye Plateau have been included in the *Loiseleurio-Vaccinietea* class (Ermakov and Cherosov 2005).

The association *Piceo obovatae-Betuletum divaricatae* occurs in the lowest part of the subalpine belt, in the contact zone with boreal forests of the *Betulo divaricatae-Laricetum cajanderi. Picea obovata* and *Larix cajanderi* are represented as scattered trees or sparse groups of trees. *Betula divaricata* together with the other subalpine shrub *Betula exilis* compose a well-developed shrub layer (cover of 20–30%). Species of dwarf shrubs and fruticose lichens predominate in the ground layer. As a whole, the community is characterized by low numbers of higher vascular plant species (8–15). On warmer southern slopes, this community is represented by a special variant with *Pinus sylvestris*.

The association *Cladonio-Pinetum pumilae* is widespread in the lower altitudinal strip of the central part of the Stanovoye Plateau where it forms a combination with the previous association at the upper part of the forest belt. The community occupies flat summits and gentle slopes of different aspects on the Plateau. It is characterized by a well-developed layer (cover of 20–70%, height of 2–7 m) of the subalpine prostrate tree *Pinus pumila*. Subalpine shrubs – *Rhododendron aureum, Betula divaricata* and *B. exilis* are typical for this layer as well. Scattered trees of *Larix cajanderi* never form a distinct canopy. Dwarf shrubs (*Vaccinium vitis-idaea, V. uliginosum, Ledum palustre, Empetrum nigrum, Loiseleuria procumbens*), as well as *Pinus pumila* and fruticose lichens are the main dominants in the community. Like in previous association, the *Cladonio-Pinetum pumilae* is characterized by a poor composition of higher vascular plants and a minor participation or even absence of typical boreal elements. On the warmest parts of southern slopes this community is represented by a special variant with *Pinus sylvestris*.

The association *Pino pumilae-Cassiopetum* occurs at the highest part of the subalpine belt where it occupies tops of mountains with shallow stony soils at altitudes of 1,200–1,400 m. The subalpine prostrate tree *Pinus pumila* and shrubs (*Rhododendron aureum, Rh. adamsii, Betula divaricata, Sorbaria pallasii)* form a shrub layer with a height of 0.4–1.2 m and a cover of 25–40%. Widespread dwarf shrubs (*Vaccinium vitis-idaea, V. uliginosum, Ledum palustre, Empetrum nigrum,*) together with typical subalpine species *Cassiope ericoides*, *Arctous alpina, Loiseleuria procumbens* predominate in the herb layer. The latter is also characterized by presence of subalpine herbs: *Aconogonon tripterocarpum, Tilingia ajanensis, Calamagrostis lapponica.* Awell-developed lichen layer includes widespread fruticose lichens and some subalpine-alpine species: *Nephroma arcticum, Asahinea chrysantha, Flavocetraria nivalis, Alectoria ochroleuca, Thamnolia vermicularis.* A variant of the association with *Rhododendron aureum* is typical of more shallow and stony soils.

In the subalpine belt of South Yakutia sometimes a fragmentary *Betula ermannii ssp. lanata* crooked forests occurs but it covers insignificant areas within the Stanovoy Range, the Sette-Daban and Suntar-Khayata Ranges and in the central part of the Verkhoyansk mountain system. The communities form a narrow strip of 50 to 100 m wide and fringe on the forest belt at its lower limit and on the tundra belt at its upper limit.

*The tundra belt* is properly developed, fringing on the subalpine, or sometimes the forest belt. The lower limit of the belt is recorded at 1,000–1,200 m a.s.l. in the middle taiga subzone and at 1,300–1,500 m a.s.l. in the South. Its upper fringe contacts the belt of epilithic lichen vegetation. The width of the tundra belt varies from 200 to 300 m, though in the mountains of South Yakutia it covers small areas. The tundra belt vegetation is predominated by dwarf shrub communities of *Dryas* growing in hedgerows (*Dryas punctata* in North–East and *D. crenulata* in South Yakutia) and polydominant dwarf shrub communities (*Dryas punctata* + *Ledum decumbens* + *Arctous alpina* + *Vaccinium minus* + *V. uliginosum* + *Cassiope tetragona* or *C. ericoides*). Vast territories, especially on North–facing slopes, are rather smooth, showing very little relief or hummocks, etc. They are occupied by the summer-green dwarf shrub tundra with a predomination of *Betula nana ssp. exilis*. In the Aldan upland and Sette-Daban Range the winter-green tundra of *Rhododendron aureum* occurs, while in the northern highlands the winter-green tundra of *Rhododendron adamsii*. On the North–facing slopes, at the bottom of circuses and hollows, as well as around snow patches the chionophilous winter-green *Cassiope* (*Cassiope tetragona* or *C. ericoides*) tundra communities are observed. Gentle slopes of intermountain depressions are covered with the dominating waterlogged tundra communities of *Eriophorum vaginatum* with fragments of *Eriophorum polystachyon* and *Carex concolor* bogs. Gentle slopes in the Far North feature the summer-green dwarf shrub tundra represented by *Salix sphenophylla, S. reticulata, S. polaris,* as well as *Hierochloe alpina* communities. Significant areas in the upper part of the tundra belt are occupied by fruticose lichen tundra communities: *Alectoria ochroleuca* or *Cetraria nivalis* + *C. cucullata* or polydominant communities (*Cetraria islandica* +

*C. cucullata* + *C. laevigata* + *Cladonia arbuscula*, etc.). On the south-facing slopes of the tundra belt stepped tundra or cryophilous steppes occur with participation or predomination of *Festuca lenensis, F. brachyphylla, Kobresia myosuroides, Carex pediformis*, etc. Along water streams and around snow-patches mesic forb and *Chamaenerion latifolium* alpine-like meadows occur. In the South–East of the Sette-Daban Range patches of the unique *Baeothryon uniflorum* tundra can be observed. The tundra belt also features widespread patches of green moss-*Sphagnum* tundra with participation of dwarf shrubs (*Cassiope tetragona, Oxycoccus microcarpus*, and *Vaccinium minus*) confined to shallow depressions on mountain slopes.

*The belt of epilithic lichens* borders on the tundra belt at its lower fringes at 500–700 m a.s.l. in the North and at 1,400–1,600 m a.s.l. in the South. The upper border lies at 2,000–2,500 m a.s.l. Alternative names for this belt used in the literature are the fell-field belt and the belt of cold stony deserts. It is represented by two types of vegetation: the foliose lichen communities predominated by *Umbilicaria sp., Parmelia sp., Cetraria hepatizon, Asahinea chrysantha*, etc., and crustose lichen communities of *Haematomma ventosum, Rhizocarpon geographicum, Sporastatia sp., Verrucaria sp., Aspicilia sp., Lecidea sp.*, etc. Both vegetation types include mosses: *Racomitrium lanuginosum, Tetralophosia setiformis, Thuidium abietinum. Andraea alpestris, Schistidium sp., Grimmia sp.*, etc.

*The nival belt* is characteristic for the highest mountain systems with a strongly pronounced modern glaciation: the Buordakh mountains, the Chersky and Suntar-Khayata Ranges. The belt lies at an altitude over 2,300–2,500 m a.s.l. It lacks a closed vegetation cover. Only small fragments of foliose and crustose lichen synusiae occur there.

It should be noted that all the belts, except for the nival one, feature an azonal vegetation type: the subalpine summer-green scrub (*Salix alaxensis, S. lanata, S. krylovii, S. boganidensis*, etc.).

The above mentioned altitudinal zonation gives the general principles but there are many deviations from the limits given. Very often the belts penetrate the adjacent belt with large tongues or fragments, and sometimes a vegetation type is found deep inside the adjacent belt.

## 3.5 Azonal Vegetation

### *3.5.1 Steppes*

#### V.I. Zakharova, M.M. Cherosov, E.I. Troeva, and P.A. Gogoleva

The steppe communities are an intrinsic, peculiar and unique component of the vegetation cover of Yakutia (Karavaev 1945, 1958, 1968; Sheludyakova 1957c; Buks 1964; Karavaev and Skryabin 1971; Sheludyakova and Skryabin 1969; Yurtsev 1981; Mirkin et al. 1985, 1992a, 1992b; Gogoleva et al. 1987; Skryabin and Karavaev 1991; Koropachinsky 1996; Zakharova 2005). The steppe vegetation is confined to the prominent parts of terraces above floodplains and to south-facing slopes of river valleys. In the North–East cryophilous steppes are common among the sparse *Larix* forests within the subalpine zone (Yurtsev 1981). Most pronounced steppe landscapes are confined to loamy soils underlain by calcareous bedrocks, while sandy alluvium and loamy sands represent unfavourable growth conditions.

The forest-steppe landscapes of Yakutia are the relics of the Late Pleistocene. The structure and composition of the relic steppes of Central and North–East Yakutia remind of periglacial steppes (Lavrenko 1981).

The Yakutian steppes are either true, meadow or cryophilous steppes of bunchgrass and rhizomatous plants (Andreyev et al. 1987), and belong to the class *CLEISTOGENETEA SQUARROSAE* Mirkin et al. 1986 ex Korolyuk 2002.

True bunchgrass steppes are very widely spread in Central Yakutia and the North–East (Indigirka and Kolyma River basins) on south-facing slopes and above-floodplain terraces of river valleys.

The eroded landscapes support bunchgrass steppe communities of the order *Stipetalia krylovii* Kononov et al. 1985 and feature a low α-diversity (from 2 to 13 species per 100 $m^2$) and low cover values (20–50%). The loose soils on the slopes of the Lena, Yana, Aldan, Viluy, and Indigirka Rivers valleys support *Agropyron cristatum* and *Elytrigia villosa* steppes. In Central Yakutia these communities are composed of *Artemisia jacutica, Veronica incana, Delphinium grandiflorum, Phlox sibirica,* etc. *Psathyrostachys* steppe communities are common in the Lena River valley and the Yana River basin. On the eroded steep slopes of valleys and alases they form pure grass stands of *Psathyrostachys caespitosa*, sometimes with *Artemisia jacutica* as a co-dominant. The forb component of the eroded steppe landscapes often consists of *Alyssum obovatum, A. lenense, Eritrichium sericeum, Chamaerodos erecta, Androsace septentrionalis*. On more fixed substrates these communities are enriched with other xerophytic grasses (*Stipa capillata, S. krylovii, Festuca lenensis, Koeleria cristata*, and sometimes *Cleistogenes squarrosa*).

The *Stipa* steppe communities are characteristic for stabilized landscapes with decreasing erosion processes. They cover insignificant areas in the valleys or on the south-facing slopes of large rivers. They characteristically contain *Stipa capillata, S. krylovii, Koeleria cristata, Festuca lenensis, Pulsatilla flavescens, Artemisia frigida,* etc. (Ivanova and Perfilyeva 1972).

The forb-bunchgrass steppes and forb-grass meadow steppes growing on soils with a properly developed sod, belong to the order *Festucetalia lenensis* Mirkin in Gogoleva et al. 1987. The communities of this order feature a higher α-diversity (20 species on average). The forb-bunchgrass steppes of Central Yakutia are predominated by *Festuca lenensis* with participation of *Helictotrichon krylovii, H. schellianum, Stipa spp., Onobrychis arenaria, Veronica incana, Potentilla nivea, Carex duriuscula, Potentilla bifurca* and *Thymus spp*. It is assumed that the *Festuca* steppes of Central Yakutia represent a secondary formation in a digressive series of the steppe landscapes. In case of anthropogenic impact (overgrazing, recreational load) they replace the original *Stipa-Festuca* steppes or the four-grass steppe communities (Ivanova 1981). Indeed, studies of the early twentieth century (Dolenko 1913; Abolin 1929) show the predominance of *Festuca-Stipa* communities on the above-floodplain terraces of the Middle Lena River (Yakutsk vicinity), while presently only fragments of these communities occur there. Besides, on the slopes of the Middle Lena River valley the impoverished variant of the four-grass steppes occur with *Cleistogenes squarrosa, Koeleria cristata, Stipa krylovii,* and *Festuca lenensis*.

Along with the *Festuca lenensis* formations, the *Koeleria cristata* steppes also represent degraded landscapes. The first develop on light soils with proper drainage, while the latter occur on heavier substrate. The *Koeleria* formations are typically pooer in species, and often contain ruderal species (*Lepidium densiflorum, Lappula squarrosa*).

Forb-grass meadow steppes combine the features of forb meadows and forb-*Festuca* steppes. They are characteristic for the landscapes with better moisture conditions: depressions of the terraces above the floodplain, upper floodplain levels, the mesoxerophytic belts of alases, and as forest edge communities. This is reflected in high values of α-diversity (up to 50 species per 100 $m^2$) and cover (50–90%). The floristic basis of the meadow steppes of Central Yakutia is provided by xerophytes and mesoxerophytes (*Pulsatilla flavescens, Carex pediformis,Bromopsis pumpelliana, B. korotkiji, Agrostis trinii, Aster alpinus, Lychnis sibirica, Dianthus versicolor, Euphorbia discolor,* etc.), though among the forbs mesophilous plants are well represented, such as *Veronica longifolia, Sanguisorba officinalis, Geranium pratense*, etc.

In the Lena-Amga Interfluve very distinct relic forest steppes are common representing *Betula pendula* forests with a xerophilous undergrowth (*Festuca lenensis* and *Poa botryoides* with participation of *Lychnis sibirica, Galium verum, Sanguisorba officinalis, Phlox sibirica,* and numerous mesophilous forbs) (Sheludyakova and Skryabin 1969; Skryabin and Karavaev 1991).

The rich forb-grass meadow steppes of the mountainous North–East are also well represented in river valleys (*Bromopsis pumpelliana, Pulsatilla patens s.l., Delphinium grandiflorum, Heteropappus biennis, Castillea rubra, Poa botryoides,* etc.) and they also form forest edge communities (*Helictotrichon krylovii, Calamagrostis purpurascens, Arnica iljinii, Pulsatilla patens s.l.*, etc.). *Carex pediformis* meadow steppes occur in the Upper Kolyma and the Middle Indigirka Rivers basins. The communities also include *Poa glauca, Eremogone tschuktschorum,* and *Dianthus repens.*

Fragments of relic cryophilous (or hemicryophytic) steppes are observed in the Lower Kolyma River basin (α-diversity 3–4 species per 100 $m^2$, cover 20–40%). They are confined to the pingo (hydrolaccolith) tops and lake sides in the lowland waterlogged tundra. The characteristic species of the cryophilous steppes are the *Carex spaniocarpa* with participation of *Trisetum spicatum, Potentilla hookeriana, Armeria scabra, Polemonium boreale, Arnica iljinii, Koeleria asiatica, Astragalus alpinus,* etc. The moss-lichen layer is rather pronounced there and includes *Polytrichum piliferum, Cetraria cucullata, C. islandica, Cladonia cocciferra, Alectoria ochroleuca, Dactylina arctica* (Yurtsev 1981). Data on these communities are very scarce; but the floristic composition allows us to attribute them to the meadow steppes due to their proper moisture conditions that impede the development of bunchgrass vegetation. Anthropogenic activity and invasion of tundra species resulted in gradual degradation of the cryophilous steppe landscapes (Andreyev and Perfilyeva 1975; Andreyev and Galaktionova 1981; Koropachinsky 1996).

The rhizomatous *Carex* steppes represent strongly degraded landscapes with a predomination of *Carex duriuscula*. They cover rather large territories in the more densily populated areas of Central Yakutia and the Middle Indigirka River basin. In the digression series they follow upon the *Festuca* and *Koeleria* steppe formations under conditions of permanent and intensive overgrazing (About 10 species per 100 $m^2$, cover up to 50%). The communities also may include an insignificant amount of suppressed *Artemisia commutata, A. jacutica, Agrostis trinii, Poa botryoides, Lychnis sibirica, Galium verum,* etc. Cherosov et al. (2005) attribute the *Carex duriuscula* communities with a pronounced anthropogenic component (*Artemisia jacutica, Lappula squarrosa, Lepidium densiflorum, Leptopirum fumarioides*) to the synanthropic class *ARTEMISIETEA VULGARIS* Lohmeyer et al. in R. Tx. 1950 (alliance *Artemisio-Caricion duriusculae* (Gogoleva et al. 1987) Czerosov (2005), order *Onopordetalia acanthii* Br.-Bl. et R. Tx. ex Klika et Hadac 1944). The communities with a predomination of both *Carex duriuscula* and the natural xerophilous component (*Artemisia commutata, Koeleria cristata, Veronica incana,* etc.) should be considered as steppes of the alliance *Festucion lenensis* of the *CLEISTOGENETEA SQUARROSAE,* however.

Petrophytic steppes are an inherent component of mountainous arid landscapes. They are confined to calcareous bedrock and are observed all over Yakutia. Skryabin (1976) described the *Elytrigia jacutorum* formation for the lower reaches of the Tokko River (200 m a.s.l., the Near-Lena Plateau). These petrophytic communities inhabit south-facing slopes with a moderate inclination (15–20°). The soils are shallow and have a significant admixture of detritus. The α-diversity is 20–25 species per 100 $m^2$, and cover values reach 70%. *Elytrigia jacutorum* forms various associations with the steppe xerophytes (*Thesium refractum, Artemisia commutata, Silene repens, Euphorbia discolor, Carex pediformis, Galium verum,* etc.) and with some participation of petrophytes (*Youngia tenuifolia, Artemisia santolinifolia, Thalictrum foetidum, Orostachys spinosa* and *Juniperus sibiricus*). Similar vegetation was described from steep slopes (40–45°) in the Buotama River mouth, the Lena Pillars area (the Middle Lena River), and the Maya River valley (South–East Yakutia) (Zakharova 2005). Contrary to the communities of South–West Yakutia, they feature participation of *Stipa capillata, Ephedra monosperma, Helictotrichon schellianum,* etc. K.A. Volotovsky (Koropachinsky 1996) described steppes of the *Elytrigia jacutorum* formation from the Aldan Tableland (280–400 m a.s.l.) on gentle slopes (25–30°). The forb-*Elytrigia jacutorum* association features isolated and suppressed trees of *Pinus sylvestris* and *Populus tremula*. The shrub layer is represented by *Juniperus davurica, Spiraea dahurica,* and *Pentaphylloides fruticosa*. The herb layer (35–40 species per 100 $m^2$, 20–30%) contains *Elytrigia jacutorum* as well as *Artemisia gmelinii (A. santolinifolia), Carex pediformis, C. trautvetteriana, Dendranthema calciphilum, Gypsophila patrinii, Phlojodicarpus sibiricus, Polygala sibirica, Potentilla nivea, Thymus serpyllum s.l., Youngia tenuifolia, Saussurea hypargyrea,* etc.

The carbonate slopes of the Middle Lena River (300–200 km south of Yakutsk) carry petrophytic variants of the steppes of the *Stipetalia* order, i.e. the communities of *Stipa krylovii* and *Psathyrostachys juncea* with a petrophytic component (*Artemisia santolinifolia, Youngia tenuifolia, Patrinia rupestris, Orostachys malacophylla, Thalictrum foetidum,* etc). The richness is 15 species per 100 $m^2$, cover 20–30%. Only here *Krascheninnikovia lenesis* and *Artemisia martjanovii* are found.

The forb-bunchgrass steppes of the northern mountainous regions (the Yana and Indigirka River basins) are very similar to those of Central Yakutia, though less diverse in species composition and they include a cryophilous component, such as *Calamagrostis purpurascens, Dracocephalum palmatum,* etc. Besides, with minor exceptions they are all confined to dry heated slopes with loamy mountain chestnut soils and detritus. Cold-resistant mountain steppe plants (*Helictotrichon krylovii, Poa botryioides, Festuca lenensis, Artemisia bargusinensis*) may form communities as high as 1,100 m a.s.l. S.Z. Skryabin considers the steppe communities of the Yana and Indigirka River basins as a special geographical northern variant of extrazonal mountain steppes (Skryabin and Konorovsky 1975). The Yana River basin features *Festuca lenensis* steppes ($\alpha$-diversity 8–25 species per 100 $m^2$, cover 30–80%) including *Artemisia bargusinensis, Galium verum, Eritrichium sericeum, Lychnis sibirica, Poa botryoides,* and *Pulsatilla flavescens* (Fig. 3.15). The endemic species *Potentilla tollii* grows strictly within the Yana River basin and is also a constituent part of the *Festuca lenensis* steppes.

The same region is also well known for its relic *Stipa krylovii* communities (*Artemisia bargusinensis, Ephedra monosperma, Carex pediformis, Thalictrum foetidum*) that remained beyond the Polar Circle since the Late Pleistocene. The steppe communities of the Indigirka River basin usually lack *Festuca lenensis*. In the subalpine belt of the mountains it is replaced by *F. auriculata*. Most common steppe formations there are represented by the coenoses of *Koeleria cristata* (south-facing gentle slopes) and *Poa botryoides* (landscapes with a colder microclimate, often on slopes at 700–800 m a.s.l.). Only this region features a tetraploid population of *Artemisia frigida,* and also contains *Eremogone meyeri, Smelowskia alba, Thymus indigirkensis, Oxytropis sheludjakovae,* etc. (Yurtsev 1981).

The North–East Yakutia also features a distinct type of *Helictotrichon krylovii* steppes developing on mountain chernozem soils and forming forest edge communities (Karavaev 1958, Skryabin 1968). Besides *Helictotrichon,* the core species of these communities are *Carex duriuscula, Agropyron cristatum, Festuca lenensis, Poa botryoides, Lychnis sibirica,* etc. In the Middle Indigirka River basin *Helictotrichon* steppes may cover large areas on terraces above the floodplain.

All Yakutian steppes are characterized by the occurrence of *Xanthoparmelia camschadalis (Parmelia vagans)* that may cover up to 8%.

The Yakutian steppes play a significant role in agriculture being used as pastures and, rarely, as hayfields.

**Acknowledgments** The work was partly supported by RFBR 08-05-00747 (2008–2010) Investigation of genesis, floristic and entomofaunistic relations of relic steppe ecosystems of Central and South–West Yakutia

**Fig. 3.15** At the top of a convex stony slope with the *Festuca lenesis* steppe formation. *Larix cajanderi* forest is seen in the valley in the background

## *3.5.2 Rock Vegetation*

### E.I. Troeva and E.G. Nikolin

The plant communities of rocks and screes are one of the least studied vegetation types of Yakutia. Most detailed studies were conducted in South Yakutia (Ivanova 1973, 1976; Volotovsky 1992; Volotovsky and Kuznetsova 1998), as well as in separate regions of North–West (Lukicheva 1963a, b) and North–East (Nikolin 1990) Yakutia.

Substrate characteristics are the significant factor that determines the qualitative features of the coenoses of rocky and stony landscapes. In Yakutia, the petrophytic vegetation occurs basically on two substrate types: magmatic (granite, basalt, traprock, cimberlite, etc.) and sedimentary (limestones, dolomites, sandstones, clay slates, etc.) bedrocks, both as weathering-resistant rocks and as coarse or fine detritus on slopes. Additional factors determining petrophytic vegetation are aspect and inclination of slopes, local climate and cryogenic processes.

Calcareous bedrocks do widely occur all over Yakutia, representing Cambrian sediments (lime-stones and dolomites) of various ages. Being softer, the lime-stones are easily affected by cryogenic erosion. Owing to this, they have a flaglike structure and have detritus, while the harder dolomites appear as rocks.

The petrophytic vegetation on calcareous bedrocks is most pronounced on the slopes of the narrow, canyon-like, river valleys that cut the plateaus. Such landscapes are characteristic for the Indigirka, Upper Lena and Upper Aldan Rivers' basins (Fig. 3.16). Within the plateau of North–West Yakutia calcareous bedrocks occur on the flat-hilly relief of watersheds with gentle slopes composed of sandy-carbonate substrate and steep dolomite slopes (Lukicheva 1963a, b).

**Fig. 3.16** Calcareous slopes of the Tokko River valley (South–West Yakutia)

The carbonate eluvium and slide-rocks of lime-stones, dolomites, dolomitic lime-stones carry shallow heavy soils. They are frozen soddy-carbonate, soddy-carbonate podzolized, and soddy-podzolic types with sufficient resources of nutrients (Petrova 1971). As opposed to acid soils, nitrogen is mineralized faster there, while P, Fe, Mn and heavy metals are less accessible for plants (Larcher 1978). High content of calcium serves as a limiting factor for certain plant species, so the basis of the communities on calcareous bedrocks is provided by obligate and optional calciphils. On the one hand, because of their heavy structure the soils retain moisture which makes them potentially cold. On the other hand, the shallow profile, the strong stoniness, and crevices in the underlying bedrock provide proper conditions for drainage. This explains the rather warm and dry characteristics of the south-facing slopes. Owing to this, xerophilous and thermophilous species find better growth conditions there as compared to silicate bedrocks. The abovementioned features of the calcareous substrate determine the peculiar combination of species of various ecology and altitudinal-latitudinal confinement (Volotovsky 1992; Volotovsky and Kuznetsova 1998). Thus, one community includes both mesophytes and xerophytes, arctic alpine and steppe species. For instance, the following tree species with a widely different ecology may grow together: *Pinus sylvestris, Picea obovata, Larix spp.* (More or less developed tree stands on calcareous substrate are described in the Section 3.5.8).

Generally, the plant communities on screes are characterized by suppressed trees or they have no trees at all. The stony substrate does not allow the roots to penetrate deeply, thus forcing the trees to keep their roots in dry and very shallow loose soils. Within a phytocoenosis the trees grow as isolated specimens. These are, besides the abovementioned conifers, *Populus tremula, Salix spp.,* and *Betula fruticosa.*

Plant communities on screes are affected by substrate mobility, amount of soil and moisture content. High mobility of the substrate is induced by cryogenic processes. Patches of bare ground may occupy up to 70–80% of the area. The calcareous slopes of river valleys with inclinations up to 50° support coenoses with 3 to 30% of cover abundance values, rarely reaching 50%. Steeper slopes lack vegetation, since the stony ground constantly moves, impeding soil stabilization.

The aspect of slopes is a very important factor for community formation. On the less heated North-facing slopes of the Tokko and Amga Rivers (the near-Lena and Lena-Aldan Plateau respectively, South Yakutia) *Carex-Dryas* tundroid formations occur (300 m a.s.l.). Similar communities were described by Lukicheva (1963a,b) from North–West Yakutia at a height of 400–500 m a.s.l. In southern regions the tundroids are common on screes with a very low content of soil, while the *Carex-Dryas* communities in the North–West develop exclusively on the loam of calcareous bedrocks. In both cases the loose substrate impedes the growth of trees or even shrubs. At the same time, in the Upper Amga River basin isolated specimens of *Pinus pumila* were recorded at the altitude of 261 m a.s.l. Growing under unfavourable conditions, the shrub is much suppressed and is characterized by a prostrate shape and a curved trunk of less than 1 m high.

The richness of the tundroid communities is 15–17 species per a plot (25 $m^2$). They are made up of prostrate dwarf shrubs, taproot rosette herbs (with a polycapitate caudex or a cushion form) (Yurtsev 1981). The plants of these communities belong to the arctic alpine element: *Dryas spp., Carex glacialis,* etc. An edificatory role is played by *Dryas punctata, D. viscosa, D. crenulata* with cover-abundance values of about 25%. At the same time, the sod forming species are the sedges that inhabit the loam accumulation among the stones: *Carex melanocarpa, C. glacialis, C. algida, C. sabynensis,* covering 30%. The same cover-abundance value is characteristic for *Rhododendron adamsii.* The forb component is not abundant, represented by constant *Braya siliquosa, Pinguicula alpina, Zigadenus sibiricus, Toffieldia coccinea* (North–West), and *T. cernua* (South), *Sassurea hypargyrea* (South). Ferns also occur at high constancy in the tundroid communities on calcareous screes: *Dryopteris fragrans, Woodsia glabella* in the South and *Cystopteris fragilis, Woodsia glabella,* and *Gymnocarpium continentale* in the North–West. Lichens (*Alectoria ochroleuca, Cetraria nivalis, C. cucullata)* are not numerous and are confined to the sodded parts of the landscape.

Warmer slopes are the habitats for the forb-*Dryas* "tundra-steppe", characterized by a combination of petrophytic steppe species and arctic alpine xerophytes. Malyshev (1965) described the "tundra-steppes" of the East Sayan Mountains, where they are confined to calcareous outcrops of light-exposed slopes. Similar communities were described by K.A. Volotovsky (1992) in the same landscapes of the Aldan River valley (the Aldan table-land) at the altitude of 300–400 m a.s.l. E.I. Troeva recorded "tundra-steppes" in the Lena-Aldan Plateau (the Upper Amga River) on the North-facing slopes with fine detritus (190–250 m a.s.l.) and in the Near-Lena Plateau (the Tokko and Chara Rivers) on the slopes of various aspects (unpublished data). The stony component of the slopes varies from large fragments with a minimal amount of soil to fine detritus with rather developed loamy soils. Some slopes feature pronounced solifluction processes due to seasonal thawing of the shallow permafrost. Typical petrophytes for the "tundra-steppe" communities are *Pulsatilla patens s.l., Thymus serpyllum s.l., Dendrantema zawadskii, Phlojodicarpus sibiricus, Euphorbia discolor, Youngia tenuifolia.* The arctic alpine species are the same as for the tundroid vegetation: *Dryas spp., Arctous alpina, Toffieldia cernua,* etc.

Due to the mobile substrate of coarse talus and a lack of soil, the vegetation occurs in patches. The edificatory plant is also the pioneer *Dryas*, among which herbs and lichens find appropriate conditions for growth. The cover-abundance value of *Dryas* is up to 80% within a patch. Being a pioneer vegetation, the "tundra-steppe" communities of coarse detritus are poor in species composition compared to those on fine screes. Their α-diversity is 12–15 and 20–25 species per plot, respectively. The structure of the "tundra-steppe" coenoses on fine detritus is more complicated. Isolated trees of *Pinus sylvestris* and *Picea obovata* often occur there. The shrub layer (cover 5%) consists of *Juniperus sibirica* and *Pentaphylloides fruticosa*. The cover of the herb layer is about 30%. On the south-facing slopes *Toffieldia cernua, Zigadenus sibirica, Arctous* grow poorly and give way to steppe species: *Potentilla nivea, Thymus serpyllum s.l., Thalictrum foetidum, Elytrigia jacutorum, Ephedra monosperma, Orostachys malacophylla* (the Near-Lena Plateau), O. *spinosa* (the Lena-Aldan Plateau). Lichens may cover 10%. The horizontal structure of the "tundra-steppe" communities on fine detritus is more or less even on dry slopes and spotty on moist substrate due to permafrost thaw out. The patches of vegetation are confined to relatively coarse detritus, where proper drainage is available, while the wet and cool soil remains bare. The "tundra-steppes" represent a relic vegetation preserved from the Pleistocene glaciation.

Owing to proper heating and the shallow stony soils, the rock outcrops, flagstones and screes of south-facing slopes are inhabited by communities of thermophilous xerophytes. This vegetation type lacks *Toffieldia cernua, Zigadenus sibiricus* and other arctic alpine and boreal montane species. The core of the petrophytic steppoids of South Yakutia consists of *Elytrigia jacutorum* (which is dominant), *Artemisia santolinifolia, Thymus serpyllum s.l., Scorzonera, Ephedra monosperma, Orostachys malacophylla* (the Near-Lena Plateau), and O. *spinosa* (the Lena-Aldan Plateau).

The association *Artemisia santolinifolia-Elytrigia jacutorum* (Fig. 3.17) occurs on light-exposed slopes of the Tokko and Upper Amga River valley at an altitude of 150–200 m a.s.l. The inclination varies from 50° to 90° (rock outcrops). Cover varies from 5–10 to 40%. The cover-abundance values of the dominating *Elytrigia jacutorum* and *Artemisia santolinifolia* range from 5 to 35%. They

**Fig. 3.17** The association *Artemisia santolinifolia-Elytrigia jacutorum* on calcareous flagstones (the Tokko River valley)

contain about 10 species per a plot (25 $m^2$). Rather rigorous ecological conditions determine the scarcity of such communities. Other constant species are *Youngia tenuifolia, Campanula rotundifolia, Orostachys malacophylla, O. spinosa,* and *Potentilla nivea.* Shrubs are represented by *Juniperus sibirica, Pentaphylloides fruticosa,* and *Rubus sachalinensis*. In the Indigirka River basin similar communities occur on steep (30° to 50°) south-facing slopes with a stony mobile substrate and primitive mountain chernozem soils. The vegetation is very thin (20–30%) predominated by *Elytrigia jacutorum, Artemisia santolinifolia* and *A. commutata,* sometimes with minor participation of *A. tanacetifolia* and *Veronica incana* (Skryabin 1968).

On more sodded substrate the association *Artemisia santolinifolia-Elytrigia jacutorum* is enriched by forb species. Increased values of α-diversity and cover make it possible to consider this a transition from the steppoid communities to the forb-*Artemisia santolinifolia-Elytrigia jacutorum* petrophytic steppe. For a description of this association we refer to Section 3.5.1.

Vegetation on cliffs occupies small ledges and crevices, where shallow soil accumulates. Besides the communities of the abovementioned association, the cliffs are inhabited by coenoses with a predomination of *Orostachys spp, Poa botryoides*. In the territory of the Natural Park "Lena Pillars" and the Sinyaya River basin unique communities of *Redowskia sophiifolia* occur, an endemic species that is recorded solely from this area (Sosina and Isaev 2003).

Ledges of carbonate rocks in North–West Yakutia support *Dryas* communities with the co-dominants *Saxifraga spinulosa* and *Gypsophylla sambukii,* growing on moss mats. *Pulsatilla patens, Astragalus alpinus, Oxytropis adamsiana* and other forbs are also constant there.

Acid and neutral bedrocks in North–West Yakutia are represented by igneous traprocks at an altitude of 600–700 m a.s.l. (Lukicheva 1963a). The level watershed areas are covered with large blocks of trap eluvium. Steep slopes are covered with screes, while gentle slopes have shallow soils with an admixture of detritus.

The association *Vaccinium vitis-idaea-uliginosum+Ledum palustre* was described from the Alyy-Sene River basin (the Anabar River basin). Its cover ranges from 10 to 20–50%. The dominating shrub species are *Dushekia fruticosa* and *Betula fruticosa*, as well as the less abundant *Juniperus sibirica, Rosa acicularis, Rubus sachalinensis, Ribes rubrum, Spiraea media, Salix jennisseensis, S. kolymensis,* and *Betula exilis*. The dominating plant of the dwarf shrub-herb layer is *Ledum palustre*. Other constant species are *Vaccinium vitis-idaea* and *Empetrum nigrum*. The herbs may form the thick synusiae of *Linnaea borealis, Limnas stelleri.* Low cover-abundance values are characteristic for *Carex globularis, Festuca supina, Equisetum scirpoides, Chamaenerium angustifolium, Pyrola incarnata, Dryopteris fragrans, Potentilla inquinans,* etc.

The lichens inhabit stones. Most abundant are the crustose and tubular forms, though *Stereocaulon paschale, Cladoina rangiferina, C. alpestris, C. amaurocrea, C. sylvatica,* and *Peltigera aphtosa* are also numerous.

The mosses grow between stones. They are mainly *Dicranum spp. Hylocomium splendens, Rhytidium rugosum, Aulacomnium spp., Tomentypnum nitens,* etc. Epilithic mosses are *Andreaea ruprestris, Grimmia sp.* which are not characteristic for calcareous outcrops.

Steep slopes support the same association though with less cover owing to the high mobility of substrates, less and shallower soils and deteriorated moisture conditions.

Flagstones of clay slates in North–East Yakutia (Nikolin 1990) support peculiar communities with low cover values (15–25%). Constant species are *Corydalis gorodkovii, Astragalus propinquus, Gorodkovia jacutica, Aconogon ochreatum, Dianthus repens, Dracocephalum palmatum, Thymus serpyllum, Chamerion latifolium, Gypsophylla sambukii, Rosa acicularis*, etc. Numerous cliff ledges in the mountains of the Verkhoyansk Range are inhabited by *Spiraea dahurica, Saxifraga oppositifolia, S. nivalis, S. serpyllifolia, S. tenuis, Sedum cyaneum, Rhodiola rosea subsp. borealis, Youngia tenuifolia, Androsace gorodkovii, Oxytropis nigrescens, Carex rupestris, Kobresia myosuroides, Draba cinerea, D. parvisiliquasa*, etc. On talus of large blocks the following species grow: *Ribes fragrans, R. triste, Juniperus sibiricus, Dryopteris fragrans, Aquilegia sibirica, Arabis turczaninowii, Erysimum pallasii, Campanula rotundifolia ssp. langsdorffiana, Urtica angustifolia, Polemonium boreale*, rarily *Ranunculus grayi, Dracocephalum palmatum* (or *D. stellerianum* in the East), *Salix berberifolia, S. tschuktschorum, S. requrvigemmis, Rhododendron adamsii,* etc.

The ledges of south-facing rocks represent fodder lands for a marmot (*Marmota camtschatica*). The vegetation is generally composed of the stepped tundra communities (*Dryas punctata, Salix polaris, Cassiope tetragona, Saxifraga nelsoniana, S. funstonii,* as well as *Kobresia myosuroides, Festuca brachyphylla, F. lenensis, Carex rupestris*), while in the marmot affected areas they are modified by vigorously developed rhizomatous grasses (*Bromopsis pumpelliana, Helictotrichon dahuricum*) and forbs (*Cerastium maximum, Potentilla stipularis, Arnica iljinii, Delphinium chamissonis, Myosotis asiatica, Sileni repens*) (Nikolin 1988).

### *3.5.3 Vegetation of Sandy Landscapes*

**E.G. Nikolin**

The ancient alluvial sands are very characteristic for the central regions of Yakutia. Being the remnants of former river processes they are found on upper terraces of the large tributaries of the Lena River. Spread by the wind, sandy landscapes form isolated "islands" among the taiga. Their Yakut name is *"tukulan"* (pronounced as *too-koo-lan*). As noted in previous studies (Karavaev and Skryabin, 1971), the modern state of these landscapes is partly the result of fires which exterminated the vegetation on sandy soils. This is also often proved by the large amount of wood flinders (mainly remains of *Pinus sylvestris*) scattered on the surface (Fig. 3.18). Such sandy soils occur mostly on prominent elements in the relief on both banks of the Lena River.

Though on the right bank (the Lena-Amga Interfluve region) these soils are covered by forests of *Pinus sylvestris* with an insignificant participation of *Larix cajanderi,* and with a ground cover composed of *Arctostaphylos uva-ursi* or lichens (mostly *Cladonia spp.*). Sometimes such vegetated sands fringe the alas depressions becoming a substrate for herbaceous xeromesophytic communities. Only small patches of loose sands are found upstream of Olyokminsk, in the Lena-Buotama Interfluve stretching downstream far to the North. But the *tukulan*-rich region is the Viluy River basin, the left tributary of the Lena River. The total area of *tukulans* makes up 50 thousand ha (Andreyev et al. 1987). The largest sandy landforms have their own names. The *tukulans* where studied by Kuznetzov (1927) and Rabotnov

**Fig. 3.18** Loose sands of the Munduguchchu *tukulan*

**Fig. 3.19** The elphin form of *Pinus sylvestris* with spreading crown at a tukulan's edge

(1935). Besides, some data on these natural landscapes were recorded by Zolnikov (1954a, 1958), Karavayev and Skryabin (1971), Galaktionova (Andreyev et al. 1987) and also reviewed in the collective monograph "Vegetation of the Viluy River basin" (Galaktionova et al. 1962). Most researches find that the *tukulans* very much resemble sandy deserts both in landform features (vast areas of almost bare sand; dunes up to 6 m high with a steep crumbling crest; hot, withering wind in sunny weather) and plant adaptations. The prevalent plants are low trees with a spreading crown, creeping forms of shrubs and dwarf shrubs (Fig. 3.19).

*Pinus sylvestris*, for example, acquires sometimes a bowl-like prostrate form very characteristic for *Pinus pumila*. The herbaceous vegetation typically consists of species with long rhizomes which produce clearly discernible tiers of additional roots on their shoots as sand covers them (Fig. 3.20). The roots of many plants are covered with sand cases "glued" by root secretion.

The leaves of most species are rough, convolute or narrow, with very dense indumentum. Some taxa grow as cushions. Many species propagate by rhizomes forming a single organism with an appearance of a mono-dominant plant community which covers an area up to 10–20 $m^2$ (Karavayev and Skryabin 1971). However, the fact that the *tukulans* are located in the taiga zone where climatic conditions are very distinct from those of the arid subtropics, and humidity is sufficient due to over-permafrost waters, suggests that the *tukulan* flora has developed convergent features, which are only similar to those of desert plants. The characteristic feature of the *tukulans* which distinguish them from deserts is their location close to large water bodies (usually lakes) which are fed from thawing permafrost underlying the sands. The flora of the sandy landforms draws from the surrounding communities, though some taxa have developed distinctive features and are endemic: *Koeleria karavajevii, K. skrjabinii, Festuca karavaevii, F. skrjabinii, Thymus sergievskjae,* etc. The total number of the psammophyte flora counts less than 100 species of higher plants (65–70 species according to Karavayev and Skryabin (1971)). Not all

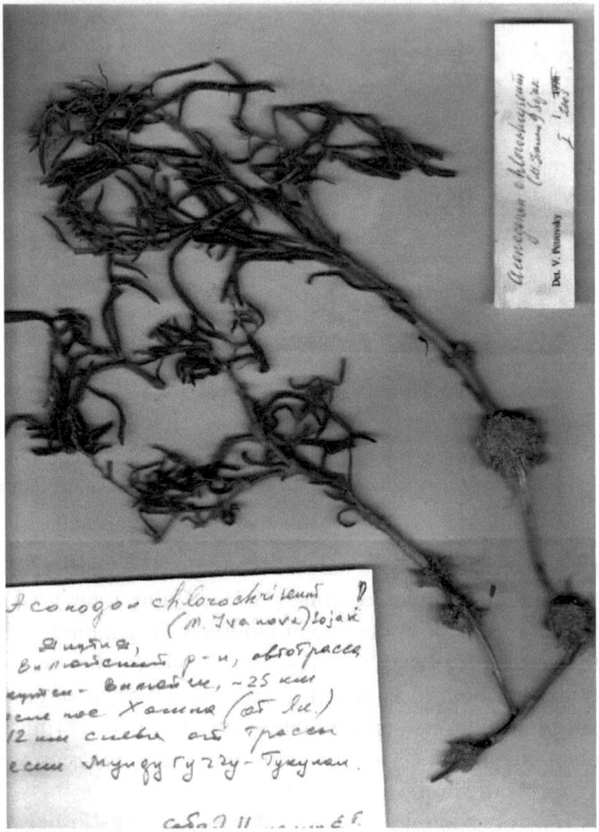


**Fig. 3.20** Tiers of additional roots of *Aconogonon chlorochryseum*

the large sandy formations in Yakutia have been investigated. The fact that they are isolated from each other by forests suggests that every *tukulan* has its own features. Thus, a recent study of the Mundguchchu-Tukulan yielded *Aconogon chlorochryseum,* a species which has been previously considered as endemic of the Chitinskaya Oblast.

The *tukulans* of the Viluy River basin, like those in the Lena-Amga Interfluve, are being overgrown with pine forests with *Arctostaphylos* and lichens forming the ground layer. Their herb-dwarf shrub layer consists of *Carex vanheurckii, Lychnis sibirica subsp. samojedorum, Arctostaphylos uva-ursi, Selaginella rupestris*, sometimes *Vaccinium vitis-idaea, Ledum decumbens, Loiseleria procumbens,* etc. The most characteristic mosses are *Polytrichum piliferum, Dicranum spp*. The lichen cover is composed of *Cladonia amaurocraea, C. arbuscula, C. cornuta, C. rangiferina, C. stellaris, Cetraria cucullata, C. islandica, C. nivalis,* etc.

The landscapes with a thin sand layer and with shallow permafrost occurrence are characterized by frost fractures. In such cases the fractures in the *Pinus sylvestris* forests become overgrown with mesohydrophyllous vegetation including *Larix gmelinii, Betula divaricata, B. exilis, Ledum decumbens, Chamaedaphne calyculata, Vaccinium uliginosum, Carex globularis,* etc. Their moss-lichen layer consists of *Dicranum elongatum, Pleurozium schreberi, Ptilidium ciliare, Sphagnum fuscum, Peltigera spp.* Dwarf shrub – *Sphagnum* bogs may also occur on sands. Sometimes *Betula pendula* may grow on semi-loose sands. The special *tukulan* communities in the Linde and Khoruonka Rivers basins are scrub of *Pinus pumila*. This is an isolated population with a distinct ecological optimum located on the left bank of the Lena River while the nearest habitat area of this species is located on the right bank of the Lena River within the Verkhoyansk Range. Common plants that form small communities or clones on loose sands are *Antennaria dioica, Artemisia pubescens* (sand populations of this species possess certain morphological peculiarities that allow some researches to consider it as a different species *A. karavajevii*), *Calamagrostis epigeios, Carex argunensis, C. ericetorum, C. korshinskyi, C. melanocarpa, Elytrigia repens, Ephedra monosperma, Koeleria cristata subsp. seminuda, Leymus littoralis, Rumex graminifolius, Saxifraga bronchialis, Silene jeniseensis subsp. popovii,* and *Thymus serpyllum*. Species which are not characteristic of *tukulan* communities but nevertheless are found there are *Campanula rotundifolia, Dianthus repens, Minuartia verna, Rosa acicularis,* etc. The *tukulans* belong to the category of especially valuable, unique landscapes of Yakutia requiring additional investigations and protection.

## *3.5.4 Vegetation of Saline Landscapes*

**M.M. Cherosov**

Central and to some extent also South–West Yakutia features the presence of salt and alkaline landscapes with halophytic vegetation (Gogoleva et al. 1987; Mirkin et al. 1992a, b). This phenomenon is unique for high latitudes. It can be explained by the complex effect of the following factors: an arid climate, abundant depressions without drainage, but the most important factor is the presence of a perennially frozen ground. These factors together serve as a natural barrier impeding the movement of salts and mineral-rich water from the upper soil horizons to lower ones. The most typical salinization in Yakutia is provided by chlorides, sulphates or both in various proportions (Elovskaya and Konorovsky 1978; Elovskaya 1987).

Typical halophytes are not numerous in Yakutia, though they form various communities along with glycohalophytic species. The dominant species of the salt and alkali landscapes of continental Yakutia are *Puccinellia tenuiflora, P.hauptiana, Suaeda corniculata, Salicornia europaea, Glaux maritima,* and *Halerpestes salsuginosa*. Species as *Elytrigia repen, Saussurea amara,* and *Knorringia sibirica* also often become a constituent part of the halophilous communities.

*Puccinellia tenuiflora* has a very wide phytocoenotic range and grows in habitats from light to strong salt concentration in the soils. The species is more characteristic for mesic communities of the alas ecosystems (alliance *Puccinellion tenuiflorae*, class *ASTERETEA TRIPOLII* Westhoff et Beeftink ap. Beeftink 1962) where it often forms monodominant grass stands. Similar but salt-free habitats are covered by mesic forb-grass meadows with a predominance of *Hordeum brevisubulati, Poa pratensis*, and other mesophilous species (*Thalictrum simplex, Sanguisorba officinalis, Vicia cracca,* etc.). They belong to the alliance *Hordeion brevisubulati* of the *MOLINIO-ARRHENATHERETEA* R. Tx. 1937 em Mirkin et Naumova 1986. In the Lena-Amga Interfluve the *Puccinellia* meadows may cover up to 70% of the whole alas area. Apart from the fact that *Puccinellia* indicates saline conditions, it forms valuable hayfields for cattle and winter pastures for horses.

In moderately dry saline habitats (mesoxerophytic belts of alases, prominent parts of floodplains) we find stepped meadows (*Artemisia commutata, Carex duriuscula, Hordeum brevisubulatum*) (see Section 3.5.7), though with participation of *Puccinellia tenuiflora* and *Saussurea amara*.

Wet saline soils are covered by vegetation of the association *Puccinellio-Alopecuretum arundinacei*. In fact, these are wet meadows of *Alopecurus arundinaceus* though with a clear diagnostic role of the species of the alliance *Puccinellion tenuiflorae* (*Puccinellia tenuiflora, Saussurea amara, Elytrigia repens*).

Soil trampling by livestock promotes the capillary rise of soil water. Consequently it evaporates, and salt accumulates near the soils surface, or sometimes even covers the surface with white incrustations. These soils, characterized by concentrated salts, are the habitats for communities of the class *THERO-SALICORNIETEA* Tx. in Tx. et Oberd. 1958. They are characterized by a prominent role of annual succulent halophytes (*Suaeda corniculata, Salicornia europaea*). These two species are the only components of the association *Suaedetum corniculatae* (Fig. 3.21). These communities are confined to the most saline soils of depressions (in alases or river terraces).

**Fig. 3.21** Degraded landscapes on the Middle Lena River terrace with oozed salts and vegetation represented by the associations of the class *THERO-SALICORNIETEA*

The *Artemisio jacuticae-Suaedetum corniculatae,* another association of the class, is characteristic for dryer and less saline conditions (terraces above the floodplain). Besides the diagnostic and dominant species, the communities may include *Salicornia europaea, Thellungiella salsuginea, Peucedanum salinum*, etc.

In the *Suaedo-Puccinellietum tenuiflorae* we find, apart from *Puccinellia tenuiflora* and *Suaeda corniculata,* glycogalophytic species such as *Taraxacum ceratophorum, Artemisia mongolica, Plantago major* or *P. media*, but they are never abundant. This vegetation is confined to soils with sulfate salinization.

The communities of the association *Puccinellio-Salicornietum europeae* occur on the soils of the chloride-sulphate salinization type under mesic moisture conditions.

The communities of the *Elytrigio-Glaucetum maritimi* form the vegetation on soils with chloride salinization and with some influence of sulphates under excessive moisture conditions.

These associations are arranged along an increasing moisture gradient and ion solubility. Thus, the easily soluble chlorides are characteristic for moist habitats. In contrast, the sulphate ions are accumulated in dryer soils. Mesic ecotopes possess both ions in various proportions.

Thus, the alas vegetation of saline soils represents the whole complex of halophytic communities. Under floodplain conditions (along the Lena and Viluy Rivers) salinization is not so common. Usually, saline soils occur in the middle or upper levels of the floodplain, where irregular flooding and high evaporation rates favour the accumulation of salt at the soil surface. The halophytic vegetation is the same as in the alases, though less pronounced. The soils of the terraces above the floodplain, however, very often feature salinization, and there solontchaks cover rather large areas, especially in populated areas.

## *3.5.5 Riparian, Lake-Side and Aquatic Vegetation*

**N.K. Sosina and E.I. Troeva**

Yakutia takes one of the leading places in Russia in the number of lakes and rivers. The general number of the rivers of the Republic is over 700,000 with a total length of about 2 million km. The major waterway of Yakutia, the Lena River (4,400 km long) is one of the largest rivers in the world. Its tributaries, the Viluy (2,650 km), the Aldan (2,273 km), and the Olyokma (1,436 km) Rivers are comparable in size with the largest European rivers. The basins of these rivers cover 65% of the territory of the Republic. Other large rivers of Yakutia are the Kolyma (2,129 km), the Indigirka (1,726 km), the Anabar (939 km) and the Yana (872 km) Rivers.

The total number of Yakutian lakes with an area over 1 hectare (ha) is 708,844 with a cumulative area of 7,399,300 ha, or 2.4% of the territory of the Republic (the global proportion of lakes in the world is 1.8%). The largest Yakutian lakes are Mogotoevo (323 km$^2$, North–east of the Indigirka River mouth), Nerpichye (237 km$^2$, the lower reaches of the Kolyma River), Ozhogino (157 km$^2$, the Middle

Indigirka Lowland), and Nidjili (119 km$^2$, the lower reaches of the Viluy River). The deepest lakes are Greater Toko and Lesser Toko (about 70 m deep, the Tokinsky Stanovik Range). Most lakes are drainless and contain fresh water. Mineralized lakes are Kempendyay and Abalakh.

The lakes of Yakutia have various origins: tectonic, glacial (glacial-erosion, glacial-accumulative, water-glacial), river (as a result of a change in a river course; these floodplain lakes are connected to rivers during floods), maritime (lagoons and laidas at coastlines and river mouths), caldron (or thermokarst, as a result of ground ice thawing), and artificial. The thermokarst and floodplain lakes are most widespread, representing a typical element of the Yakutian landscape. They are abundant both in the coastal plains and in river valleys throughout the whole territory of Yakutia (Korzhuev 1965).

*Vegetation of lake shores and slow streams*. As regards their riparian vegetation, the lakes of Central and South Yakutia can be divided into three groups (Trufanova 1967): group 1 represents the lakes which nearly completely lack a shallow shore. The banks are low, swamped, and composed of peaty soils. The lakes are 4–5 m deep. The water is dark-brownish. Their vegetation consists of a narrow belt containing *Phragmites australis, Menyanthes trifoliata, Typha latifolia,* etc.; group 2 includes shallow (2–4 m) lakes with low and swamped banks. In winter these lakes get frozen very deeply. As a result, high concentrations of $H_2S$ accumulate in the water and cause mass mortality of fishes. These lakes are characterized by a rich vegetation. Plants occupy all the shallow parts of the lake. The rims are covered by *Scolochloa, Carex* or *Equisetum-Carex* stands. In the water body *Myriophyllum* and *Utricularia* freely float. During relatively short periods of time the lakes overgrow and get swamped; group 3 represents lakes with a moderately developed rim of vegetation. Most important are *Potamogeton, Phragmites, Butomus, Sagittaria* and *Sparganium.*

The large lakes that gradually get deeper from their shore to their centre due to a very gently down-sloping lake bottom, especially in the alas landscapes, show concentric belts of vegetation. The over-wetted soils near the water body are inhabited by various *Carex* species (*Carex vesicata, C saxatilis, C. driandra,* etc.) and/or tall grasses (*Glyceria triflora, Scolochloa festucacea, Beckmannia syzigachne,* etc.*)*. The belt of shallow water vegetation (from the lake rim up to 1 m depth) supports large *Carex spp, Equisetum fluviatile, Cicuta virosa, Persicaria amphibia, Sagittaria natans, Sparganium spp*. Between these plants *Lemna trisulca* or *L. minor* may cover water surface. Sometimes they form a continuous cover in small stagnant ponds. The shallow water vegetation is followed by a belt community of tall hydrophytes: *Phragmites australis, Scirpus spp.,* and *Typha spp.* The freely floating plants of *Ceratophyllum, Myriophyllum,* and *Utricularia spp*. are also common there. Sometimes this belt is characterized by associations which contain *Tephroseris palustris* or *Acorus calamus*. The deeper water masses are inhabited by *Potamogeton spp.,* and in some places by *Nymphaea tetragona* and *Nuphar pumila*. During dry periods, when the lakes desiccate, *Scolochloa* and *Phragmites,* or even *Scirpus lacustris* may occur under moderate moisture conditions (Permyakova 1961; Usanova 1961; Galaktionova et al. 1962; Trufanova 1967; Karavaev and Skryabin 1971).

The riparian vegetation of young oxbow lakes is more scanty compared to alas lakes. Later in the succession, they start to be vigorously overgrown. The rims and shallow water are covered by thick, often monospecific stands of large *Carex* species (*Carex aquatilis, C. rhynchophysa, C. rostrata, C. driandra, C. vesicata,* etc.), or *Equisetum fluviatile, E. arvense,* or *Glyceria triflora.* Deeper in the water *Myriophyllum spicatum, Utricularia vulgaris, Ceratophyllum demersum,* and *Hippuris vulgaris* appear. The shallow floodplain lakes feature *Hippuris-Ceratophyllum* communities on oozy clayey-sapropel grounds, or *Lemna-Ceratophyllum* coenoses on sapropel substrate (Egorova et al. 2006). Very often *Sagittaria natans* plays a primary role in communities at a depth of 1.5 m, along with *Nymphaea tetragona* and *Nuphar pumila.* Deeper water is inhabited by *Potamogeton spp.* (Karavaev and Skryabin 1971). In the Middle Lena River valleys the rims of oxbow and floodplain lakes are covered by *Scolochloa festucacea* with *Acorus calamus* and *Carex spp.*

With time, the accumulation of plant remains leads to a gradual shoaling of lakes. As a result the hydrophytes disappear from the shallow rims and move to the deeper parts of the lake. Gradually, a lake becomes completely overgrown, transforming into a swamp and later into a *Carex* meadow.

Overgrowing lakes with steep slopes are often characterized by quagmires composed of a thick mat of *Carex chordorrhiza, Menyanthes trifoliata, Comarum palustre, Calla palustris,* and some other plants. As the process continues the water mirror reduces in area and, finally, almost disappears (Karavaev and Skryabin 1971).

The hydrophytic vegetation of the valley lakes of the northern taiga zone is scarce and monotonous. In the Middle Kolyma River valley not all lakes possess a vegetated rim (Trufanova 1972a,b). The vegetation may form isolated "islands" along the lake side or occur in a narrow belt getting into the water as far as 1 m deep. The surrounding tree and shrub vegetation approach the lake rims. The rim vegetation consists mainly of *Equisetum fluviatile, Menyanthes trifoliata, Arctophila fulva, Carex rhynchophysa, C. rostrata, C. vesicata,* or *Comarum palustre.* Less abundant are *Cicuta virosa, Naumburgia thyrsiflora, Rumex aquaticus, Caltha arctica, Hippuris vulgaris,* etc. In the water *Potamogeton spp., Myriophyllum verticillatum, Geratophyllum demersum, Sagittaria natans,* etc. occur. In some lakes littoral or freely floating quagmires of *Menyanthes trifoliata* are observed. Aquatic mosses, such as *Calliergon giganteum, Drepanocladus vernicosus, D. aduncus, D. sendtneri,* etc. favour quagmire formation, speeding up overgrowing process of the lake. The aquatic mosses usually grow in shallow water bodies, in bays and gulfs, where they predominate. The overgrowing of a lake starts from the bottom. First, mosses form separate clumps within submerged vegetation. Shallow lakes are characterized by high rates of overgrowing thanks to annual accumulation of dead plants. With time, mosses displace the underwater vegetation, and lakes turn into swamps.

The most common communities of hydrophytic vegetation in the forest-tundra and tundra zones are monospecific coenoses of *Arctophila fulva* and *Carex concolor* (Tyrtikov 1955; Nosova 1964; Perfilyeva et al. 1991). They are widespread also in the northern taiga. *Arctophila* is characteristic for large and small lakes and microdepressions without drainage, while *Carex concolor* communities are confined to the polygonal-ridged microcomplexes of the southern arctic tundra. The arctic tundra subzone also features communities of *Pleuropogon sabinii, Dupontia fisheri, Alopecurus alpinus*. The subarctic tundra is characterized by communities of *Hippuris vulgaris* and *H. lanceolata, Equisetum fluviatile, Ranunculus pallasii, Menyanthes trifoliata,* or *Sparganium borealis.* Communities of *Potamogeton* are observed in the thermokarst lakes of the Kolyma River valley (Trufanova and Galaktionova 1975). Communities of *Batrachium* and especially *Ranunculus gmelinii* are more widespread and confined to shallow water bodies. The coenoses of *Arctophila fulva* are best studied (Shvedchikov 1974a, b, 1975; Andreyev et al 1978; Matveyeva 1993; etc.). They inhabit the deepest (up to 70 cm) littoral parts of lakes. Strips of this community vary in width. Closer to the rim, mats of *Menyanthes trifoliata* are found with an admixture of *Arctophila fulva, Equisetum fluviatile, Utricularia vulgaris* and *Comarum palustre.* Very often *Menyanthes trifoliata* may form a continuous carpet covering the whole surface of wide and shallow lakes. Shallow lakes also support *Sparganium minimum*, or *Hippuris vulgaris (*with participation of *Caltha palustris),* forming an outer belt of vegetation. Peat sedimentation as a result of the annual accumulation of dead plants yields an even more accelerated shallowing and swamping of a water body. At a depth of 10 cm communities of *Equisetum fluviatile* with participation of *Arctophila fulva, Comarum palustre, Carex aquatilis, C. saxatilis, Eriophorum scheuchzeri,* and *E. polystachion* are observed. The moss layer is represented by *Drepanocladus revolvens* and *Scorpidium scorpiodes.*

*Riparian vegetation.* The lower floodplain of many rivers is formed by pebbly and sandy alluvium. Young floodplains of rivers and streams are characterized by *Equisetum arvense* or *E. fluviatile* communities. *Equisetum fluviatile* forms wide strips along the banks of small rivers. *Equisetum* may grow on substrates of any mechanical structure, from clayey-muddy to medium-sized pebbles. The *Equisetum* communities typically have a very plain structure and a very scanty species composition. Besides the dominating *Equisetum arvense*, with cover abundance values 30 to 100%, only *Carex aquatilis, Eleocharis palustris* or *Deschampsia borealis* occur. The stage of *Equisetum arvense* or *E. fluviatile* lasts only a very short in time. On properly drained soils it is followed by grass communities (*Equisetum-Beckmannia, Beckmannia-Poa* or *Carex-Calamagrostis* meadows), and on heavy substrates by *Carex* (*Carex acuta, C. aquatilis, C. Diandra,* etc.) meadows. In both cases *Equisetum* remains but in insignificant amounts (Ivanova 1961; Galaktionova et al. 1962).

The vegetation of carbonate pebble alluvium has a distinct character. During floodings the water stream has a strong mechanical action on the vegetation growing on pebbles. The abiotic conditions are thus very unfavourable there. This determines the scanty species composition and cover abundance values of such communities, representing the initial succession stage of floodplain overgrowing. The pebble communities of North–western Yakutia were studied by Lukicheva (1963a, b). Due to strong erosive effects of the rivers, the alluvium in that region consists of coarse detrital sediments. The following species are observed there: *Allium schoenophrasum, Angelica decurrens, Salix saxatilis,* etc. The next stage is characteristic for elevated spits, where loamy sand mingles with pebbles. In summer this substrate dries up completely after the spring tides, and the spits become covered by a rich and thick vegetation. The communities consist of *Dryas grandis, Thesium refractum, Astragalus alpinus, Thymus spp., Aster alpinus, Sanguisorba officinalis,* etc.

In South Yakutia, the Near-Lena and Aldan Plateaus, the lower floodplains consist of carbonate flagstones covered by small-sized (5–10 cm) pebble alluvium. The communities on this substrate are characterized by the presence of erosiophilous, often annual or biennual species (*Setaria viridis, Chenopodium sp., Corispermum sp., Polygonum aviculare,* etc.). The *Setaria-Corispermum* formation (provisional name) occurs on the banks of the Tokko, Upper Amga, and Middle Lena Rivers.

The *Setaria-Sanguisorba* association is a typical example described from the Upper Amga River (unpublished data of E. Troeva). The cover of these communities varies from 5 to 40%, the average height of the upper layer is 50–60 cm, species richness is 20–25 per 25 $m^2$. The communities are composed of two layers. The lower layer consists of prostrate species (*Setaria viridis, Potentilla anserina, Persicaria amphibia*), that are adapted to withstand regular flooding and wave action. The upper layer consists of *Sanguisorba officinalis* and *Aconogonon ochreatum*, featuring symmetrical rounded shapes. (Fig. 3.22). This is probably explained by the low cover abundance values and uniform light colour of the substrate, as a result of which a plant is evenly exposed to sun and reflected radiations. Other constant species are *Stachys aspera, Inula britannica, Equisetum arvense* and *Allium schoenoprasum.*

On elevated parts of the floodplains the pebble alluvium includes sand and sandy loam. Absence of wave action and the presence of a mixed substrate favour the development of a richer vegetation. Similar to formations of the North–West, these landscapes support thick and diverse communities composed of mesophytic and xeromesophytic species. On the Near-Lena and Aldan Plateaus the communities are represented by the forb-*Sanguisorba* meadow formation. They are characterized by

**Fig. 3.22** Rounded and compact shape of *Aconogonon ochreatum* on the carbonate pebble alluvium

average cover values of 40%, an average height of 60 cm, and a species diversity of 20–25 species per 25 $m^2$, sometimes reaching over 30 species. The dominant species is *Sangisorba officinalis*. Constant species are *Vicia cracca, Allium scoenoprasum, Galium verum, Lupinaster pentaphyllus, Ranunculus propinquus, Galium boreale*, etc. The mesophytic *Thalictrum simplex*, *Vicia cracca* and *Allium scoenoprasum* co-exist with steppe species such as *Koeleria sp., Dianthus versicolor or Potentilla nivea* that penetrate the communities from the stepped meadows or meadow steppes from terraces above the floodplain (Savvinov et al. 1992).

In the Middle Lena River (200–300 km South of Yakutsk) similar landscapes are occupied by the forb-*Medicago falcata* communities that are very similar in ecology to the above-mentioned *Sanguisorba* meadows (unpublished data of E. Troeva). The edificatory species there is *Medicago falcata* characterized by cover abundance values reaching 50%. Besides *Medicago,* constant species are *Setaria viridis, Artemisia scoparia* and *Potentilla bifura. Equisetum arvensis* and *Corispermum sp.* are also characteristic. These communities also include such xerophytic species such as *Thymus pseudoaltaicus, Poa transbaicalica, Euphorbia discolor*, or sometimes even *Stipa krylovii* that penetrate from the steppe communities on elevated terraces and watershed areas on which the floodplain transgressed

The aquatic vegetation of South–West Yakutia is dominated by *Butomus umbellatus* and *Potamogeton spp.* with an admixture of *Lemna, Myriophyllum, Callitriche hermaphroditica*. The *Phragmites australis-Potamogeton* community with participation of *Persicaria amphibian* and *Ranunculus repens*is also common. Rarely communities of *Nuphar pumila* occur. On the Pilka River (right tributary of the Upper Lena River) and the Khamra River (left tributary of the Upper Lena River) *Fontinalis antipyretica* is abundant. *Potamogeton lucens* is rarer (Egorova et al. 2006).

## 3.5.6 Bogs

### A.A. Egorova

Bogs are landscapes characterized by excessive and stagnant moisture conditions, an acid reaction of the soil, and processes of peat formation. In Yakutia they are associated with the widespread development of thermokarst processes. As regards feeding type, the bogs of Yakutia are distinguished into two types: the upland bogs and lowland bogs. Generally, the bogs and tundra-bog complexes occupy over half of the territory of the tundra zone and 1.2% of the boreal zone (Andreyev et al. 1987). A very small number of publications is dedicated to Yakutian bogs. The bogs and tundra-bog complexes of the tundra zone are best studied (Sheludyakova 1938, 1948; Tyrtikov 1955, 1958; Petrovsky 1959, 1962; Dobretsova 1962; Schelkunova 1970; Andreyev and Perfilyeva 1980; Boch 1975, 1978; Andreyev et al. 1987; Labutin et al. 1985; Skryabin and Karavaev 1991; Perfilyeva et al. 1991, etc.). Some data on forest bogs are given in the works by Drobov (1927), Rabotnov (1939), Sheludyakova (1957b), Neinshtadt and Nikonov (1958), Ivanova 1961; Galaktionova et al.(1962), Lukicheva (1963a); Karavaev and Skryabin 1971; Kats 1971; etc.

*The lowland bogs* occur in the tundra zone and are confined to the concave elements of the relief. Their characteristic feature is gleization of the soils and low rates of peat formation (the peat layer does not exceed 25–35 cm) due to the shallow occurrence of frozen grounds, low summer temperatures and late dates of seasonal thawing of the permafrost. Water logging in the tundra zone is favoured by low evaporation rates.

The lowland bogs are most common in maritime lowlands and in lake depressions. Besides, bogs are a constituent part of the polygonal-ridged and hummocky-hollow tundra-bog complexes that are widespread in the valleys and deltas of the rivers, as well as in watershed landscapes disturbed by thermotkarst. They are most characteristic for the Indigirka and Kolyma Rivers basins (North–East Yakutia).

The tundra lowland bogs are represented by grass vegetation. There are communities with participation of *Sphagnum*, but they do not play any edificatory role (Perfilyeva et al. 1991; Galkina 1956). The composition of dominant species changes from the North southwards. Within the strip of northern arctic tundra (the Novosibirskie Isl.) the dominant species are *Dupontia fischeri, Carex concolor, Arctophila fulva, Eriophorum polystachion,* and *E. medium* (Alexandrova 1963). In the southern arctic tundra *Dupontia fischeri* disappears and is replaced by *Eriophorum scheuchzeri*. In the northern Subarctic the bog communities are supplemented by *Carex chordorrhiza* and *C. rotundata.* The southern Subarctic bogs feature polydominant communities of *Carex chordorrhiza, C. concolor, C. rariflora,* and *Eriophorum polystachion* (Perfilyeva et al. 1991).

*Arctophila fulva* bogs are common along tundra river banks, lake sides, as well as in microdepressions, and are characterized by a vary scanty species composition (5 to 7 species). *Arctophila* covers 10 to 40%. The moss layer is represented by *Drepanocladus exannulatus,* and *Sphagnum sguarrosum.*

Bogs with *Eriophorum polystachion* develop in the shallow water of lake rims, depressions in maritime terraces and river deltas, and especially in the polygonal-ridged microcomplexes. They are the richest communities containing 18 to 61 species of higher vascular plants and mosses. Besides *Eriophorum polystachion* and *Carex concolor, Dupontia fischeri* is common there. Cover values of the moss layer reach 30–60%, sometimes forming an almost continuous cover.

The shoreline east of the Indigirka River features watered bog communities of *Eriophorum scheuchzeri*. The species composition includes 7 to 15 moss and higher vascular plant species. The herb layer may include *Carex concolor* and *Arctophila fulva*. The moss layer is thick, sometimes forming a continuous carpet: *Drepanocladus exannulatus, Polytrichum jensenii, Sphagnum sguarrosum, S. fimbriatum*, etc.

Habitats that dry up deteriorate the growth conditions of the dominant grasses and the species composition of mosses and higher vascular plants increases (even shrubs may appear). This is very characteristic for the subarctic part of the Lena River Delta, where drying up of the polygonal-ridged tundra-bogs is observed (Labutin et al. 1985; Perfilyeva et al. 1991). They feature communities without mosses in moderately watered polygons; communities with hydrophilous mosses, which are saturated with water; transitional communities with hygrophilous and mesophilous mosses; and, finally tundra communities in case of complete drying up of the substrate. A similar series is inherent to the grass bogs with *Carex concolor* on properly drained soils, while on heavy soils the hydrophilous green moss stage is followed by the *Sphagnum* communities, missing the hygrophilous stage (Perfilyeva et al. 1991).

Bogs and tundra-bog complexes are also characteristic for the subzone of the near-tundra forests. However, there they represent the initial stage of forest succession (Tyrtikov 1958): polygonal-ridged tundra-bog complexes – polygonal-hollow forest-bog complexes – hummocky-hollow forest-bog complexes.

*The upland bogs* are common for the forest zone and do not occur in the tundra zone and in the near-tundra forest landscapes.

In the forest zone bogs have a limited distribution due to the arid climate, pronounced mountainous topology, and widespread carbonate and saline soils. However, as small patches, they are scattered all over the Republic. Very seldom, relatively large areas of upland bogs are found in level areas, where lakes are characterized by intensive peat formation processes. Diversity and areas occupied by bogs in the forest zone of Yakutia decrease from the North southwards and from the East westwards.

The floristic composition of forest bogs is similar to those of the Subarctic tundra. They also contain *Betula exilis, Ledum palustre, Vaccinium vitis-idaea, V. uliginosum,* etc. The herb layer is predominated by *Eriophorum polystachyon, Carex limosa, C. chordorrhiza,* etc. The shrub layer's cover reaches 20–30%.

According to the classification of M.S. Boch and V.V. Mazing (1979), the bogs of the forest zone are distinguished into 3 types based on the dominating eco-biomorphs: the hydrophilous grass bogs, the hydrophilous moss bogs of plains, and the mesohydrophilous grass bogs of foothills.

The first type is characteristic for river valley terraces above the floodplain and watershed areas with stagnant moisture conditions. The characteristic species are *Carex chordorrhiza, C. karoi, C. lithophila.* In the bogs shrubberies of *Salix myrtilloides* or *Spiraea salicifolia* occur. The moss layer of such bogs is weakly developed and is represented by *Drepanocladus vernicosus. Eriophorum polystachion* bogs occur in river valleys and foothills. The moss layer is composed of *Calliergon giganteum, Drepanocladus spp.,* and sometimes *Sphagnum fuscum* and *S. warnstofrrii.* The peat horizon is as thick as 0.5–1 m.

The second type is represented by the green moss and *Sphagnum* bogs occurring in watersheds in the thermokarst hollows as wide as 100–300 m, and rarely in river valleys. In watershed areas of the North–West *Carex*-green moss bogs are common with predomination of *Carex chordorrhiza* in the herb layer. The continuous moss layer consists of *Drepanocladus intermedius, Dr. exannulatus, Calliergon giganteum,* etc. (Lukicheva 1963a). The *Carex*-green moss bogs of river valleys differ in species composition (Ivanova 1961; Karavaev and Skryabin 1971; Kats 1971). The dominant sedge is *Carex limosa* with participation of *C. rostrata.* The dwarf shrub layer is represented by *Chamaedaphne calyculata* and *Andromeda polifolia.* The moss layer is composed of *Drepanocladus uncinatus, D. vernicosus* and *Calliergon giganteum. Sphagnum* bogs are widespread in the Olenyok,

Zhigansk, Viluy, Upper Viluy and Uchur-Olyokma geobotanical districts, characterized by sandy sediments. The bogs there are confined to watershed areas and river valleys occupying up to 1 ha in area. These landscapes have an uneven microrelief. The tree layer consits of isolated trees of *Larix spp., Pinus sylvestris* or *Betula spp.* as high as 1–4 m. The dwarf shrub-herb layer is dominated by *Ledum palustre, Vaccinium vitis-idaea, Andromeda polifolia, Rubus chamaemorus* and *Oxycoccus microcarpus* in watersheds or *Chamaedaphne calyculata* in river valleys. The moss layer is represented by *Sphagnum fuscum, S. lenense* and *S. warnstorfii* in watersheds and is predominated by *S. obtusum* in river valleys. The peat layer reaches 20 to150 cm.

The third type occurs in the foothills of the Verkhoyansk Range (100 m a.s.l.) (Andreyev et al. 1987). It represents the *Carex-Eriophorum* community with participation of shrubs. The formation of this bog type is explained by higher amounts of precipitation and permanent run-off from the slopes. The bog is confined to depressions and is surrounded by the *Larix* forests of the green moss type on prominent parts of the landscape. The bog represents the hummocky microrelief (each hummock is 20–25 cm high and 25–30 cm wide). Suppressed trees of *Larix* occur as high as 5–6 m. Shrubs are *Salix pyrolifolia* and *Betula exilis* with an admixture of *Salix myrtilloides*. The dwarf shrub-herb layer is predominated by *Eriophorum vaginatum, Carex appendiculata, Vaccinium uliginosum, V. vitis-idaea,* and *Ledum palustre*. The moss layer is represented mainly by *Aulacomnium turgidum*.

The bogs in Yakutia have no economic significance. Having alternative sources of energy, peat is not used as a fuel. The grass bogs and swamped larch sparse forests represent good pastures for horses and reindeer.

### 3.5.7 Meadow Vegetation

**M.M. Cherosov, P.A. Gogoleva, and E.I. Troeva**

The meadows of Yakutia represent a special phenomenon in that they reflect the regional features and uniqueness of Yakutian nature. On the other hand, the boreal flora, which makes up most of the meadow coenoflora, determines the similarity of Yakutian meadows with their analogues in other parts of the world, including the well-studied meadows of Western Europe.

The meadows of Yakutia have been studied since the beginning of the twentieth century. In 1901 A. Cajander travelled all along the Lena River and later published his observations (Cajander 1903, 1904). Later on, the meadow vegetation was studied by a number of expeditions: of the so called Resettlement Department in 1912, of the Academy of Sciences of the USSR (1920s–1930s), of the People's Commissariat for land use (since 1932), etc. They yielded both scientific and industrial descriptions of the meadows of various regions of Yakutia. These works were either descriptive or had an application-oriented character. The second half of the twentieth century is characterized by more fundamental investigations, focused on succession, ecology and the geographical principles of meadows distribution. The general results of those studies are given in the works by M.N. Karavaev (1965), Karavaev and Skryabin (1971), Andreyev et al. (1975), Skryabin and Karavaev (1991). The syntaxonomy and ecology of the meadows of the Lena River floodplain are given in the monograph by Kononov (1982). The alas meadows of Central Yakutia were studied by Gogoleva et al. (1987) and Ivanov et al. (2004). The classification of the meadow vegetation of the Lena River floodplain and Central Yakutian alases using the floristic-sociological approach was published abroad (Mirkin et al. 1985, 1992a, b). Korolyuk et al. (2005) conducted an analysis of the growing conditions of the main syntaxa of meadow vegetation of Central Yakutia using the authors' ecological indices. However, most data on Yakutian meadow vegetation was obtained based on the traditional ecological-dominance approach.

Some meadow types have names of local origin, which are also used in the scientific literature, e.g. the alas and small river valley meadows. The alas meadows represent the vegetation of near-lake depressions in forested watershed areas that appear as a result of thermokarst processes. The main conditions of alas formation include the presence of level area, ice lenses underground, and anthropogenic disturbance of the environment (fires, cuttings). Alas vegetation forms about half of the Yakutian meadows. The small river meadows are herbaceous communities of erosion valleys of rivulets, which often dry up or have a variable course. The most widespread meadow types in Yakutia are the floodplain, alas, and small river valley types.

The high diversity of the present meadow vegetation is determined by a number of factors:

- The meadows may be primary or secondary in origin. Primary meadows are formed as pioneer vegetation on various alluviums of rivers, lakes, or seas and on erosion valleys created by small rivers. Secondary meadows occupy large areas as a result of human activity (shrubbery and forest elimination, lake drainage, etc.). If the human activity is stopped, the secondary meadows are replaced by the original vegetation. Too strong anthropogenic effects lead to severe degradations of meadows.
- The combination of the arid steppe elements, penetrating as far as the Arctic Ocean's coastline, and the cryogenic tundra elements, occurring even in the central regions of Yakutia, is a characteristic feature of the Yakutian vegetation, including meadows. The period of the so called "tundra-steppes" is reflected in the composition of meadows.
- The topology of Yakutia supports the presence of both lowland and mountainous types of meadows.
- And finally, in their contact with bogs, forest, steppe, alpine, and anthropogenic communities, the meadow coenoses show the influence of diverse ecological factors.

In Yakutia mesic meadows are not common, giving way to the waterlogged meadows (mainly in northern regions) and the stepped meadows (in central, south–western and mountainous North–eastern regions where steppe landscapes are widespread).

Here we do not mention the maritime meadows of the Arctic zone, since they have been described above (see Section 3.3.1), and neither the near-icing meadows due to lack of data.

*The northern taiga subzone.* The waterlogged areas are occupied by communities of moist and swamped meadows composed of *Arctophila fulva, Carex concolor, Eriophorum polystachion, Calamagrostis langsdorffi, C. neglecta.* Tussock-forming *Carex spp.* are very common there, especially *Carex juncella.* The tussocks may reach a height up to 1 m, serving as a substrate and a source of mineral nutrition for other plants. One of such species is *Calamagrostis holmii.* Being a rather mesophilous plant, it finds proper growth conditions on the *Carex juncella* tussocks in the waterlogged area.

In the Yana and Indigirka River basins within the Verkhoyansk mountain system the meadows acquire a distinct feature due to the wide occurrence of polygonal cryogenic relief. The stepped meadows of the subzone, as of the other regions of Yakutia, are genetically related to the steppes. Apart from the North–East, this phenomenon is also common in Central and South–West Yakutia, though isolated stepped elements occur all over the boreal zone of Yakutia, characterized by a local lack of moisture, e.g. on the slopes of river valleys. The development of the steppe element is favoured by the arid ultra-continental climate of Yakutia. The coenoflora of stepped meadows includes such species as *Agrostis trinii, Bromopsis pumpelliana, Poa botryoides, Koeleria cristata, Carex duriuscula, Linum perenne, Potentilla nivea, Lychnis sibirica, Pedicularis venusta,* etc.

The middle floodplains support mesic meadows with a predomination of *Calamagrostis holmii, Carex enervis, Carex minuta.* The lower floodplains and depressions of the upper floodplains are occupied by swamped meadows with participation of *Eriophorum polystachion* in a background of hygromesophilous forb species.

In mountains the meadows cover insignificant areas in floodplains and around icings. The subalpine and alpine tundra belts feature mesic meadows on sandy-gravel alluvium with a prevalence of *Leymus interior, Bromopsis pumpelliana,* and *Helictotrichon dahuricum.* These communities include also higher vascular species and mosses that are typical of the tundra. The riparian vegetation of small rivers and streams represents a complex of bogs, tundra communities and swamped meadows of *Carex concolor.* The terraces above the floodplain within the alpine tundra belt are occupied by the short-grass meadows of *Carex media, Carex misandra, Poa pratensis, Alopecurus alpinus, Kobresia myosuroides, Ranunculus propinquus,* etc. The communities of *Equisetum variegatum* are economically important as they are valuable pastures for horses that significantly put on weight when they feed on them just before winter.

In the northern taiga of West Yakutia (the Olenyok and Anabar River basins) the meadow communities represent basically narrow strips of riparian vegetation (see Section 3.5.5). There are also the stepped meadows that are sometimes contact communities with insignificant (small) steppe areas.

*The middle taiga subzone.* For this subzone the meadows of lowland Central Yakutia are described. The meadows there cover the largest areas, occupying:

– river floodplains,
– alases,
– erosion valleys of small rivers that surround the alas region of the Lena, Amga and Viluy Interfluves and are situated away from the large rivers,
– forest (waterless valley, or upland) meadows.

As in many other regions of the world, the floristic basis of the meadows of Central Yakutia is provided by species of *Poacea* and *Cyperaceae*, forming mono- or polydominant communities. The most competitive species are *Hordeum brevisubulatum, Alopecurus arundinaceus, Puccinellia tenuiflora, Agrostis trinii, Agrostis gigantea, Calamagrostis langsdorffi, Calamagrostis neglecta, Carex juncella, Carex schmidtii, Carex acuta, Carex appendiculata, Carex aquatilis,* etc. Forb species are usually less abundant (*Vicia cracca, Lupinaster pentaphyllus, Lathyrus pilosus, Lathyrus pratensis, Geranium pratense, Veronica longifolia, Thalictrum minus, Silene repens, Galium verum, G. boreale, Achillea millefolium, A. asiatica, Heracleum dissectum*, etc.).

A distinct feature of the meadow vegetation of the subzone is its xerophytic character, which is even more pronounced then in the northern taiga. The floristic peculiarity of stepped meadows is determined by *Agrostis trinii, Bromopsis pumpelliana, Koeleria cristata, Poa botryoides, Delphinium grandiflorum, Linum perenne, Festuca lenensis,* and *Anemone sylvestris* in combination with the mesophilous *Hordeum brevisubulatum, Puccinellia tenuiflora, Poa pratensis, Elytrigia repens, Thalictrum simplex, Galium verum, Sanguisorba officinalis,* etc. Depending on the amount of precipitation during part of the vegetation period, one of these species groups gets the advantage as regards growth conditions, resulting in either a meadow or a steppe character of a community. The stepped meadows cover the largest areas in the Lena River valley and in the upper vegetation belts of alases. The communities of the mesic and stepped communities belong to the class *MOLINIO-ARRHENATHERETEA* according to the floristic-sociological criteria of vegetation classification.

Another distinct feature of the meadows of the middle taiga subzone is their halophytic character. The upward movement of soil moisture with dissolved salts in July-August under conditions of lack of drainage (alas depressions, drying up small rivers) results in salt accumulation in the upper horizons. These saline soils of primary or secondary (anthropogenic) origin support halophytic meadow communities, especially vigorous under conditions of a pronounced variability in moisture regimes. The grasses are *Hordeum brevisubulatum, Alopecurus arundinaceus, Puccinellia tenuiflora,* and *Puccinellia hauptiana*. The latter two species play an edificatory role. The sedges are *Carex reptabunda* in mesic habitats and *C. atherodes* and *C. lithophila* under moist conditions. Indicator species of salinization are the halophytes and glycohalophytes *Glaux maritima, Potentilla anserina, Knorringia sibirica, Saussurea amara, Salicornia europaea*, composing the second layer of a grass stand. The saline meadows are most characteristic for the alases of Central Yakutia and above-floodplain terraces of the Lena River, and are absent in the small river valleys due to proper drainage and lower values of summer temperatures (see Section 3.5.4). The communities of saline meadows belong to the class *ASTERETEA TRIPOLII*.

*The floodplain meadows* are characterized by higher moisture levels and are less dependent on annual climatic conditions. The floodplains support a high variety of meadow types: from stepped to swamped ones that occur at all floodplain levels (upper, middle and lower).

*An alas* in Yakut means "a meadow surrounded by forested mountains". As has been mentioned before, the alases are a unique landscapes induced by thermokarst processes. The vegetation of alases is composed of steppe, meadow, hygrophytic and aquatic communities. An alas goes through a series of developmental stages. Each stage is characterized by a peculiar combination of communities.

A typical alas landscape is arranged into four concentric soil-vegetation belts. The first belt represents the hydrophilous communities with a predomination of *Glyceria triflora, Scolochloa festucacea, Scirpus lacustris, Tephroseris palustris, Phragmites australis, Typha latifolia,* etc. on frozen peaty-bog soils. The second belt consists of the communities of swamped and moist meadows with participation of *Carex juncella, Carex lithophila, Calamagrostis neglecta, Beckmannia syzigachne, Alopecurus arundinaceus, Agrostis stolonifera,* etc. on frozen meadow-bog soils (Fig. 3.23). The third belt represents the frozen chernozem-meadow soils with mesic meadows with a predomination of Hordeum *brevisubulatum* or *Puccinellia tenuiflora*, the fourth belt is composed of stepped meadows of *Elytrigia repens Artemisia commutata, Carex duriuscula* and other species on frozen meadow-chernozem soils. The south-facing slopes are occupied by steppe coenoses of *Stipa krylovii, Festuca lenensis, Carex pediformis,* and *Pulsatilla flavescens*, while the North-facing slopes are covered by the larch forest of *Larix gmelinii,* or *L. cajanderii.*

**Fig. 3.23** The meadow communitiy of the association *Alopecuretum arundinacei* representing the mesohygrophilous belt of an alas vegetation

Not all alases feature this clear belt differentiation. The belts may vary in width. Sometimes, a lake may be absent, or, on the contrary, fill the whole depression. The sizes of alases also vary greatly from several tens of metres to several kilometres. Sometimes several alases may join together by means of small "grass" rivers, forming complexes, the so called "alas valleys".

Taking alas vegetation as an example the approach of sigma-syntaxonomy was developed for distinguishing territorial units (horizontal structure) of vegetation cover (Gogoleva et al. 1987). The basic unit is a sigma-association representing a number of syntaxa that characterize a certain type of landscape. For instance, the sigma-association *Artemisio commutatae-Hordeetum/Puccinellietum* comprises the vegetation of dry alases of the Lena-Amga Interfluve, where most area is covered by stepped meadows. The *Artemisio rupestris-Hordeetum/Scolochloetum festucaceae* covers the alases of the Lena-Viluy Interfluve that are more watered and less salinized. The *Caricetum juncellae/Scolochloetum festucaceae* comprises the alases without pronounced xerophytic belt.

*The small river valley* meadows are formed on watersheds and cover small areas. Due to inconstant, often drying up, water courses and pronounced tall grass vegetation, they have acquired the local name "grass rivers". The main feeding source of such rivers is melting snow and rains. The soil that is waterlogged in spring retains moisture till mid summer or even longer. An additional source of moisture is the thawed permafrost. The melt water is characterized by low temperatures and deposits very little sludge, poor in mineral elements (Zolnikov 1954b). Such conditions hamper swamping and favour the formation of meadow vegetation.

The narrow rivers (up to 300 m wide) are fringed by the communities of *Equisetum fluviatile, Carex appendiculata,* or *Beckmannia syzigachne. Carex juncella* forms large tussocks as high as 70–80 cm with a density of 1–2 tussocks per $m^2$. Moving away from the river bed, the tussocks reduce in size, and their number increases up to 4–5 per $m^2$. The tussocks support *Calamagrostis langsdorffii,* which becomes more vigorous and "moves" to the soil in dryer places. Such communities also contain *Poa palustris, Beckmannia syzigachne, Alopecurus arundinaceus, Hordeum brevisubulatum, Caltha palustris, Acorus calamus, Ranunculus gmelinii, Ranunculus propinquus, Vicia cracca,* and *Lathyrus pilosus.* They belong to the class *CALAMAGROSTETEA LANGSDORFFII* Mirkin in Achtjamov et al. 1985 (Fig. 3.24).

Under even dryer conditions *Carex schmidtii* becomes dominant replacing *C. juncella,* gradually forming meadows of *Carex schmidtii, Calamagrostis neglecta, Poa pratensis, Hordeum brevisubulatum,* and other species of mesic habitats. The mesic meadows belong to the *MOLINIO-ARRHENATHERETEA*. At early stages of the development of small river meadows the communities are characterized by *Poa palustris, Caltha palustris, Thalictrum minus, Sanguisorba officinalis*, that later give way to competitively stronger species. With time, shrubs start invading the small river valley meadows (*Spiraea salicifolia, Betula fruticosa*).

The small river meadows may provide with up to 30% of the fodder for cattle and horse breeding. On the other hand, tussocks hamper proper haying, so such

**Fig. 3.24** Rich meadow with *Calamagrostis langsdorffii* in a water-logged area (Central Yakutia)

meadows are used as hayfields only during the years with unfavourable growth conditions.

*The forest (waterless valley) meadows* are formed on the prominent elements of the relief where deforestation took place (cuttings, fires). These meadows are scattered among the forest tracts in watershed areas and on forested slopes of river valleys. They possess the features of previous forest communities: young growth of tree species, mosses, *Arctostaphylos uva-ursi, Vaccinium vitis-idaea, Limnas stellerii,* etc. Often the only moisture source for such communities is atmospheric precipitation.

The forest meadows develop in various habitats, from waterlogged to xeric ones. This determines the floristic composition of each community. These meadows belong to the *MOLINIO-ARRHENATHERETEA*.

The forest meadows persist under regular haying. Otherwise they quickly overgrow with shrubs (*Spiraea media, S.salicifolia, Rosa acicularis, Lonicera altaica,* etc.) and are followed by the appropriate succession stages.

Meadows also occur in South and South–West Yakutia. They are comparable to those of Central Yakutia, though have distinct features. For instance, halophilous communities are not common in South and South–West Yakutia. They contain species with a more southern distribution: *Festuca pratensis, Phleum pratense, Trifolium pratense, Glechoma hederacea, Lathyrus pratensis*. The most widespread are meadows with *Festuca pratensis*, and less common are communities with a predomination of *Alopecurus arundinaceus, Equisetum fluviatile*, as well as stepped meadows.

In the mountainous regions of South Yakutia (the Upper Aldan River basin) meadows occupy insignificant areas, most of them situated in floodplains. They are composed of *Bromopsis inermis, Festuca rubra, Alopecurus pratensis, Trifolium pratense, Phleum pratense,* etc.

Thus, the latitudinal range of meadow distribution is rather wide, but in each zone the meadows acquire distinct features.

The nival meadow communities developing under perennial snow cover are characteristic for the subalpine and tundra belts. Hydrological and thermal conditions are the key factors determining their peculiar composition and structure. The following associations are observed: pure thickets of *Ranunculus sulphureus* grow on very wet melkozem surrounding melt water pools in flat cirque bottoms; *Sibbaldia procumbens* with an admixture of *Carex tripartita, Minuartia biflora,* etc. and a more or less developed moss layer on slopes covered with detritus or shallow soils; *Saxifraga merckii* with an admixture of *Luzula pallescens* and *Carex aterrima* and a weakly developed moss layer are observed rarely on gentle south-facing slopes; the most common forb-*Salix*-moss communities with *Salix turczaninowii* grow straight under snow cover between large stones, on shaded crests or in damp shady bottoms of narrow and deep intermountain hollows. The moss layer is almost continuous and multi-specific, represented mostly by pioneer species preferring bare substrates. Such meadow communitites are very similar to those growing in the mountain systems of Siberia, Yurtsev (1964) mentioned their affinity to alpine meadows of mountains with a well developed alpine belt.

The studied sub-alpine meadows of South Yakutia are represented by four associations depending on moisture conditions: *Festuca*+forb, *Calamagrostis*+forb, forb+*Calamagrostis*+*Carex* and forb+*Geranium*. The *Festuca*+forb meadows (dominated by *Festuca altaica* and *Geranium albiflorum*) are characteristic for the most xeric, slightly prominent stony parts of trough valleys. Its moss layer (*Hylocomium splendens, Pleurozium schreberi* with an admixture of *Brachythecium reflexum*) covers 60% of soil surface. The dealluvial gentle slopes under North–east-facing cliffs with a sufficient moisture supply are covered with the dense *Calamagrostis*+forb communities dominated by *Calamagrostis langsdorfii, Aruncus asiatica, Polemonium coeruleum* and *Carex pallida.* They are characterized by a continuous though thin moss layer of mainly *Hylocomium splendens.* The flat parts of cirque bottoms with poor drainage are the habitats for forb+*Calamagrostis*+*Carex* meadows (*Carex podocarpa, C. pallida, Calamagrostis spp.).* Its moss layer covers only 5% of the soil surface and is dominated by *Sanionia uncinata.* The forb+*Geranium* meadows form narrow communities along erosional hollows on mountian slopes. The featuring species are *Geranium albiflorum, Trollius aldanensis* and *Carex podocarpa.* The continuous thin moss layer is dominated by *Hylocomium pyrenaicum* and *Dicranum spadiceum* and *Sanionia uncinata* is usually common. The gentle, mainly south-facing slopes are the habitats for large grass meadows (dominated by *Aquilegia glandulosa* or *Veratrum oxysepalum)* growing along stream banks.

Like other communities in Yakutia, the meadow coenoses are characterized by a low $\alpha$-diversity (very rarely a coenoflora may number 40–50 species), though they are peculiar in species combination (especially under variable conditions). Many regions and meadows types occurring here still require proper investigation.

### *3.5.8 River Valley Complexes*

**A.P. Efimova, I.F. Shurduk, L.V. Kuznetsova, and A.P. Isaev**

Vegetation of river valleys represents an intrazonal complex comprising coniferous (*Larix cajanderi, L. gmelinii, L. sibirica, Picea obovata, P. ajanensis*, rarely *Abies sibirica,* and *Pinus sibirica*) and deciduous (*Populus suaveolens, Chosenia arbutifolia, Betula pendula, B. pubescens,* arborescent *Salix* species) forests, shrubberies (frutescent *Salix spp., Alnus hirsuta, Betula fruticosa, B. nana ssp. exilis, B. Divaricata,* etc.), mixed forests, meadows, bogs, riparian and aquatic vegetation. River valley vegetation characteristics are directly related to relief, character and intensity of alluvial processes and valley landform peculiarities, i.e. developed floodplains, river terraces and other landscape units of alluvial origin.

The lowland character is inherent to the following Yakutian rivers: the Lena River, its large tributaries (the Viluy River, the middle and lower reaches of the Aldan River, the lower and partly middle reaches of the Olyokma River), the Amga River, as well as such northern rivers as the Alazeya River, the Anabar River (middle and lower reaches), the lower reaches of the Olenyok and Yana Rivers, and lower and middle reaches of the Indigirka and Kolyma Rivers. The following rivers partly possess a mountainous character: the Olyokma River (upper and partly middle reaches), its large tributary the Tokko River, the Aldan River (upper and partly middle reaches), its large tributaries the Timpton, Uchur, Upper and Middle Maya and Algolma Rivers, the upper reaches of the Viluy River, the upper reaches of the Anabar and Kolyma Rivers and the upper and partly middle reaches of the Olenyok, Yana and Indigirka Rivers.

*The Lena River* is one of the ten largest rivers on Earth. From the upper part of its middle reaches to the delta it flows through Yakutia. The upper part of its middle reaches has a narrow valley limited by the firm Pre-Cambrian and Cambrian bedrocks that are highly resistant to erosion. The sandy-pebble alluvium along the river banks and on islands bears narrow strips of *Salix viminalis* and *S. dasyclados* shrubberies. The middle reaches of the Lena River are characterized by loose Cretaceous and Jurassic sediments. There the valley suddenly expands up to 25 km. The valley landscape is predominated by natural meadows used as hayfields, as well as by agricultural lands, fallows, and pastures. The forests are communities of *Salix, Betula, Picea obovata, Larix,* and *Pinus sylvestris*. There are also shrubberies of various typology and structure. The scientific data on the forest and shrub vegetation of the Middle Lena River are summarized in the following works (Cajander 1903, 1904; Abolin 1929; Dolenko 1913; Drobov 1927; Sheludyakova 1957a; Kononov 1982; Savvinov and Kononov 1981; Timofeyev and Shurduk 1996; Gogoleva 1996; Perfilyeva and Shurduk 1999; Efimova 2001, 2003; Efimova and Shurduk 1998, 2001a, b, 2003, 2005; Efimova et al. 2003, 2004, 2005; Shurduk et al. 2005). According to the opinion of many specialists (Abolin 1929; Kononov 1971; Savvinov and Kononov 1981; etc.), the predominating meadow vegetation in the Middle Lena River valley has a secondary origin as a result of large-scaled cutting since the times when man first occupied the valley. The three-century-long process of urbanization, land cultivation and cattle grazing has yielded the almost complete elimination of forests, especially in the most urbanized sites: the vicinities of Pokrovsk (the Erkeeni Valley), the vicinities of Yakutsk (the Tuymaada Valley), and the Namsky region (the Ensieli Valley). In the most populated areas the forest cover has been reduced to 5–10 (15)%. The highest values of forest cover (over 30%) are characteristic for the less urbanized sites of the valley (Andreyev et al. 1987).

The floodplain of the Middle Lena River features a very high phytocoenotic diversity (Fig. 3.25). The fresh silty-sandy alluvium of river banks and islands are covered by vast communities of *Salix viminalis* and *S. dasyclados* with an admixture of *S. udensis* and *S. triandra*. Depending on conditions and age of the ecotopes they form various associations. The successful spread of *S. viminalis* and *S. dasyclados* in the Lena River floodplain became possible thanks to their phenological rhythm that has been established during their microevolution and is perfectly adapted to the river regime and the arid zonal conditions. The essence of this adaptation is that the period when the seeds are flying-off coincides with the period of exposure of the fresh alluvium substrates after the spring tide. Arriving on the wet silty-sandy surface, the seeds germinate and have time to develop roots penetrating sufficiently deep into wet ground before the arid season begins.

The communities of *S. udensis* and *S. triandra* form thickets of 5–6 m high on fresh silty-sandy alluvium of the lower floodplain. They are limited in area. Very rarely the temporary streams in the lower and middle floodplain may support pure or mixed communities of *Salix viminalis, Alnus hirsuta* with a very dense undergrowth of *Swida alba* (ass. *Alnetum hirsutae swidetosum)* (Efimova 2003).

The upper floodplain (in inter-ridge hollows, along the channels and banks of oxbow lakes on frozen soddy-meadow soils) supports forb and mixed shrub communities of the alluviophobic *Salix bebbiana* and *S. pyrolifolia*, sometimes with an admixture of *S. abscondita*. They represent the successional stage following upon the pioneer alluviophilous communities. At sites with sufficient moisture and proper drainage (the slopes of banks of oxbow lakes and channels) *Alnus hirsuta* often participates in the abovementioned communities. Depending on specific ecological conditions, the following species may act as dominants and co-dominants in the herb layer: *Geranium pratense, Anemonidium dichotomum,* hygromesophilous and mesophilous grasses and *Carex spp*. The appearance of *Pyrola spp.*, indicates the beginning of forest formation and self-seeding of conifers.

With time, the coenoses of alluviophobic *Salix* species are replaced by forb and mixed shrub forests of *Betula pendula*. As a rule, these primary forests follow the elongated patches of former *Salix* communities and can be found exclusively on the upper floodplain (Efimova 2001). Later on, they are replaced by zonal coniferous forests of *Picea obovata* and *Larix* (Efimova 2000; Efimova and Shurduk 2001a, 2003; Efimova et al. 2004).

Peculiar *Salix* and *Salix-Alnus* coenoses form a micro-belt within the Middle Lena River valley on steep erosion slopes of streams and ravines (*S. viminalis, S. dasyclados, S. traindra, S. pyrolifolia, S. pseudopentandra,* and *Alnus hirsuta).*

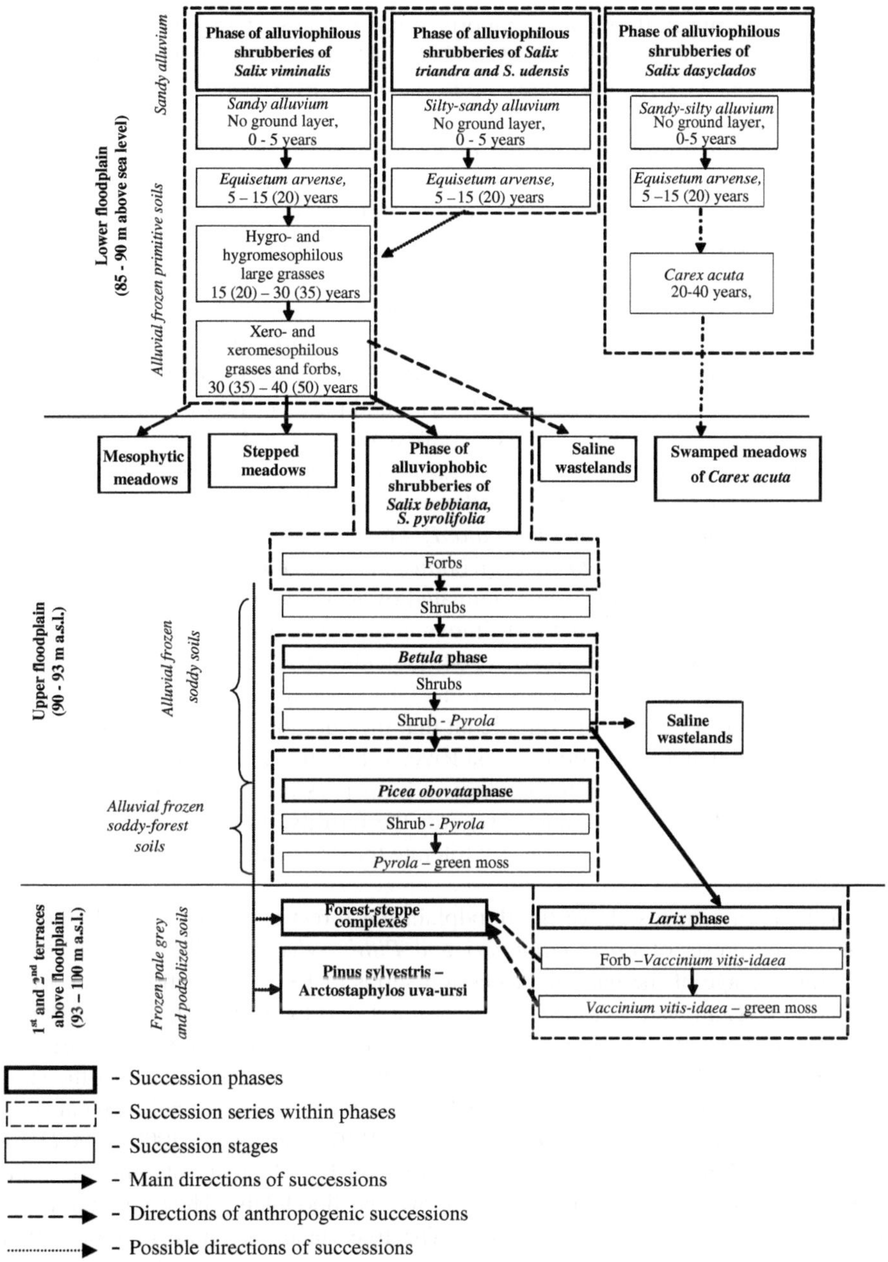


**Fig. 3.25** Scheme of succession dynamics of forest and shrub vegetation of the Middle Lena River's modern valley

Sometimes the upper floodplain supports small patches of secondary park communities of *Salix viminalis* and *S. dasyclados* at sites with a sandy or silty-sandy substrate deposited over the soddy soils occupied by ruderal fallow or natural meadow vegetation.

In the upper and upper part of the middle reaches of the Lena River forests of *Picea obovata* occur along the Lena River and along small river banks, represented by the herb-green moss type. In the Middle Lena River valley *Picea obovata* forests are observed on the upper floodplain along channel banks and as small patches on terraces above the floodplain on concave or flat relief elements. As a whole, the forests are characterized by a complex composition of the 2–3-layered tree stand, a high tree canopy coverage and a specific composition of the herb-dwarf shrub and moss-lichen layers. The tree stands typically have medium to high values of crown closure, are medium-productive, and pure or often with an admixture of *Betula pendula* and *Larix spp*. The weak understory is formed of *Rosa acicularis, Ribes glabellum,* and *Lonicera altaica.* The herb-dwarf shrub layer is thin and is represented by forest species *Lathyrus humilis, Viola mauritii* and *Saussurea dubia,* taiga sciophytes (*Pyrola incarnata, P. dahurica,* etc.), dark coniferous forests species *Trientalis europaea, Petasites frigidus,* and *Adoxa moschatellina.* The well-developed moss-lichen layer is predominated by *Rhytidium rugosum, Hylocomium splendens, Aulacomnium palustre* (Efimova et al. 2004; Shurduk et al. 2005). The earlier publications (Cajander 1903, 1904; Dolenko 1913; Chugunov 1955) report vast areas covered by *Picea* forests on the 1st terrace above the floodplain of the Middle Lena River valley. It is significant that they do not occur on the 2nd terrace above the floodplain of the Middle Lena River valley. In our opinion, such a limited distribution is determined by the ecological properties of *Picea*: this species has a very narrow ecological amplitude as regards the fertility, drainage and temperature of soils. In contrast to *Larix*, it does not tolerate the drastically worse soil conditions of the 2nd terrace.

On the river terraces above the floodplain the forest cover is formed by coniferous zonal forests of *Larix cajanderi* and *Pinus sylvestris*. They represent the final climax stage of the natural succession series of forest vegetation in the Lena River valley (Efimova et al. 2005). The original development of *Larix* tree stands on terraces above the floodplain is related basically to the shallower occurrence of permafrost, and, subsequently, by modified hydrothermal conditions of the substrates and the development of frozen pale grey (transitional) and pale typical soils (Pozdnyakov 1963, 1983, 1986; Abaimov et al. 1997; Samsonova 2000). The *Larix* forests in the Middle Lena River valley grow on flat elements of relief or wide shallow depressions of the 2nd terrace above the floodplain. They are characterized by a limited typological structure. The tree stands are pure or more often include an admixture of *Betula pendula, Picea obovata* or *Pinus sylvestris.* They are 1–2-layered, medium-closed, often of secondary post-fire origin. Usually the reproduction of *Larix* is not satisfactory. The *Larix* forests almost always contain *Betula pendula* in the second layer, and *S. bebbiana, Rosa acicularis,* often *Salix taraikensis, S. abscondita, S. brachypoda, and S. pyrolifolia* in the understory. The herb-dwarf shrub layer is characterized by constant or dominant *Vaccinium vitis-idaea, Arctostaphylos uva-ursi, Arctous erythrocarpa, Lathyrus humilis,* etc.

The moss-lichen layer is dominated by *Rhytidium rugosum, Cladina rangiferina, C. arbuscula, Peltigera aphthosa,* etc.

The *Pinus syvestries* forests are confined to weakly podzolized sandy soils of relic elevations of the 2nd terrace above the floodplain. The tree stands are pure or rarely contain *Larix cajanderi* and *Betula pendula.* They are low- or medium-closed, often of secondary post-fire origin. The understory is practically absent or contains isolated *Rosa acicularis* and *Spiraea media.* The herb-dwarf shrub layer is dominated by the dwarf shrubs *Vaccinium vitis-idaea* and *Arctostaphylos uva-ursi.* The herb component is represented by forest grasses, *Carex* spp., steppe xerophytes and mesoxerophytes (*Festuca jacutica, Carex macroura, C. conspissata, Lychnis sibirica, Phlox sibirica, Pulsatilla flavescens, Dianthus versicolor, Veronica incana,* etc.). The ground layer is formed by lichens: *Cladonia rangiferina, C. stellaris, Cetraria laevigata, Peltigera didactyla,* but mosses also occur: *Rhytidium rugosum, Ceratodon purpureus* (Efimova et al. 2005).

The shrub formations of the Middle Lena River valley are mainly mono-dominant communities of *Swida alba, Spiraea salicifolia, Crataegus dahurica, Salix brachypoda,* and *Rosa acicularis* characterized by their limited distribution on the upper floodplain and terraces above the floodplain. They have a high density and a plane vertical structure. In the upper part of the middle reaches there are, besides the abovementioned coenoses, communities of *Padus avium, Cotoneaster melanocarpus* and other species. Their ground layer is dominated by conspicuous forbs: *Lilium pensilvanicum, L. martagon, Trollius asiaticus, Aconitum spp., Delphinium spp.,* etc.

Within the taiga zone the Lower Lena River features *Larix* forests, while the river and channel banks are covered by shrubberies mainly of *Salix viminalis*, and the lower floodplain supports meadows of *Calamagrostis langsdorffii* (Andreyev et al. 1987).

Within the Lena River basin there are numerous small rivers with valleys as wide as 0.5–1.5 km. On the left bank of the Lena River the small rivers flow on the ancient alluvial plain and possess lowland characteristics and have more or less developed valleys. They support the yernik shrubberies of *Betula fruticosa, B. nana ssp. exilis* and *Larix cajanderi – Vaccinium vitis-idaea* forests alternating with meadows of *Calamagrostis langsdorffii* and tussock meadows of *Carex juncella + C. appendiculata* (Andreyev et al. 1987). Communities of *Betula nana ssp. exilis – Carex spp.* are common in low swamped landscapes, while *B. nana ssp. exilis – Eriophorum vaginatum – Aulacomnium palustre* coenoses are characteristic for depressions of small river valley complexes.

*The Viluy River* is a left large tributary of the Lena River. Data on the vegetation cover of its basin are not numerous (Abolin 1929; Utkin 1958; Cheremkhin 1961; Galaktionova et al. 1962; Scherbakov 1992; Poiseyeva and Mironova 2000; etc.). The valley of the Viluy River is less developed then that of the Lena River. Its middle reaches are characterized by a more or less pronounced floodplain. The river banks support narrow strips of shrubberies of *Salix viminalis, S. dasyclados,* rarely of *S. udensis,* that are generally similar to those of the Lena River. The upper floodplain is occupied by forests of *Picea obovata* – green moss and *Larix – Vaccinium vitis-idaea* (80% of the total area of the floodplain), as well as by small patches of *Betula* forests. In the lower reaches of the Viluy River its valley is wider and is indented by numerous channels and oxbow lakes. Here herb *Larix* forests predominate (85% of the area) (Andreyev et al. 1987). The valleys of small rivers support yerniks of *Betula fruticosa* and *B. exilis*.

*The Yana and Adycha Rivers*. The fresh sandy-pebble and sandy alluvium of the lower floodplain of islands and narrow spits support pure or mixed shrubberies of *Salix boganidensis, S. udensis,* and *S. schwerinii.* As a rule, such communities have an elongated shape following the configuration of fresh alluvial sediments. These pioneer coenoses may include *S. dasyclados* as an insignificant admixture. Being intrazonal vegetation, these *Salix* shrubberies are characterized by a structure similar to that of the alluviophilous vegetation of the Middle Lena River: the vertical and horizontal structure is very plane, the understory is absent, and the herb layer is thin, composed of *Equisetum arvense, E. pratense, Calamagrostis langsdorffii, C. lapponica,* etc. In the upper floodplain, experiencing sporadic spring tides, the sandy loamy and loamy soils are covered by *Larix* – green moss forests and mixed coenoses with a dominance of *Salix boganidensis* with participation of *S. udensis, S. schwerinii, S. dasyclados,* and rarely *Populus suaveolens.* The understory is absent, only isolated specimen of *Betula fruticosa, Rosa acicularis* and *Ribes dikuscha* may occur. The herb layer is composed of grass and *Carex* species. The moss layer has a mosaic horizontal structure. Communities of *Populus suaveolens* and *Populus suaveolens – Chosenia arbutifolia* were not recorded there.

There are mountain rivers with temporary catchments that flow into the Yana and Adycha Rivers in shallow inter-mountain depressions. Their valleys support dwarf shrub tundra coenoses of a rather complicated composition. A narrow valley complex was described as wide as 7–8 (10) m at an altitude 750 m a.s.l. The width of the riverbed was up to 2 m in eastward direction, its slope inclinating at 10–20°. The highest records of rising water levels are 1–2 m, up to 3 m in some places. All along the river valley the following species are abundant: *Ledum palustre* and co-dominant *Arctous alpina* and *Aconopogon tripterocarpum*. Constant species are *Betula exilis*, *Empetrum nigrum*, and *Calamagrostis lapponica*. Other common species are *Salix pulchra, S. tschuktschorum, Bistorta major subsp. ellipticum, Calamagrostis holmii, C. langsdorffii, Carex holostoma, Vaccinium vitis-idaea subsp. minus, V. uliginosum, Rubus arcticus, Artemisia arctica, Tephroseris integrifolia,* etc. In non-flooded areas isolated specimens of *Salix bebbiana* occur. In some areas pure *Calamagrostis* meadows are observed as riparian communities. The soil cover abounds with various hygrophilous mosses and lichens.

*The Kolyma River* can be considered a mountain-lowland river in its middle reaches and a lowland river in its lower reaches due to the orographical conditions of the basin. The alluvial processes of its valley in the middle reaches have their own distinct features and differ from those in the valleys of lowland and typical mountain rivers. The pebble and pebble-sandy sediments of the Upper and Middle Kolyma river (up to the confluence of the Korkodon River, 64° 43′ N, 154° 58′E) are replaced by pebble-sandy alluvium (the Krokodon River confluence through the Zyryanka River confluence 65° 49′ N, 150° 46′E) and sandy alluvium downstream the Zyryanka River confluence. At the same time, patches of pure pebble are found throughout the whole right bank of the river. The characteristics of alluvial sediments determine the peculiarities of the syngenesis of the vegetation cover, and the forest vegetation in particular.

In the lower floodplain upstream the Zyryanka settlement there are no stable herbaceous coenoses (Fig. 3.26). Sometimes *Equisetum arvense* forms thick mono-dominant stands on silty substrate. Open herb communities are gradually replaced by different pioneer tree species depending on the fractional structure of the alluvial sediments.

The pebble and pebble-sandy alluvium without silt are overgrown by *Chosenia arbutifolia*. Communities with *Populus suaveolens, Salix rorida,* and *S. schwerinii* are confined to substrates with an increased content of silt fraction in the alluvium, but they do not withstand strong silting. The shrubberies of *Salix rorida* and *S. schwerinii*, being relatively stable ecosystems, function during 2–3 generations and then transform into various communities depending on moisture conditions: *Larix cajanderi* forests, rarely *Duschekia fruticosa – Equisetum pratense* with berry-bearing shrubs (*Ribes dikuscha, R. triste, Rosa acicularis*) and young growth of *Larix cajanderi*. The initial, riparian, communities in the series of *Larix* forests are represented by *Larix cajanderi – Calamagrostis langsdorffii, Larix cajanderi – Equisetum pratense*, and *Larix cajanderi – Pyrola asarifolia + forb*. These are most productive forests in the Kolyma River basin attributed to the growth class III or rarely IV. They are confined to silty-sandy ecotopes, seldom flooded by high water. The wood reserves of ripe and overripened forests is 200–300 $m^3$ per 1 ha. Beyond the floodplain conditions the succession process is finished by *Larix cajanderi – Vaccinium vitis-idaea*. Later on, they gradually transform into *Larix cajanderi – Vaccinium uliginosum – Aulacomnium palustre* and *Larix cajanderi – Sphagnum spp.* with all characteristic features of the forests of the northern taiga sub-zone.

The valley of the Middle Kolyma River, downstream of the Zyryanka settlement, contains vast lowlands with abundant bogs and lakes. The *Chosenia* and *Populus* communities gradually disappear there, and *Salix rorida* is replaced by *S. udensis*. The productivity of *Larix* forests is distinctly lower here and their typological composition is also different. The valley vegetation consists mainly of *Larix* forests. The confinement of the basic vegetation types of the Kolyma River valley to various topologies is given in simplified schemes (Figs. 3.27 and 3.28).

In the floodpain the young alluvial sediments are inhabited by an *Equisetum arvense* meadow, subsequently replaced by the strips of *Salix* shrubberies of the *Equisetum arvense* and *E. pratense* types. On the upper part of the modern floodplain, which is not subject to regular flooding, the process of forest formation is observed. The *Larix cajanderi – Duschekia fruticosa – Equisetum pratense* and *Larix cajanderi – Duschekia fruticosa + Rosa acicularis + Ribes triste – Equisetum pratense* are most common types there. The terraces above the floodplain support narrow strips (several tens of meters) of the most productive forests of *Larix cajanderi – Pleurozium schreberi*. Moving away from the river bed, depressions in the valley and the adjacent plateaus of watersheds are characerized by worse hydrothermal conditions of the soils. Owing to this, these areas are covered by low productive

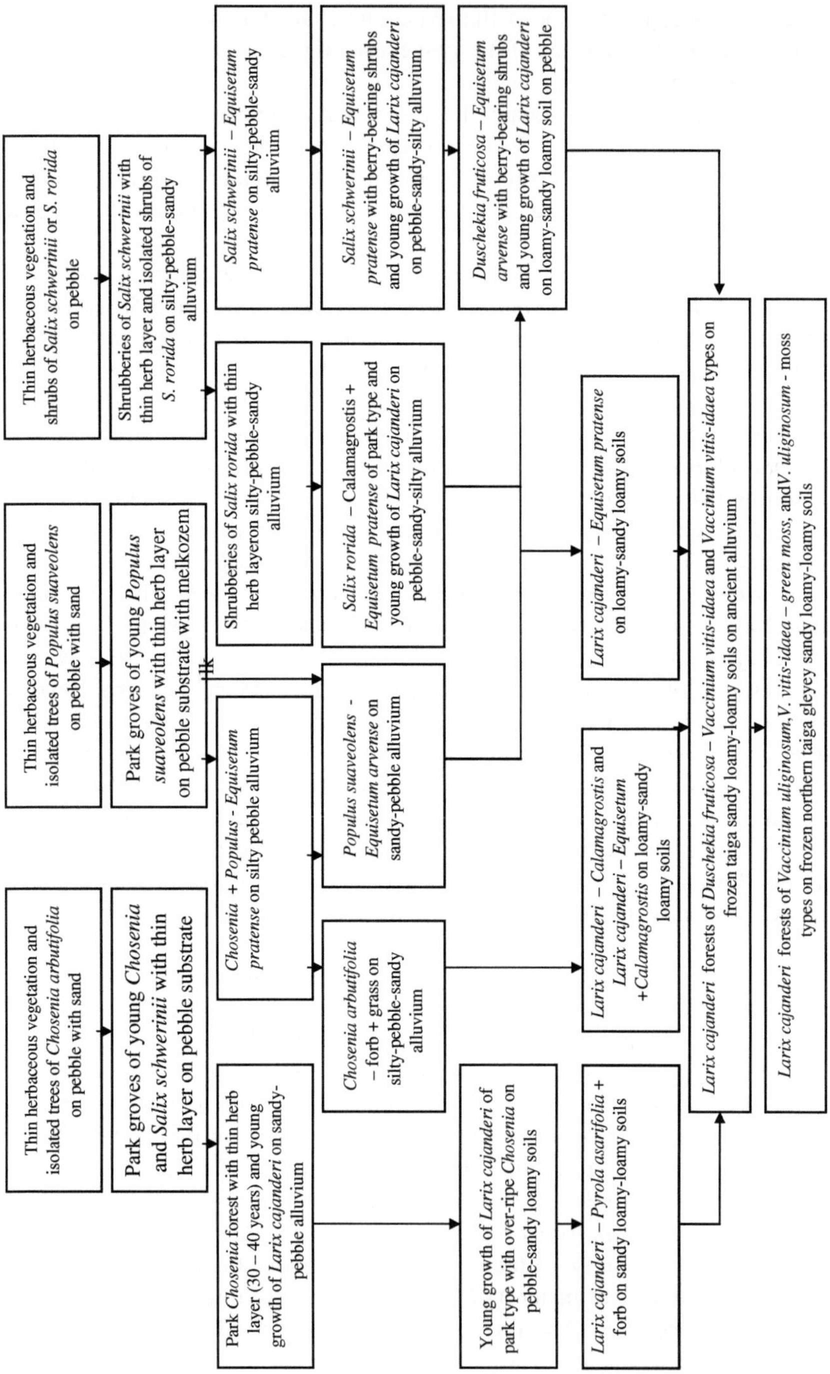


**Fig. 3.26** The scheme of ecogenesis of tree-shrub vegetation in the Kolyma River valley within the Verkhnekolymsk Region

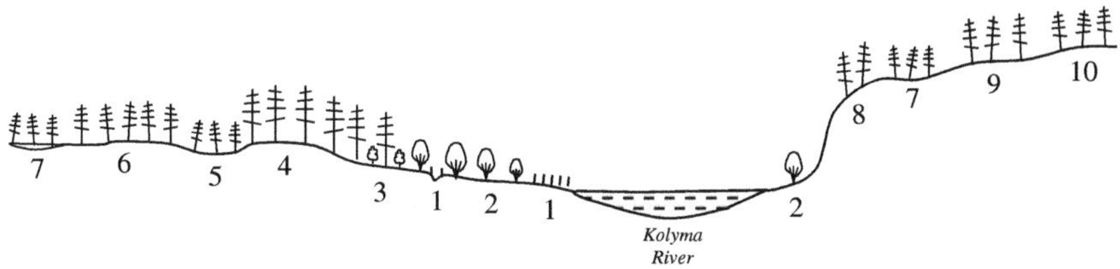


**Fig. 3.27** The cross-section of the Kolyma River valley in its middle reaches: (*1*) *Equisetum arvense* meadow, (*2*) thickets of *Salix udensis* and *S. schwerinii*, (*3*) *Larix cajanderi – Duschekia fruticosa*, (*4*) *Larix cajanderi – Vaccinium vitis-idaea – Pleurozium schreberi*, (*5*) *Larix cajanderi – Vaccinium uliginosum*, (*6*) *Larix cajanderi – Vaccinium uliginosum – Aulacomnium palustre*, (*7*) *Larix cajanderi – Ledum palustre – Sphagnun s.l.*, (*8*) *Larix cajanderi – Pleurozium schreberi*, (*9*) *Larix cajanderi – Vaccinium vitis-idaea – Ptilidium ciliare*, (*10*) *Larix cajanderi – Ledum palustre – Cetraria cucullata + Cladonia rangiferina*

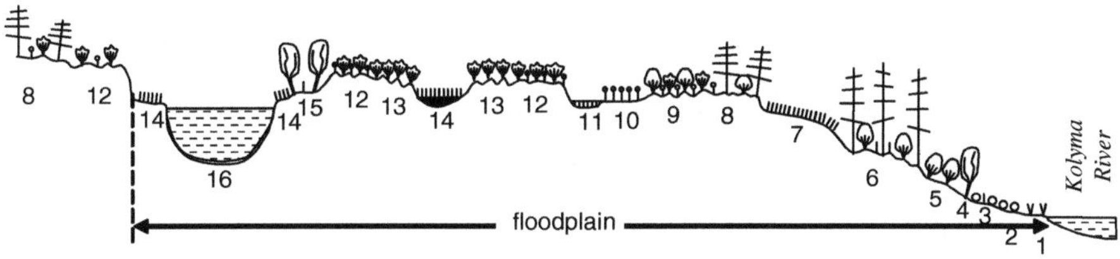


**Fig. 3.28** Generalized vegetation profile of the left bank of the Kolyma River in the vicinity of the town of Srednekolymsk: (*1*) Open herb groupings, (*2*) *Equisetum arvense* meadow, (*3*) *Equisetum arvense – Agrostis stolonifera – Alopecurus arundinaceus* meadow, (*4*) *Salix udensis – Equisetum pratense*, (*5*) *Duschekia fruticosa – Equisetum pratense*, (*6*) *Larix cajanderi – Duschekia fruticosa – forb + Calamagrostis langsdorffii*, (*7*) forb + *Bromopsis* stepped meadow, (*8*) *Larix cajanderi – Salix pulchra + Duschekia fruticosa – Vaccinium uliginosum*, (*9*) *Salix pulchra + Duschekia fruticosa – Carex spp. + Vaccinium uliginosum*, (*10*) *Carex rhynchophysa + Eriophorum polystachion* bog, (*11*) *Carex capitata – Calliergon bog*, (*12*) *Salix pulchra* – green moss – *Vaccinium uliginosum*, (*13*) *Salix pulchra – Carex appendiculata*, (*14*) *Arctophila fulva* and other swamped meadows, (*15*) *Salix udensis + Rosa acicularis – Equisetum pratense + Calamagrostis langsdorffii*, (*16*) thermokarst lake

*Larix cajanderi – Ledum palustre + Vaccinium uliginosum* forest. The waterlogged mesodepressions support *Larix cajanderi – Carex spp. + Vaccinium uliginosum* forest. Further from the river, the terraces above the floodplain are inhabited by *Larix cajanderi – Duschekia fruticosa – Vaccinium vitis-idaea, Larix cajanderi – Vaccinium vitis-idaea,* and *Larix cajanderi – Vaccinium vitis-idaea* – green moss forests. At sites with a poor drainage the *Larix cajanderi – Vaccinium uliginosum* – green moss communities of growth class V occur; they are replaced by the *Larix cajanderi* – green moss and *Larix cajanderi – Sphagnum spp.* forests of growth class Va or Vb under even worse drainage conditions.

Where ridges of the Yukagir Tableland penetrate the Kolyma River valley, most drained and properly heated parts of flat summits and slopes are covered by the *Larix* forests of dwarf shrub-lichen, *Vaccinium vitis-idaea – Ptilidium ciliare* and other similar types. The swamped areas support the *Larix cajanderi – Ledum palustre – Sphagnum spp.* forest.

In its lower reaches the Kolyma River flows within the near-tundra forest subzone and tundra zone. The pebble alluvium in floodplains is covered by monodominant *Equisetum arvense* communities forming a solid or discontinuous cover. *Carex*, grass and forb species also occur. The young alluvial deposits are inhabited mainly by shrubberies of *Salix udensis* and *S. schwerinii*. Gentle silty banks of channels are overgrown by *Arctophila fulva* or *Equisetum arvense*.

On sites with sufficient moisture we find communities of *Calamagrostis langsdorffii* reaching 70–120 (150) cm in height and 60–75% coverage in mesic habitats and up to 90–100% under conditions of excess moisture. The monodominant grass stands may include insignificant admixtures of *Carex juncella, Hordeum jubatum, Poa pratensis, Chamaenerion angustifolium, Rubus arcticus, Lactuca sibirica, Potentilla norvegica, P. stipularis,* etc. The species composition is dependent on moisture conditions. The mesic meadows feature mesophytes, while the waterlogged communities contain hygro- and hydrophytes. The low, level parts of the floodplain support widespread tussock meadows of *Carex juncella.* With time a thick green moss layer develops resulting in deterioration of soil ventilation. The tussock *Carex juncella* meadow is replaced by the transitional stage of *Carex – Calamagrostis* coenoses (*Calamagrostis langsdorffii, Carex juncella, C. bochemica, Eriophorum polystachion*, etc). In the upper floodplain grass bogs with *Carex concolor* and *Eriophorum polystachion* are common. The coenoses of *Arctophila fulva* cover significant areas along the river banks and on drained lake sites (Andreyev and Perfilyeva 1980). The river valleys also commonly contain polygonal-ridged tundra-bogs with thickets of *Duschekia fruticosa, Salix pulchra, S.glauca.* The ridges are overgrown by *Eriophorum vaginatum,* while the polygonal depressions support bogs of the *Carex concolor* and *C. chordorrhiza* in complex with *Salix pulchra* and *S. alaxensis*, grass meadows of *Eriophorum polystachion, Arctophila fulva* and *Equisetum arvense – Poa alpigena.*

The river network of South Yakutia entirely belongs to the Lena River basin. The largest rivers, the Vitim, Olyokma, Aldan, Timpton and Uchur Rivers, have a typical mountain character with swift currents, significant fluctuations in the water level, numerous rapids and rifts, as well as icings. Almost all the rivers near the southern limits of Yakutia originate in its territory (the Udokan, Stanovoy, Zverev and Tokinsky stanovik Ranges) with the exception of the Olyokma River that enters Yakutia in its middle stream. Following the high dynamism of the floodplain regime of the mountain rivers the vegetation complex consists here of very dynamic, constantly developing successional series.

Basically, the vegetation cover of the river valleys of the South Yakutian mountain regions consists of pure *Larix* (*Larix gmelinii, L. cajanderi*) taiga alternating with open mires. The pioneer riparian vegetation is formed by grasses, *Carex* species and forbs, and sometimes the monodominant coenoses of *Chamaenerion latifolium* occur on pebble alluvium. Characteristic species of the young sediments of lower floodplain are *Deschampsia sukatschewii, Astragalus alpinus, A. schelichovii, Oxytropis adamsiana, Artemisia bargusinensis, Agrostis clavata,*

*A. trinii, Calamagrostis langsdorffii,* etc. There are about 150 pioneer species in the vegetation of the overgrown floodplain (see also Section 3.5.5). The subsequent successional course depends on the hydrological peculiarities of a river and, consequently, on alluvial characteristics, drainage level, etc. Along slow currents in sandy depressions dense thickets of *Salix schwerinii* and *S. udensis* occur lacking the ground layer. The fastest growing tree species in Yakutia is *Chosenia arbutifolia* that forms groves of the forb + *Calamarostis* type on sandy-pebble alluvium.

*The Tokko River* originates from the glacial lake of the same name situated at an altitude of about 1,400 m a.s.l. At the beginning of its course downway the river dodges along the bottom of the glacial valley between the randomly scattered moraine hills. The river banks support narrow strips of *Salix* communities (*Salix schwerinii, S. udensis, S. krylovii,* etc.) growing among coarse and medium-sized talus. Small (up to 15 m) trees of *Populus suaveolens* and *Chosenia arbutifolia* seldomly occur. Levelled, strongly sodded and moist sites are occupied by *Larix* woodlands of the *yernik* type with a lot of green mosses. Regardless of aspect, the gentle stony slopes are covered by *Larix* woodlands of the dwarf shrub – lichen and lichen types. The understory is composed of *Pinus pumila, Duschekia fruticosa, Betula divaricata* and *Rhododendron aureum.* As slope inclination increases *Larix* disappears and *Pinus pumila* increases. The tree-line communities, especially along the stream banks, may often include the *Betula ermannii ssp. lanata* forests with a properly developed herb layer (*Calamagrostis langsdorffii, Carex pallida, Viola biflora,* sometimes *Vaccinium vitis-idaea* and *V. myrtillus*). The upper border of the forest belt lies at 1,200 m a.s.l., but isolated trees of *Larix gmelinii* along the stream banks are observed as high as 1,500 or even 1,600 m a.s.l. In the river head area (the subalpine belt) the predominating shrubberies of *Pinus pumila* grow in a complex with *Rhododendron aureum, Salix krylovii, S. saxatilis,* as well as fragments of large herb meadows (*Aquilegia glandulosa, Veratrum oxysepalum, Carex aterrima*), nival meadows (*Saxifraga merckii, Luzula palescens, Carex aterrima*), etc. Downstream the Tokko River crosses a tectonic depression. Meandering there through a wide floodplain, it forms numerous oxbow lakes overgrown by large *Carex* and *Eriophorum* species (*Eriophorum russeolum, Carex acuta, C. vesicata,* etc.). The same landscape supports *Picea* forests of the herb – green moss and *Oxalis* – green moss types. The islands are characterized by *Picea* forests lacking a ground layer. The sandy-pebble alluvium is the best substrate for the development of *Populus suaveolens* and *Chosenia arbutifolia* forests of the shrub type (*Salix rorida, Sorbaria sorbifolia* and *Rosa acicularis*) with a weakly developed herb-dwraf shrub layer and absence of a moss layer.

Further downstream the most of the areas of the 1st and 2nd terraces above the floodplain are occupied by *Larix* forests of the dwarf shrub-green moss type on slightly podzolized loamy, heavy loamy or stony-loamy soils. These forest communities include *Picea obovata* and *Betula pendula* in tree layer, and the herb-dwarf shrub layer is predominated by *Vaccinium uliginosum* or *Ledum palustre*. The understory consists of the constant *Betula divaricata, Duschekia fruticosa* and *Rosa acicularis,* and sometimes includes *Swida alba, Sorbaria sorbifolia, Sorbus sibirica, Lonicera edulis, Pinus pumila,* etc. The loose screes of the steep slopes of the valley support sparse *Larix* forests with a weakly developed understory of *Juniperus sibirica, Betula divaricata, Spiraea media* and *Rosa acicularis.* The well-heated and properly drained sites are occupied by *Larix + Pinus sylvestris* and *Pinus sylvestris* forests of the *Vaccinium vitis-idaea, Vaccinium vitis-idaea + Arctostaphylos uva-ursi* or *Arctostaphylos uva-ursi* types with participation of *Rhododendron dauricum* and *Duschekia fruticosa.*

The moist sites of the river valleys and watershed slopes support dark coniferous forests composed of several tree species, mainly of *Picea obovata* and *Abies sibirica* in various proportions with presence of *Pinus sibirica, P. sylvestris, Betula pendula, Populus suaveolens, P. tremula* and *Sorbus sibirica.* The understory is poor and is confined to "windows" in the closed canopy: *Sorbaria sorbifolia, Swida alba, Rhododendron aureum, Rosa acicularis,* seldom suppressed specimens of *Ribes nigrum* and *Rubus sachalinensis.* The ground layer is typical for the dark coniferous taiga: *Oxalis acetosella, Linnaea borealis, Mitella nuda, Circaea alpina, Goodyera repens, Trientalis europaea, Vaccinium vitis-idaea, Orthilia obtusata,* and *Pyrola rotundifolia.* Ephemeral streams and low relief elements are characterized by *Athyrium filix-femina, Smilacina dahurica,* and *Ranunculus repens.* Depending on moisture conditions, the forest ranges from the mesic shrub – *Oxalis* type to the damp *Oxalis* and fern + *Oxalis* – green moss type (Kuznetsova and Ivanova 2001).

*The Aldan River* flows into the Lena River from the East. In its lower and middle stream it has a lowland character, while the upper reaches possess mountain features. Like its large tributaries, the Amedichi, Uchur and Timpton Rivers, the Aldan River originates from the North-facing slopes of the Stanovoy Range. It flows through the wide depression composed of Jurassic bedrocks stretched along the Stanovoy Range's foothills. This territory is occupied by wet *Larix* woodlands of the shrub – dwarf shrub – *Sphagnum* type in a complex with tussock *Carex* meadows with participation of low montane *Salix* species and forbs (*Trollius sibiricus, Calamagrostis langsdorffii, Rubus arcticus, Veronica longifolia,* etc.). The wide terrace-like elements in the relief, as well as the very gentle slopes of various aspects, feature *yernik Larix* woodlands and *yerniks* of *Betula divaricata.* The herb-dwarf shrub layer is predominated by hygrophilous dwarf shrubs, *Carex, Eriophorum* and *Sphagnum* species. At the foot of the slopes *Hypnum* peat bogs form in the hollows. This bog type consists of large hollows covered by *Hypnum* mosses and scattered *Carex limosa* and *Betula exilis.* The hollows are surrounded by dwarf *Salix* species, *Triglochin palustris, Majanthemum bifolium, Andromeda polifolia* and *Eriophorum russeolum.* The ridges are also overgrown by mosses and contain *Ledum palustre, Vaccinium uliginosum, Oxycoccus microcarpus, Chamaedaphne calyculata, Trollius sibiricus, Calamagrostis langsdorffii, Rubus arcticus,* etc. The hillocky peat bogs are confined to the mountain river valleys. The hillocks appear in the central parts of a bog reaching 4–5 m in height and up to 50 m in diameter. The deep hollows between hillocks are overgrown by *Hypnum* mosses. The vegetation is similar to that of the abovementioned hollow peat bog (Elenevsky 1933).

The river valleys have no developed floodplain, and only in meanders the wet *Carex*, grass-*Carex* or grass meadows occur (mainly of *Calamagrostis langsdorffii*). The pebble alluvium is overgrown by forbs and *Salix* species.

In the lowland sections of the rivers the lower floodplain is occupied by shrubberies of *Salix viminalis, S. dasyclados* and rarely *S. udensis.* In the upper levels of the floodplain herb *Betula pendula* forests grow. The terraces above the floodplain are dominated by *Larix* forests of the *Vaccinium uliginosum + V. vitis-idaea* – green moss type, sometimes with participation of *Pinus sylvestris, Betula pendula* and *Picea obovata.* The *Pinus sylvestris* forests also occur there rarely (Andreyev et al. 1987).

*The Timpton River*. The meadow vegetation of its floodplain forms a narrow strip of large forb species with participation of shrubs (*Heracleum dissectum, Sanguisorba officinalis, Tanacetum vulgare, Cacalia hastata,* etc.) (Povarnitsyn 1932a; Galaktionova and Permyakova 1964; Andreyev et al. 1975, 1987).

The forest cover of the upper and middle reaches of the Timpton River consists of *Picea ajanensis* and *P. obovata* forests. They grow along the river banks, at the upper limit of the forest belt and in intermountain depressions. *Picea ajanensis* is strictly confined to crystalline bedrocks all along the water front of the Timpton River, on slope ecotopes (intermountain depressions, along springs), as well as at the upper border of the forest belt on steep slopes with a well-developed snow cover. This tree species typically forms pure tree stands, however *Betula ermanni subsp. lanata* often participates in its communities. The following types of *Picea ajanensis* forests are most characteristic for Yakutia: *Hylocomium*, *Vaccinium vitis-idaea* – green moss and fern – green moss. They cover large areas in the upper part of the forest belt. Very rarely the *Pinus pumila*, forb and *Sphagnum* types are observed. The understory of the *Picea ajanensis* forests usually includes *Duschekia fruticosa*, *Pinus pumila, Betula divaricata, Rhododendron aureum, Vaccinium vitis-idaea, Linnaea borealis,* etc. (Sokolov 1923; Povarnitsyn 1932b, 1933; Tyulina 1956, 1962; Scherbakov 1975; Volotovsky 1993, 1994). The *Picea obovata* forests form narrow communities on the floodplain. On the 1st terrace above the floodplain they are replaced by *Larix* woodlands with a rich cover of moss-lichens, lichens and dwarf shrubs *(Vaccinium uliginosum, V. vitis-idaea, Ledum palustre).* In waterlogged areas they are represented by the *Chamaedaphne calyculata* and *Sphagnum* types. On the 2nd terrace similar *Larix* communities alternate with *Pinus sylvestris* – *Larix* woodlands of the dwarf shrub – lichen or dwarf shrub – moss – lichen types growing in dryer habitats (Tyulina 1957).

Following the Archaean crystalline bedrocks, the forests of *Betula ermanni subsp. lanata* often descend from the mountain forest belt into the valley down to the river bed. They usually feature a properly developed herb layer with predomination of mesophilous forest-meadow and meadow forb species: *Calamagrostis langsdorffii, Carex pallida, Viola biflora,* and sometimes *Vaccinium vitis-idaea* and *V. myrtillus* (Rabotnov 1936; Tyulina 1956, 1959, 1962; Volotovsky and Chevychelov 1991; Volotovsky 1994).

The acid metamorphic or igneous bedrocks support *Larix* forests with a well developed moss-lichen cover, pure or with an admixture of other tree species. The herb-dwarf shrub layer is predominated by either *Ledum palustre, Vaccinium uliginosum,* or *V. vitis-idaea* depending on moisture conditions. The plant communities on acid bedrocks have low α- and β-diversities.

In the middle reaches of the Timpton River small fragments of *Betula pendula* and *Populus tremula* forests can seldom be observed. They are usually of secondary post-fire origin and participate as admixtures in *Larix* forests. The banks of large rivers and islands in the Timpton River basin can be covered by vast shrubberies of *Salix viminalis,* sometimes of *Populus suaveolens*, *Chosenia arbutifolia* and *Alnus hirsuta* that develop on pebble alluvium. The small rivers support thickets of *Salix schwerinii*, *Spiraea salicifolia* and *Betula divaricata* (Andreyev et al. 1987).

The alluvium of the rivers of the Tokinsky Stanovik Range consist of coarse pebble and boulders and are the habitat of unique *Salix cardiophylla* communities. The pebble alluvium covered by a thick sand layer up to 25–30 cm is covered by the *Populus suaveolens* forests or *Alnus hirsuta* shrubberies in the lower parts of the floodplain. All the abovementioned communities composed of deciduous tree species exist as long as one generation of trees. Even in early stages of the development of these communities the understory may contain young growth of *Larix* and *Picea* that replace the pioneer species during the subsequent successional stages. In valleys at 730 m a.s.l. or below *Picea ajanensis* and *P. obovata* forests grow. The *Picea obovata* communities are confined to the upper floodplains and terraces above the floodplain. They usually form narrow strips bordering the *Salix* or *Populus suaveolens* communities. The *Picea obovata* forests are replaced by *Larix* forests higher in the valleys. On calcareous bedrocks *Picea obovata* often penetrates the higher altitude zones often forming the tree-line. The region is characterized by the following types of *Picea obovata* forests: shrub – *Equisetum*, *Pyrola*, shrub – *Equisetum* + *Vaccinium vitis-idaea* – green moss. Rarely open herb communities occur. The forests of *Picea ajanensis* are confined to the crystalline bedrocks both in watershed areas and valleys. They descend from the mountains as narrow communities along the stream banks and narrow depressions. *Betula ermanni ssp. lanata* often participates in the *Picea ajanensis* tree stands. The most characteristic types are *Hylocomium*, *Vaccinium vitis-idaea* – green moss, and fern – *Hylocomium.* The banks of streams with swift currents may support *Picea ajanensis* – *Pyrola* or *Picea ajanensis* – *Calamagrostis* + forb communities. In the valleys near the upper limit of the forest belt the *Picea ajanensis* communities are represented by the following types: *Pinus pumila* – *Vaccinium vitis-idaea* or *Ledum palustre* – *Sphagnum* under stagnant moisture conditions and *Lycopodium* – *Sphagnum* in habitats with proper drainage.

## 3.6 Anthropogenic Vegetation

### *3.6.1 Ruderal Vegetation*

**M.M. Cherosov, P.A. Gogoleva, B.N. Pestryakov, and S.I. Mironova**

The ruderal vegetation of Yakutia consists of the communities of 5 traditional classes adopted in European geobotany, as well as by 2 classes that are characteristic for the northern territories of North Eurasia and are almost not present in Europe.

The class *ARTEMISIETEA VULGARIS* Lohmeyer et al. in R. Tx. 1950. The class and its order *Onopordietalia acanthi* comprise the communities of biennual and perennial hemicryptophytes representing various stages of synanthropic vegetation or natural vegetation with a significant participation of ruderal species in xeric habitats.

The diagnostic species of the class in Yakutia are *Arctium tomentosum, Bromopsis inermis, Carduus crispus, Lamium album, Leonurus quinquelobatus, Melilotus albus, M. officinalis, Tanacetum boreale* and *T. vulgare*. The diagnostic species of the order are *Astragalus danicus, Carduus nutans, Descurainia sophia, Festuca ovina, Hyoscyamus niger, Lappula squarrosa, Oberna behen, Rumex thyrsiflorus,* and *Sisymbrium officinale*.

In the territory of Yakutia the communities of this class are characterized by a significant participation of steppe species of the class *CLEISTOGENETEA SQUARROSAE* Mirk. et al. 1986 ex Korolyuk 2002, since they replace the steppe vegetation after disturbance.

There are two alliances distinguished within the class: *Artemisio-Caricion* (Gogoleva et al. 1987) Czerosov 2005 and *Dauco-Melilotion* Gors 1966 (Cherosov et al. 2005). The communities of the *Artemisio-Caricion duriusculae* develop as a result of anthropogenic and zoogenic impact on steppes and stepped meadows, while the vegetation of the *Dauco-Melilotion* replaces the true and saline meadows of mesic habitats. Growing in residential areas and representing degraded fodder lands, these communities are used in agriculture as a food source for cattle. Some of them serve even as verdure decorating settlements.

The class *POLYGONO ARENASTRI – POETEA ANNUAE* Rivas-Martinez 1975 corr. Rivas-Martinez et al. 1991 comprises the coenoses of compacted soils and is formed by annual plants. The vegetation of the class has a very wide distribution, covering North America, Eurasia, and, probably, the neo-tropical regions of the Southern Hemisphere (Ishbirdin 2001). The communities are strictly confined to landscapes exposed to regular recreation and grazing, and characterized by trampled soils. This feature determines the diagnostic species of the class. The diagnostic species of the class in Yakutia are *Amoria repens, Capsella bursa-pastoris, Lepidium densiflorum, Lepidotheca suaveolens, Plantago major, Plantago media, Poa annua, Polygonum aviculare, Polygonum humifusum, Potentilla anserina, Sagina procumbens* and *Stellaria media.*

The vegetation develops in a retrogressive direction. Representing a certain stage in a successional series, the communities of this class may stay unchanged for a long time in case of permanent anthropogenic impact. Cessation of human activity induces the recovery process.

Despite a low productivity (100–200 kg per ha), the vegetation of this class is used as pastures. Besides, the communities of the *POLYGONO ARENASTRI – POETEA ANNUAE* are characterized by high cover abundance values. This helps reducing the dust level in settlements, thus improving the environmental conditions in residential areas.

The class *EPILOBIETEA ANGUSTIFOLII* R. Tx. et Preising in R. Tx. 1950 represents the secondary communities of post-forest areas, developing as a result of both anthropogenic and natural factors, such as fires, cuttings, windfall, or after wood pest impacts (Isaev and Cherosov 1996). The diagnostic species of the class are *Chamaenerion angustifolium* and *Rubus saxatilis.*

The habitat conditions favour the formation of communities composed of eutrophic helophytes (light demanders) with a r-life strategy (Whittaker 1975), i.e. they develop from the soil seed bank under favourable growth conditions. Elimination of strong competitors favours the mass development of species with a ruderal strategy and even the penetration of weeds into these communities. If at least 10% of the species in a coenoflora is synanthropic we may consider the communities as ruderal. The synanthropic groups often found in vegetation of the class *EPILOBIETEA* are species of the classes *STELLARIETEA* R. Tx., Lohmeyer et Preising in R. Tx. 1950 and partly *MATRICARIO-POETEA* Ishbirdin 2002. On the other hand, however, they are not typical synanthropic communities, since, as mentioned before, their development may be induced by natural processes as well. Two associations are distinguished within the class: the *Calamagrostio epigeiosi-Chamerietum angustifolii* and the *Calamagrostio neglectae-Chamerietum angustifolii.* The first association develops in dry habitats and is characterized by xerophilous species (*Aegopodium alpestre, Bromopsis pumpelliana, Chamaenerion angustifolium, Calamagrostis epigeios* and *Pulsatilla flavescens*). The second association is confined to mesic conditions (*Calamagrostis neglecta, Chamaenerion angustifolium*).

The communities of the *EPILOBIETEA ANGUSTIFOLII* start to develop in the first year after a fire or cutting and remain for a rather long period until the succession towards forest makes headway. The obvious sign of a transition of the vegetation from *EPILOBIETEA* to a forest class is the appearance of stable species combinations of forest communities and the disappearance or reduction of synanthropic or meadow species.

The communities of the *EPILOBIETEA* are used by a man. In case of limited areas of hayfields, such grass stands in the taiga are cut for hay or used as pastures. Besides, elimination of a tree layer favours vigorous growth of a dwarf shrub-herb layer, including such food species as *Fragaria orientalis, Vaccinium vitis-idaea*, etc.

Such post-forest vegetation is also important for nature conservation. The properly developed synanthropic community represents a shelter for the better development of the species taking part in subsequent forest recovery.

The class *MOLINIO-ARRHENATHERETEA* Tx. 1937 em Mirk. et Naum. 1986 comprises the meadows used for hayfields and pastures. The natural meadows of the class are properly described in Section 3.5.7. Their synanthropic derivatives develop in residential areas and are represented by the degraded meadows and initial serial communities. Constant species of such communities are *Agrostis stolonifera, Potentilla anserina, Poa pratensis, Trifolium pratense, Amoria repens*, etc. The studied communities that occur in Yakutia belong to the order *Potentillo-Polygonetalia* R. Tx. 1947. It is represented by the following diagnostic species: *Agrostis stolonifera, Elytrigia repens, Equisetum arvense, Inula britannica, Juncus compressus, Persicaria lapathifolia, Potentilla anserina, Potentilla norvegica, Rorippa palustris* and *Sisymbrium officinale.*

Surrounding settlements, the vegetation of this class is also used as pastures for cattle and horses.

The class *BIDENTETEA TRIPARTITI* Tx., Lohm et Prsg. in Tx. 1950 occurs in the temperate zone from Europe through Japan. The communities of this class grow on eroded lake sides. The diagnostic species of the class are *Alopecurus alpinus, Bidens radiata, Bidens tripartita, Chenopodium glaucum, Chenopodium rubrum, Epilobium palustre, Erysimum cheiranthoides, Potentilla anserina, Ranunculus repens, Ranunculus sceleratus, Rorippa palustris* and *Rumex maritimus.*

The synanthropic landscapes of Yakutia feature distinct communities as well (Cherosov 2005). The most peculiar classes are the *PUCCINELLIO-HORDEETEA JUBATI* Mirkin in Gogoleva et al. 1987 and the *MATRICARIO-POETEA ARCTICAE* occurring in the northern taiga, forest tundra and tundra.

The communities of the class *PUCCINELLIO-HORDEETEA JUBATI* contain *Puccinellia hauptiana* (or *P. tenuiflora* in the habitats where it plays a significant role in the surrounding natural environment), *Hordeum jubatum*, and *Saussurea amara*. The coenoses are characterized by low α-diversity. The communities of the association *Puccinellio-Hordeetum jubati* occur under normal and insufficient moisture, while those of the *Beckmannio-Hordeetum jubati* occur in waterlogged landscapes. The habitats occupied by this vegetation show a slight to moderate salinization. This determines the presence of the halophytes. The coenoses of the *PUCCINELLIO-HORDEETEA JUBATI* are used for grazing.

The class *MATRICARIO-POETEA ARCTICAE* comprises the anthropogenic (waste areas in settlements, road slopes) and erosiophilous (various alluvium types, breaking river banks and lake sides) vegetation of the Holarctic. The diagnostic species of the class in Yakutia are *Polygonum humifusum, Artemisia tilesii, Calamagrostis neglecta, Calamagrostis purpurea, Deschampsia caespitosa, Descurainia sophioides, Equisetum arvense, Equisetum pratense, Eriophorum scheuchzeri, Festuca ovina, Festuca rubra, Poa alpigena, P. arctica, Tanacetum bipinnatum, Tephroseris palustris* and *Tripleurospermum hookeri.*

In northern Eurasia communities of anthropogenically disturbed tundra landscapes are very common (Gruzdev 1987, 1996; Sumina 1994; Ishbirdin et al. 1999; Ishbirdin 2001, etc.). The dominant species are *Poa alpigena, Tripleurospermum hookeri, Deschampsia caespitosa*, etc.

The syntaxonomical approach to the study of anthropogenic Arctic communities was used by Hadac (1989); Pestryakov (1992); Gruzdev and Martynenko (1994); Sumina (1994); Ishbirdin et al. (1999); Ishbirdin (2001), etc.

The class consists of two orders: 1. *Phippsio-Cochleariopsietalia* Hadac 1989 (the communities of the northern Holarctic), composed of two alliances: the *Cochleariopsion groenlandicae* Hadac (1989) and the *Poion glauco-malacanthae* Sumina 1994; 2. *Chamerio-Betuletalia nanae* Khusainov et Ishbirdin in Khusainov et al. 1989 with one alliance: the *Chamerio-Matricarion hookerii* Ishbirdin et al. 1996. The studied communities occur in the continental parts of Eurasia. We assume that the Novosibirskie Islands and the shoreline of the Arctic Ocean may feature vegetation of the order *Phippsio-Cochleariopsietalia* Hadac 1989.

The peculiarities of synanthropic communities are determined both by anthropogenic impact (intensity and character of anthropogenic or zoogenic load) and natural factors (salinization, moisture conditions, geographical location). This is discussed in details in works discussing synanthropic vegetation (Cherosov 2005; Cherosov et al. 2005).

### *3.6.2 Weed Vegetation*

**M.M. Cherosov and N.P. Sleptsova**

As one may think, the severe nature of Yakutia does not favour the development of agriculture, and particularly, crop production. Consequently, there should not be the phenomenon of weed vegetation in Yakutia.

The history of plant cultivation in Yakutia is as long as 300 years. Presently, 60% of all crops consumed by the residents of the Republic are provided by local production. Considering the short growing season and arid climate, this is a rather significant achievement. Under the conditions of the Yakutian climate and soils the following crops are cultivated: cereals (*Hordeum vulgare, Avena sativa, Triticum cereale, Secale cereale*) and vegetables (*Solanum tuberosum, Brassica oleracea, Cucumis sativus, Solanum lycopersicum, Beta vulgaris, Daucus spp.*). Recent agricultural experience even has proved the successful growth of water-melons (*Citrullus lanatus*) provided the summer is very hot.

Crop farming at an industrial scale under the conditions of the rigorous northern climate was aroused by the remoteness of the Republic and the imperfect transport network. Thus, hundreds of thousands of hectares (97% are situated in Central Yakutia) are allotted for plant production and, consequently, for weed vegetation development.

The complete list of weed species and a classification of the weed vegetation based on ecological-floristic criteria are given in the works by N.P. Sleptsova with her co-authors (Sleptsova 1984; Sleptsova et al. 1985; Sleptsova and Kononov 1986; Mirkin et al. 1988; Cherosov et al. 2005).

Despite the small areas of arable lands (compared to the traditional arable regions of Russia), the diversity of weed communities in Yakutia is rather high. For Central and South–West Yakutia seven associations are distinguished of the alliance *Sperguio-Oxalidion* Gors in Oberd. et al. 1967 (class *STELLARIETEA MEDIAE* R. Tx., Lohmeyer et Preising in R. Tx. 1950).

The diagnostic species for Yakutia are *Agrostis stolonifera, Amaranthus retroflexus, Brassica campestris, Brassica juncea, Chenopodium album, Convolvulus arvensis, Crepis tectorum, Elytrigia repens, Fallopia convolvulus, Galeopsis bifida, Geranium sibiricum, Lamium purpureum, Oberna behen, Senecio vulgaris, Setaria viridis, Sonchus arvensis, S. oleraceus* and *Urtica urens.*

Weed vegetation is characterized by the prevalence of annual plants (therophytes) in communities representing the initial stages of recovery succession in disturbed landscapes. The appearance of the species of the subsequent successional stages indicate that those communities belong to the class *ARTEMISIETEA VULGARIS* or to other synanthropic classes depending on substrate and local climate characteristics, and other factors.

The class *STELLARIETEA MEDIAE* clearly differs from the other synanthropic classes by the significant contribution of synanthropic species and the very negligible role of apophytes (native plants that have left their natural habitats and passed spontaneously to artificial sites).

In most cases the arable lands of Yakutia are characterized by insufficient moisture conditions (for comfortable crop farming rather dry habitats are used for ploughing, i.e. stepped meadows or meadow steppes). The moisture deficit is compensated by irrigation. However, due to political and economical events, which have taken place during the past decades in Russia, reclamation works were significantly reduced. This determines the prevalence of xeromesophytes in the weed coenoflora. Besides, due to widespread saline soils, the weed vegetation often contains halophytic plants.

The α-diversity of Yakutian weed communities is much lower than those of other regions of Russia. For instance, the associations of weed vegetation of Bashkiria, Ural or South Siberia may comprise 50–60 species. Yakutian weed communities feature 17 to 25, sometimes 32 species.

The associations of weed vegetation are distinguished based on the complex of abiotic factors determining the growth conditions. The factor irrigation plays an important role in the development of a weed vegetation.

The communities of the association *Elytrigio-Chenopodietum albi* are most widespread and typical for Central Yakutia. They are confined to depressions (middle belts of alases, inter-ridge depressions in terraces above the floodplains). The communities of the association *Aveno-Lappulatum squarrosae* develop on fertile soils on properly heated, prominent elements of landscapes under conditions of water deficiency. The association *Cannabido-Melilotetum officinalis* is typical for lands with spring cereals, while the association *Polygonetum tomentoso-amphibii* was described from irrigated arable lands in the Viluy River basin. The communities of the association *Potentillo anserinae Chenopodietum glaucii* are also common for Central Yakutia, particularly for the arable lands in the Lena and Aldan River valleys under conditions of sufficient heat and moisture. The diagnostic species include *Setaria viridis*, *Chenopodium glaucum* on saline soils, and *Potentilla anserina* in moist depressions. The association *Chenopodio-Chamerietum angustifolii* comprises the weed communities of post-forest arable lands (3–15 years after forest elimination). The association *Fagopyro tatarici-Stellarietum media* was described from South Yakutia (Aldan region), characterized by a milder climate and sufficient moisture conditions.

The nature of Yakutia, covering the tremendous territory from the islands in the Arctic Ocean to the mountain systems in the South, has influence on ruderal and weed vegetations that are traditionally considered to be determined by human activity. The synanthropic vegetation has a very important conservational significance, participating in recovery of landscapes that are disturbed as a result of mining industry, especially for overgrowing open pits.

# References


Abaimov AP, Prokushin SG, Zyryanova OA, Kaverzina LN (1997) Osobennosti formirovaniya i funktsionirovaniya listvennichnykh lesov na merzlotnykh pochvakh. Lesovedeniye 5: 13–23 – *Features of formation and functioning of larch forests on frozen soils* (in Russian)

Abolin RI (1929) Geobotanicheskoy i pochvennoye opisaniye Leno-Viluyskoy ravniny. In: Trudy komissii po izucheniyu Yakutskoy ASSR 10. Leningrad – *Geobotanical and soil description of the Lena-Viluy Plain* (in Russian)

Alexandrova VD (1960) Flora sosudistykh rasteniy Bolshogo Lyakhovskogo (Novosibirskiye ostrova). Botanichesky zhurnal 45(11): 1687–1693 – *The flora of vascular plants of Bolshoy Lyakhovsky Island (Novosibirskie Isalnds)* (in Russian)

Alexandrova VD (1961) Vliyaniye snezhnogo pokrova na rastitelnost v arkticheskoy tundre. In: Alexandrova VD, Kuvaev VB, Tikhomirov BA et al. (ed) Materialy po rastitelnosti Yakutii. Izd-vo LTA, Leningrad – *Influence of snow cover on vegetation in the Arctic tundra* (in Russian)

Alexandrova VD (1962) O podzemnoy structure nekotorykh rastitelnykh soobschestv arkticheskoy tundry na ostrove Bolshom Lyakhovskom. In: Problemy botaniki. Izd-vo AN SSSR, Moscow-Leningrad 6: 148–160 – *On the underground structure of some plant communities of the Arctic tundra on Bolshoy Lyakhovsky Island* (in Russian)

Alexandrova VD (1963) Ocherk flory i rastitelnosti ostrova Bolshogo Lyakhovskogo. In: Novosibirskiye ostrova. Morskoy transport, Leningrad – *Essay on the flora and vegetation of Bolshoy Lyakhovsky Island* (in Russian)

Alexandrova VD (1970) Dinamika mozaichnosti rastitelnykh soobschestv pyatnistykh tundr v arkticheskoy Yakutii. In: Mozaichnost rastitelnykh soobschestv i eyo dinamika. Vladimir – *Dynamics of mosaicity of plant communities of spotted tundra in Arctic Yakutia* (in Russian)

Alexandrova VD (1977) Geobotanicheskoye raionirovaniye Arktiki i Antarktiki. Nauka, Leningrad – *Geobotanical regionalization of the Arctic and Antarctic* (in Russian)

Andreyev VN (1954) Prodvizheniye lesnoy rastitelnosti v tundru v svyazi s zaschitnymi svoistvami lesoposadok na Severe. Botanichesky zhurnal 39(1): 28–47 – *Advancement of forest vegetation into the tundra in relation to protecting properties of tree-plantings in the North* (in Russian)

Andreyev VN (1956) Zaseleniye tundry lesom v sovremennuyu epokhu. In: Rastitelnost Krainego Severa SSSR i eyo osvoyeniye 1. Izd-vo AN SSSR, Moscow-Leningrad – *Forest invasion into the tundra in the modern period* (in Russian)

Andreyev VN (1975) Olenyi pastbischa Severnoy Yakutii. In: Botanicheskiye issledovaniya v Yakutii. Izd-vo YaF SO AN SSSR, Yakutsk – *Reindeer pastures in north Yakutia* (in Russian)

Andreyev VN, Galaktionova TF (1981) Relictovye tundrostepnye gruppirovki v Nizhnekolymskoy tundre. In: Andreyev VN (ed) Rastitelnost Yakutii i eyo okhrana. Izd-vo YaF SO AN SSSR, Yakutsk – *The relic tundra-steppe groups in the Lower Kolyma tundra* (in Russian)

Andreyev VN, Nakhabtseva SF (1974) Podzony yakutskoy tundry. In: Biologicheskiye problemy Severa. 4th Symposium. Issue 3. Izd-vo YaF SO AN SSSR: 40–45 – *Subzones of the Yakut tundra* (in Russian)

Andreyev VN, Perfilyeva VI (1975) Fragmenty stepnoy rastitelnosti na Severo-Vostoke Yakutii. In: Byulleten NTI: Biologicheskiye problemy Severa. Izd-vo YaF SO AN SSSR, Yakutsk: 14–16 – *The fragments of steppe vegetation in North–East Yakutia* (in Russian)

Andreyev VN, Perfilyeva VI (1980) Rastitelnost Nizhnekolymskoy tundry. In: Rastitelnost i pochvy subarkticheskoy tundry. Nauka, Novosibirsk – *Vegetation of the Lower Kolyma River tundra* (in Russian)

Andreyev VN, Galaktionova TF, Govorov PM, et al. (1978) Sezonnaya i pogodovaya dimanika fitomassy v subarkticheskoy tundre. Nauka, Novosibirsk – *Seasonal and annual dynamics of phytomass in the Subarctic tundra* (in Russian)

Andreyev VN, Galaktionova TF, Mikhalyova VM et al. (1975) Luga Yakutii. Nauka, Moscow – *Meadows of Yakutia* (in Russian)

Andreyev VN, Galaktionova TF, Perfilyeva VI, Scherbakov IP (1987) Osnovnye osobennosti rastitelnogo pokrova Yakutskoy ASSR. Izd-vo YaF SO AN SSSR, Yakutsk – *Basic features of vegetation cover of the Yakutian ASSR* (in Russian)

Andreyev VN, Nakhabtseva SF, Perfilyeva VI (1976) Tipy tundr Yakutii. In: Prirodnye resursy Yakutii, ikh ispolzovaniye i okhrana. Knizhnoye izd-vo, Yakutsk – *Types of tundra of Yakutia* (in Russian)

Belyaeva NV (1975) Klassifikatsiya kormovykh ugodiy khrebta Cherskogo. In: Botanicheskiye issledovaniya v Yakutii. Izd-vo YaF SO AN SSSR, Yakutsk – *Classification of fodder lands of the Chersky Range* (in Russian)

Birkengof AL (1932) Kratky predvaritelny otchyot o lesoissledovatelnykh rabotakh, provedennykh Indigirskim gidrilogicheskim otryadom Yakutskoy ekspeditsii AN SSSR v 1929–1930. In: Trudy SOPS AN SSSR 6. Izd-vo AN SSSR, Leningrad – *Brief preliminary report on forest investigation conducted by the Indigirka hydrological group of the Yakut expedition of the AS USSR in 1929–1930* (in Russian)

Boch MS (1975) Bolota nizovyev reki Indigirki (v predelakh tundrovoy zony). In: Flora, sistematika i filogeniya rasteniy. Kiev – *Bogs of the lower reaches of the Indigirka River* (*within the tundra zone*) (in Russian)

Boch MS (1978) Dostizheniya sovremennogo bolotovedeniya. In: Botanika 2: Sovremennoye bolotovedeniye i tundrovedeniye. Nauka, Moscow – *Achievements of modern bog science* (in Russian)

Boch MS, Mazing VV (1979) Ekosistemy bolot SSSR. Nauka, Leningrad – *Bog ecosystems in the USSR* (in Russian)

Borisova NI, Popova NP (1987) Ob allelopaticheskoy aktivnosti nekotorykh lesnykh rasteniy. In: Ekologo-biologicheskiye osnovy lesovodstvennykh mer v Yakutii. Izd-vo YaF SO AN SSSR, Yakutsk – *On allelopathic activity of some forest plants* (in Russian)

Bosikov NP (1991) Evolyutsiya alasov Tsentralnoy Yakutii. Izd-vo IMZ SO RAN, Yakutsk – *Evolution of alases of Central Yakutia* (in Russian)

Boychenko AM (1987) Proizrastaniye sosny obyknovennoy na severe areala v Yakutii. In: Ekologo-biologicheskiye osnovy lesovodstvennykh mer v Yakutii. Izd-vo YaF SO AN SSSR, Yakutsk – *Growth of Pynus sylvestris at the northern limits of its distribution area in Yakutia* (in Russian)

Boychenko AM (1992) Nekotorye aspekty plodonosheniya kedrovogo stlanika v Aldanskom nagorye. In: Botanicheskiye issledovaniya v kriolotozone. Collection of articles. Izd-vo YaNTs SO RAN, Yakutsk – *Some aspects of Pinus pumila fruiting in the Aldan Tableland* (in Russian)

Boychenko AM, Isaev AP (1992) Voprosy vosstanovleniya pritundrovykh lesov vblizi severnoy granitsy ikh rasprostraneniya v Yakutii. Lesnoy zhurnal 4: 33–37 – *Problems of restoration of near-tundra forests at the northern limits of their distribution in Yakutia* (in Russian)

Boychenko AM, Isaev AP (1996) Lesa doliny reki Kolymy v srednem techenii. In: Problemy ekologii Yakutii (Biogeographicheskiye issledovaniya). Izd-vo YaGU, Yakutsk – *The forests of the Kolyma River valley in its middle reaches* (in Russian)

Buks II (1964) O nekotorykh novykh mestonakhozhdeniyakh fragmentov stepnoy rastitelnosti v Yakutskom Zapolyarye. Byulleten Moskovskogo obschestva ispytateley prirody 69 (2): 135–137 – *On some new fragments of steppe vegetation in the Yakut Arctic* (in Russian)

Buks II (1966) Listvennichnye redkolesya na severe Leno-Olenekskogo mezhdurechya. Byulleten MOIP. Otdel Biologiya 71 (4): 71–82 – *The sparse larch forests in the North of the Lena-Olenyok Interfluve* (in Russian)

Cajander AK (1901) Lenan rannoilta. Luonnon Ystävä 5: 162–164

Cajander AK (1902) Die Wälder im Lena-Thale. Compt Rend Congr Natur Méd du Nord, Helsingfors sect 7: 41–42

Cajander AK (1903) Beiträge zur Kenntniss der Vegetation der Alluvionen des nordlichen Eurasiens. I. Die Alluvionen des unteren Lena-Thales. Act Soc Sc Fenn 32 (1): 1–182

Cajander AK (1904) Studion uber die Vegetation des Urwaldes am Lena-Fluss. Act Soc Sc Fenn 32 (3): 1–40

Cajander AK, Poppius RB (1903) Eine naturwissenschaftliche Reise im Lena-Thal. Fennia 19(2): 1–44

Cheremkhin SS (1961) Lesa verkhnego techeniya reki Viluy. In: Materialy o lesakh Yakutii 7. Izd-vo AN SSSR, Moscow: 243–259 – *Forests of the Upper Viluy River* (in Russian)

Cherosov MM (2005) Sinantropnaya rastitelnost Yakutii. YaF izd-va SO RAN, Yakutsk – *Synanthropic vegetation of Yakutia* (in Russian)

Cherosov MM, Sleptsova NP, Mironova SI et al. (2005) Sintaksonomiya sinantropnoy rastitelnosti Yakutii. YaF Izd-va SO RAN, Yakutsk – *Syntaxonomy of the synanthropic vegetation of Yakutia* (in Russian)

Chugunov BV (1955) Osnovnye tipy lesov Lenskogo raiona YaASSR i lesovozonovleniye v nikh. In: Trudy Instituta biologii YaF SO AN SSSR. Colletion of articles. Issue 1: 147–174 – *Basic forest types of the Lensk Region of the Yakut Autonomous Soviet Socialist Republic and reforestation there* (in Russian)

Cwynar LC, Ritche JC (1980) Arctic steppe-tundra: a Yukon perspective. Acc Adv Sci 208: 1375–1377

Danilova NS (ed) (2005) Raznoobrazie rastitelnogo mira Yakutii. Izd-vo SO RAN, Novosibirsk – *Diversity of the vegetation world of Yakutia* (in Russian)

Desyatkin RV (1984) Pochvy alasov Leno-Amginskogo mezhdirechya. Izd-vo YaF SO AN SSSR, Yakutsk –*Soils of alases of the Lena-Amga Interfluve* (in Russian)

Dobretsova LA (1961) Yernikovye formatsii v basseyne reki Viluya. In: Materialy po rastitelnosti Yakutii. Leningrad – *The yernik formations in the Viluy River basin* (in Russian)

Dobretsova LA (1962) Rastitelnost Zhiganskogo raiona. In: Nauchnye soobscheniya. Issue 8: Biologia. Knizhnoye izd-vo, Yakutsk – *Vegetation of the Zhigansk Region* (in Russian)

Dolenko GI (1913) Dolina reki Leny bliz Yakutska. In: Predvaritelny otchyot po organizatsii i issledovaniyu rabot po issledovaniyu pochv Aziatskoy Rossii v 1912 godu. Saint-Petersburg – *The Lena River valley in Yakutsk vicinity* (in Russian)

Drobov VP (1927) Kratky otchyot o rastitelnosti Leno-Aldanskogo plato. In: Materialy komissii po izucheniyu Yakutskoy ASSR. Issue 8. Izd-vo AN SSSR, Leningrad – *Brief report on the vegetation of the Lena-Aldan Plateau* (in Russian)

Efimova AP (2000) Suktsessionny ryad lesnoy rastitelnosti poimy Yakutskogo i Namskogo uchastkov doliny Sredney Leny. In: Ekologiya 2000: estafeta pokoleniy (Proceedings of 1st Inetrantional school-workshop on ecology). MGUL, Moscow: 61–63 – *Successional series of forest vegetation of the floodplain of the Yakutsk and Namsky sections of the Middle Lena River valley* (in Russian)

Efimova AP (2001) Klassifikatsiya lesnoy rastitelnosti poimy Sredney Leny s primeneniem "blok-metoda". In: Proceedings of scientific conference of students and young scientists of the Republic of Sakha (Yakutia) "Lavrentyevskiye chteniya". Yakutsk: 101–102 – *Classification of forest vegetation of the Middle Lena River's floodplain using the "block-method"* (in Russian)

Efimova AP (2003) Klassifikatsiya lesnykh soobschestv nizkoy poimy Yakutskogo i Namskogo uchastkov Sredney Leny. In: Proceedings of 7th scientific conference of students and young scientists of the Republic of Sakha (Yakutia) "Lavrentyevskiye chteniya" Volume 3. Yakutsk: 30–34 – *Classification of forest communities of the lower floodplain of the Yakutsk and Namsky sections of the Middle Lena River* (in Russian)

Efimova AP, Shurduk IF (1998) Ivnyaki doliny reki Leny v eyo srednem techenii. In: Problemy botaniki na rubezhe 20–21 vekov 1. Saint-Petersburg: 249 – *Willow forests of the Lena River valley in its middle reaches* (in Russian)

Efimova AP, Shurduk IF (2001a) Rol ivnyakov v dinamike rastitelnogo pokrova poimy Tuymaady i ikh prirodnoye znacheniye. In: Timofeev PA, Shmatkova LE (ed) Flora i fauna osobo okhranyaemykh prirodnykh territoriy Respublikanskoy sistemy Ytyk Kere Sirder. Kuduk, Yakutsk – *Role of willow communities in the dynamics of vegetation cover of the floodplain of the Tuymaada valley and their natural significance* (in Russian)

Efimova AP, Shurduk IF (2001b) Vliyaniye antropogennykh faktorov na zaschitnye funktsii poymennykh lesov Sredney Leny. In: Ekologicheskaya bezopasnost reki Leny: Monitoring, prirodnye i tekhnogennye kataklizmy. Collection of articles. Yakutsk: 82–83 – *Effect of anthropogenic factors on the protective function of floodplain forests of the Middle Lena River* (in Russian)

Efimova AP, Shurduk IF (2003) Osobennosti pervichnogo lesoobrazovatelnogo protsessa v levoberezhnoy poime Sredney Leny. In: Botanicheskiye issledovaniya v Aziatskoy Rossii. Proceedings of 11th Congress of the Russian Botanical Society. Volume 2. Izd-vo "Azbuka", Barnaul: 353–354 – *Features of primary forest forming process in the left-bank floodplain of the Middle Lena River* (in Russian)

Efimova AP, Shurduk IF (2005) Osnovnye cherty dinamiki lesov doliny Sredney Leny. In: Lesopolzovaniye, ekologiya i okhrana lesov: fundamentalnye i prikladnye aspekty. Proceedings of International Conference. STT, Tomsk: 77–79 – *Basic features of forest dynamics of the Middle Lena River valley* (in Russian)

Efimova AP, Shurduk IF, Popova LT, Koryakina GA (2004) K voprosu o elovykh lesakh Namskogo uchastka doliny Sredney Leny. Collection of articles. In: Problemy izucheniya rastitelnogo pokrova Yakutii. NIPK "Sakhapoligrafizdat", Yakutsk: 48–56 – *On the question of Picea forests of the Namsky section of the Middle Lena River valley* (in Russian)

Efimova AP, Shurduk IF, Isaev AP(2005) Dinamika lesnoy i kustarnikovoy rastitelnosti doliny Sredney Leny. Nauka i obrazovaniye 2(38): 9–15 – *Dynamics of forest and shrub vegetation of the Middle Lena River valley* (in Russian)

Efimova AP, Shurduk IF, Sivtseva TA, Nikitin IA (2003) K klassifikatsii lesnykh soobschestv nizkoy poimy Yakutskogo i Namskogo uchastkov sredney Leny. In: Isaev AP (ed) Problemy botanicheskikh i lesovodstvennykh issledovaniy v RS (Ya) i Finlyandii. Proceedings of Sakha-Finnish Conference dedicated to 100th Anniversary of Cajander's expedition along the Lena. Izd-vo YaF SO RAN, Yakutsk: 96–101 – *On the classification of the forest communities of a lower floodplain of the Yakutsk and Namsky sections of the Middle Lena River* (in Russian)

Egorova AA, Vasilyeva II, Stepanova NA, Fesko NN (1991) Flora tundrovoy zony Yakutii. Izd-vo YaNTs SO RAN, Yakutsk – *The flora of the tundra zone of Yakutia* (in Russian)

Egorova AA, Zakharova VI, Safronov RR, Sosina NK (2006) Nelesnaya rastitelnost Lenskogo raiona. In: Volpert YaL (ed) Pochvy, rastitelny i zhivotny mir Yugo-Zapada Yakutii. Nauka, Novosibirsk – *Non-forest vegetation of the Lensk Region* (in Russian)

Elenevsky RA (1933) Priroda Aldanskogo kraya. In: Prirodnye resursy Yuzhnoy Yakutii v svyazi s sotsialisticheskoy rekonstruktsiey selskogo khozyaistva. Selkhozgiz, Moscow-Leningrad – *The nature of the Aldan territory* (in Russian)

Elovskaya LG (1987) Klassifikatsiya i diagnostika merzlotnykh pochv Yakutii. YaF SO AN SSSR, Yakutsk. – *Classification and diagnostics of frozen soils of Yakutia* (in Russian)

Elovskaya LG, Konorovsky AK (1978) Rayonirovanie i melioratsiya merzlotnykh pochv Yakutii. Nauka, Novosibirsk. – *Regionalization and melioration of frozen soils of Yakutia* (in Russian)

Ermakov NB, Cherosov MM (2005) Differentiation of the *VACCINIO-PICEETEA* and *LOISELEURIO-VACCINIETEA* in mountains of Yakutia. Annali di Botanica 5: 15–28

Ermakov NB, Cherosov MM, Gogoleva PA (2002) Classification of ultracontinental boreal forests in central Yakutia. Folia Geobotanica et Phytotaxonomica 37: 419–440

Ermakov NB, Dring J, Rodwell J (2000) Classification of continental hemiboreal forests of north Asia. Braun-Blanquetia 28: 1–131

Fyuodorov PP, Nikolaev AN, Desyatkin AR (2007) Vyyavleniye gidrotermicheskogo rezhima pochv na radialny prirost listvennitsy v Tsentralnoy Yakutii. Proceedings of the Conference "Novye metody dendroekologii". Irkutsk – *Study on the effect of the hydrothermal regime of soils on radial growth of larch trees in Central Yakutia* (in Russian)

Galaktionova TF, Permyakova VA (1964) Luga srednego techeniya reki Aldana. In: Issledovaniya pochvennykh, rastitelnykh i faunisticheskikh resursov. Knizhnoye izd-vo, Yakutsk – *The meadows of the Middle Aldan River* (in Russian)

Galaktionova TF, Dobretsova LA, Permyakova AA, Usanova VM (1962) Rastitelnost basseyna reki Viluy. Proceedings of the Institute of Biology YaF SO AN SSSR 8. Moscow-Leningrad – *Vegetation of the Viluy River basin* (in Russian)

Galkina EA (1956) Sphagnovye bolota. In: Geobotaicheskaya karta SSSR 1:4000000. Izd-vo AN SSSR, Moscow-Leningrad – *Sphagnum bogs* (in Russian)

Giterman RE (1963) Etapy razvitiya chetvertichnoy Yakutii i ikh znacheniya dlya stratigraphii. In: Trudy GIN AN SSSR 78. Moscow – *Developmental stages of the Quaternary of Yakutia and their significance for stratigraphy* (in Russian)

Giterman RE (1985) Istoriya rastitelnosti severo-vostoka SSSR v pliotsene i pleistotsene. Nauka, Moscow – *The history of vegetation of the North–eastern USSR in the Pliocene and Pleistocene* (in Russian)

Giterman RE, Golubeva LV, Zaklinskaya ED et al. (1968) Osnovnye etapy razvitiya rastitelnosti Severnoy Azii v antropogene. In: Trudy GIN AN SSSR. Moscow – *Main stages of development of vegetation of North Asia in the Anthropogene* (in Russian)

Gogoleva PA (1996) Klassifikatsiya rastitelnosti Yakutii. In: Problemy ekologii Yakutii 1: Biogeographicheskiye issledovaniya. Collection of articles. Izd-vo YaGU, Yakutsk: 21–32 – *Classification of vegetation of Yakutia* (in Russian)

Gogoleva PA, Kononov KE, Mirkin BM, Mironova SI (1987) Sintaksonomiya i simfitosotsiologiya rastitelnosti alasov Tsentralnoy Yakutii. Izd-vo Irkutskogo GU, Irkutsk – *Syntaxonomy and symphytosociology of alas vegetation of Central Yakutia* (in Russian)

Gorodkov BN (1946) Problemy paleogeographii chetvertichnogo perioda. Trudy Instituta geographii AN SSSR 37: 348–349 – *Problems of paleogeography of the Quaternary* (in Russian)

Gorodkov BN (1956) Rastitelnost i pochvy ostrova Kotelnogo (Novosibirsky Arkhipelag). In: Rastitelnost Krainego Severa SSSR i eyo osvoyeniye. Izd-vo AN SSSR, Moscow-Leningrad 2: 7–132 – *Vegetation and soils of Kotelny Island (Novosibirsky Archipelago)* (in Russian)

Gruzdev BI (1987) Antropogennaya transformatsiya vidovogo sostava rastitelnykh soobschestv Bolshezemelskoy tundry. In: Ekologo-tsenoticheskoye i floristicheskoye izucheniye fitotsenozov Evropeyskogo Severa SSSR. Collection of articles. Syktyvkar: 58–66 – *Anthropogenic transformation of species composition in plant communities of the tundra in Bolshaya Zemlya* (in Russian)

Gruzdev BI (1996) Sinantropnaya rastitelnost tekhnogennykh uchastkov Vostochno-Evropeyskikh tundr Rossii. Proceedings of 3rd International Conference "Osvoeniye severa i problemy rekultivatsii". Syktyvkar: 34–35 – *Synanthropic vegetation of technogenic sites in the East European tundra of the USSR* (in Russian)

Gruzdev BI, Martynenko VA (1994) Rastitelny pokrov tekhnogennykh uchastkov na evropeyskom Severe. In: Proceedings of 2nd International Conference "Osvoeniye severa i problemy rekultivatsii". Syktyvkar: 34–35 – *Vegetation cover of technogenic sites in the European North* (in Russian)

Hadac E (1989) Notes on plant communities of Spitsbergen. Folia Geobotanica et Phytotaxonomica 24(2):131–169

Hopkins DM (1967) The Cenozoic history of Beringia. A synthesis. In: Bering Land Bridge. Stanford university press, Standford

Hughes MK, Vaganov EA, Shiyatov SG, et al. (1999) Twentieth-century summer warmth in northern Yakutia in a 600-year context. The Holocene 5: 603–608

Isaev AP, Cherosov MM (1996) Soobschestva klassa *EPILOBIETEA ANGUSTIFOLII Tx. et Prsg.* in Tx. 1950 v Yakutii. In: Problemy ekologii Yakutii 1: Biogeographicheskiye issledovaniya. Collection of articles. Izd-vo YaGU, Yakutsk: 21–32 – *Communities of the class EPILOBIETEA ANGUSTIFOLII Tx. et Prsg. in Tx. 1950 in Yakutia* (in Russian)

Isaev AP, Mikhalyova LG (ed) (2000) Les i vechnaya merzlota: osobennosti sostava i struktury lesov merzlotnogo regiona, problemy ratsionalnogo vedeniya khozyaystva i okhrany. Izd-vo YaGU, Yakutsk – *Forest and permafrost: composition and structural peculiarities of the forests of the permafrost region, problems of rational economy and conservation* (in Russian)

Isaev AP, Borisov ZZ, Chikidov II (2004) Napravleniye i khod estestvennogo zarastaniya gornykh otvalov v basseyne srednego techeniya reki Molodo. In: In: Ekologicheskaya bezopasnost pri razrabotke rossypnykh mestorozhdeniy almazov. Collection of articles. Sakhapoligrafizdat,

Yakutsk: 154–160 – *Direction and process of natural overgrowing of mine dumps in the basin of the Middle Molodo River* (in Russian)

Ishbirdin AR (2001) Ekologo-geograficheskiye zakonomernosti formirovaniya sisantropnykh flor i rastitelnosti selitebnykh territoriy Rossii. Abstract of Thesis for Doctor Degree. Moscow – *Ecological-geographical principles of the formation of synanthropic flora and vegetation of populated areas of Russia* (in Russian)

Ishbirdin AR, Khusainov AF, Mirkin BM (1999) Tekhnogennaya suktsessionnaya sistema rastitelnosti mestorozhdeniya "Medvezhye" i upravleniye vosstanovitelnymi protsessami. Byulleten MOIP (Biologiya) 104(1): 90–98 – *Technogenic successional system of vegetation of deposit "Medvezhye" and management of recovery processes* (in Russian)

Ivanov AA, Mironova SI, Savvinov DD (2004) Alasnye luga Leno-Viluyskogo mezhdurechya Tsentralnoy Yakutii pri razlichnykh rezhimakh ispolzovaniya. Nauka, Novosibirsk – *The alas meadows of the Lena-Amga Interfluve of Central Yakutia under various modes of usage* (in Russian)

Ivanova MM (1973) O priurochennosti rasteniy k gornym porodam na Stanovom nagorye. Botanichesky zhurnal 58(9): 1252–1260 – *On the confinement of plants to the bedrocks on the Stanovoy tableland* (in Russian)

Ivanova MM (1976) Osobennosti rastitelnosti na izvestsoderzhaschikh porodakh Stanovogo nagorya. Botanichesky zhurnal 65(5): 675–682 – *Characteristics of vegetation on calcareous bedrocks of the Stanovoy tableland* (in Russian)

Ivanova VI (1961) Kratky ocherk rastitelnosti srednego techeniya reki Olenyok. In: Alexandrova VD, Kuvaev VB, Tikhomirov BA et al. (ed) Materialy po rastitelnosti Yakutii. Izd-vo LTA, Leningrad – *Brief review of the vegetation of the Middle Olenyok River* (in Russian)

Ivanova VP (1967) O stepnoy rastitelnosti v doline sredney Leny. In: Uchyonye zapiski YaGU. Seria biologicheskikh i khimicheskikh nauk. Issue 7: 11–19 – *On the steppe vegetation in the Middle Lena River valley* (in Russian)

Ivanova VP (1981) Tipchakovye stepi – odin iz etapov pastbischnoy digressii rastitelnosti v doline reki Leny. In: Andreyev VN (ed) Rastitelnost Yakutii i eyo okhrana. Izd-vo YaF SO AN SSSR, Yakutsk – *Fescue steppes as one of the stages of vegetation degradation of pastures in the Lena River valley* (in Russian)

Ivanova VP, Perfilyeva VI (1972) Sokhranit kovylnye stepi Yakutii. In: Priroda Yakutii i eyo okhrana. Knizhnoye izd-vo, Yakutsk – *To protect the Stipa steppes of Yakutia* (in Russian)

Karavaev MN (1945) Kratky analiz flory stepey Tsentralnoy Yakutii (predvaritelnoye soobscheniye). Botanichesky zhurnal 30 (2): 62–76 – *Brief analysis of the steppe flora of Central Yakutia (preliminary report)* (in Russian)

Karavaev MN (1955) K voprosu o geobotanicheskom raionirovaniyi tayozhnoy zony Yakutii. Vestnik MGU 5(8): 109–115 – *On the question of geobotanical regionalization of the taiga zone of Yakutia* (in Russian)

Karavaev MN (1958) Fragmenty reliktovykh stepey s *Helictotrichon krylovii* (N. Pavl.) Henrard v Yakutii. Botanichesky zhurnal 43(4): 481–489 – *Fragments of relic steppes with Helictotrichon krylovii (N. Pavl.) Henrard in Yakutia* (in Russian)

Karavaev MN (1965) Rastitelny pokrov Yakutii. In: Gerasimov IP (ed) Yakutia. Nauka, Moscow – *Vegetation cover of Yakutia* (in Russian)

Karavaev MN (1968) *Agropyron karavaevii* P. Smirn. kak odin primer amerikano-aziatskikh stepnykh svyazey. Botanichesky zhurnal 53 (10): 1457–1461 – *Agropyron karavaevii P. Smirn. as an example of the American-Asian steppe relations* (in Russian)

Karavaev MN, Dobretsova LA (1964) Kratky ocherk rastitelnosty doliny reki Nery v eyo nizhnem techenii (basseyn verkhney Indigirki). Botanichesky zhurnal 49(11): 1544–1559 – *Brief review of the vegetation of the Nera River valley in its lower reaches (the Upper Indigirka River basin)* (in Russian)

Karavaev MN, Skryabin SZ (1971) Rastitelny mir Yakutii. Yakutknigoizdat, Yakutsk – *Vegetation world of Yakutia* (in Russian)

Karpel BA, Medvedeva NS (1977) Plodonosheniye listvennitsy daurskoy v Yakutii. Nauka, Novosibirsk – *Fruiting of Larix dahurica in Yakutia* (in Russian)

Karpov NS (1980) O priroste listvennitsy Gmelina v anabarskoy tundre. In: Byulleten NTI Biologicheskie problemy Severa. Izd-vo YaF SO AN SSSR, Yakutsk – *On the growth of Larix gmelinii in the Anabar tundra* (in Russian)

Kartushin VM (1963) O rastitelnosti ostrova Bennetta, Novosibirskiye ostrova. In: Trudy AANI 224. Leningrad: 177–179 – *On the vegetation of Bennett Island, Novosibirskie Islands* (in Russian)

Kats NYa (1971) Bolota zemnogo shara. Nauka, Moscow – *Bogs of the Earth* (in Russian)

Kildyushevsky ID (1960) Formirovanie rastitelnosti yugo-vostochnoy chasti Verkhoyanskogo khrebta v raznykh usloviyakh sredy. Problemy botaniki 5: Materialy po izucheniyu flory i rastitelnosti vysokogoriy. Moscow-Leningrad: 62–71 – *Formation of vegetation in the southeastern part of the Verkhoyansk Range under various environmental conditions* (in Russian)

Komarov VL (1926) Vvedeniye v izucheniye rastitelnosti Yakutii. Trudy komissi po izucheniyu Yakutskoy ASSR. Vol 1. Leningrad – *Introduction to the vegetation study of Yakutia* (in Russian)

Kononov KE (1971) Poimennye luga Sredney Leny. Knizhnoye izd-vo, Yakutsk – *The floodplain meadows of the MiddleLena River* (in Russian)

Kononov KE (1982) Luga poimy reki Leny (ekologo-fitostenotichesky anakiz). Knizhnoye izd-vo, Yakutsk – *The meadows of the Lena River floodplain (ecological-phytocoenotic analysis)* (in Russian)

Korolyuk AYu, Troeva EI, Cherosov MM et al. (2005) Ekologicheskaya otsenka flory i rastitelnosti Tsentralnoy Yakutii. Izd-vo YaNTs SO RAN, Yakutsk – *Ecological assessment of the flora and vegetation of Central Yakutia* (in Russian)

Koropachinsky IYu (ed) (1996) Zelyonaya kniga Sibiri: redkiye i nuzhdayuschiesya v okrane rastitelnye soobschestva. Nauka, Novosibirsk – *Green Book of Siberia: Rare and protection-requiring plant communities* (in Russian)

Korzhuev SS (1965) Vody. In: Gerasimov IP et al. (ed) Yakutia. Nauka, Moscow-Leningrad – *Waters* (in Russian)

Kozhevnikov YuP, Ukraintseva VV (1996) Tundrostepi pleistotsena: argumenty "za" i "protiv". Izvestiya RAN (geographiya) 3: 96–110 – *The tundra-steppes of the Pleistocene: arguments pro and contra* (in Russian)

Kryuchkov VV (1968) Kochkarnye tundry. Botanichesky zhurnal 53(12): 1716–1730 – *Hummocky tundra* (in Russian)

Kuminova RV (1936) Ocherk o rastitelnosti Aldanskogo raiona YaASSR. In: Trudy Tomskogo Universiteta 90: 1–56 – *Review of the vegetation of the Aldan Region of the YaSSR* (in Russian)

Kuvaev VB (1955) Luga i pastbischa Olyokminskogo Raiona. In: Trudy Instituta Biologii. Izd-vo AN SSSR, Moscow-Leningrad 1: 103–135 – *Meadows and pastures of the Olyokminsk Region* (in Russian)

Kuvaev VB (1956) Rastitelnost Vostochnogo Verkhoyanya. Rastitelnost Krainego Severa i eyo osvoyeniye 2: 133–186 – *Vegetation of the East Verkhoyansk Region* (in Russian)

Kuvaev VB (1961) K vydeleniyu poyasa kholodnykh kamenistykh pustyn v gorakh Severa Evrazii. Botanichesky zhurnal 46(3): 337–347 – *On distinguishing a belt of cold stony deserts in the mountains of North Eurasia* (in Russian)

Kuvaev VB (1964) Botaniko-geographichesky profil cherez levoberezhye Amura i Yuzhnuyu Yakutiyu. Botanichesky zhurnal 49(4): 511–522 – *Botanical-geographical profile through the left bank of the Amur River and South Yakutia* (in Russian)

Kuznetsov SS (1927) Barkhannaya oblast v Yakutskom krae. Priroda 10: 785–790 – *Dune region in the Yakutian territory* (in Russian)

Kuznetsova LV, Ivanova EI (2001) Elovo-pikhtovye lesa Olyokmo-Charskogo nagorya. In: Timofeev PA, Shmatkova LE (ed) Flora i fauna osobo okhranyaemykh prirodnykh territoriy Respublikanskoy sistemy Ytyk Kere Sirder. Kuduk, Yakutsk – *Picea-Abies forests of the Olyokma-Chara upland region* (in Russian)

Labutin YuV, Perfilyeva YuV, Revin YuV et al. (1985) Rastitelny i zhivotny mir delty reki Lena. Izd-vo YaF SO AN SSSR, Yakutsk – *Plant and animal world of the Lena River Delta* (in Russian)

Larcher W (1978) Ekologiya rasteniy. Mir, Moscow – *Ecology of plants* (in Russian)

Lavrenko EM (1981) O rastitelnosti pleistotsenovykh periglyatsialnykh stepey SSSR. Botanichesky zhurnal 66(3): 313–327 – *On the vegetation of the Late Pleistocene periglacial steppes of the USSR* (in Russian)

Lazarev PA, Tirskaya NF (1975) Selerikanskaya iskopaemaya loshad i usloviya eyo suschestvovaniya (po rezultatam sporo-pyltsevogo analiza). In: Palinologicheskiye materialy k stratigraphii osadochnykh otlozheniy Yakutii. Knizhnoye izd-vo, Yakutsk – *The Selerikan fossil horse and its living conditions (based on results of spore-pollen analysis)* (in Russian)

Lukicheva AN (1960) Rastitelny pokrov kak indikator kimberlitovykh trubok. Geologiya i geofizika 11: 35–48 – *Vegetation cover as an indicator of a kimberlite pipe* (in Russian)

Lukicheva AN (1963a) Rastitelnost severo-zapada Yakutii i eyo svyaz s geologicheskim stroeniem mestnosti. Izd-vo AN SSSR, Moscow-Leningrad – *The vegetation of North–West Yakutia and its relation to the local geological structure* (in Russsian)

Lukicheva AN (1963b) Rastitelnost verkhovyev reki Muny (severo-zapadnaya Yakutia) i vliyaniye na neyo noveyshey tektoniki. Botanichesky zhurnal 48 (3): 328–340 – *The vegetation of the upper reaches of the Muna River (North–West Yakutia) and the influence of recent tectonics on it* (in Russian)

Malyshev LI (1965) Vysokogornaya flora Vostochnogo Sayana. Nauka, Moscow-Leningrad – *The alpine flora of the east Sayan Mountains* (in Russian)

Matveyev IA, Nikolaev ME, Sivtsev PD et al. (ed) (1989) Atlas sleskogo khozyaistva Yakutskoy ASSR. GUGK, Moscow – *Atlas of agriculture of the Yakutian ASSR* (in Russian)

Matveyeva IP (1993) Pushitsa mnogokoloskovaya i arktofila ryzhevataya v nizovyakh reki Kolymy. Izd-vo YaNTs SO RAN, Yakutsk – *Eriophorum polystachion and Arctophila fulva in the lower reaches of the Kolyma River* (in Russian)

Matveyeva NV (1980) Dve poezdki na ostrov Bolshoy Begichev (kratky ocherk flory i rastitelnosti). Botanichesky zhurnal 65(11): 1543–1559 – *Two trips to Bolshoy Begichev Island (brief review of the flora and vegetation)* (in Russian)

Medvedeva NS (1971) Plodonosheniye listvennitsy daurskoy na severe Yakutii. In: Issledovaniya rastitelnosti i pochv v lesakh Severo-Vostoka SSSR. Knizhnoye izd-vo, Yakutsk – *Fruiting of Larix dahurica in North Yakutia* (in Russian)

Medvedeva NS (1973) Morfogenez generativnykh organov listvennitsy daurskoy v Severnoy Yakutii. In: Polovaya reproduktsiya khvoinykh. Nauka, Novosibirsk – *Morphogenesis of generative organs of Larix dahurica in north Yakutia* (in Russian)

Mikhailov IS (1963) Evolyutsiya pochv i rastitelnogo pokrova severo-vostochnoy chasti ostrova Faddeyevsky, Novosibirskiye ostrova. In: Trudy AANI 224. Leningrad: 121–132 – *Evolution of soils and vegetation cover of Faddeyevsky Island, Novosibirskie Islands* (in Russian)

Mirkin BM, Gogoleva PA, Kononov KE (1985) The vegetation of central Yakutian Alases. Folia Geobotanica et Phytotaxonomica 20: 345–396

Mirkin BM, Kononov KE, Gogoleva PA, Burtseva EI, Naumova LG (1992a) The floodplain grasslands of the middle Lena-River I: general characteristic and ordination. Folia Geobotanica et Phytotaxonomica 27: 225–245

Mirkin BM, Kononov KE, Gogoleva PA, Burtseva EI, Naumova LG (1992b) The floodplain grasslands of the middle Lena-River II: classification. Folia Geobotanica et Phytotaxonomica 27: 247–300

Mirkin BM, Solomesch AI, Khaziakhmetov RM (1988) Ruderalnye soobschestva kak istochnik rastitelnykh resursov. Rastitelnye resursy 24(2): 10–17 – *Ruderal communities as a source of plant resources* (in Russian)

Naurzbaev MM, Vaganov EA, Sidorova OV (2003) Izmenchivost prizemnoy temperatury vozdukha na severe Evrazii po dannym tysyacheletnikh drevesno-koltsevykh khronologiy. Kriosfera Zemli 7(2): 84–91 – *Variability of surface air temperature in North Eurasia based on the thousand-year-old tree ring chronologies* (in Russian)

Neinshtadt MI, Nikonov MN (1958) Torfyanye mestorozhdeniya Yakutskoy ASSR. In: Problemy Severa: Trudy Soveta po izucheniyu proizvoditelnykh sil. Izd-vo AN SSSR, Moscow – *Peat deposits in the Yakutian ASSR* (in Russian)

Nikitina NV (2004) Formirovaniye temnokhvoino-kedrovykh lesov v Yakutii. In: Problemy izucheniya rastitelnogo pokrova Yakutii. NIPK "Sakhapoligrafizdat", Yakutsk – *Formation of dark coniferous Pinus sibirica forests in Yakutia* (in Russian)

Nikolaev AN (2000) Drevesno-koltsevye khronologii kak otrazheniye izmeneniy klimata. In: Isaev AP. Mikhalyova LG (ed) Les i vechnaya merzlota: osobennosti sostava i struktury lesov merzlotnogo regiona,problemy ratsionalnogo vedeniya khozyaystva i okhrany. Izd-vo YaGU, Yakutsk – *Tree ring chronologies as a reflection of climate change* (in Russian)

Nikolaev AN, Fyodorov PP (2004) Vliyaniye klimaticheskikh faktorov i termicheskogo rezhima merzlotnykh pochv Tsentralnoy Yakutii na radialny prirost listvennitsy i sosny (na primere statsionara "Spasskaya Pad"). Lesovedeniye 6: 1–11 – *Influence of climatic factors and thermal regime of frozen soils on radial growth of larch and pine trees (case study of the Research station "Spasskaya Pad")* (in Russian)

Nikolaev AN, Naurzbaev MM, Yeger IV, et al. (2001) O vozraste listvennitsy v srednetayozhnoy podzone Yakutii. In: Timofeev PA, Shmatkova LE (ed) Flora i fauna osobo okhranyaemykh prirodnykh territoriy Respublikanskoy sistemy Ytyk Kere Sirder. Kuduk, Yakutsk – *On the age of larch trees in the middle taiga sub-zone of Yakutia* (in Russian)

Nikolin EG (1986) Dinamika rastitelnosti baidzharakhov v vysokoshirotnoy Arktike. In: Proceedings of the 11th All-Union Symposium "Biologicheskiye problemy Severa" 2. Yakutsk: 22–24 – *Dynamics of baidjarakh vegetation in the high-latitude Arctic* (in Russian)

Nikolin EG (1988) Reliktovye stepnye gruppirovki v gorakh Verkhoyanskogo khrebta. In: Vinokurov PM, Germogenov NI, Tazlova RS (eds) Proceedings of the 7th Yakutian scientific-practical conference of young scientists and specialists 3. Yakutsk: 42–43 – *The relic steppe groupings in the mountains of the Verkhoyansk Range* (in Russian)

Nikolin EG (1990) Flora i rastitelny pokrov Tsentralnogo Verkhoyanya. Abstract of Thesis of Candidate Degree. Novosibirsk – *The flora and vegetation cover of the Central part of the Verkhoyansk region* (in Russian)

Nikolin EG (1992) Obschiye cherty structurnoy organizatsii rastitelnosti Tsentralnogo Verkhoyanya. In: 10 Vsesoyuznoye soveschaniye po izucheniyu flory i rastitelnosti vysokogoriy. Proceedings. TsSBS SO RAN, Novosibirsk: 106 – *Common features in the structural organization of vegetation of the Central part of the Verkhoyansk region* (in Russian)

Nosova LI (1964) Ocherk tundrovoy i lesotundrovoy rastitelnosti mezhdurechya Yany i Omoloya. Botanichesky zhurnal 49 (5): 661–668 – *Review of the tundra and forest-tundra vegetation of the Yana-Omoloy Interfluve* (in Russian)

Perfilyeva VI (1977) Osnovnye cherty rastitelnosti gornoy chasti basseyna reki Yany. In: Teoreticheskiye i prikladnye problemy biologii na Severo-Vostoke SSSR. Collection of articles. Izd-vo YaF SO AN SSSR, Yakutsk: 85–94 – *Basic features of the vegetation of the mountainous part of the Yana River basin* (in Russian)

Perfilyeva VI, Rykova YuV (1975) Arkticheskaya tundra v ustye reki Chukochyey. In: Botanicheskie issledovaniya v Yakutii. Izd-vo YaF SO AN SSSR, Yakutsk – *The Arctic tundra in the Chukochya River mouth* (in Russian)

Perfilyeva VI, Shurduk IF (1999) Rastitelnost ostrovov Ust-Viluyskogo zakaznika. In: Timofeyev PA, Ivanova EI, Isaev AP (ed) Flora i rastitelnost Yakutii. Collection of articles. FIPS, Moscow: 78–82 – *Vegetation of islands of the Ust-Viluysky Reserve* (in Russian)

Perfilyeva VI, Egorova AA, Stepanova NA (1981) K flore Kondakovskogo ploskogorya (Severo-Vostochnaya Yakutia). In: Andreyev VN (ed) Rastitelnost Yakutii i eyo okhrana. Izd-vo YaF SO AS SSR, Yakutsk – *On the flora of the Kondakovskoye Plateau (North–eastern Yakutia)* (in Russian)

Perfilyeva VI, Teterina LV, Karpov NS (1991) Rastitelny pokrov tundrovoy zony Yakutii. Izd-vo YaNTs SO RAN, Yakutsk – *Vegetation cover of the tundra zone of Yakutia* (in Russian)

Permyakova AA (1961) Rastitelnost alasov Suntarskogo raiona Yakutskoy ASSR. In: Alexandrova VD, Kuvaev VB, Tikhomirov BA et al. (ed) Materialy po rastitelnosti Yakutii. Izd-vo LTA, Leningrad – *The vegetation of alases of the Suntar Region of the Yakut ASSR* (in Russian)

Pestryakov BN (1992) Ekologo-fitotsenoticheskaya kharakteristika rastitelnosti doliny reki Yany (severo-vostok Yakutii) i protsessy eyo sinantropizatsii. Abstract of Thesis for a Candidate Degree. Dnepropetrovsk – *Ecological-phytocoenological characteristic of the vegetation of the Yana River valley (North–east Yakutia) and processes of its synanthropization* (in Russian)

Petrova EI (1971) Pochvy Yuzhnoy Yakutii. Yakutskoye knizhnoye izd-vo, Yakutsk – *The soils of South Yakutia* (in Russian)

Petrovsky VV (1959) O structure rastitelnykh assotsiatsiy poligonalnykh bolot v nizovyakh reki Leny. Botanichesky zhurnal 44(10): 1500–1509 – *On the structure of plant associations of polygonal bogs in the lower reaches of the Lena River* (in Russian)

Petrovsky VV (1962) Kompleksnye assotsiatsii v rastitelnom pokrove tundry i lesotundry. In: Problemy botaniki. Izd-vo AN SSSR, Moscow-Leningrad – *Complex associations in vegetation cover of tundra and forest-tundra* (in Russian)

Petrovsky VV, Korolyova TM (1979) K flore delty reki Kolymy. Botanichesky zhurnal 63(1): 19–31 – *On the flora of the Kolyma River Delta* (in Russian)

Petrovsky VV, Korolyova TM (1980) K flore poberezhiy Vostochnogo-Sibirskogo morya. Botanichesky zhurnal 64(1): 13–26 – *On the flora of the East-Siberian Sea coast* (in Russian)

Petrovsky VV, Zaslavskaya TM (1981) K flore pravoberezhya reki Kolymy bliz eyo ustya. Botanichesky zhurnal 65(5): 1520–1528 – *On the flora of the Kolyma River's right bank near its mouth* (in Russian)

Pivnik SA (1958) Ekologo-biologicheskiye osobennosti kedrovogo stlanika (*Pinus pumila* (Pall) Rgl) v osnovnykh rastitelnykh assotsiatsiyakh severo-zapadnoy chasti ego areala. Abstract of thesis for a Candidate Degree. Leningrad – *Ecological-biological peculiarities of Pinus pumila (Pall) Rgl in basic plant associations in the North–western part of its distribution area* (in Russian)

Poiseyeva SI, Mironova SI (2000) Sintaxonomiya rastitelnosti reki Viluy. In: Nauka v Yakutii: dostizheniya i problemy. Collection of articles. Izd-vo SO RAN, Novosibirsk: 147–151 – *Syntaxonomy of the vegetation of the Viluy River* (in Russian)

Povarnitsyn SA (1932a) Lesa doliny reki Aldan ot Tommota do ustya reki Notory. In: Lesnye resursy Yakutii. Izd-vo AN SSSR, Leningrad – *The forests of the Aldan River from Tommot town through the Notora River mouth* (in Russian)

Povarnitsyn SA (1932b) Lesa Leno-Aldanskogo vodorazdela po Sanyyakhtakhskoy trope. In: Lesnye resursy Yakutii. Izd-vo AN SSSR, Leningrad – *The forests of the Lena-Aldan watershed along the Sanyyakhtakh pathway* (in Russian)

Povarnitsyn SA (1933) Lesa doliny reki Aldan ot Tommota do ustya reki Uchura. In: Trudy Instituta po izucheniyu lesa. Izd-vo AN SSSR, Leningrad 1: 155–231 – *The forests of the Aldan River from Tommot town through the Uchur River mouth* (in Russian)

Pozdnyakov LK (1946) Sortimentnye i tovarnye tablitsy dlya daurskoy listvennitsy. In: Tekhnicheskiye pravila otpuska lesa. Yakutsk – *Assortment and commodity tables for Larix dahurica* (in Russian)

Pozdnyakov LK (1948) O khode rosta daurskoy listvennitsy Verkhoyanskogo raiona Yakutskoy ASSR. Doklady AN SSSR 60(2): 301–304 – *On the growth process of Larix dahurica in the Verkhoyansk Region of the Yakutian ASSR* (in Russian)

Pozdnyakov LK (1956) O vliyanii rastitelnosti na glubinu letnego ottaivaniya pochvy. In: To the 75th Anniversary of Academician VN Sukachev. Collection of articles on geobotany, silvics, paleogeography and floristics. Izd-vo AN SSSR, Moscow–Leningrad – *On the influence of vegetation on the depth of seasonal thawing of soils* (in Russian)

Pozdnyakov LK (1961a) Lesa verkhnego techeniya Yany. In: Trudy Instituta biologii YaF SO AN SSSR. Izd-vo AN SSSR, Moscow 7: 162–242 – *The forests of the Upper Yana River* (in Russian)

Pozdnyakov LK (1961b) Listvennichnye i sosnovye lesa verkhnego Aldana. Izd-vo AN SSSR, Moscow – *The Larix and Pinus sylvestris forests of the Upper Aldan River* (in Russian)

Pozdnyakov LK (1962) O vliyanii rasteniy zhivogo pokrova na prorastaniye semyan i razvitiye vskhodov daurskoy listvennitsy. In: Problemy botaniki: voprosy botanicheskoy geografii, geobotaniki i lesnoy biogeotsenologii 6. Izd-vo AN SSSR, Moscow-Leningrad: 308–311 – *On the influence of plants of the living cover on seed germination and young growth development of Larix dahurica* (in Russian)

Pozdnyakov LK (1963) Gidroklimatichesky rezhim listvenichnykh lesov Tsenrtalnoy Yakutii. Izd-vo AN SSSR, Moscow – *Hydroclimatic regime of larch forests of Central Yakutia* (in Russian)

Pozdnyakov LK (1969) Lesa Yakutskoy ASSR. In: Lesa SSSR: Lesa Urala, Sibiri i Dalnego Vostoka. Nauka, Moscow – *The forests of the Yakut ASSR* (in Russian)

Pozdnyakov LK (1975) Listvennitsa daurskaya. Nauka, Moscow – *Larix dahurica* (in Russian)

Pozdnyakov LK (1983) Estestvennoye izrezhivaniye i prirost listvennichnikov v Tsentralnoy Yakutii. Lesovedeniye 6: 28–34 (in Russian) – *Natural thinning and growth of larch forests in Central Yakutia* (in Russian)

Pozdnyakov LK (1986) Merzlotnoye lesovedeniye. Nauka, Novosibirsk – *Permafrost silvics* (in Russian)

Prakhov NI (1957) Osnovnye elementy rastitelnosti Verkhoyanskogo khrebta. Trudy Instituta biologii YaF SO AN SSSR. Izd-vo AN SSSR, Moscow 3: 39–67 – *Basic elements of the vegetation of the Verkhoyansk Range* (in Russian)

Protopopov AV (2005) Formirovaniye flory listvennichnykh lesov Yano-Indigirskogo mezhdurechya. Abstract of Thesis for a Candidate Degree. Yakutsk – *Development of the flora of larch forests of the Yana-Indigirka Interfluve* (in Russian)

Rabotnov TA (1935) Landshafty peschanykh obrazovaniy v nizovyakh Vilyuya. Zemlevedeniye 37(4): 321–338 – *Landscapes of sand formations in the lower reaches of the Viluy River* (in Russian)

Rabotnov TA (1936) Obzor rabot po izucheniyu rastitelnosti Yakutskoy ASSR (k 200-letiyu botanicheskogo izucheniya Yakutii). Botanichesky zhurnal 26(6): 727–738 – *Review of publications concerning investigation of the vegetation of the Yakutian ASSR (to the 200th anniversary of botanical investigation of Yakutia)* (in Russian)

Rabotnov TA (1937) Rastitelnost vysokogornogo poyasa basseyna verkhovyev rek Aldana i Timptona. Izvestiya RGO 69(4): 585–605 – *Vegetation of the high-mountian belt of the Upper Aldan and Upper Timpton Rivers' basins* (in Russian)

Rabotnov TA (1939) O bolotakh Tsentralnoy Yakutii. Izvestiya geograficheskogo obschestva 71(4): 5447–550 – *On marshlands of Central Yakutia* (in Russian)

Rabotnov TA (1940) Fiziko-geografichesky ocherk Aldanskogo okruga YaASSR. In: Materialy zemelno-vodnogo ustroistva raionov Krainego Severa 6. Moscow – *Physiographical characteristics of the Aldan Region of the YaSSR* (in Russian)

Safronov FG, Safronov AF (ed) (2007) Entsiklopediya Yakutii: Rastitelny mir 2. Izd-vo YaNTs SO RAN, Yakutsk – *Encyclopedia of Yakutia: Plant World* (in Russian)

Safronova IN (1980) K flore ostrova Kotelny (Novosibirskiye ostrova). Botanichesky zhurnal 65(4): 544–551 – *On the flora of Kotelny Island (Novosibirskie Islands)* (in Russian)

Safronova IN (1982) Materialy k flore mysa Svyatoy Nos (Arkticheskaya Yakutia). Botanichesky zhurnal 67(12): 1660–1669 – *Data on the flora of the Svyatoy Nos Cape (Arctic Yakutia)* (in Russian)

Samsonova VV (2000) Mnogoletnyaya merzlota na Zemnom share i v Yakutii. In: Isaev AP, Mikhalyova LG (ed) Les i vechnaya merzlota: osobennosti sostava i struktury lesov merzlotnogo regiona, problemy ratsionalnogo vedeniya khozyaystva i okhrany. Izd-vo YaGU, Yakutsk – *Permafrost on Earth and in Yakutia* (in Russian)

Savvinov DD, Kononov KE (1981) Teplovoy balans lugovoy rastitelnosti i klimat merzlotnykh poymennykh pochv. Nauka, Novosibirsk – *Heat balance of meadow vegetation and climate of frozen floodplain soils* (in Russian)

Savvinov DD, Savvinov GN, Tyaptirgyanov MM et al. (1992) Ekologiya verkhney Amgi. Izd-vo YaNTs SO RAN, Yakutsk – *Ecology of the Upper Amga River* (in Russian)

Schelkunova RP (1970) Rastitelnost tundrovoy zony Yakutskoy ASSR. Thesis for a Candidate Degree. Irkutsk – *Vegetation of the tundra zone of the Yakutian ASSR* (in Russian)

Scherbakov IP (1962) Lesnye resursy Yakutii i ikh ispolzovaniye. Knizhnoye izd-vo, Yakutsk – *Forest resources of Yakutia and their utilization* (in Russian)

Scherbakov IP (1964) Tipy lesa Yuzhnoy Yakutii. In: Pozdnyakov LK (ed) Lesa Yuzhnoy Yakutii. Nauka, Moscow – *The forest types of south Yakutia* (in Russian)
Scherbakov IP (1965) O vosstanovlenii lesnoy rastitelnosti na Krainem Severnom predele Evrazii. Izvestiya SO AN SSSR (Biologicheskiye i meditsinskiye nauki) 2(8): 54–61 – *On the recovery of forest vegetation in the Far North of Eurasia* (in Russian)
Scherbakov IP (1975) Lesnoy pokrov Severo-Vostoka SSSR. Nauka, Novosibirsk – *Forest cover of the North–east of the USSR* (in Russian)
Scherbakov IP (1992) Lesa verkhnego i srednego techeniya reki Viluy. In: Botanicheskiye issledovaniya v kriolotozone. Collection of articles. Izd-vo YaNTs SO RAN, Yakutsk – *The forests of the Upper and Middle Viluy River* (in Russian)
Sheludyakova VA (1938) Rastitelnost basseyna reki Indigirki. Sovetskaya botanika 4–5: 43–79 – *Vegetation of the Indigirka River basin* (in Russian)
Sheludyakova VA (1948) Rastitelnost severo-vostoka Yakutii. In: Doklady na pervoy nauchnoy sessii Yakutskoy bazy AN SSSR.Gosizdat, Yakutsk – *Vegetation of North–eastern Yakutia* (in Russian)
Sheludyakova VA (1957a) Kratky ocherk lugov doliny reki Leny. In: Trudy Instituta Biologii YaF SO AN SSSR. Izd-vo AN SSSR, Moscow 3: 127–138 – *Brief review of meadows of the Lena River valleys* (in Russian)
Sheludyakova VA (1957b) Materialy k izucheniyu torfyanikov. Trudy TGU 141: 130–139 – *Materials for investigation of peatlands* (in Russian)
Sheludyakova VA (1957c) Stepnaya rastitelnost Yakutskogo Zapolyarya. In: Trudy Instituta Biologii YaF SO AN SSSR. Izd-vo AN SSSR, Moscow 3 – *The steppe vegetation of the Yakut Arctic* (in Russian)
Sheludyakova VA (1961) Tabunnoye soderzhaniye loshadey (tebenyovka v Verkhoyanskom Raione Yakutskoy ASSR). In: Poimennye luga Krainego Severa i priyomy ikh ispolzovaniya i uluchsheniya 10. Knizhnoye izd-vo, Norilsk – *Herd keeping of horses (winter pasturage in the Verkhoyansk Region of the Yakutian ASSR)* (in Russian)
Sheludyakova VA, Skryabin SZ (1969) Charany – landshafty lesostepnogo tipa v Tsentralnoy Yakutii. Botanichesky zhurnal 54(4): 539–556 – *Charans, the forest-steppe landscapes in Central Yakutia* (in Russian)
Shurduk IF, Efimova AP, Popova LT, Kolesova L (2005) Sostav i structura elovykh lesov doliny Sredney Leny. Proceedings of All-Russian Conference "Prirodnaya i antropogennaya dinamika nazemnykh ekosistem". Izd-vo Irkutskogo GU, Irkutsk: 480–482 – *Composition and structure of Picea forests of the Middle Lena River valley* (in Russian)
Shvedchikov GV (1974a) K ekologii rastitelnykh soobschestv *Arctophila fulva* v nizovyakh Kolymy. Botanichesky zhurnal 59 (3): 385–393 – *On the ecology of Arctophila fulva plant communities of the lower reaches of the Kolyma River* (in Russian)
Shvedchikov GV (1974b) Zavisimost prirosta i fenologicheskogo razvitiya arktofily ot ekologicheskikh usloviy. Biologicheskiye problemy Severa 3. Izd-vo YaF SO AN SSSR, Yakutsk: 118–121 – *Response of growth and phenological development of Arctophila to ecological conditions* (in Russian)
Shvedchikov GV (1975) *Arctophila fulva* (Trin) Anders. kak tsenozoobrazovatel gigrofitnykh soobschestv na severe Yakutii. Abstract of Thesis for a Candidate Degree. Leningrad – *Arctophila fulva (Trin) Anders. as a builder of hygrophytic communities in the north of Yakutia* (in Russian)
Sidorova OV, Naurzbaev MM (2002) Klimaticheskiy otklik derevyev listvennitsy, proizrastayuschikh na verkhney granites lesa i nadpoimennoy terrase v nizovyakh reki Indigirki. In: Osnovnye zakonomernosti globalnykh i regionalnykh izmeneniy klimata i prirodnoy sredy v pozdnem kainozoye Sibiri 1. Novosibirsk: 391–401 – *Climatic feedback of larch trees growing at the upper forest limits and in a terrace above the floodplain in the lower reaches of the Indigirka River* (in Russian)
Sidorova OV, Naurzbaev MM (2005) Rekonstruktsiya temperatury vozdukha za posledniye 2000 let po dannym godichnykh kolets derevyev listvennitsy vostoka Taimyra i severo-vostoka

Yakutii. Sibirskiy ekologichesky zhurnal 1: 51–60 – *Air temperature reconstruction for the last 2000 years based on the data of larch tree rings from eastern Taimyr and North–eastern Yakutia* (in Russian)

Skryabin SZ (1968) Stepnaya rastitelnost v srednem techenii Indigirki. Abstract of Thesis for a Candidate degree. Yakutsk – *The steppe vegetation in the Middle Indigirka River* (in Russian)

Skryabin SZ (1976) Fragmenty gornoy petrofitnoy stepi v nizovye reki Tokko (Yuzhnaya Yakutia). Botanichesky zhurnal 61(12): 1748–1751 – *The fragments of a mountainous petrophytic steppe in the lower reaches of the Tokko River (South Yakutia)* (in Russian)

Skryabin SZ, Karavaev MN (1991) Zelyony pokrov Yakutii. Knizhnoye izd-vo, Yakutsk – *The green cover of Yakutia* (in Russian)

Skryabin SZ, Konorovsky KA (1975) Rastitelnost i pochvy stepey srednego techeniya Indigirki. In: Botanicheskiye materially po Yakutii. Izd-vo YaF SO AN SSSR, Yakutsk – *Vegetation and soils of the steppes of the Middle Indigirka River* (in Russian)

Sleptsova NP (1984) O nekotorykh osobennostyakh segetalnoy rastitelnosti TSentralnoy Yakutii. Botanichesky zhurnal 69(8): 1070–1073 – *On some peculiarities of weed vegetation of Central Yakutia* (in Russian)

Sleptsova NP, Kononov KE, Must NM (1985) *Elytrigio – Chenopodion* all. nova – novy soyuz segetalnoy rastitelnosti Tsentralnoy Yakutii. In: Antropogennye protsessy v rastitelnosti. Collection of articles. Ufa: 69–75 – *Elytrigio – Chenopodion all. nova is a new alliance of weed vegetation in Central Yakutia* (in Russian)

Sleptsova NP, Kononov KE (1986) Segetalnaya rastitelnost v Tsentralnoy Yakutii. In: Mirkin BM (ed) Klassifikatsiya rastitelnosti SSSR c ispolzovaniem floristicheskikh kriteriev. Izd-vo MGU, Moscow – *Weed vegetation in Central Yakutia* (in Russian)

Sochava VB (1933a) Estestvennye kormovye ugodya tundrovoy zony Yakutii. Sovetskoye olenevodstvo 2: 47–118 – *Natural fodder lands of the tundra zone of Yakutia* (in Russian)

Sochava VB (1933b) Tundry basseyna reki Anabary. Izvestiya gosudarstvennogo geographicheskogo obschestva 65 (4) – *The tundra of the Anabar River basin* (in Russian)

Sochava VB (1934a) Botaniko-geographicheskiye podzony v zapadnykh tundrakh Yakutii. In: Proceedings of the 1st All-Union Congress of Geographers. Leningrad – *Botanical-geographical sub-zones of the western tundra of Yakutia* (in Russian)

Sochava VB (1934b) Rastitelnye assotsiatsii Anabarskoy tundry. Botanichesky zhurnal 19 (3) – *Plant communities of the Anabar tundra* (in Russian)

Sochava VB (1957) Taiga na severo-vostoke Sredne-Sibirskogo ploskogorya. Botanicheskiy zhurnal 42 (9): 1408–1415 – *Taiga in the northeastern Middle Siberian Plateau* (in Russian)

Sokolov FV (1923) Ocherk yugo-vostochnoy chasti Yakutiskoy oblasti. Izvestiya RGO. 1919–1923. Saint-Petersburg 55 (1): 181–236 – *Essay on the south–eastern part of the Yakutian Region* (in Russian)

Sosina NK, Isaev AP (2003) Novye nakhodki *Redowskia sophiifolia (Brassicaceae)* v Yakutii. In: Isaev AP (ed) Problemy botanicheskikh i lesovodstvennykh issledovaniy v RS (Ya) i Finlyandii. Proceedings of Sakha-Finnish Conference dedicated to 100th Anniversary of Cajander's expedition along the Lena. YaF SO RAN, Yakutsk: 81–83 – *New records of Redowskia sophiifolia (Brassicaceae) in Yakutia* (in Russian)

Sosina NK, Zakharova VI (2004) K izucheniyu rastitelnosti basseyna reki Muny (nizhnee techeniye reki Leny). In: Ekologicheskaya bezopasnost pri razrabotke rossypnykh mestorozhdeniy almazov. Collection of articles. Sakhapoligrafizdat,Yakutsk: 228–237 – *On the study of vegetation of the Muna River basin (lower reaches of the Lena River)* (in Russian)

Stark SK (1933) Lesa Verkhnekolymskogo raiona. Sovetskaya Aziya, Moscow – *The forests of the Upper Kolyma Region* (in Russian)

Stepanov GM (1981) Iskusstvennoye lesovosstanovleniye na garyakh Severnoy Yakutii. Lesnoye khozyaystvo 12: 59–60 – *Artificial recovery of burnt-out forests of North Yakutia* (in Russian)

Stepanov GM (1987) Osobennosti estestvennogo lesovozobnovleniya na garyakh Severnoy Yakutii. In: Ekologo-biologicheskiye osnovy lesovodstvennykh mer v Yakutii. Izd-vo YaF SO

AN SSSR, Yakutsk – *Peculiarities of natural recovery of burnt-out forests of North Yakutia* (in Russian)

Stepanov GM (1988) Temperaturny rezhim merzlotnykh pochv na garyakh severnoy Yakutii. Lesovedeniye 5: 67–71 – *Temperature regime of frozen soils in burnt-out forests of North Yakutia* (in Russian)

Sumina OI (1975) Rastitelnost baidjarakhov ostrova Kotelnogo (Novosibirskiye ostrova). Botanichesky zhurnal 60(9): 1311–1319 – *Vegetation of baidjarakhs of Kotelny Island (Novosibirskie Islands)* (in Russian)

Sumina OI (1986) Dopolneniye k flore i rastitelnosti ostrova Kotelnogo i Zemli Bunge (Novosibirskiye ostrova). Botanichesky zhurnal 71(7): 903–911 – *Supplement to the flora and vegetation of Kotelny Island and Zemlya Bunge (Novosibirskie Islands)* (in Russian)

Sumina OI (1994) Sozdaniye edinogo kompyuternogo banka dannykh po antropogennoy rastitelnosti arktiki v tselyakh izucheniya antropogennykh izmeneniy tundrovykh ekosistem. In: Proceedings of the 2nd International Conference "Osvoeniye severa i problemy rekultivatsii". Syktyvkar: 34–35 – *Creation of a common computer database on the anthropogenic vegetation of the Arctic for the investigation of the anthropogenic transformation of tundra ecosystems* (in Russian)

Tarabukina VG, Savvinov DD (1990) Vliyaniye pozharov na merzlotnye pochvy. Nauka, Novosibirsk – *Fire effect on frozen soils* (in Russian)

Tikhomirov BA, Shtepa VS (1956) K kharakteristike lesnykh forpostov v nizovyakh reki Leny. Botanichesky zhurnal 41(8): 1107–1122 – *On the characteristics of the forest outposts in the lower reaches of the Lena River* (in Russian)

Timofeyev PA (1976) O vzaimootnosheniyakh rasteniy listvennichnykh lesov Tsentralnoy Yakutii. Proceedings of the Conference "Struktura i dinamika rastitelnogo pokrova". Moscow: 101–103 – *On plant interrelationships in larch forests of Central Yakutia* (in Russian)

Timofeyev PA (1979) Vliyaniye vodnykh vytyazhek iz rasteniy travyano-kustarnichkovogo yarusa listvennichnykh lesov na vozobnovleniye listvennitsy daurskoy. Ekologiya 4: 89–91 – *Effect of water extraction of plants from the herb-dwarf shrub layer of larch forests on the recovery of Larix dahurica* (in Russian)

Timofeyev PA (1980) Lesa Yakutii. Knizhnoye izd-vo, Yakutsk – *The forests of Yakutia* (in Russian)

Timofeyev PA (1990) O roli allelopaticheskogo faktora v zhizni rasteniy. Izvestiya SO AN SSSR (Biologicheskiye nauki) 1: 129–134 – *On the role of an allelopathic factor in plant life* (in Russian)

Timofeyev PA (2003) Lesa Yakutii: sostav, resursy, ispolzovaniye i okhrana. Izd-vo SO RAN, Novosibirsk – *The forests of Yakutia: composition, resources, utilization and protection* (in Russian)

Timofeyev PA, Protopopov AV (1996) Listvennichnye lesa basseyna reki Yany v eyo srednem techenii. In: Problemy ekologii Yakutii (biogeographicheskiye issledovaniya). Izd-vo YaGU, Yakutsk – *Larch forests of the Yana River in its middle reaches* (in Russian)

Timofeyev PA, Shurduk IF (1996) Sostoyaniye i voprosy okhrany lesov doliny Tuymaady. In: Problemy ekologii Yakutii: biogeograficheskiye issledovaniya. Collection of articles. Izd-vo YaGU, Yakutsk: 101–105 – *Condition and protection issues of the forests of the Tuymaada valley* (in Russian)

Timofeyev PA, Isaev AP, Scherbakov IP et al. (1994) Lesa srednetayozhnoy podzony Yakutii. Izd-vo YaNTs SO RAN, Yakutsk – *The forests of the middle taiga subzone of Yakutia* (in Russian)

Tolmachev AI (1932) Flora tsentralnoy chasti Vostochnogo Taimyra. Trudy Polyarnoy Komissii AN SSSR 8: 126 – *The flora of the central part of East Taimyr* (in Russian)

Tolmachev AI (1954) K istorii vozniknoveniya i razvitiya temnokhvoynoy taigi. Izd-vo AN SSSR, Moscow-Leningrad – *On the history of the origin and development of the dark coniferous taiga* (in Russian)

Tomskaya AI (1981) Palinologiya kainozoya Yakutii. Nauka, Novosibirsk – *Palynology of the Cainozoic of Yakutia* (in Russian)

Tomskaya AI (2000) Kormovaya baza mamonta v pozdnem pleistotsene Yakutii. Yakutsk – *Fodder base of mammoths in the Late Pleistocene in Yakutia* (in Russian)

Trufanova ER (1967) Tsvetkovye rasteniya vodoyomov Yakutii i ikh khozyaistvennoye ispolzovaniye. In: Lyubite i okhranyaite prirodu Yakutii. Yakutskoye knizhnoye izd-vo, Yakutsk – *The flowering plants of the water bodies of Yakutia* (in Russian)

Trufanova ER (1972a) Nekotorye svedeniya ob isplzovaniyi ondatroy vodnoy i pribrezhno-vodnoy rastitelnosti Kolymskikh ozyor. Yakutskoye knizhnoye izd-vo, Yakutsk – *Some data on the use of aquatic and rim vegetation of the Kolyma lakes by musk-rat* (in Russian)

Trufanova ER (1972b) Rastitelnotst ozyor srednego techeniya reki Kolymy. In: Pochvennye i botanicheskiye issledovaniya v Yakutii. Yakutskoye knizhnoye izd-vo, Yakutsk – *The vegetation of lakes of the Middle Kolyma River* (in Russian)

Trufanova ER, Galaktionova TF (1975) Rastitelnost vodoyomov v nizovyakh reki Kolymy. In: Byulleten NTI. Biologicheskiye problemy Severa. Izd-vo YaF SO AN SSSR, Yakutsk: 16–17 – *The vegetation of the water bodies of the lower reaches of the Kolyma River* (in Russian)

Trufanova ER, Egorova AA, Karpov NS (1981) Ostrovki drevesnoy rastitelnosti v Anabarskoy tundre. In: Proceedings of the 4th All-Union Symposium "Biologicheskiye problemy Severa" 1. Syktyvkar – *Fragments of tree vegetation in the Anabar tundra* (in Russian)

Tyrtikov AP (1955) Rastitelnost nizovyev reki Yany. Byulleten MOIP (Biologiya) 60(5): 135–146 – *Vegetation of the lower reaches of the Yana River* (in Russian)

Tyrtikov AP (1958) Nekotorye svedeniya o rastitelnosti nizovyev reki Indigirki. Byulleten MOIP (Biologiya) 63(1) – *Some data on vegetation of the lower reaches of the Indigirka River* (in Russian)

Tyulina LN (1956) Na ozere Toko i severnom sklone Stanovogo Khrebta (kratky geobotanichesky ocherk). In: Akademiku VN Sukachevu k 75-letiyu so dnya rozhdeniya. Izd-vo AN SSSR, Moscow-Leningrad – *On Toko lake and North-facing slopes of the Stanovoy Range (brief geobotanical review)* (in Russian)

Tyulina LN (1957) Ocherk lesnoy rastitelnosty verkhnego techeniya reki Aldan. In: Trudy Instituta Biologii YaF SO AN SSSR. Izd-vo AN SSSR, Moscow 3: 83–138 – *Essay on the forest vegetation of the Upper Aldan River* (in Russian)

Tyulina LN (1959) Lesnaya rastitelnost srednego i nizhnego techeniya reki Yudomy i nizovyev Mai. Izd-vo AN SSSR, Moscow – *The forest vegetation of the Middle and Lower Yudoma River and lower reaches of the Maya River* (in Russian)

Tyulina LN (1962) Lesnaya rastitelnost sredney i nizhney chasti basseina reki Uchur. Izd-vo AN SSSR, Moscow-Leningrad – *The forest vegetation of the middle and lower parts of the Uchur River basin* (in Russian)

Ukraintseva VV (2002) Rastitelnost i klimat Sibiri epokhi mamonta. Izd-vo Polikom, Krasnoyarsk – *Vegetation and climate of Siberia during the mammoth epoch* (in Russian)

Usanova VM (1961) K voprosu klassifikatsii alasov Tsentralno-Yakutskoy ravniny. In: Alexandrova VD, Kuvaev VB, Tikhomirov BA et al. (ed) Materialy po rastitelnosti Yakutii. Izd-vo LTA, Leningrad – *On classification of alases of the Central Yakutian Plain* (in Russian)

Utkin AI (1958) Nekotorye osobennosti rasprostraneniya kornevykh system drevesnykh porod v kholodnykh pochvakh. In: Soobscheniya Instituta lesa AN SSSR. Collection of articles. Moscow: 64–71 – *Some features of root system distribution of tree species in cold soils* (in Russian)

Utkin AI (1961) Kedrovy stlanik na severo-zapadnoy okraine areala i istoriya ego rasprostraneniya. In: Trudy Instituta lesa i drevesiny: Voprosy lesovodstva i lesovedeniya 50: 104–119 – *Pinus pumila at the northwestern border of its distribution area and the history of its distribution* (in Russian)

Utkin AI (1965) Lesa Tsentralnoy Yakutii. Nauka, Moscow – *The forests of Central Yakutia* (in Russian)

Utkin AI (1976) Lesnye biogeotsenozy kriogennoy oblasti kak spetsifichnye sistemy. Ekologiya 3: 15–22 – *Forest biogeocoenoses of the cryogenic region as specific systems* (in Russian)

Vaganov EA, Arbatskaya MK, Shashkin AV (1996) Istoriya klimata i chastota pozharov v tsentralnoy chasti Krasnoyarskogo Kraya 2. Dendrokhronologicheskiy analiz svyazi izmenchivosti priroste derevyev, climata i chatoty pozharov. Sibirskiy ecologicheskiy zhurnal 1: 19–28 – *History of climate and fire frequency in the central part of Krasnoyarsk Territory 2. Dendrochronological analysis of the correlation between tree growth variability, climate and fire frequency* (in Russian)

Vaskovsky AP (1959) Kratky ocherk rastitelnosti, climata i khronologii chetvertichnogo perioda v verkhovyakh rek Kolymy, Indigirki i na severnom poberezhye Okhotskogo morya. In: Lednikovy period na territroii Evropeyskoy chasti SSSR i Sibiri. Izd-vo MGU, Moscow – *Brief description of vegetation, climate and chronology of the Quaternary period in the upper reaches of the Kolyma and Indigirka Rivers, as well as at the northern coast-line of the Sea of Okhotsk* (in Russian)

Volotovsky KA (1992) Osobennosti rastitelnogo pokrova na karbonatnykh porodakh Aldanskogo nagorya. In: Botanicheskiye issledovaniya v kriolitozone. Izd-vo YaNTs SO RAN, Yakutsk: 80–91 – *Characteristics of the vegetation cover on calcareous bedrocks of the Aldan tableland* (in Russian)

Volotovsky KA (1993) Tipologiya i lesotaksatsionnye parametry ayanskikh elnikov na khrebte Tokinsky Stanovik. In: Proceedings of scientific-practical conference of young scientists and post-graduate students. Yakutsk: 93 – *Typology and forest inventory parameters of forests of Picea ajanensis on the Tokinsky Stanovik Range* (in Russian)

Volotovsky KA (1994) Lesa iz eli ayanskoy. In: Lesa srednetayozhnoy podzony Yakutii. Izd-vo YaNTs SO RAN, Yakutsk – *Forests of Picea ajanensis* (in Russian)

Volotovsky KA, Chevychelov AP (1991) Kamennoberyozovye lesa Yakutii. Botanichesky zhurnal 76(6): 39–47 – *The forests of Betula ermanii in Yakutia* (in Russian)

Volotovsky KA, Kuznetsova LV (1998) Novye danye o estestvennom mezhrodovom gibride *Sorbocotoneaster pozdnjakovii (Rosaceae)*. Botanichesky zhurnal 83(1): 94–103 – *New data on the natural intergeneric Sorbocotoneaster pozdnjakovii (Rosaceae)* (in Russian)

Whittaker RH (1975) Communities and Ecosystems. Macmillan Publ Co, New York.

Yarovoy MI (1939) Rastitelnost basseyna reki Yany i Verkhoyanskogo khrebta. Sovetskaya botanika 1: 21–40 – *Vegetation of the Yana River basin and the Verkhoyansk Range* (in Russian)

Yurtsev BA (1959) Vysokogornaya flora gory Sokuydakh i eyo mesto v ryadu gornykh flor arkticheskoy Yakutii. Botanichesky zhurnal 44(8): 1171–1177 – *Alpine flora of Sokuydakh Mountain and its position in the series of montane floras of Arctic Yakutia* (in Russian)

Yurtsev BA (1961) K kharakteristike podzony severo-tayozhnykh listvennichnikov v zapadnoy chasti basseina reki Yany. In: Alexandrova VD, Kuvaev VB, Tikhomirov BA et al. (ed) Materialy po rastitelnosti Yakutii. Izd-vo LTA, Leningrad – *On characteristics of the subzone of the northern taiga larch forests in the western part of the Yana River basin* (in Russian)

Yurtsev BA (1962) Botaniko-geograficheskiye nablyudeniya u severnogo predela rasprostraneniya listvennitsy na reke Olenyok. In: Problema botaniki. Izd-vo AN SSSR, Moscow-Leninrad – *Botanical-geographical observations at the northern limits of the larch distribution area on the Olenyok River* (in Russian)

Yurtsev BA (1964) Botaniko-geograficheskiy ocherk Indigirskogo sklona gornogo uzla Suntar-Khayata. In: Rastitelnost zarubezhnykh stran. Nauka, Moscow-Leningrad – *Botanical-geographical essay on the Indigirka slope of the Suntar-Khayata Ridge* (in Russian)

Yurtsev BA (1968) Flora Suntar-Khayata: Problemy istorii vysokogornykh landshaftov Severo-Vostoka Sibiri. Nauka, Leningrad – *The flora of Suntar-Khayata: The issues of the history of alpine landscapes of North–East Siberia* (in Russian)

Yurtsev BA (1974) Stepnye soobschestva chukotskoy tundry i pleistotsenovaya "tundrostep". Botanichesky zhurnal 59(4): 484–501 – *The steppe communities of the Chukchi tundra and the Pleistocene "tundra-steppe"* (in Russian)

Yurtsev BA (1981) Reliktovye stepnye kompleksy Severo-Vostochnoy Azii: problemy rekonstruktsii kriokseroticheskikh landshaftov Beringii. Nauka, Novosibirsk – *The relic steppe complexes of the North–East Asia: problems of reconstruction of the cryoxeric landscapes of Beringia* (in Russian)

Zakharova VI (2005) K flore petrophitnykh stepey okhranyaemykh territory. In: Cherosov MM, Gogoleva PA, Simonova SV (ed) Flora i rastitelnost kriolitozony 2: Rastitelnost kriolitozony. Izd-vo YaNTs SO RAN, Yakutsk – *On the flora of the petrophytic steppes of protected territories* (in Russian)

Zakharova VI, Isaev AP, Ivanova EI, Sosina NK, Mikhalyova LG, Chikidov II (2004) Rastitelny pokrov, flora i mikobiota srednego techeniya reki Molodo. In: Ekologicheskaya bezopasnost pri razrabotke rossypnykh mestorozhdeny almazov. Collection of articles. Sakhapoligrafizdat, Yakutsk: 133–142 – *Vegetation cover, flora and mycobiota of the Middle Molodo River* (in Russian)

Zolnikov VG (1954a) Relief i pochvoobrazuyushiye porody vostochnoy poloviny Tsentralnoy Yakutii. In: Materialy o prirodnykh usloviyakh i selskom khozyaistve Tsentralnoy Yakutii. Izd-vo AN SSSR, Moscow – *Relief and soil-forming bedrocks of the eastern part of Central Yakutia* (in Russian)

Zolnikov VG (1954b) Za ratsionalnoye ispolzovaniye zemel v selskom khozyaistve Tsentralnoy Yakutii. Knizhnoye izd-vo, Yakutsk – *On rational land use in agriculture of Central Yakutia* (in Russian)

Zolnikov VG (1958) Pochvenno-landshaftnye raiony Zapadnoy Yakutii. In: Razvitiye proizvoditelnykh sil Zapadnoy Yakutii v svyazy s sozdaniem almazodobyvayuschey promyshlennosti 3: Prirodnye usloviya i selskoye khozyaistvo. Knizhnoye izd-vo, Yakutsk – *Soil-landscape regions of West Yakutia* (in Russian)

# Chapter 4
# Vegetation and Human Activity

**M.M. Cherosov, A.P. Isaev, S.I. Mironova, L.P. Lytkina, L.D. Gavrilyeva, R.R. Sofronov, A.P. Arzhakova, N.V. Barashkova, I.A. Ivanov, I.F. Shurduk, A.P. Efimova, N.S. Karpov, P.A. Timofeyev, and L.V. Kuznetsova**

## 4.1 Human Activity and the Ecological Situation in Yakutia

**M.M. Cherosov and A.P. Isaev**

The territory of Yakutia also experiences ecological problems even though the human impact is less than in many other regions of the world. They are aggravated by the following specific factors.

1. The presence of perennially frozen grounds, which are continuous in the northern and central parts of the Republic and disjunctive in the south.
2. Vast areas are occupied by such unstable landscapes as forest tracts bordering on the tundra zone (northern taiga) and the subalpine belt (mountain sparse forests), and steppe communities. Even an insignificant anthropogenic impact may cause negative processes in those landscapes.
3. Low rates of biogeochemical processes in landscapes due to the prolonged cold period. This results in a very slow decomposition of contaminants (10–15 times slower than in the forest-steppe and steppe zones).

Besides, the relative overpopulation of Yakutia should also be taken into consideration. Though characterized by a vast territory and low population total (less than one million), some regions of the Republic are characterized by high densities, 4 times exceeding that of Alaska and 40 times of the Canadian North.

The Yakutian nature strongly suffers from various industrial activities. They are ore-dressing and processing enterprises, coal, oil and gas extraction, timber industry, agriculture, power industry, transport networks. According to the information of the Ministry of Nature Conservation, over 7,000 ha of reindeer pastures are destroyed annually, the lichen fodder resources being reduced by 2.5–3%. The productivity of hayfields has been reduced by a factor 1.5–2 during the last 40 years.

M.M. Cherosov (✉)
Institute for biological problems of the cryolithozone, 41 Lenin Ave., Yakutsk, 677980, Russia
e-mail: cherosov@mail.ru

E.I. Troeva et al. (eds.), *The Far North: Plant Biodiversity and Ecology of Yakutia*, Plant and Vegetation 3, DOI 10.1007/978-90-481-3774-9_4,

Every year forest fires destroy hundreds of thousands of hectares. Millions of tons of waste water are discharged into rivers; hundreds of thousands of tons of pollutants are ejected into the atmosphere. The territory of the Republic features such processes as water erosion, secondary salinization, permafrost degradation induced by construction works, forest destruction, etc.

The economy of most of the Yakutian regions has an agricultural basis, growing traditional crops (barley, oat, wheat, potato, etc.). Hundreds of thousands of hectares are arable lands.

There are five ecologically stressed regions recognized in Yakutia:

1. The *Viluy region,* due to diamond-mining activity, power industry, and agriculture. Since 1956, for over 50 years the diamond mining enterprises have discharged waste waters into the natural water bodies contaminating them with toxic substances such as salts of thallium, strontium, arsenic and mercury. Construction of the Viluy Power Station has caused the submergence of tens of millions of tons of organic mass, which started to release phenols, hydrogen sulphide, and carbon dioxide. The sulphide rocks under the water reservoir release zinc and copper. The liquid waste drain from the cattle farms, mineral fertilizers and pesticides also contribute to the negative ecological situation. The Viluy River is characterized by exceeding the maximum permissible concentrations (MPC) of oil products, as well as of phenols, by a factor 5 and that of copper by a factor 3.
2. The *South Yakutian region,* due to power and mining industries. Two basic contamination sources are the coal strip-pit and gold ore extracting and processing enterprises. The Aldan River experiences destructive influences of the gold mining activity. The surface waters of the Neryungri, Chulman and Timpton Rivers are in even a worse condition.
3. The *North-Eastern region,* mostly due to mining industry. The fishery basis of this region has been crippled. The spawning areas of such marketable fish species as Siberian Cisco, Broad Whitefish and Humpback Whitefish have been eliminated. In the Adycha and Omoloy Rivers the content of mercury reaches 7 MPCs, and phenols 5–8 MPCs. The phenol content in the Indigirka River 4 times exceeds the MPC. The River is considered to be moderately polluted. At the same time some of its tributaries contain critical amounts of the following elements: copper up to 39 MPCs and zinc up to 7 MPCs. A similar situation is observed in the Kolyma River.
4. The *Yakutsk region* suffers from very strong anthropogenic impacts. The poor quality of the air is caused by dense traffic, industrial enterprises, and a high population density. On some days the excess of the MPC values for the following substances may reach: over 5 for hydrogen sulphide and dust, 17 for nitrogen dioxide, 4 for ammonia, 8 for carbon dioxide. The lakes of Yakutsk are in a bad condition as well, containing 11 MPCs of phenol. The highest concentration of pollutants is observed in the central and northern parts of the city. The Lena River is polluted with phenols and oil products coming from river transport, timber enterprises, and municipal engineering.

5. The *Trans-Lena regions* (Megino-Kangalas, Churapcha, Tatta, Ust-Aldan regions) experience lack of water, both of drinking and other water. These regions are also characterized by intensive agricultural activities.

Such vigorous, careless and ecologically unsafe human activities have no doubt a negative impact on the vegetation cover of Yakutia, which is as sensitive and vulnerable as all other ecosystems throughout the North.

The spatial variability in the natural conditions of the vast territory of Yakutia causes a high diversity of plant communities. This implies that a geographical approach is needed when selecting restoration methods for the ecosystems that are disturbed by human activity or natural disasters. Cleaning and subsequent revegetation of disturbed lands should be based on knowledge of the peculiarities of succession in each geographical region, on unprejudiced and comprehensive assessments of natural restoration status of the specific plant communities in the area, on economical propriety, as well as on the social and ecological significance of the effort.

The following disturbed lands are subject for re-vegetation or other restoration:

1. post-fire forest tracts lacking plants able to produce seeds;
2. anthropogenically disturbed areas (various cuttings) characterized by the replacement of commercially valuable tree species by less valuable ones, continuous overgrowing with herbs and shrubs, strongly littered areas, etc.;
3. technogenically disturbed lands (coarse sceletal grounds);
4. insect affected tree stands;
5. areas with a thermokarst, thermoerosive, or erosive topography, as well as the areas exposed to prolonged floods by water from melted snow;
6. residential areas including settlements situated in alases, etc.;
7. tukulans (northern sand deserts), if moving, sands fixation is required.

## 4.2 Technogenic Transformation of Vegetation

**S.I. Mironova and M.M. Cherosov**

The global scale of the negative environmental impact of mining activities emphasizes the barest necessity of nature conservational measures. This problem is of special urgency in the world's northern regions. Over half (54%) of the technogenically disturbed lands of Russia are situated in the territories of Siberia and the Far East. Of these, 20% of the prospected mineral deposits are situated in the regions of perennially frozen grounds (Kryuchkov 1973).

Yakutia is famous for its mineral deposits, ranking first in Russia in mining diamonds, antimony and tin. The beginning of the mining activity in Yakutia dates back to the 1920s, when large gold deposits were discovered in the South of the republic (Aldan Region). The diamond placer and pipe deposits "Mir", "Aikhal" and "Udachny" have been mined since the 1950s in the territory of the Mirny Region

(West Yakutia). Open pit mining of coal in its largest deposit (the "Neryungrinsky" strip-pit) has been conducted in the territory of the Neryungry Region (South Yakutia) since the 1970s. In recent years the gas and oil deposits are rigorously exploited in the Southwest of the republic (Lensk and Olyokminsk Regions).

Open pit mining implies the removal of the soil-vegetation cover requiring tens or hundreds of years for recovery under the condition of permafrost. Therefore, the study of the mechanisms of the formation of vegetation cover in disturbed areas should have priority and is of practical significance.

Using ecological-floristic classification and ordination techniques we have revealed the whole technogenic succession system (TSS) of vegetation dynamics in technogenically disturbed lands (Mironova 2000). The concept "succession system" was proposed by Razumovsky (1981) and involves all serial and climax communities of a specific natural region. Kucherov (1995) successfully used this concept to describe the total diversity of anthropogenic changes. We suggest that for the regions of intensive technogenic disturbances the epithet "technogenic" should proceed the term "succession system", because all natural successions that take place in those regions are strongly affected by anthropogenic impact. So TSS is a set of serial communities that are formed within a natural-technogenic landscape, i.e. as a part of the whole succession system (Mironova 2000).

Two major succession systems were shown to be characteristic for vegetation overgrowing diggings in West and South Yakutia. Substrate characteristics (mine wasting dumps, coarse mother beds, and dredged deposits of finer composition) and location of vegetation (on a dump itself or between mine dumps) reveal different subsystems of succession.

The succession systems are similar to R. H. Wittaker's idea of "superclimax", when serial and climax communities are formed based on one and the same species, approaching the model of "initial floristic composition" by F. Eagler, stating that almost the whole species composition that participate in the succession appear during the first stage, and later only the ratio of the participating species changes.

The β-diversity was estimated on basis of the number of lower units (communities and association variants or associations and subassociations in cases when variants were not distinguished). The study revealed that technogenic disturbance does not cause the development of absolutely new types of synanthropic vegetation. It consists of serial communities containing species from the natural vegetation. This allows to forecast the recovery process and build the TSS (Table 4.1). In the studied regions they are represented by stable combinations of ruderal communities, mostly belonging to associations of the classes *ARTEMISIETEA VULGARIS, PUCCINELLIO-HORDEETEA JUBATI* (in the South) and *PUCCINELLIO-HORDEETEA JUBATI* and *MATRICARIO-POETEA ARCTICAE* (in the North).

The β-diversity appeared to be higher in all TSSs at medium stages of succession. Later on, it may remain stable, increase or decrease depending on the strength of the secondary technological pressure on a certain subsystem. This should be taken into account when working out the projects of biological reconstruction of disturbed lands in West and South Yakutia.

**Table 4.1** Change of β-diversity of technogenic succession systems of West Yakutia

| Syntaxa | Age classes | | | | |
|---|---|---|---|---|---|
| | I | II | III | IV | V |
| D.c. *Chenopodium album* + *Artemisia jacutica.* *[Artemisietea]* | | ++ | | | |
| D.c. *Polygonum aviculare* + *Artemisia jacutica.* *[Artemisietea]* | | ++ | | | |
| D.c. *Hordeum jubatum* + *Artemisia jacutica.* *[Artemisietea]* | | ++ | | | |
| *Chamerio-Hordeetum jubati typicum v. Chenopodium album* | | ++ | +++ | ++ | + |
| D.c. *Rubus sachalinensis* + *Chamaenerion angust.* *[Epilobietea.]* | | + | ++ | ++ | + |
| *Puccinellio-Hordeetum typicum v. Chenopodium album* | | + | ++ | ++ | + |
| *Chamerio-Hordeetum jubati typicum v. Artemisia mongolica.* | | | + | ++ | ++ |
| *Chamerio-Hordeetum jubati typicum* | | | + | + | ++ |
| *Puccinellio-Hordeetum jubati typicum* | | | + | + | +++ |
| *Puccinellio-Hordeetum typicum v. Elymus kronokensis* | | | | + | ++ |
| D.c. *Salix viminalis* + *Chamaenerion angustifolium* *[Epilobietea]* | | | | + | ++ |
| D.c. *Betula fruticosa* + *Chamaenerion angustifolium* *[Epilobietea]* | | | | | + |
| β-diversity | | 6 | 6 | 8 | 9 |

+++ communities prevailing in area; ++ communities occupying no less than 30% of the area; + communities occupying an insignificant area. D.c. – derived communities.

## 4.3 Forest Fires

### L.P. Lytkina and A.P. Isaev

Fires are a natural phenomenon in boreal forests (Utkin 1965; Sannikov 1983; Furyaev 1996; etc.). That is why the normal ecological functioning of the forests is determined by fires. Their effect on ecosystem condition and dynamics has

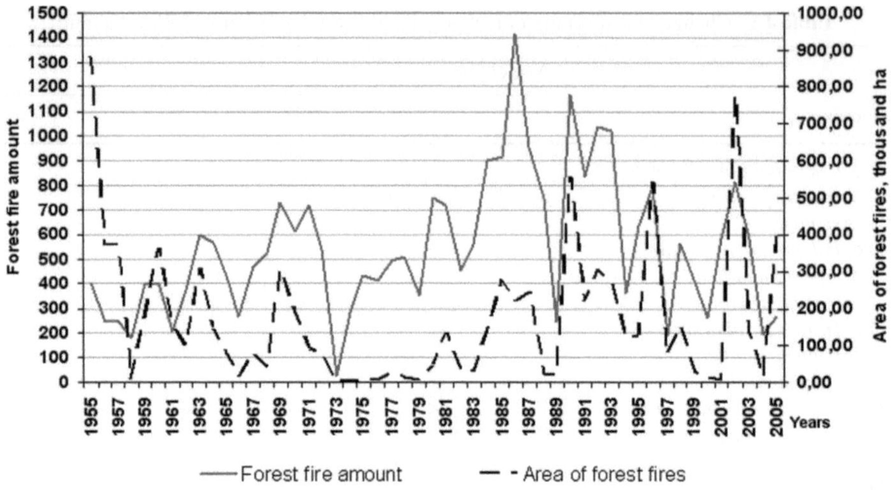


**Fig. 4.1** Statistics of forest fires in Yakutia

become stronger during the last decades because the proportion of anthropogenic fires increased. Over 8.2 thousand fires were recorded in Yakutia during the last 10 years, covering 1.3 million ha and resulting in 20 million $m^3$ of burned out timber (Fig. 4.1).

An analysis of the forest fire statistics reveals that the primary cause of forest fires in Yakutia is thunderstorms in July-August (over 50%). These are so called "dry thunderstorms", very characteristic for Yakutia. A thunderbolt induces ignition of the ground cover and dead wood. Lightning-caused fires usually affect the largest areas of forest tracts owing to their remoteness and difficulty to access the location. A little less then half of all fires are induced by a man, especially in densely populated and industrial regions. Their amount increases during the intensified recreation periods (agricultural work, spring and autumn hunting, outdoor leisure, harvesting berries and mushrooms, etc.).

The effects of fires on the components of forest ecosystems (tree stand, living ground layer, microclimate, soil) under conditions of permafrost are very widely covered in literature (Chugunova 1964; Scherbakov and Chugunova 1960, 1961; Utkin 1965; Tarabukina and Savvinov 1990; Timofeyev et al. 1994; Abaimov et al. 1998; Isaev and Mikhalyova 2000; Lytkina 2005; Lytkina and Protopopova 2006, etc.).

In the forests of Yakutia ground (surface) fires are most common. Quick spring fires destroy last year's herbal vegetation, and partly also the fresh tree waste and other dry plant remains, having a minor negative effect on the forest communities. Settled ground fires, however, are more dangerous because of the complete elimination of lower layers, i.e. herbage, shrubs, and young trees. Tree stands remain untouched for 10–90%. After the total or partial destruction of the litter and soil-vegetation cover the heat, moisture and permafrost conditions of soils change drastically. For example, in Central Yakutia the early post-fire succession stages are characterized by the following indices: compared to intact forest, the soil temperature at a depth of 30 cm rose up to 4.3–6.2° C, the soil moisture content increased by a factor 2.3, the depth of the seasonally thawed layer increased 0.3–0.8 m (Lytkina 2005). However, the most significant changes take place during the first 10 years after the fire (Chugunova 1964; Utkin 1965; Tarabukina and Savvinov 1990; Isaev 1993; Timofeyev et al. 1994; Isaev and Mihkalyova 2000; Lytkina 2005, etc.). All parameters of microclimatic and soil conditions (soil temperature, seasonally thawed layer depth, moisture content) directly depend on the degree of overgrowth of a fire-site or the development of the insulating capacity of the living ground layer. Fire induced changes in habitat conditions stabilize after 20–25 year of succession (Fig. 4.2).

The forest recovery process is determined by a number of factors. Foremost are the biological and ecological properties of *Larix spp., Pinus sylvestris* and *Betula pendula*. Being the pioneer tree species in Yakutia, they may participate in the formation of the initial recovery stage. This can be explained by their high seed production, good seed germination, and the highly adaptive potential of the species. Larch, for instance, is well adapted to growing under conditions of an arid climate and shallow occurrence of permafrost. *Larix cajanderi* is characterized by the remarkable feature of pouring out seeds in late summer in the year of ripening.

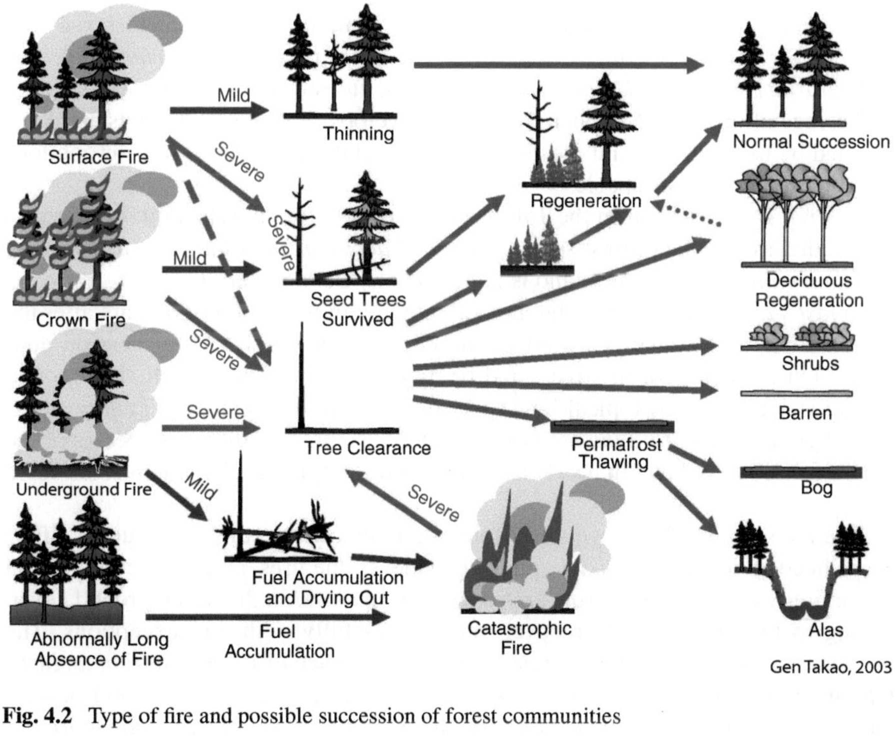


**Fig. 4.2** Type of fire and possible succession of forest communities

Being covered with needle litter and snow, the seeds have the advantage to fully use the spring moisture. So the young sprouts have time to strengthen before the summer drought is settled. However, this feature of the *Larix cajanderi* has a negative effect on the food availability for animals, resulting in rather low biodiversity of the larch taiga.

Secondly, the first years after a fire are characterized by the establishment of favourable conditions for forest vegetation recovery: completely destroyed litter, soil enriched with ashy elements, increased moisture of the upper soil horizons due to inflow from lower horizons, etc.

Most specialists recognize several stages of post-fire vegetation succession. The following stages are characteristic for larch forests (Timofeyev et al. 1994):

- the *initial herb stage.* Lasts 0–8 (10) years. There are two phases in this stage: an open herb community with participation of pioneer species and a closed herb vegetation (i.e. no new species enters the community); they are mainly forbs-grass communities. Right after a fire has happened, the fire-site becomes inhabited by the pioneer *Chamaenerion angustifolium* and the pioneer mosses *Ceratodon purpureus* and *Marchantia polymorpha.* Gradually, they give way to grasses (*Calamagrostis langsdorffii, Poa pratensis, Puccinellia tenuiflora*) with some forb species (*Lupinaster pentaphyllus, Crepis tectorum, Euphrasia jacutica,* etc.). The first shoots of tree species appear (*Larix, Betula*).
- the *shrub stage* lasts the next 10–15 years after a fire. Several years later the herb layer becomes thinner. *Betula fruticosa* and *B. divaricata* appear to serve as a shelter for young growth of tree species. The forest mosses *Polytrichum strictum, Pleurozium schreberi, Hylocomium splendens,* etc. substitute the pioneer species. The vegetation cover composition further develops. The humid habitats are dominated by hygrophytes and mesohygrophytes, while dryer ones carry xerophilous and mesoxerophilous species.
- the *birch stage* starts from the 15th year at the earliest and lasts until 40–45 years after the fire. Typical forest species appear (*Vaccinium vitis-idaea, V. uliginosum, Ledum palustre*). The tree stand is dominated by *Betula pendula.*
- the *last stage* includes birch-larch and larch young forests (40–60 years after the fire). Birch is a constituent part of the dense larch stands. As succession goes on birch is displaced, the tree stand being composed of pure larch. The ground layer is dominated by typical forest species (*Vaccinium uliginosum, Ledum palustre, Pyrola incarnata, Linnaea borealis,* etc.). The major edificator species of the dwarf shrub-herb layer in nearly all Yakutian forest types is *Vaccinium vitis-idaea.*

The recovery of *Pinus sylvestris* forests has its own peculiarities due to differences in edaphic conditions, the quick character of the fires, and the predomination of the fire-resistant dwarf shrub *Arctostaphylos uva-ursi* in the living ground cover. Owing to this, pine forests recover very successfully with rare exceptions. The first succession stages in pine forests feature dominance of *Chamaenerion angustifolium* and *Chelidonium majus,* as well as xerophilous forbs, grasses and pioneer mosses (*Ceratodon purpureus, Funaria hygrometrica, Marshantia polymorpha*).

*Arctostaphylos uva-ursi* also quickly recovers thanks to patches of this dwarf shrub remaining after the fire. Shrubs (*Rosa acicularis, Dushekia fruticosa, Salix spp.*, *Cotoneaster melanocarpa,* etc.) appear 3–5 years after the fire, and the pioneer mosses are replaced by *Polytrichum juniperinum* and *P. piliferum* 5–10 years after the fire.

The recovering ability of forests varies; it is being rather satisfactory in the lowland forests of South Yakutia, but weak or very poor in the North or in mountainous regions. Considering the poor recovery rates of the undergrowth and ground layer, especially in Central Yakutia, forest fires serve as a factor stimulating the recovery process (Utkin 1965; Pozdnyakov 1974; Timofeyev et al. 1994).

During the last years, the effectiveness of the fire fighting and prevention measures has been greatly reduced, as is apparent from the considerable areas affected by fire. Moreover, considering the low population density and the undeveloped transport network in the mountain taiga, the forest fires in the most part of Yakutia (especially in the mountain taiga region) are not extinguished. Fire control using aircraft is conducted only for an insignificant area of the republic, due to cost – benefit considerations and the impossibility to cover the whole territory of Yakutia by preventive measures.

## 4.4 Timber Felling

**A.P. Isaev**

The forest recovery process in cuttings, especially at initial stages, is prolonged and greatly differs from that of fire-sites owing to the specificity of the conditions after cutting. Fire-sites are characterized by a complete elimination of the vegetation and organic horizon of the soil, while in cuttings only tree stands are removed, since the effect of the machinery on soil and vegetation cover in winter (the main season of timber felling) is insignificant: the soil is frozen and the living ground layer is protected by snow.

It has been mentioned before, that larch, pine and birch often become the pioneer species in fire-sites. However, this is usually not true for succession series in cuttings due to the absence of favourable conditions for the growth of young shoots: a developed ground cover is a serious competitor for tree sprouts. That is why more or less favourable conditions appear in disturbed sites.

Modern timber cutting processes have changed the course of forest recovery. The use of machinery raised the urgent problem of safety of young growth of primary renewal. Preservation of young trees is important both for rapid formation of the generation of a new tree stand in a cutting, and for its positive effect on environmental microclimatic conditions, since it reduces the processes of thermokarst, water-logging and soil sodding. Proceeding from the fact, that 10–20 year-old larch growing in open ecotopes is able to bear seeds in productive years (Karpel and Medvedeva 1977) the role of young trees for dissemination in cuttings is obvious.

At present, tree felling in the middle taiga of Yakutia implies the cutting method called clear-patch cutting (Pobedinsky 1965), since in forest tracts trees occur in patches of specific age groups (Pozdnyakov 1975) that are as large as 2–5 ha. Near clear fellings are also common, where trees with trunks of 12–16 cm in diameter, and imperfect and other subquality trees remain untouched. To preserve the undergrowth the narrow-swath method of cutting is applied, mainly by enterprises that fell timber at an industrial scale. The width of swaths varies from 15 to 20 m in Central Yakutia and to 30–40 m in Southwest Yakutia. This method allows the preservation of 65–70% of the young trees situated in groups, when the groups remain untouched between strip roads. Unsystematic cuttings reduce the preservation of young trees leaving untouched only 20–40% or even less.

Preservation of the undergrowth also depends on the season of cutting. It is much higher if felling is conducted in winter time. Also, cutters often mainly fell the tall and elder trees and as a consequence the remaining stand consists of smaller and younger trees.

Cutting that involves the use of harvesters (both wheeled and caterpillar-mounted) that collect the stems has a heavily damaging impact. The main reason of their negative effect is their short boom that forces them to closely approach each cut tree. This results in destruction of up to 80% of the young growth. Young trees as high as 0.4–0.5 m are eliminated almost completely. Besides, use of the machinery strongly litters up the cuttings, especially in winter time, as wood remains are only pressed into snow without being crushed by the wheels or caterpillars. Cleaning up of such cuttings requires high costs. Thus, the stem-collecting machinery strongly impedes further forest renewal in the cuttings, especially in summer time.

Drastic changes in the growth conditions of forests yield suppression, faultiness (imperfect stems) and often partial death of young growth. This problem for the middle taiga subzone of Yakutia is covered in the publications of L.K. Pozdnyakov (1963, 1975), Gavrilova (1967, 1969), Savvinov (1969, 1971, 1972, 1973, 1976), Scherbakov et al. (1977), Isaev (1992, 1993, 2000).

Timber felling induces considerable change in the radiation balance in cuttings. The amount of light quantity in a cutting of forbs-*Vaccinium vitis-idaea* larch forest is increased by a factor two and in *Vaccinium vitis-idaea* larch forest by three as compared to intact forests, where most of the light (54.8 and 69.5% respectively) is intercepted by the tree canopy. Each unit of crown closure or canopy density corresponds to a 8–9% decrease in light intensity (Scherbakov et al. 1977; Isaev 1993). The light intensity at ground level is also impeded by the undergrowth and dwarf shrub-herbaceous cover, reaching only 30% of its maximal (above the forest canopy) value (Isaev 1993).

Temperature conditions in cuttings are considerably harder for plant growth, then in intact forests, especially as regards temperature extremes. In cuttings, in summer, the surface air is warmer at day-time and colder at night, especially at daybreak. Therefore, night frosts on the surface of continuous cuttings are stronger and occur more often, then in intact forest, affecting the development of certain plants. Increased heating induces changes in soil temperature conditions. Generally, it is "warmer" in a cutting then in a forest. At that, soil temperature conditions greatly depend on the degree to which the living ground layer and litter is intact. Their removal results in a rise of the average daily soil temperature. The best soil heating conditions lead to a thickening of the active horizon and an increase of the seasonally thawing depth of frozen grounds. Subsidence of the thermokarst in cuttings of the middle taiga subzone occurs very rarely, but the conditions of the frozen grounds change considerably after elimination of the tree stand. According to Fyodorov (1985) the depth of the seasonally thawed layer under *Vaccinium vitis-idaea* larch forests increases after cutting from 1.2–1.5 to 1.8–2.2 m. When the cutting is overgrown, the depth of the seasonally thawed layer decreases. It is interesting that the seasonally thawed depth of frozen ground under young growth can be thinner then that under a mature tree stand. The reason is the higher density of the canopy of young trees that impedes the heating of the soil surface compared to mature tree stands with a medium or low canopy density.

Water conditions are also subject to change in cuttings. A disturbed ground layer (live or dead) results in an increase in moisture content due to water entry from lower saturated horizons at the increased depth of the seasonally thawed layer and reduced transpiration rates of the disturbed ground layer. Later on, like in fire-sites (Tarabukina and Savvinov 1990), moisture content reduces greatly due to increased evaporation (better soil heating conditions and decreased water entry from lower horizons).

The formation and transformation of vegetation cover in cuttings follow certain natural laws, which are reflected in the forest recovery process. After removal of the tree stand, shrubs, dwarf shrubs and herbs start growing thickly and hamper seed germination, growth and development of the tree species. The development of the vegetation cover in larch forests under conditions of the middle taiga subzone of Yakutia was studied in general by Chugunov (1955, 1961), Utkin (1960), Scherbakov and Chugunova (1961); Vipper (1964, 1973). Detailed study was conducted by Mikhalyova (1971, 1974, 1975, 1977), Mikhalyova and Chugunova (1971), and Isaev (1991, 1993).

Figure 4.3 illustrates the overgrowing of a cutting in *Vaccinium vitis-idaea* larch forests, the most typical forest type of the middle taiga subzone. Cuttings with a removed tree stand and other severely damaged layers of the forest community represent relatively open phytocoenoses allowing free access of new species. *Chamaenerion angustifolium* is the most active species growing in cuttings of this forest type. In modern cuttings with a disturbed soil cover of 30–40% (or even more) the projective cover of this species can be considerable. This is similar to the fire-affected forest recovery process, when the mineralized substrate is covered with abundant *Chamaenerion angustifolium*. However, in spite of high vigour of certain species the vegetation cover is predominated by typical forest plants, participants of the ground layer of cut forest: *Vaccinium vitis-idaea, Lathyrus humilis, Arctous erythrocarpa, Vaccinium uliginosum,* etc. Some forest species disappear from coenoses, such as *Maianthemum bifolium* and *Pyrola incarnata*. Besides the obvious simplification of the stratification (eliminated tree stand, strongly damaged subordinate layers), the structural modifications are also apparent in a complication of the horizontal structure. A system of mosaic microgroupings is formed, which is

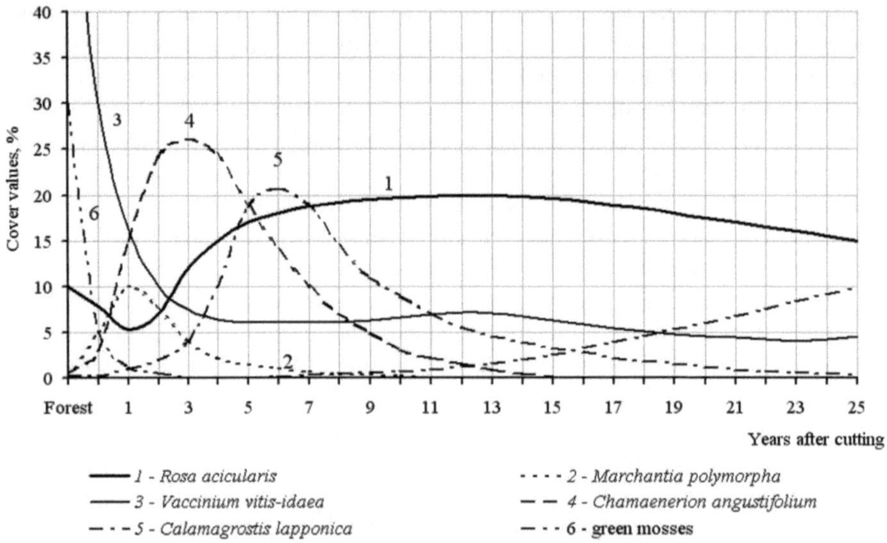


**Fig. 4.3** Cutting overgrowth in *Larix-Vaccinium vitis-idaea* forests

a general characteristic for all early stages of the recovery processes (Borman and Likens 1979). The stronger the cutting-induced damage, the more motley the vegetation cover. Later on, the mosaic pattern of the vegetation cover in cuttings is even more expressed. Though *Vaccinium vitis-idaea* is still abundant, its projective cover steadily reduces (up to 10%). The one dominating species is *Chamaenerion angustifolium*, but grasses also start to spread (*Festuca spp., Agrostis trinii, Calamagrostis langsdorffii, C. lapponica, Bromopsis*, *Poa sibirica, P. pratensis*). Sedges also participate in gradually overgrowing the cutting (*Carex vancheurckii, C. melanocarpa, C. pallida,* etc.), as well as *Luzula rufescens*. The herbal species composition becomes more diverse as it is supplemented by non-forest species (*Galium verum, Lychnis sibirica, Crepis tectorum,* etc.), and an irregular herbal layer is formed. Grasses and forbs now take the lead in the herbal layer, while *Chamaenerion angustifolium,* as a typical explerent, gradually loses its dominant position. As the density of the vegetation increases, the plant community shifts from open to closed, the competitive relations between plants strengthen, and the possibility of invasion of new species becomes limited. Starting from 3 to 5 years, the number of species invading the plant communities sharply reduces and becomes stable by the 8th–10th year. At the same time, the more or less settled coenosis is formed with a relatively stable composition and structure. The cuttings in *Vaccinium vitis-idaea* larch forests feature a relatively weak shrub growth. The most vigorous species are *Betula fruticosa* and *Salix spp.,* while birch starts to establish during the first years after cutting.

With further development of the birch-larch tree stand, the species diversity, projective cover and abundance of some species decrease. The projective cover of the dwarf shrub-herb layer in a 30-year old cutting does not exceed 30–40%. Such a cutting often is a combination of dense birch-larch young growth lacking a

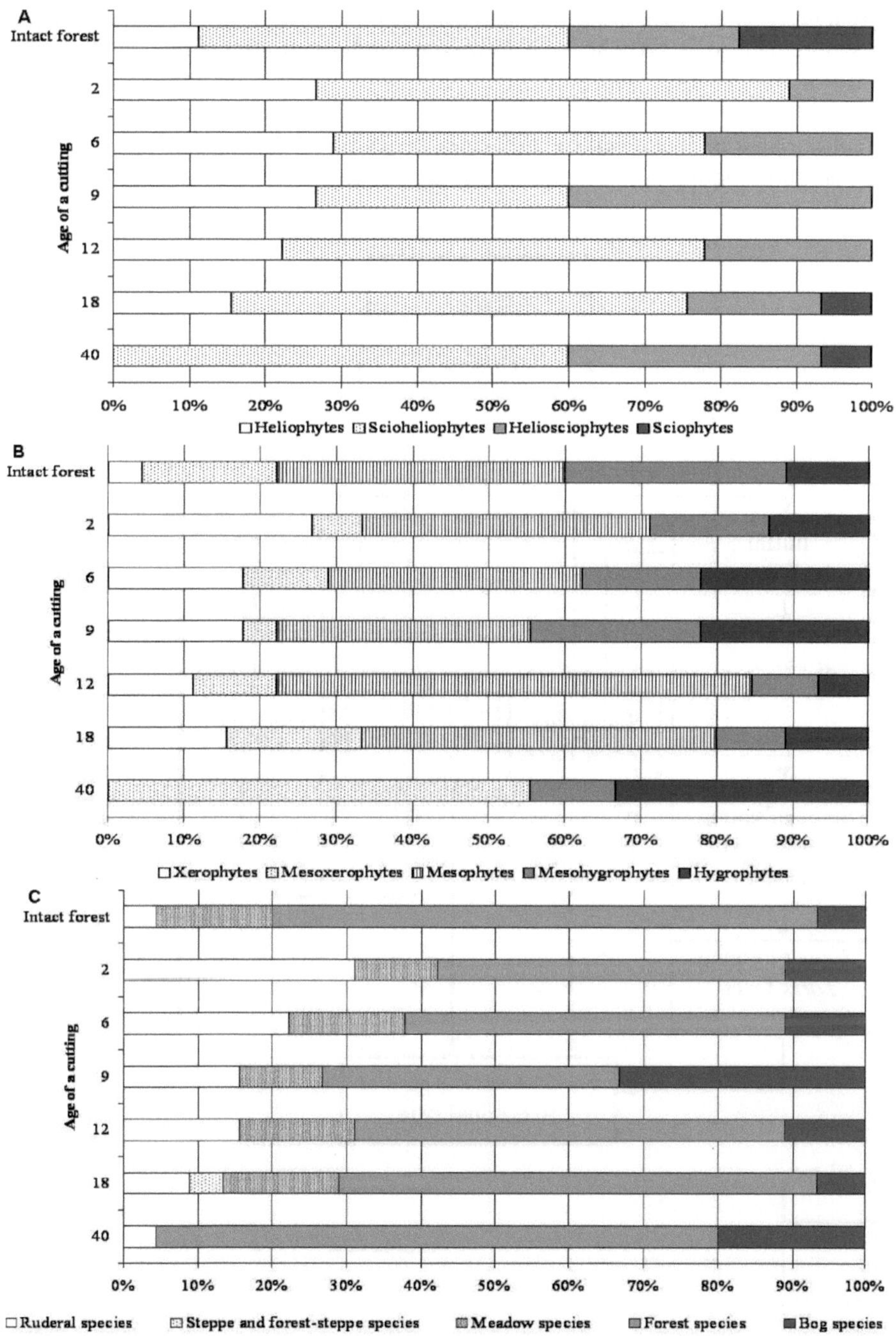


**Fig. 4.4** Dynamic series of ecological-coenotic spectra of a dwarf shrub-herb layer of *Larix cajanderi* forests as regards light (**a**) and moisture (**b**) conditions, as well as coenotic structure (**c**)

herbal layer on 65–70% of its area, and patches of a herbal vegetation. Participation of subordinate trees and shrubs is also common though not significant (less then 5–10%).

The change in species composition determined by peculiar ecological conditions in cuttings can be illustrated by a dynamic series of ecological-coenotic spectra of a dwarf shrub-herbal layer (Fig. 4.4). During the first years after cutting the species

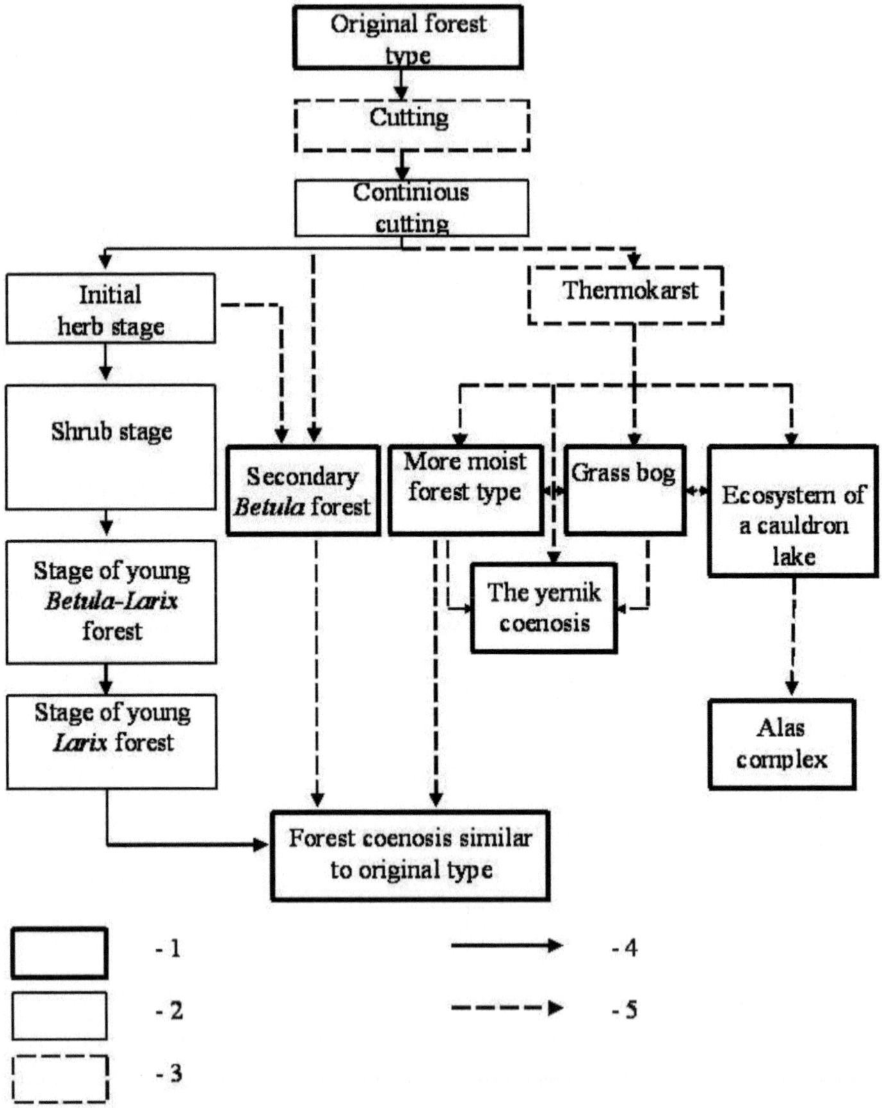


**Fig. 4.5** Scheme of vegetation dynamics in cuttings 1 – Original forest type and derivative phytocoemoses; 2 – succession stages; 3 – main factors influencing successional course; 4 – most typical directions of succession; 5 – possible directions of succession

of open ecotopes acquire strong positions. As regards the hygromorphs in young cuttings, xerophytes increase their positions in communities, hygrophilous species retain their portion unchanged in the course of the succession, while the species of intermediate moisture requirement reduce in number. With the development of the forest vegetation, the compositions of coenomorphs, heliomorphs, and hygromorphs gradually return to their original state.

Thus, forest recovery in cuttings is rather successful. Direction and course of this process may abruptly change, influenced by possible thermokarst manifestations. In this case the secondary succession represents not an ordinary recovery process, but a peculiar "primary" one, called quasi-primary by Rabotnov (1983). This may result in the formation of both moist types of larch forests, *yernik* formations, grass bogs, or cauldron lakes (Utkin 1960; Tyrtikov 1969; Pozdnyakov 1983, 1986). A scheme of the vegetation dynamics in cuttings is given in Fig. 4.5.

## 4.5 Natural Fodder Lands of Central Yakutia

### *4.5.1 Hayfields and Pastures*

**L.D. Gavrilyeva R.R. Sofronov, A.P. Arzhakova, N.V. Barashkova, and I.A. Ivanov**

A thousand years of experience of cattle- and herd horse-breeding in Yakutia indicates the rather long-term process of developing the natural fodder lands for pastures and hayfields in the enormous territory of Northeast Eurasia. During that period the ancestors of the Yakuts first settled in the Middle Lena River region and the Lena-Amga Interfluve (Basharin 1956).

The extreme climatic conditions (cold winters lasting nine months, short droughty summers, cold soils, and perennially frozen grounds) forced the nomadic migrants to change their lifestyle, switching to settled cattle-breeding. During their centuries-old history the Yakuts have worked out the system of scattered stock keeping in small alases and in river valleys. The size of the livestock pool was determined by the capability to maintain them in droughty years. Herds of horses and cattle grazed on the rather vast grasslands of the three valleys of the Middle Lena River (Tuymaada, Erkeeni, Enseli), as well as in the valleys of the Tatta and Amga Rivers and in taiga alases.

Exploring Siberia in the seventeenth century, the Russians found the established settled stock-breeding economy of the Yakuts in the Middle Lena region and the Lena-Amga Interfluve (Central Yakutia). With increased livestock of cattle and horses, practically all the suitable hayfields and pastures had been set aside by the first quarter of the nineteenth century. An economic description of grassland management as an agricultural sector of Yakutia was first made by V.N. Poryadin in 1926, who participated in the expedition of the Academy of Sciences of the USSR aiming to study the agriculture of Yakutia. At that time the condition of the grasslands did not cause anxiety, since no sign of their degeneration due to anthropogenic pressure was obvious.

At present, significant areas of hayfields and pastures of Central Yakutia are in imperfect condition as a result of the absence of land improving arrangements, water-logging and remoteness.

The best natural fodder lands of Central Yakutia are represented by the following types of grasslands of various topologies: floodplain, alas and lakeside, upland, and small river valley meadows (Kononov et al. 1979).

*Floodplain meadows* cover the largest areas in the Lena River floodplain, as well as the rivers of the second and third order: the Viluy, Aldan, Amga, and Tatta. The average productivity is 10–18 centners per ha (c/ha) (= 1000–1800 kg per ha) of hay, comprising 37% of its gross production.

*Shortly flooded* (*stepped*) *meadows* are confined to high levels of a central floodplain. In some years they are flooded when ice jams are formed in a river causing the water level to rise. The soils are stepped-meadow, stratified, carbonate loamy and loamy sandy. The floristic composition of stepped meadows includes up to 40–50 species. They are mainly grasses *Agrostis trinii, Hordeum brevisubulatum, Elytrigia repens, Bromopsis korotkiji, Koeleria gracilis;* forbs *Artemisia jacutica, Galium verum, Saussurea amara,* and sedges (*Carex duriuscula*) with an occurrence of legumes *Lupinaster pentaphyllus, Vicia cracca, Medicago falcata,* etc. The productivity of shortly flooded meadows is low, 3–4 c/ha of hay.

*Medium flooded* (*true*) *meadows* occupy the largest part of a central floodplain, as well as parts of the lower and upper floodplains, exposed to flooding for 3–4 weeks. The soils are frozen alluvial soddy soils, with a high content of humus (from 10 to 11% in the upper horizons to 0.4–0.7% in the lower horizons at a depth of 20–30 cm). The vegetation is predominated by mesoxerophytes. The average productivity is 10–12 c/ha of hay. Various types of grass communities (*Hordeum brevisubulatum, Elytrigia repens, Agrostis gigantea*) are of special economic significance, giving a productivity up to 10–15 c/ha of hay.

*Long flooded* (*swamped*) *meadows* develop under conditions of prolonged flooding and are characteristic for floodplain depressions. The soils are frozen alluvial peat-gley soils. The species diversity is low with hygromesophytes as edificators: *Glyceria triflora, Equisetum palustre, Calamagrostis langsdorffii, Scolochloa festucacea, Carex disticha, C. acuta.* The communities with *Calamagrostis* are characterized by highest productivity, thus representing valuable hayfields. They typically cover large areas though often they are strongly tussocky.

By gross hay production the floodplain meadows rank second after the alas meadows.

*Small river* (*taiga river*) *valley meadows* represent a special variety of floodplain meadows and are situated in the valleys of small taiga rivers. By gross fodder production they take the third place (26%) and are mainly used as summer pastures for cattle and winter pastures for horses. The soils they grow on vary widely: frozen meadow-bog, muddy-bog, peaty, peaty-bog, and frozen soddy-meadow types. The vegetation is dominated by *Calamagrostis langsdorffii, Carex schmidtii* and *C. juncella,* the latter two species forming tussocks of various height (10–60 cm) in cold and overwetted habitats. The dryer areas are covered with bluegrass communities with a prevalence of *Poa pratensis,* occurrence of *Alopecurus arundinaceus*, *Bromopsis inermis*, *Vicia cracca,* and forbs.

*Upland meadows* occupy the drained and flood-free sectors of the valleys and dry ravines. They also may represent secondary communities on stubbed out lands and in post-residential areas. These meadows are used as summer and early spring pastures providing 1% of the fodder. The vegetation consists of forbs-*Elytrigia repens* communities with a prevalence of *Elytrigia repens.* Abundant forbs are *Sanguisorba officinalis, Thalictrum simplex* and *Galium verum.* Sometimes sedge-forbs or sedge-*Elytrigia repens*-forbs meadows occur with participation of *Carex duriuscula.*

*Alas and near-lake meadows* are most typical for the Lena-Amga Interfluve and the Viluy River basin. They provide with up to 36% of the Yakutian hay making with a productivity of 4–13 c/ha depending on certain weather condition during the year. These meadows are situated in depressions of various origins, mostly of thermokarst.

The vegetation basically consists of the grasses *Hordeum brevisubulatum, Elytrigia repens* and *Puccinellia tenuiflora* and forbs *Artemisia mongolica, A. jacutica* and *Salicornia perennans.* Various *Cyperaceae* grow all along the moisture gradient from xerophytic (*Carex duriuscula*) to aquatic (*Scirpus lacustris*) habitats. Alas grasslands are characterized by a concentric distribution of plant communities around a lake. Basically, three belts are distinguished: the upper, middle and lower. The upper belt represents the habitat for xerophytes: *Stipa capillata, Elytrigia repens, Agrostis trinii, Koeleria gracilis.* The middle belt is covered with communities of *Hordeum brevisubulatum, Sanguisorba officinalis, Thalictrum simplex, Poa pratensis* and *Puccinellia tenuiflora.* The latter species serves as an indicator for this belt. *Alopecurus arundinaceus, Calamagrostis langsdorffii, Beckmannia syzigachne, Glyceria triflora, Scirpus lacustris* grow in the lower belt of alases. Depending on weather conditions during the growing season, the alas grasslands feature pronounced perennial and seasonal dynamics, so that the belt communities may change in width and spatial position each year, and strongly vary both in productivity and nutritional value of the forage.

Alases of river origin occur more rarely, and mainly occur in the Viluy River basin. They are confined to taiga river valleys and represent nearly flat, hardly noticeable depressions of a roundish or elongated form. Sometimes they appear as a result of drainage and drying out of taiga lakes.

#### 4.5.1.1 Hayfields

From the total area of hayfields of Central Yakutia flooded hayfields comprise 17%, of them 13% being clean (i.e. featuring no shrubs, tussocks, litter, etc.); 59% go for upland meadows, of them 5% being improved and 51% clean. The recent 30–40 years are characterized by a tendency of decreasing productivity of the natural fodder lands. Following the data of the State Committee of Statistics of the Republic of Sakha (Yakutia), the hay harvesting rates are characterized by a 30–40% reduction, and hayfield productivity has decreased 2.2–2.5 times (Desyatkin 1996). By January 1, 2007, the livestock numbered 259.2 thousand heads, including 103.2 thousand milk cows, and 134.5 thousand horses. Natural hayfields and pastures occupy 718 thousand and 794 thousand ha respectively, providing only 40–50% of the required forage protein. Only 62% of the hayfields were used in 2007, since about 270 thousand ha were flooded, deserted, or were hard to access due to their remoteness. Meadow swamping rates increase every year, comprising now 24% of the total hayfield area. Due to shrub invasion and presence of tussocks, swamped meadows are used only for 50–60%.

Figure 4.6 shows that since the 1960s (the period of large collective farms establishment) the hayfield productivity in Yakutia has declined, even though their area has increased.

The traditional stock-breeding system that has established in Yakutia during the recent decades does not imply a clear division of grasslands into hayfields and pastures, resulting in strong degradation of the fodder lands near the settlements in Central Yakutia (especially in the Lena-Amga Interfluve). Another major reason of grassland degradation is soil compression as a result of the technogenic activity of various vehicles. This first refers to agricultural machinery, tractors and their trailers. The modern hay harvesting methods imply multiple passages of machinery. As a result, their wheels and caterpillars damage 10–15% of the total hayfields area (Denisov and Prokopyev 1979).

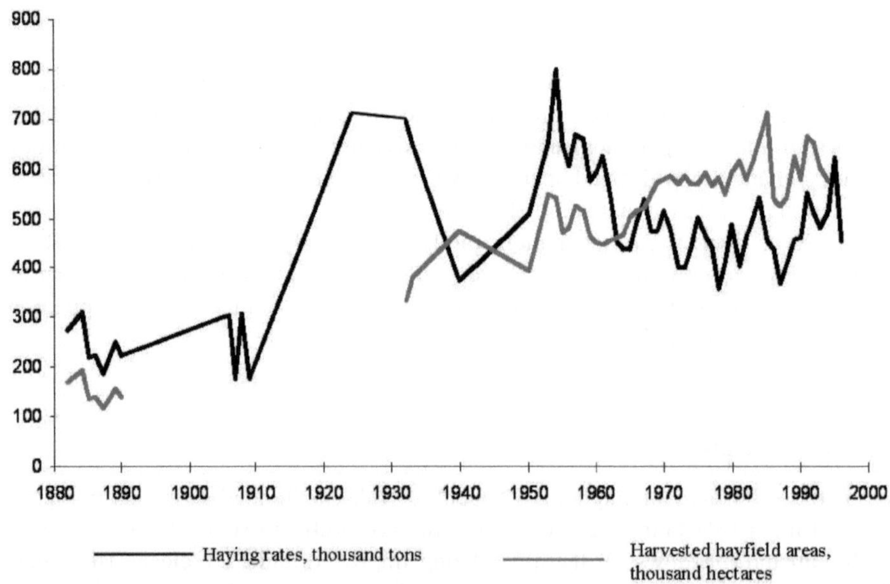


**Fig. 4.6** Haying rates and hayfield areas in Yakutia for the period of 1882–1996 (based on the analysis of literature data)

A special classification of alas vegetation (fodder types) was elaborated to be used for Yakutian fodder land characterization (Kononov et al. 1979; Gogoleva et al. 1987; Gavrilyeva 1998; Gerasimova and Gavrilyeva 2001; Savvinov et al. 2005). The following is a list of vegetation fodder types best suitable for haying.

Forbs-*Bromopsis* type. The communities of this type grow mainly in forest edges, on the slopes of *baidjarakh*s and *bulgunnyakh*s. At that, only shaded eastern-, western-, and northern-facing slopes are used as hayfields. The grass stand is usually dense (projective cover up to 90%), high (average height 70–80 cm) and is characterized by a rich species composition with predomination of *Bromopsis korotkiji, Pulsatilla flavescens, Veronica incana,* etc. The average productivity is 14–15 c/ha.

*Elytrigia* type. These communities are common in upper and middle alas belts. The projective cover is 60%, the height of the grass stand is 40–60 cm. The productivity varies significantly depending on weather conditions and amounts to 10.4 c/ha on average. In spring and autumn such communities serve as pastures for cattle. The vegetation is dominated by *Elytrigia repens*. Other characteristic grass species are *Hordeum brevisubulatum, Poa stepposa, P. pratensis* and *Agrostis trinii.* Also *Carex duriuscula* and forbs (predomination of *Artemisia commutata* and *Saussurea amara*) grow there.

*Puccinellia* type. Mesic habitats support the domination of *Puccinellia tenuiflora* (80–90%) with a projective cover up to 80% and an average height up to 70 cm. The average productivity is 22.0 c/ha, and also depends on yearly weather conditions and soil salinity. Other community components are scarcer: *Hordeum brevisubulatum, Alopecurus arundinaceus, Knorringia sibirica, Saussurea amara, Glaux maritima.* Litter is absent in this type.

The *Hordeum* type is not common in alases and occupy the middle mesic belts. The grass stand is dense (95%), with an average height of 90 cm, and a productivity of 20.7 c/ha. The most representative species of this type are the grasses *Hordeum brevisubulatum, Poa pratensis, Elytrigia repens,* etc.; the forbs *Sanguisorba officinalis, Thalictrum simplex, Geranium pratense, Gentiana macrophylla, Silene repens, Galium verum,* etc.; as well as legumes. The meadows represent high quality hayfields.

The *Alopecurus type* is widely distributed in the lower vegetation belts. It is characterized by a dense (95%) and high (70 cm) grass stand with a productivity of 25.3 c/ha. The dominant species are *Alopecurus arundinaceus, Poa pratensis, P. palustris, Agrostis stolonifera* and *Carex lithophila.*

The *Carex-Calamagrostis* type is confined to overwetted lake-sides (lower belts of alases). The communities of this type feature tussocks formed by *Carex juncella.* Projective cover is 80%, height 90 cm, productivity 19.8 c/ha. Dominant species are *Carex juncella* and *Calamagrostis neglecta*; co-dominants are *Alopecurus arundinaceus, Agrostis stolonifera, Carex lithophila, C. atherodes* and *C. disticha.* These communities contain litter.

As mentioned before, there are no alases nowadays that would be used strictly for hay harvesting. In early spring and in autumn they serve as pastures (Gavrilyeva 1998).

#### 4.5.1.2 Pastures

Traditionally, grazing lands are situated around residential areas and summer cattle farms. Like hayfields, the main pastures of Central Yakutia represent grassland vegetation of alases (55% of their total area), floodplains (up to 23%), taiga river valleys (up to 17%), and uplands (up to 4%), including forest communities. The topological location of the grasslands determines the seasonality of grazing. Thus, spring pastures are situated on floodplain ridges and upland meadows, in the upper belts of alases, and on steep slopes with steppe vegetation. Summer pastures (grazing from early June onwards) represent inter-ridge depressions of above-floodplain terraces, alas belts of variable moisture conditions with *Hordeum brevisubulatum* and *Alopecurus arundinaceus,* as well as sedge and swamped vegetation.

The Soviet period's enlargement and consolidation of collective farms, established near large settlements, has led to a reduction in the total pasture area due to desertion of remote fodder lands. In contrast, the nearby grazing lands have become overstocked by an even smaller amount of livestock (Sofronov and Karpov 1999; Nikolaeva and Sofronov 2001). The area of trampled, salinized, and low-productive pastures totals now 230 thousand ha in Central Yakutia. Overgrazing results in exhaustion of the vegetation cover and compression of the upper soil layer, yielding a complex of ecological transformations. Increased soil solidity raises capillary pressure. This leads to rising salt concentrations in saline soils, the appearance of trampling-induced hillocks on wet soils, and the transformation of dry habitat soils of light mechanical structure into dust (Kononov et al. 1979; Mironova and Poiseeva 1996). All these signs of pasture degradation are especially pronounced in alases, since most settlements and cattle-breeding husbandries are situated in these peculiar landforms (Denisov et al. 1983).

The following are the basic fodder types of pasture vegetation (Kononov et al. 1979; Gogoleva et al. 1987; Gavrilyeva 1998; Gerasimova and Gavrilyeva 2001; Savvinov et al. 2005).

The *Stipa-Koeleria* type is characteristic for valley and *baidjarakh* slopes. Its projective cover is 80%, its average height 70 cm and its average productivity 8.8 c/ha. Major species are *Stipa krylovii, S. capillata, Festuca lenensis, Poa stepposa*, *Carex duriuscula,* etc. This type is used as spring pasture.

The *Koeleria-Carex duriuscula* type is situated in the upper xeric belts of alases and on above-floodplain terraces, indicating moderate pasture pressure. About 50% of the species composition goes for sedges (*Carex duriuscula*). The average projective cover is 30%, the average height 10 cm, and its productivity 8.5 c/ha.

With increased grazing load the previous type transforms into the *Carex duriuscula* type, characterized by a lower productivity (4.4 c/ha), and average height (up to 5 cm). The projective cover remains unchanged making up 30% on average. Grasses disappear or decrease in abundance in the community, 30–50% of the grass stand being represented by *Carex duriuscula.* Ruderal forb species (*Artemisia jacutica, Lepidium apetalum, Chenopodium album,* etc.) on the contrary, strengthen their position, comprising 35–60% of the grass stand.

The *Puccinellia* type differs from its hayfield variant both by a significant biomass reduction (up to 6.4 c/ha) and a change in floristic composition. *Puccinellia tenuiflora* loses its dominating position, giving way to *Elytrigia repens, Poa pratensis* and *Agrostis stolonifera.* Grasses make up 20–80% of the grass stand. The proportion of forbs (*Taraxacum ceratophorum, Plantago media*, sometimes *Descurainia sophia, Lepidium densiflorum*) is also increased.

Increased grazing induces replacement of the *Puccinellia* type by the *Potentilla anserina* type with a projective cover up to 60%, an average height of 5 cm and an average productivity of 8.3 c/ha. Grasses are suppressed and give way to creeping and rosette forbs (*Potentilla anserina, Plantago media, Knorringia sibirica,* etc.).

The *Alopecurus* type also significantly reduces the proportion of grasses under grazing conditions in favour of forbs (*Potentilla anserina, Armoracia sisymbrioides, Persicaria amphibia*). Species number is lower, the projective cover makes up 60%, the average height is 40 cm, and the productivity 11.8 c/ha. In wetter habitats with increased grazing pressure the *Alopecurus* type transforms into the *Eleocharis-Alopecurus* and *Eleocharis-Potentilla* types, representing two different stages of pasture degradation.

The *Eleocharis-Alopecurus* type features dominance of *Eleocharis palustris, Sparganium emersum, Carduus crispus, Armoracia sisymbrioides* or *Knorringia sibirica;* and the grasses *Alopecurus arundinaceus* and *Beckmannia syzigachne* occupy 10–60% of the grass stand. The average projective cover is 70%, and productivity 13.8 c/ha. Water-logging and trampling-induced hillocks in the lower alas belts are also characteristic for this fodder type.

The *Eleocharis-Potentilla* type represents communities with a dominance of *Eleocharis palustris* and *Potentilla anserina.* Grasses (*Alopecurus arundinaceus, Puccinellia tenuiflora*) almost disappear comprising only 5% of the grass stand. The latter is dense (up to 95%), its average height is 30 cm, and its productivity up to 27.6 c/ha.

The grazing variant of the *Carex-Calamagrostis* type has a high density of tussocks and predominance of *Carex juncella* and *Calamagrostis langsdorffii.* Compared to hayfields, the average height decreases to 40 cm and the productivity to 14.1 c/ha or lower. From year to year, much litter is accumulated between the tussocks (10–15%).

Applying the ecological principles of the rational use of natural fodder lands allows both to stabilize the negative processes and increase the productivity of hayfields and grazing lands.

### 4.5.2 Grazing Effect on Forest Communities

**I.F. Shurduk and A.P. Efimova**

The valley forests also suffer from trampling and grazing. This results in a change in the relative size of the different ecological-coenotical species groups; in a decrease in the abundance of dominants and subdominants; in a reduced vitality of plants, in a short-time increase of the biodiversity, particularly of synanthropic species, followed by its eventual decline. Thus, light and medium disturbance of the willow and birch communities induced by grazing are characterized by the appearance of ruderal and pasture species. Lightly disturbed vegetation communities feature the appearance and strong growth of vegetatively propagating plants with creeping sprouts and stolons (*Amoria repens, Potentilla anserina*) or short rhizomes (*Poa pratensis, Festuca*). These species are accompanied with weeds and halophytic species such as *Armoracia sisymbrioides*, *Artemisia jacutica*, *Chenopodium album*, *Descurainia sophia*, *Erysimum cheiranthoides*, *Fallopia convolvulus*, *Glaux maritima*, *Saussurea amara* and *Sonchus arvensis*. At present, among the woods of *Salix bebbiana* growing in the Middle Lena valley, the anthropogenic variant of the association is distinguished as *Salicetum bebbianae pyroloso-fruticosum* var. *amoriosa repentis*, characterized by the dominance of *Amoria repens* in the herb layer. Grasses (*Poa pratensis, Festuca*) prevail under medium grazing pressure. The last stages of pasture degradation are characterized by trampling-tolerant species, such as *Polygonum aviculare* and *Plantago major.* Strongly disturbed communities are characterized by suppressed herbs of low vitality, while the shrub layer (undergrowth) becomes completely degraded as is apparent from the separate dead standing plants or thin shrubbery. The upper layer is characterized by a thin canopy, partly with dead leaves, a degraded vitality, presence of dead trees and many broken stem parts. In some willow communities growing in inter-ridge depressions a strong deformation of the ground surface is observed, often with trampling-induced hillocks.

In the valley coniferous forests with *Picea obovata* and *Larix cajanderi* anthropogenic post-pasture disturbance is visible in a change in the percentage of forest-meadow-ruderal species. The abundance of typical forest dwarf shrubs (*Vaccinium vitis-idaea, Arctostaphylos uva-ursi,* etc.) and other forest species (*Arctous alpina ssp. erythrocarpa, Linnaea borealis, Pyrola spp., Orthilia, Moehringia lateriflora, Lathyrus humilis*) declines, and disappears completely with a further increase of grazing pressure. The role of pasture species and weeds (similar as for the deciduous forests of the valleys) becomes stronger. The shrub layer becomes degraded and of low vitability. First the hygromesophilous, then the mesophilous species disappear (*Spiraea salicifolia, Salix pyrolifolia*), later followed by *Ribes glabellum*. The xeromesophilous and mesoxerophilous shrubs (*Rosa acicularis, Spiraea media, Salix bebbiana*) are more tolerant. Later lichens and mosses are eliminated. In the second above-floodplain terrace of the Middle Lena, characterized by uncontrolled recreation activity and overgrazing, the anthropogenic variant of *Pinus sylvestris* forest association was classified as *Pinetum mixoherboso-arctostaphylosum substepposum* var. *festucosa jacuticae*. In the dwarf shrub-herb layer *Arctostaphylos uva-ursi* is completely eliminated, yielding to grazing-tolerant dense-tussock grasses (*Festuca jacutica, Poa spp.* etc.) and sedges.

Haying has different effects on the valley forest communities then grazing. Annual mowing of the adjacent meadow grass-stands also destroys the young, vegetative and reproductive specimens of willows and birches. As a result, the potential advance of willow and birch communities into meadow is limited.

Long-term anthropogenic pressure leads to transformations in the natural succession series of forest and shrubbery vegetation of the Middle Lena valley. In most parts of the anthropogenically affected landscapes of the valley the succession series turns to predominance of meadow-swamp coenoses, as well as to saline bare grounds and desertification. Complete elimination of forests on the above-floodplain terraces in the vicinity of Yakutsk (Tuymaada valley) has resulted in a disruption of the succession process due to lack of seed sources and changes in soil conditions. The succession series has stopped at the willow stage of *Salix bebbiana* and *S. pyrolifolia,* and partly at the birch stage. The forested areas in the most disturbed 1st and 2nd above-floodplain terraces of the Tuymaada valley are reduced to 10–15% of their original area (Timofeyev and Shurduk 1996; Timofeyev et al. 1992). The development of forest vegetation in the floodplain of the Nam ulus area (the Middle Lena River valley) has stopped at the birch stage as the natural dynamics are disturbed by overgrazing. As mentioned before, light and medium anthropogenic pressure results in the appearance of peculiar anthropogenic variants of forest associations. Further increase of overgrazing is assumed to lead to convergence of all forest types to one anthropogenic association.

Deforestation, which in the Middle Lena valley has taken place since the eighteenth century, has lead to soil deterioration and stirring up of cryogenic processes in soils with shallow occurrences of ground ice and iced grounds (Elovskaya and Petrova 2001). In the 2nd above-floodplain terrace, which almost completely lacks forests, the concave elements of mesorelief of thermokarst origin bear peculiar phytocoenocomplex-microseries developed at places previously occupied by oxbow lakes and swamps. They represent a complex of phytocoenoses of various overgrowing stages of those depressions. They are usually of a rounded or oval form and are limited in area. However, these microseries are very widespread and along with the surrounding stepped meadows, *Artemisia-Stipa* steppes and fallows form a quasi-climax forest-steppe complex of anthropogenic origin. These ecological microseries mainly develop according to the following scheme: *Equisetosum fluviatile-Caricetum acutae* → *Spiraetum salicifoliae-Caricetum appendiculatae* → *Salicetum brachypodae-pyrolifoliae-Caricetum vesicariae* → *Salicetum bebbianae mixoherbosum* → *Betuletum mixoherbosum.* As a whole, the phytocoenocomplexes feature a high floristic diversity. The terminal stage of these microseries are birch-larch communities or stepped meadows (in case of overgrazing). At present, these phytocoenocomplexes, representing different degrees of degradation, experience strong recreational activities. However, they now are very special communities with a rich species diversity, being "oases" among stepped and overgrazed landscapes and requiring purposeful protection.

Thus, the anthropogenic transformation of forest vegetation of the Lena River valley is expressed in changes in floristics, dominance and edificator composition, in ecocoenomorphic structure, in reduced plant vitality, in simplification of the vertical structure, in disturbance of the dynamic series and in a deviating succession. Profound changes are irreversible. Owing to this, purposeful conservational measures are necessary for the protection of the valley's forest cover. The valley forests do not undergo industrial-scale timber cutting and thus have an invaluable stabilizing effect and are of social significance. Their nature conservational role is many-sided, including water and permafrost protection and erosion prevention. Pasture use without pasture rotation should be prohibited. It should be forbidden to further exploit the degraded lands and they should be submitted to surface improvement. Monitoring investigations in the valley are required to study the degradation-recovery series in over-grazed communities. The recreational activity in the territory should be partially managed on a commercial basis.

## 4.6 Anthropogenic Effects on Tundra and Forest-Tundra Vegetation

**N.S. Karpov**

### *4.6.1 Grazing Effect*

Grazing is one of the most ancient factors that have a great influence on vegetation. In the tundra zone any sector of the tundra has experienced reindeer grazing pressure to a certain extent. We recognize four stages of reindeer pasture degradation that takes place in the tundra (Karpov 1989).

*The initial or original vegetation stage* (*low grazing pressure*). Reindeer use such places very rarely, and generally not more than once every 10 years. The vegetation cover shows no traces of grazing. No damage of the lichen cover is also observed. Such areas are situated between the migratory routes and the permanent stands of the reindeer breeders.

*The slightly changed original vegetation stage* (*moderate grazing pressure*). Reindeer grazing or only run-through in the snow-free period results in trampled and compressed or broken lichen thallus totaling 10–20% of the total phytomass of moderately grazed areas. Reduction in lichen abundance values is followed by an increased role of green mosses, grasses or even dwarf willows in the Hypoarctic subzone. On elevated dry landforms the participation of dwarf shrubs increases.

*The disturbed original vegetation stage* (*high grazing pressure*). The stage is characterized by complete destruction of the lichen cover, increased species diversity and cover values, and increased phytomass of grasses tolerant to grazing. In the tundra-bog complexes the dwarf shrubs degrade but, in contrast, become abundant in hummocky tundra. Thermokarst phenomena are very common there.

*The eliminated original vegetation stage* ( *overgrazing*). This is common for the areas surrounding nomad camps. Such landscapes are characterized by complete elimination of the vegetation, with bare ground occupying 50–60% of the total area. The vegetation recovers in 8–10 years, similar to that of the previous stage.

Fodder reserve dynamics were studied in the pastures of the Indigirka River basin in relation to effects of reindeer grazing (Fig. 4.7). The diagram indicates a decrease of the lichen productivity rates and an increase of green fodder productivity in the 2nd and 3rd stages (long-term grazing).

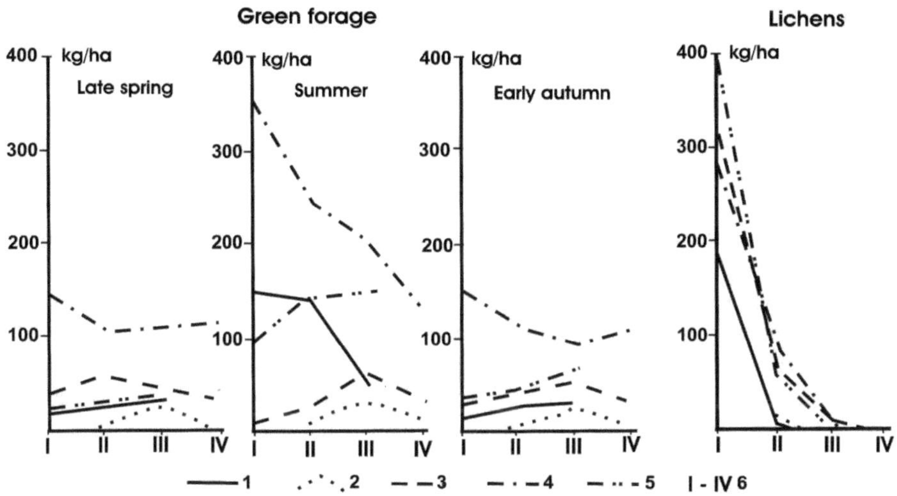


**Fig. 4.7** Fodder productivity dynamics influenced by grazing Tundra-bog complexes: 1 – flat frost mound; 2 – polygonized. Tundra landscapes: 3 – northern tussocky; 4 – south tussocky; 5 – hummocky; 6 – degradation stages

However, the south tussocky tundra shows a declining productivity of green fodder due to trampling of tussocks of *Eriophorum*. The decreased phytomass of *Eriophorum* is compensated by grasses only in the 3rd stage. In summer the fodder mass in flat frost mound tundra-bog complexes also tends to sharply decrease owing to trampling and shrub utilization for firewood, the leaves of these shrubs being a constituent part of green fodder. Thus, under intensive long-term grazing conditions, lichen pastures that could be used all-year-round are transformed into late spring and summer grazing areas. Although this provides sufficient food for reindeer in snow-free periods, it is altogether a negative development because the reindeer pastures of the tundra zone of Yakutia thus start to lack a lichen cover. According to Andreyev (1980), the permanent decline rates of the lichen forage productivity in the Ust-Yansky and Anabarsky farms reached 4.0%, while in the Olenegorsky farm it was 2.5% (Karpov 1984). Despite the opinion that reindeer can manage without lichens, the fodder is best assimilated if it consists of both lichens and green plants. The major cause of the elimination of lichens is the tradition of domestic reindeer grazing around a camp. This is normal in case of a small private herd but not acceptable for large gatherings of animals leading to uneven pasture pressure and elimination of the vegetation cover encircling the camps (Karpov 1988). There are no observations of similar phenomenon in the wild reindeer of the Arctic tundra which in summer concentrate in large (tens of thousands head) herds trying to protect themselves against heat and mosquitoes.

Generally, it is possible to reduce the grazing effect through controlled pasture pressure. This can be reached by fulfillment of a complex of measures: by involving all the pastures according to their grazing capacity and by carefully utilizing the pastures in accordance with their seasonal purpose.

### 4.6.2 Use of Shrubs for Fuel

Due to lack of tree vegetation, the reindeer breeders, hunters and fishermen living in the tundra zone use shrubs as a fuel. This is especially characteristic for reindeer breeders living a nomadic life. *Betula exilis* is most preferred since it burns well both in dry and wet states. The scale of firewood gathering from this species can be estimated from the sharp decrease in its abundance in tundra around the permanent stands of reindeer breeders. The reindeer breeders state that they move northward as far as this *yernik* species occurs. Besides *Betula exilis,* the shrubs of *Salix pulchra* growing along stream banks are used for this purpose. Though in contrary to the birch, the latter is used only in a dry state.

Apart from the fuel purpose, these shrubs serve as a litter underlying reindeer skins, protecting them from damp tundra soils or snow. Before moving to another place, the litter is gathered into a pile, so that next time it can be used as firewood for cooking. In order to decrease the rates destroying the shrubs for firewood, the reindeer breeders should be provided with sufficient wood, liquefied gas, or fuel oil.

### 4.6.3 Fires in Tundra

Due to the natural characteristics of the tundra (high moisture content in soils and air, widespread bog complexes), fires in this zone are very rare and affect small areas. They are caused mainly as a result of human activity. Tundra fires appear to have their own typical features. Fire effects on the tundra vegetation are ambiguous. Thus, Galaktionova and Neustroeva (1980) proved that autumn burning in non-lichen communities may facilitate plant growth because it takes away the litter. Wein and Bliss (1973) also state positive effects of burnings, as it increases the *Eriophorum* productivity in the tussocky tundra of North America. The same was found by Polozova (1986) in western Chukotka. Bely (1974) reported to a certain extent a rejuvenation of tundra pastures (grass stand improvement) after burning of non-palatable dwarf shrubs and large herbs, dead plants, and the moss cover. On the other hand, by destroying fodder shrubs and lichens, fire reduces their phytomass and deteriorates the forage quality. The length of the post-fire recovery period is not long for non-lichen vegetation, while for lichens it lasts for many decades.

### 4.6.4 Transport Influence

Studies on the vegetation and soil covers of the tundra zone of Yakutia revealed that caterpillar transport causes significant damages. Most intensive traffic is observed near settlements and in mine working areas, the vicinities of which are disfigured by numerous roads. The regions of the most rigorous mining activity are in a most deplorable state.

An assessment of caterpillar vehicle impact on the vegetation in the southern Hypoarctic tundra was conducted in the vicinity of Chokurdakh in frost mound tundra and polygonized tundra-bog (Andreyev and Perfilyeva 1979). Investigation included testing of two models of carriers (GAZ-71 and GTT) with specific pressures on the ground of 0.17 and 0.20 g/cm$^2$ and track widths 25 and 54 cm, respectively. Even a single passing of the rovers over the vegetation appeared to cause 20–30% damage to the lichens. Passing three times in dry weather and six times in wet weather, results in 100% damage of the lichens. Bare ground with a complete elimination of vegetation was observed after 12 times repeated traffic in wet weather, while in polygonal grass bogs only 3–6 times are enough, depending on the speed of the vehicles. By the end of summer, the permafrost level was 6–10 cm deeper than in intact areas. Tracks may induce erosion on slopes or thermokarst ponds in hollows. Monitoring studies of natural regrowth on caterpillar tracks showed extremely long rehabilitation periods for the restoration of the original vegetation or even its impossibility. V.N. Andreyev and V.I. Perfilyeva showed two types of regrowth in ten years after the experiment with the carriers: sedge-grass and *Eriophorum* types. Properly drained places are overgrown by grasses and sedges, and over-wetted landforms by rhizomatous *Eriophorum* species. A similar picture is observed in other northern regions in Siberia (Tvorogov 1988).

To summarize the whole impact of a caterpillar cross-country vehicle on the soil-vegetation cover of the tundra zone, we may say that its biological resources and economical potential get reduced, while the thermokarst and erosion processes are, on the contrary, stimulated. Based on investigations the Government of the Republic in 1978 issued a directive on the regulation of caterpillar transport traffic in the northern regions of the Republic. The prohibition of traffic outside roads and accepted routes has played a significant role in the conservation of tundra pastures and of the soil-vegetation cover of the tundra zone as a whole.

The vegetation along winter roads is disturbed as well. Shtyrev et al. (1979) reported that after two years of exploiting the winter road Zeleny Mys – Bilibino (292 km) the top soil layer in the track was destroyed. Later it was transformed into a gully of 5 m deep and 10 m wide at soil surface. After the traffic was stopped, 40 km (13.7%) of the road appeared to be destroyed. Vegetation on winter roads suffers greatly from cleaning the roads from snowdrifts by bulldozers. Our investigations in May 1986 on the winter road Tumat – Kazachye revealed that 5–10 bulldozer passings through places most covered with snow are enough to almost completely eliminate the vegetation cover.

## 4.7 Herb Use for Medicine and Food

**P.A. Timofeyev, L.V. Kuznetsova, and A.P. Isaev**

The importance and value of non-arboreal resources (medicinal and food plants, fungi, industrial crops) are even presently underestimated. For instance, 1 ha of forest in Central Yakutia can produce no more than 100–120 m$^3$ of timber, and no further economic benefit can be expected for the following 120–140 years. Yet, 15–18 tons of *Vaccinium vitis-idaea* can be harvested from the same area during this period. Converting these values in money means, that long term berry harvesting appears to be more profitable than just "wood making". It can be even more beneficial if other plant resources are taken into consideration.

### *4.7.1 Food Plants*

Food plants are very common in the flora of Yakutia (Andreyev et al. 1987). They can be divided into the following groups:

1. Berry plants. They take a special place due to their common occurrence, abundance and food value. There are 45 berry species growing in Yakutia (Timofeyev and Ivanova 2007). The following forest species have food value: trees *Sorbus sibirica* and *Padus asiatica*; shrubs *Ribes spp.* (7 species), *Rubus matsumuranus, Lonicera edulis, L. altaica, Rosa spp.* (4 species); dwarf shrubs *Vaccinium vitis-idaea, V. uliginosum, V. myrtillus, Oxycoccus palustris, O. microcarpus, Empetrum nigrum*; perennial herbs *Fragaria orientalis, Rubus arcticus, R. chamaemorus, R. saxatilis, Rubus humulifolius. Vaccinium vitis-idaea, V. uliginosum, Ribes spp.,* and *Fragaria orientalis* are generally appreciated because of their abundance and taste. One hectare of *Vaccinium vitis-idaea* or *Dushekia-Vaccinium vitis-idaea* larch forests in Central Yakutia is able to produce up to 721.5 kg of *Vaccinium vitis-idaea* (Timofeyev and Petrova 1983).
2. "Nut" plants, including *Pinus sibirica* and *Pinus pumila,* are also valuable. Unfortunately, the Yakutian distribution area of *Pinus sibirica* is limited to Southwest Yakutia, while *Pinus pumila* has a wider distribution in the mountains of East, South, and, sometimes, West Yakutia.
3. The three species of arborescent birches: *Betula alba, B. pendula* and *B. ermannii*, produce juice. One tree may give up to 50 l of birch juice (Vershnyak 1978, 1987).
4. Plants, the organs of which (leaves, stems, underground organs) are used for food. There are 123 food herb species in Yakutia (Fyodorov 1999).
5. Of 240 pileate fungi species (mushrooms) 150 are edible (Petrenko and Lopatina 1978). Most of them grow in forests, significantly contributing to the richness of food plant resources (Vasilyeva et al. 1971). Fungi resources are used insufficiently, since there is no harvesting on a industrial scale.

### *4.7.2 Medicinal Herbs*

Medicinal herbs play an important role in the life of the northern dwellers. The history of the study of medicinal herbs in Yakutia dates back to 1669, when S. Epishev collected and described plants used for medical treatment. Further records of the healing flora can be found in reports and papers of numerous explorers and travellers: D.G. Messerschmidt, I.G. Gmelin, V.L. Seroshevsky, R. Maak, P. Pallas,

I.M. Petukhova, etc. In the twentieth century medicinal herbs were studied by E.M. Yaroslavsky, M.A. Gabyshev, M.N. Karavaev, A.D. Egorov, V.P. Samarin, V.V. Lebedeva, L.V. Sleptsova, etc. A.A. Makarov summarized the data on the healing power of Yakutian plants in the monograph "Biologically active substances in plants of Yakutia" (1989), in which he emphasized the promise of the local medicinal flora for practical use. The author discussed 130 species of officinal herbs, including 6 fungi species and 3 lichen species. He proved that the content of biologically active substances of the plants growing under the extreme natural-climatic conditions is higher and, thus, their pharmaceutical properties are more pronounced.

At present, 16% of the whole flora of Yakutia consists of medicinal herbs comprising 319 species of 223 genera and 73 families (Kuznetsova 1999). A recent summary of medicinal herbs includes 89 plant species and 3 fungi species (*Fomitopsis officinalis, Claviceps purpurea, Innonotus obliquus*) that are indeed used in official medicine; while 87 plants species possess similar pharmaceutical properties (Ivanov 2003); and 102 plant species and 21 fungi species are used in folk medicine (Ivanov 2005).

### 4.7.3 Resources of Medicinal and Food Plants and Fungi

Tables 4.2 and 4.3 show the estimated resources of several species of medicinal and food plants respectively. The harvestable reserves of medicinal raw materials are by

**Table 4.2** Estimated resources of several medicinal plant and fungi species in the forests of Yakutia

| | *Vaccinium vitis-idaea* | | | *Ledum palustris* | | | *Arctostaphylos uva-ursi* |
|---|---|---|---|---|---|---|---|
| | Middle taiga | Northern taiga | Total | Middle taiga | Northern taiga | Total | Total |
| Forest area containing industrial reserves of resources, million ha | 11.0 | 9.0 | 20.0 | 5.0 | 6.0 | 11.0 | 2.0 |
| Average reserves of medicinal raw material, c/ha | 9.0 | 6.0 | – | 4.0 | 3.0 | – | 8.0 |
| General reserve of medicinal herb resources, million tons | 9.9 | 5.4 | 15.3 | 2.0 | 1.8 | 3.8 | 1.6 |
| Active reserve of medicinal herb resources, million tons | 1.7 | 0.9 | 2.6 | 0.33 | 0.3 | 0.6 | 0.26 |

**Table 4.3** Estimated resources of some edible berries and fungi species in the forests of Yakutia

| | *Vaccinium vitis-idaea* | | | *Vaccinium uliginosum* | | | Fungi (mushrooms) (fresh weight) | | | |
|---|---|---|---|---|---|---|---|---|---|---|
| | | | | | | | Middle taiga | | | |
| | Middle taiga | Northern taiga | Total | Middle taiga | Northern taiga | total | Larch forests | Pine forests | Northern taiga | total |
| Forest area containing industrial reserves of resources, million ha | 11.0 | 9.0 | 20.0 | 1.8 | 2.8 | 4.6 | 40.0 | 1.0 | 25.0 | 66.0 |
| Average reserves of food plants, c/ha | 1.5 | 0.5 | | 1.0 | 1.5 | | 2.7 | 2.7 | 2.7 | |
| General reserve of food plant resources, million tons | 1.6 | 0.5 | 2.1 | 0.2 | 0.4 | 0.6 | 11.0 | 0.3 | 0.7 | 12.0 |

far not properly estimated because of their 4–5-year recovery cycle after cutting the phytomass at ground level (Bogdanova 1980, 1983; Bogdanova and Muratov 1978; Isaev and Timofeyev 1999), while the harvestable reserves of berries are equal to their gross production.

## References

Abaimov AP, Prokushkin SG, Zyryanova OA (1998) Osobennosti poslepozharnykh povrezhdeniy listvennichnykh lesov merzlotnoy zony Sredney Sibiri. Sibirskiy ekologicheskiy zhurnal 3–4: 315–323 – *Characteristic features of post-fire damages of the larch forests of the permafrost zone of Central Siberia* (in Russian)

Andreyev VN, Perfilyeva VI (1979) Vliyaniye dvizheniya gusenichnogo transporta na rastitelnost subarkticheskoy tundry. In: Proceedings of 7th All-Union Symposium "Biological problems of the North". Apatity: 22–24. – *Impact of caterpillar transport on the Subarctic tundra vegetation* (in Russian)

Andreyev VN (1980) Nekotorye dannye o delikhenizatsii tundry. In: Rastitelnost i pochvy subarkticheskoy tundry. Novosibirsk. – *Some data on tundra delichenization* (in Russian)

Andreyev VN, Galaktionova TF, Perfilyeva VI, Scherbakov IP (1987) Osnovnye osobennosti rastitelnogo pokrova Yakutskoy ASSR. Izd-vo YaF SO AN SSSR, Yakutsk – *Basic features of vegetation cover of the Yakutian ASSR* (in Russian)

Basharin GP (1956) Istoriya agrarnykh otnosheniy v Yakutii. Moscow – *The history of agrarian relations in Yakutia* (in Russian)

Bely NF (1974) Vliyaniye pozharov na olenyi pastbischa v Magadanskoy oblasti. In: Biologicheskiye problemy Severa 3: 141–146 – *Fire effect on reindeer pastures in the Magadan Region* (in Russian)

Bogdanova GA (1980) Struktura populyatsiy brusniki v sredney taiga Sibiri. In: Biologicheskiye resursy lesov Sibiri. Collection of articles. Krasnoyarsk:15–31 – *Population structure of Vaccinium vitis-idaea in the middle taiga of Siberia* (in Russian)

Bogdanova GA (1983) Osnovy ratsionalnogo ispolzovaniya brusniki – tsennogo lekarstvennogo syrya. In: Prodovolstvennye i kormovye resursy lesov Sibiri. ILiD, Krasnoyarsk: 94–101 – *Principles of rational use of Vaccinium vitis-idaea, a valuable officinal raw material* (in Russian)

Bogdanova GA, Muratov YuI (1978) Brusnika v lesakh Sibiri. Nauka, Novosibirsk – *Vaccinium vitis-idaea in the forests of Siberia* (in Russian)

Borman FH, Likens GE (1979) Pattern and process in a forested ecosystem. In: Disturbance, development and the steady state based on the Hubbard Brock ecosystem study. Springer, New York, Heidelberg, Berlin.

Chugunov BV (1955) Osnovnye tipy lesov Lenskogo raiona YaASSR i lesovozonovleniye v nikh. In: Trudy Instituta biologii YaF SO AN SSSR 1: 147–174 – *Basic forest types of the Lensk Region of the Yakut Autonomous Soviet Socialist Republic and reforestation there* (in Russian)

Chugunov BV (1961) Vozonovleniye lesa v Yugo-Zapadnoy Yakutii. In: Trudy Instituta biologii YaF SO AN SSSR 8: 260–323 –*Reforestation in Southwest Yakutia* (in Russian)

Chugunova RV (1964) Gari Yuzhnoy Yakutii i ikh lesovozobnovleniye. In: Pozdnyakov LK (ed) Lesa Yuzhnoy Yakutii. Nauka, Moscow – *Burnt-out forests of South Yakutia and their recovery* (in Russian)

Denisov GV, Prokopyev AN (1979) Intensifikatsiya ispolzovaniya i okhrana lugov Yakutii. In: Okhrana prirody Yakutii. Izd-vo YaF SO AN SSSR, Yakutsk – *Intensification of usage and protection of meadows of Yakutia* (in Russian)

Denisov GV, Streltsova VS, Nakhabtseva SF et al (1983) Rekonstruktsiya i okhrana alasnykh lugov Yakutii. Izd-vo YaF SO AN SSSR, Yakutsk – *Reconstruction and protection of alas meadows of Yakutia* (in Russian)

Desyatkin RV (1996) O nekotorykh ekologicheskikh problemakh selskokhozyaistvennogo proizvodstva Yakutii. In: Problemy ekologii Yakutii. Collection of articles. Izd-vo YaGU, Yakutsk: 112–123 – *On some ecological problems of agricultural production of Yakutia* (in Russian)

Gavrilyeva LD (1998) Pastbishnaya digressiya i ratsionalnoye ispolzovaniye rastitelnosti alasov Leno-Amginskogo mezhdurechya. Abstract of Thesis for a Candidate degree. Yakutsk – *Pasture degradation and rational use of alas vegetation in the Lena-Amga Interfluve* (in Russian)

Gerasimova ZV, Gavrilyeva LD (2001) Khozyaistvennye tipy rastitelnosti alasov Leno-Amginskogo mezhdurechya. Nauka i obrazovaniye 1: 35–40 – *Fodder types of alas vegetation of the Lena-Amga Interfluve* (in Russian)

Elovskaya LG, Petrova EI (2001) Sovremennoye sostoyaniye pochvennogo pokrova doliny Tuymaada (ekologicheskaya otsenka, ratsionalnoye ispolzovaniye i okhrana pochv). In: Ekologicheskaya bezopasnost reki Leny: monitoring, prirodnye i tekhnogennye kataklizmy. Collection of articles. Yakutsk: 41–44 – *Modern condition of soil cover in the Tuymaada valley (ecological estimation, rational use and protection of soils)* (in Russian)

Furyaev VV (1996) Rol pozharov v protsesse lesoobrazovaniya. Nauka, Novosibirsk – *Role of fires in the forest formation process* (in Russian)

Fyodorov AN (1985) Rol vyrubok v razvitii merzlotnykh landshaftov Tsentralnoy Yakutii. In: Regionalnye i inzhenernye geokriologicheskiye issledovaniya. Collection of articles. Izd-vo IMZ SO RAN, Yakutsk: 111–117 – *Role of cuttings in the development of permafrost landscapes of Central Yakutia* (in Russian)

Fyodorov IA (1999) Pischevye travy Yakutii. In: Timofeyev PA, Ivanova EI, Isaev AP (eds) Flora i rastitelnost Yakutii. FIPS, Moscow – *Food herbs of Yakutia* (in Russian)

Galaktionova TF, Neustroeva AI (1980) Vliyaniye srokov otchuzhdeniya nadzemnoy fitomassy na urozhainost pushitsy vlagalischnoy v subarkticheskoy tundre. In: Selskoye khoziaystvo Krainego Severa 7. Abstracts of 4th All-Union meeting: 86–87 – *Effect of cutting terms of on-ground phytomass on the productivity of Eriophorum vaginatum in subarctic tundra* (in Russian)

Gavrilova MK (1967) Teplovoy balans listvennichnogo lesa na Leno-Amginskom mezhdurechye. In: Gidroklimaticheskiye issledovaniya Sibiri. Nauka, Moscow – *Heat balance of larch forests in the Lena-Amga Interfluve* (in Russian)

Gavrilova MK (1969) Radiatsionny rezhim v listvennichnykh lesakh Yugo-Zapadnoy Yakutii. Lesovedeniye 1: 16–23 – *Radiation balance in the larch forests of Southwest Yakutia* (in Russian)

Gogoleva PA, Kononov KE, Mirkin BM, Mironova SI (1987) Sintaksonomiya i simfitosotsiologiya rastitelnosti alasov Tsentralnoy Yakutii. Izd-vo Irkutskogo GU, Irkutsk – *Syntaxonomy and symphytosociology of the alas vegetation of Central Yakutia* (in Russian)

Isaev AP (1991) Adaptatsiya predvaritelnogo podrosta listvennitsy k usloviyam suschestvovaniya na vyrubke. In: Ekologiya rastitelnogo mira Yakutii. Collection of articles. Yakutsk: 25–26 – *Adaptation of the preliminary growth of larch to existing conditions in clearings* (in Russian)

Isaev AP (1992) Vozobnovleniye listvennitsy na lesosekakh Yugo-Zapadnoy Yakutii. In: Botanicheskiye issledovaniya v kriolitozone. Collection of articles. Izd-vo YaNTs SO RAN, Yakutsk: 135–143 – *Larch recovery in the cuttings of Southwest Yakutia* (in Russian)

Isaev AP (1993) Listvennichnye lesa srednetayozhnoy Yakutii, lesovozobnovleniye na vyrubkakh. Abstract of Thesis for a Candidate Degree. Krasnoyarsk – *Larch forests of the middle taiga of Yakutia, reforestation in cuttings* (in Russian)

Isaev AP (2000) Estestvennoye lesovozobnovleniye na vyrublakh i garyakh. In: Isaev AP, Mikhalyova LG (eds) Les i vechnaya merzlota: Osobennosti sostava i struktury lesov merzlotnogo regiona, problemy ratsionalnogo vedeniya khozyaistva i okhrany. Izd-vo YaGU, Yakutsk – *Natural reforestation in clearings and burning-outs* (in Russian)

Isaev AP, Mikhalyova LG (eds) (2000) Les i vechnaya merzlota: osobennosti sostava i struktury lesov merzlotnogo regiona, problemy ratsionalnogo vedeniya khozyaystva i okhrany. Izd-vo YaGU, Yakutsk – *Forest and permafrost: composition and structural peculiarities of the forests of the permafrost region, problems of rational economy and conservation* (in Russian)

Isaev AP, Timofeyev PA (1999) Opyt opredeleniya uscherba nedrevesnym resursam lesa v svyazi s lesozagotovkami v Tsentralnoy Yakutii. In: Timofeyev PA, Ivanova EI, Isaev AP (eds) Flora i rastitelnost Yakutii. FIPS, Moscow – *Estimation of damaged non-arboreous forest resources as a result of timber felling in Central Yakutia* (in Russian)

Ivanov BI (ed) (2003) Atlas lekarstvennykh rasteniy Yakutii 1: Lekarstvennye rasteniya ispolzuyemye v nauchnoy meditsine. YaF Izd-va SO RAN, Yakutsk – *Atlas of officinal herbs 1: Officinal herbs used in official medicine* (in Russian)

Ivanov BI (ed) (2005) Atlas lekarstvennykh rasteniy Yakutii 2: Lekarstvennye rasteniya ispolzuyemye v narodnoy meditsine. YaF Izd-va SO RAN, Yakutsk – *Atlas of officinal herbs 2: Officinal herbs used in folk medicine* (in Russian)

Karpel BA, Medvedeva NS (1977) Plodonosheniye listvennitsy daurskoy v Yakutii. Nauka, Novosibirsk – *Fruiting of Larix dahurica in Yakutia* (in Russian)

Karpov NS (1984) Vypas i delikhenizatsiya olenyikh pastbisch v subarkticheskoy tundre Yakutii. In: Osobennosti tekhnologii selskokhozyaistvennogo proizvodstva na Krainem Severe. BNTI NIISKh Krainego Severa. Issue 51. Novosibirsk: 14–18 – *Grazing and delichenization of reindeer pastures in the subarctic tundra of Yakutia* (in Russian)

Karpov NS (1988) Pitaniye i vypas severnogo olenya v Indigirskoy tundre v bessnezhnoye vremya. In: Zoogeograficheskiye i ekologicheskiye issledovaniya teriofauny Yakutii. Collection of articles. Yakutsk: 86–95 – *Nutrition and grazing of a reindeer in the Indigirka tundra in the snow-free period* (in Russian)

Karpov NS (1989) Dinamika rastitelnosti subarkticheskoy tundry Yakutii pod vliyaniem vypasa severnykh oleney. Abstract of Thesis for a Candidate Degree. Moscow – *Vegetation dynamics of the subarctic tundra affected by reindeer grazing* (in Russian)

Kononov KE, Gogoleva PA, Burtseva EI (1979) Senokosy i pastbischa Tsentralnoy Yakutii. Knizhnoye izd-vo, Yakutsk – *Hayfields and pastures of Central Yakutia* (in Russian)

Kryuchkov VV (1973) Krainiy Sever: problemy ratsionalnogo ispolzovaniya prirodnykh resursov. Moscow. – *The Far North: problems of rational use of natural resources.*

Kucherov IB (1995) O printsipe dopolnitelnosti v geobotanike: metodologicheskiye predposylki vozniknoveniya komplementarnykh podkhodov k izucheniyu rastitelnosty. Zhurnal obschey biologii 56(4): 486–505 – *On the principle of complementarity in geobotany: methodological preconditions of appearance of complementary approaches in the study of vegetation* (in Russian)

Kuznetsova LV (1999) Ekologo-geographicheskaya kharakteristika lekarstvennykh rasteniy Yakutii. In: Timofeyev PA, Ivanova EI, Isaev AP (eds) Flora i rastitelnost Yakutii. Collection of articles. FIPS, Moscow: 129–139 – *Ecological-geographical characteristics of the officinal herbs of Yakutia* (in Russian)

Lytkina LP (2005) Dinamika rastitelnogo pokrova na garyakh listvennichnykh lesov Leno-Amginskogo mezhdurechya (Tsentralnaya Yakutia). Abstract of Thesis for a Candidate Degree. Yakutsk – *Dynamics of the vegetation cover of post-fire larch forests of the Lena-Amga Interfluve (Central Yakutia)* (in Russian)

Lytkina LP, Protopopova VV (2006) Lesnye pozhary kak ekologichesky factor formorovaniya lesov Tsentralnoy Yakutii. Nauka i obrazovaniye 2(42): 50–56 – *Forest fires as an ecological factor of forest development in Central Yakutia* (in Russian)

Mikhalyova VM (1971) Zarastaniye raznorezhimnykh vyrubok v listvennichnike olkhovnikovo-brusnichnom na yugo-zapade Yakutii. In: Issledovaniya rastitelnosti i pochv v lesakh Severo-Vostoka SSSR. Knizhnoye izd-vo, Yakutsk – *Overgrowing of cuttings under various conditions in Dushekia-Vaccinium vitis-idaea larch forests in Southwest Yakutia* (in Russian)

Mikhalyova VM (1974) Osobennosti razvitiya osnovnykh tsenoobrazovateley na vyrubkakh Yugo-Zapadnoy Yakutii. In: Materialy o lesakh Severo-Vostoka SSSR. Knizhnoye izd-vo, Yakutsk – *Developmental characteristics of basic coenotic components in cuttings in Southwest Yakutia* (in Russian)

Mikhalyova VM (1975) Vliyaniye vyrubok na razvitiye Linnei severnoy v listvennichnykh lesakh Yugo-Zapadnoy Yakutii. In: Botanicheskiye materially po Yakutii. Collection of articles. Yakutsk: 68–77 – *Effect of cuttings on development of Linnaea borealis in the larch forests of Southwest Yakutia* (in Russian)

Mikhalyova VM (1977) Razvitiye rastitelnosti na vyrubkakh. In: Formirovaniye rastitelnogo pokrova v svyazi s rubkami v lesakh yugo-zapadnoy Yakutii. Nauka, Novosibirsk – *Vegetation development in cuttings* (in Russian)

Mikhalyova VM, Chugunova RV (1971) Vosstanovleniye rastitelnosty na vyrubkakh Yugo-Zapadnoy Yakutii. Vostochno-Sibirskoye knizhnoye izd-vo, Irkutsk – *Vegetation recovery in clearings in Southwest Yakutia* (in Russian)

Mironova SI (2000) Tekhnogennye suktsessionnye sistemy rastitelnosti Yakutii (na primere Zapadnoy i Yuzhnoy Yakutii). Izd-vo SO RAN, Novosibirsk – *Technogenic succession systems of the vegetation of Yakutia (case study of West and South Yakutia)* (in Russian)

Mironova SI, Poiseyeva SI (1996) Vliyaniye vypasov na rastitelnost alasov. In: Problemy ekologii Yakutii. Izd-vo YaGU, Yakutsk: 145–150 – *Grazing effect on alas vegetation* (in Russian)

Nikolayeva MKh, Sofronov RR (2001) Produktivnost senokosnykh i pastbischnykh ugodiy alasov Tsentralnoy Yakutii. In: Teoreticheskiye i prikladnye vorposy travoseyaniya v kriolitozone. Proceedings of International Conference. Part 1. Izd-vo YaNTs SO RAN, Yakutsk: 97–103 – *Productivity of hayfields and pastures of alases of Central Yakutia* (in Russian)

Petrenko IA, Lopatina ZA (1978) Maloizvestnye syedobnye griby lesov Tsentralnoy Yakutii. In: Lesnye rastitelnye resursy Sibiri. ILiD, Krasnoyarsk – *Little-known edible fungi of the forests of Central Yakutia* (in Russian)

Pobedinsky IA (1965) Sosnovye lesa Sibiri i Zabaikalya. Nauka, Moscow – *Pine forests of Siberia and Transbaikalia* (in Russian)

Polozova TG (1986) Vliyaniye pozhara na rastitelnost yuzhnykh gipoarkticheskikh tundr na Zapadnoy Chukotke. Botanichesky zhurnal 12: 1657–1663 – *Fire effect on vegetation of the southern hypoarctic tundra in West Chukotka* (in Russian)

Pozdnyakov LK (1963) Gidroklimaticheskiy rezhim listvennichnykh lesov Tsentralnoy Yakutii. Izd-vo AN SSSR, Moscow – *Hydroclimatic conditions of the larch forests of Central Yakutia* (in Russian)

Pozdnyakov LK (1974) Voprosy kompleksnogo osvoeniya lesov Severo-Vostoka Sibiri. In: Materialy o lesakh Severo-Vostoka SSSR. Knizhnoye izd-vo, Yakutsk – *Problems of complex exploitation of the forests of Northeast Siberia* (in Russian)

Pozdnyakov LK (1975) Listvennitsa daurskaya. Nauka, Moscow – *Larix dahurica* (in Russian)

Pozdnyakov LK (1983) Les na vechnoy merzlote. Nauka, Novosibirsk – *Forest on permafrost* (in Russian)

Pozdnyakov LK (1986) Merzlotnoye lesovedeniye. Nauka, Novosibirsk – *Permafrost silvics* (in Russian)

Rabotnov TA (1983) Fitotsenologiya. Izd-vo MGU, Moscow – *Phytocoenology* (in Russian)

Razumovsky SM (1981) Zakonomernosti dinamiki biotsenozov. Nauka, Moscow – *Principles of dynamics of biocoenoses* (in Russian)

Sannikov SN (1983) Tsiklicheski-eroziyno-pirogennaya teoriya estestvennogo vozobnovleniya sosny obyknovennoy. Ekologiya 1: 3–9 – *A cyclical-erosion-pyrogenic theory of natural restocking of Pinus sylvestris* (in Russian)

Savvinov DD (1969) Vliyaniye razlichnykh stepeney rubki na temperaturny rezhim pochv yugo-zapada Yakutii. In: Collection of articles of the Yakut hydrometeorological observatory 2. Yakutsk: 45–47 – *Effect of various cuttings on soil conditions in Southwest Yakutia* (in Russian)

Savvinov DD (1971) Temperaturny rezhim i vodny rezhim pochv lesov Yakutii. In: Issledovaniya rastitelnosti i pochv v lesakh Severo-Vostoka SSSR. Knizhnoye izd-vo, Yakutsk – *Temperature and moisture conditions of the forest soils of Yakutia* (in Russian)

Savvinov DD (1972) Vliyaniye rubok i raspashki na fiziko-khimicheskiye svoistva pochv yugo-zapada Yakutii. Knizhnoye izd-vo, Yakutsk – *Effect of cuttings and ploughing on physicochemical features of the soils of Southwest Yakutia* (in Russian)

Savvinov DD (1973) Mikroklimat lesa i vyrubok Yakutii. In: Pochvennaya klimatologiya Sibiri. Nauka, Novosibirsk – *Microclimate of forests and clearings in Yakutia* (in Russian)

Savvinov DD (1976) Gidrotermicheskiy rezhim pochv v zone mnogoletney merzloty. Nauka, Novosibirsk – *Hydrothermal conditions of soils in the permafrost area* (in Russian)

Savvinov DD, Mironova SI, Bosikov NP et al. (2005) Alasnye ekosistemy: struktura, funktsionirovaniye, dinamika. Nauka, Novosibirsk – *Alas ecosystems: structure, functioning,* dynamics (in Russian)

Scherbakov IP, Chugunova RV (1960) O klassifikatsii garey v Yugo-Zapadnoy i Tsentralnoy Yakutii. Izvestiya SO AN SSSR 1: 127–136 – *On classification of burnt-out forests in Southwest and Central Yakutia* (in Russian)

Scherbakov IP, Chugunova RV (1961) Lesa Yugo-Zapadnoy Yakutii i mery sodeystviya lesovozobnovleniyu na lesosekakh i garyakh. In: Materilay o lesakh Yakutii. Izd-vo AN SSSR, Moscow – *The forests of Southwest Yakutia and measures promoting forest recovery in cuttings and burnings* (in Russian)

Scherbakov IP, Mikhalyova VM, Chugunova RV, Karpel BA (1977) Formirovaniye rastitelnogo pokrova v svyazi s rubkami. Nauka, Novosibirsk – *Development of vegetation cover in relation to cuttings* (in Russian)

Sofronov RR, Karpov NS (1999) K probleme obespecheniya productivnosty alasnykh pastbischnykh ugodiy. In: Timofeyev PA, Ivanova EI, Isaev AP (eds) Flora i rastitelnost Yakutii. Collection of articles. FIPS, Moscow: 145–149 – *On the problem of productivity of alas pastures* (in Russian)

Shtyrev IN, Volsky SG, Grigoryev KI (1979) Vliyaniye avtomobilnogo transporta na okruzhayuschuyu sredu. In: Okhrana prirody Yakutii. Knizhnoye izd-vo, Yakutsk – *Traffic effect on the environment* (in Russian)

Tarabukina VG, Savvinov DD (1990) Vliyaniye pozharov na merzlotnye pochvy. Nauka, Novosibirsk – *Fire effect on frozen soils* (in Russian)

Timofeyev PA, Isaev AP, Scherbakov IP et al. (1994) Lesa srednetayozhnoy podzony Yakutii. Izd-vo YaNTs SO RAN, Yakutsk – *The forests of the middle taiga subzone of Yakutia* (in Russian)

Timofeyev PA, Ivanova EI (2007) Yagodnye rasteniya Yakutii. Bichik, Yakutsk – *Berry plants of Yakutia* (in Russian)

Timofeyev PA, Petrova TN (1983) O zapase syrya brusniki v lesakh doliny Sredney Leny. In: Problemy osvoyeniya lesnykh resursov Sibiri i Dalnego Vostoka. Abstracts of All-Union conference. Izd-vo SO AMN, Novosibirsk: 72–73 – *On the resources of Vaccinium vitis-idaea in the forests of the Middle Lena River Valley* (in Russian)

Timofeyev PA, Shurduk IF (1996) Sostoyaniye i voprosy okhrany lesov doliny Tuymaady. In: Problemy ekologii Yakutii: biogeograficheskiye issledovaniya. Collection of articles. Izd-vo YaGU, Yakutsk: 101–105 – *Conditions and protection issues of the forests of the Tuymaada valley* (in Russian)

Timofeyev PA, Shurduk IF, Ivanov BI (1992) Sostoyaniye zelyonykh nasazhdeniy okrestnostey Yakutska. In; Chelovek i Sever: istorichesky opyt, sovremennoye sostoyaniye, perspektivy razvitiya. Proceedings of regional inter-industry scientific-practical conference. Izd-vo YaNTs SO RAN, Yakutsk: 98–102 – *Conditions for green plantings in the vicinity of Yakutsk* (in Russian)

Tvorogov VA (1988) Estestvennoye zarastaniye narushennykh uchastkov tundry v raione Yamburgskogo gazokondensatnogo mestorozhdeniya 9P-V Tazovsky. Botanichesky zhurnal 11(73): 1577–1583 – *Natural overgrowing of disturbed tundra in the region of the Yamburg gas condensate deposit 9P-V Tazovsky* (in Russian)

Tyrtikov AP (1969) Vliyaniye rastitelnogo pokrova na promerzaniye i protaivaniye gruntov. Izd-vo MGU, Moscow – *Effect of vegetation cover on freezing through and thawing through of grounds* (in Russian)

Utkin AI (1960) O estestvennom vozobnovlenii listvennitsy daurskoi v Tsentralnoy Yakutii. In: Soobscheniya laboratorii lesovedeniya. Collection of articles. Issue 2. Izd-vo AN SSSR, Moscow: 44–68 – *On natural recovery of Larix dahurica in Central Yakutia* (in Russian)

Utkin AI (1965) Lesa Tsentralnoy Yakutii. Nauka, Moscow – *The forests of Central Yakutia* (in Russian)

Vasilyeva LN, Shurduk IF, Medvedeva NS (1971) K izucheniyu syedobnykh gribov Yakutii. In: Okhrana prirody Yakutii. Vostochno-Sibirskoye knizhnoye izd-vo, Irkutsk – *On the study of edible fungi* (in Russian)

Vershnyak VM (1978) Podsochka beryozy ploskolistnoy v Tsentralnoy Yakutii. In: Lesnye rastitelnye resursy Sibiri. ILiD, Krasnoyarsk – *Tapping of Betula plathyphylla in Central Yakutia* (in Russian)

Vershnyak VM (1987) Nekotorye osobennosti podsochki beryozy v Tsentralnoy Yakutii. In: "Podsochka i pererabotka soka listvennykh porod". Proceedings of All-Union scientific-practical conference. Lvov: 26–28 – *Some peculiarities of birch tapping in Central Yakutia* (in Russian)

Vipper VN (1964) Vliyaniye travyano-kustarnichkovogo pokrova na vozobnovleniye listvennitsy daurskoi v Tsentralnoy Yakutii. In: Trudy po lesnomu khozyaistvu Sibiri. Collection of articles. Issue 8. Nauka, Novosibirsk: 122–132 – *Influence of the herb-dwarf shrub layer on the recovery of Larix dahurica in Central Yakutia* (in Russian)

Vipper VN (1973) Vliyaniye podleska i travyano-kustarnichkovogo pokrova na vozobnovleniye listvennichnykh lesov Tsentralnoy Yakutii.Nauka, Moscow – *Influence of the undergrowth and herb-dwarf shrub layer on the recovery of larch forests in Central Yakutia* (in Russian)

Wein RW, Bliss LC (1973) Changes in arctic *Eriophorum* tussock communities following fire. Ecology 54(2): 845–852.

# Chapter 5
# Insect Impact on Vegetation

**A.I. Averensky, I.I. Chikidov, and Yu.V. Ermakova**

## 5.1 Needle- and Leaf-Eating Forest Insects

**A.I. Averensky and I.I. Chikidov**

Long-term stationary investigations on the biology and ecology of the dendrophilous insects of Yakutia were carried out in the south-western, southern and central regions which are characterized by forest tracts of industrial significance. The leaf- and needle-eating chlorophages are stated to have a primary impact on major forest-forming tree species (*Larix spp., Pinus sylvestris, Betula spp., Salix spp.*) and that is aggravated by attacks of the secondary xylophages on the weakened or dying trees.

The major, dangerous and mass pest of the *Larix* forests is a Siberian silk moth (*Dendrolimus superans sibiricus* Tschetv.) affecting large areas (Fig. 5.1). Of all local tree species, *Larix*, dominating the territory of Yakutia, is the most resistant to the tree vermin species. Due to annual renewal of assimilating organs the *Larix* is able to withstand one or two needle browsings per tree. Besides, high productivity of the *Larix* allows the losses caused by insect invasions to be quickly compensated. Nevertheless, massive affection of large forest tracts accompanied with death of most trees may induce phenomena such as thermokarst processes which impede further forest growth in the affected area. In this regard, the mass reproduction of the Siberian silk moth may have a drastic effect on *Larix* forests. With high densities of larvae per a tree the numbers of this species may exceed 3 millions per 1 ha. And each is able to consume up to 430 mg of needles daily. So, its biology and role in *Larix* communities are worth being discussed in detail.

Mass outbreaks of this insect cover large forest areas and is characterized by a high concentration of the vermin (up to 15,000 and more caterpillars per tree) causing total needle devouring, death of dark conifer species, and sharp weakening of the *Larix* forest. As a result, continuous (some times partial) dead-wood is formed. Lower concentration of the pests, which means partial needle browsing

A.I. Averensky (✉)
Institute for biological problems of the cryolithozone, 41 Lenin Ave., Yakutsk, 677980, Russia
e-mail: aaver37@yandex.ru

E.I. Troeva et al. (eds.), *The Far North: Plant Biodiversity and Ecology of Yakutia*, Plant and Vegetation 3, DOI 10.1007/978-90-481-3774-9_5,

**Fig. 5.1** *Dendrolimus superans sibiricus* Tschetv.: **a** – female imago; **b** – male imago; **c** – eggs; **d** – larva; **e** – pupa

without total crown striping, results in the decrease of productivity and tree weakening. Due to this fact as well as the presence of large dead-wood areas, the forests affected by silk moths yield mass outbreaks of a stem vermin causing final death of the trees. The Siberian silk moth induces injuries to *Larix, Pinus sibirica, Abies, Picea*, and, to a lesser extent, to *Pinus sylvestris*. Its reproduction centres appear both in mixed coniferous forests and pure monospecific communities. Mass propagation of the Siberian silk moth in *Larix* forests features prolonged peaks, especially in the regions with large tracts of pure *Larix* forests. If the dark coniferous forests show a spatially pronounced differentiation of the pestholes and their expansive character, then the *Larix* forests are characterized by a mosaic type of outbreak centres (Rozhkov 1965).

For the middle taiga zone of Yakutia the Siberian silk moth is a regular species occurring between peaks evenly all over the territory. The northern border of its distribution area lies within 62′ NL (Petrenko 1965) or in the Aldan River mouth area (Ammosov 1978).

In the inter-peak periods participation of the Siberian silk moth in the entomofauna of Central Yakutia is not significant (1–3 middle- or old-aged caterpillars per tree in *Larix* forests, or the same amount of pupae per tree of the I–V growth class) (Ammosov 1978). It takes several years for a population size of 1–3 individuals per tree to reach a mass outbreak (up to 1,000–2,000 specimens per tree) (Vinokurov et al. 2001).

In the territory of Central Yakutia the Siberian silk moth gives preference to *Larix* but sometimes may feed on *Picea* and *Pinus* (Vinokurov et al. 2001). Thus, in this region the Siberian silk moth is considered as a *Larix* monophage which is characteristic for all local needle-eating *Lepidoptera* species (Kolomiets 1957; Rozhkov 1963; Ammosov 1978).

According to Ammosov (1978), both during the periods of population decreases and of mass outbreaks the Siberian silk moth in Central Yakutia prefers the mesic *Larix* forests with *Vaccimium vitis-idaea*, particularly the *Larix cajanderi*-forbs+*Arctous*+*Vaccimium vitis-idaea* (N.B. in these association connotations "–" separates layers while "+" indicates co-dominance in the same layer) (species occurrence 40%) and *Larix cajanderi-Vaccimium vitis-idaea* (28.8%) types of middle-aged forests of middle and high density, growing on gentle south-, west- and east-facing slopes, on terrace-like landscapes and moderate summits of watersheds. Sometimes (4–8%) the insect is observed in *Larix cajanderi-Ledum palustris+Vaccimium vitis-idaea, Larix cajanderi-Rhododendron+Vaccimium vitis-idaea,* and *Larix cajanderi+Pinus sylvestris-Vaccimium vitis-idaea+Arctostaphylos* forests.

The swarming period of the Siberian silk moth in Yakutia, like in other parts of its distribution area, takes place in July (Rozhkov 1963; Ammosov 1978). Under favourable conditions (absence of abundant precipitation) the butterflies move over significant distances and disperse over the suitable territories. With persistent winds from the same direction and high precipitation during the swarming period the silk moths concentrate within limited areas. Rozhkov (1965) stated that when the silk moth's population normally increase, and the weather is rainy, a mass outbreak is possible if the butterflies are concentrated in specific locations. According to a number of authors (Petrenko 1965; Vinokurov et al. 2001), the following factors favour an increase in silk moth numbers: under pro-longed drought, when the population density of the natural enemy decreases, and when the trees weaken physiologically as a result of anthropogenic impact and fires. Insect entomophages play the major role in the regulation of the number of silk moths (Ammosov 1971b, 1978) leading to a sparse population of the vermin (1–2 cartepillars per 1–2 trees) in inter-peak periods and extinguishing the pestholes during the peak periods at all developmental stages. The most effective natural enemies are egg-eaters destroying up to 90–100% of eggs in the mass outbreak locations and, eventually, bringing the best to a rapid decrease in numbers. The caterpillars and pupae of the silk moth are also, though to a lesser extent, affected by ichneumons and tachinid flies. The role of birds in the regulation of silk moth numbers is insignificant: they do not concentrate in the mass outbreak locations and consume only up to 2% of the total pest amount (Vinokurov et al. 2001). Wintering conditions constitute the most important factor determining the population number of entomophages. Shallow depths of snow cover yields mass killing of hibernating insects-entomophages and later this is also noticeable in the mass reproduction peak of the silk moth.

In the middle taiga zone of Yakutia the mass outbreak of the Siberian silk moth has a regular character. It has been recorded since the past century in a number of the pre-Lena regions (Lensky, Olekminsky, Khangalassky, Namsky Uluses), in all trans-Lena regions and in the Ust-Maisky Ulus (Southwestern Yakutia) (Fig. 5.2).

The second half of the twentieth century is notable for intensive insect epidemics recorded in the Pokrovsk (1948–1954), Namsky (1969), Amginsky, Gorny and Ust-Maisky forest areas (Table 5.1). The last peak, that took place in 1999, was the largest one ever recorded in Yakutia (Vinokurov et al. 2001).

The details of the insect epidemics of 1999–2001 are as follows. According to the data of the Forestry Department of the Republic of Sakha (Yakutia), in the early summer of 1999 the mass outbreak locations of the silk moth were recorded in the forest husbandries of the Lena-Amga Interfluve. One year later another peak occurred in the pre-Lena uluses (Khangalassky, Namsky, Yakutsk vicinity). By 2000 over 5 million ha were affected by the vermin with the pestholes occupying 500,000 ha. Those locations were characterized by affection (with injury degrees 20–100%, though strong affection prevailed) of separate stands of *Larix* forest covering tens to tens of thousands ha. Since two generations of the pest were observed, their negative activity took place throughout the whole vegetation period. In the third

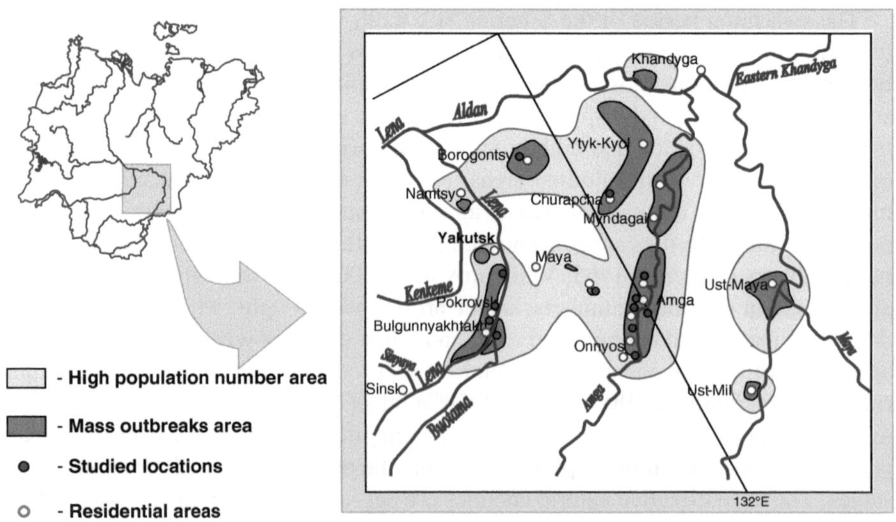


**Fig. 5.2** The map of silk moth affected *Larix* forests in Central Yakutia

**Table 5.1** The main mass outbreak locations of the Siberian silk moth in Central Yakutia (second half of twentieth century)

| Location | Peak period | Area, (in 1,000 ha) |
|---|---|---|
| Khangalassky forestry | 1948–1952 | 6.5 |
| Gorny forestry | 1965–1977 | 7.3 |
| Namsky forestry | 1969–1974 | 6.3 |
| Khangalassky forestry | 1974–1979 | 7.3 |
| Gorny forestry | 1975–1981 | 18.1 |
| Amginsky forestry | 1976–1982 | 206.2 |
| Ust-Maysky forestry | 1978–1982 | 118.6 |
| Central Yakutia | 1999–2001 | 500.0 |

year the reproduction peak entered a crisis stage, as evidenced by an abrupt decrease in population density at many examined mass outbreak locations. For instance, in 2000 the Pokrovsk mass outbreak location numbered 273–886 caterpillars and pupae per tree, while in 2001 not a single specimen was found there. The epidemic locations of the Tattinsky Ulus featured 265 specimens per a tree in 2000 and only 1–12 in 2001 (Vinokurov and Isaev 2002).

The sanitary state of the examined forests was satisfactory as a whole. Many trees were suppressed, numerous dead standing trees were observed in senile, and even in ripening and mature tree stands. The basic reasons of forest weakening are fire-caused injuries (mainly after spring agricultural burning), cattle trampling, and increased recreational load. The Siberian silk moth mass outbreak locations adjoined the settlements along the Lena, Amga, and Tatta Rivers which serve as an "ecological gutter" facilitating distribution of many animal species, including insects. Originally, before the silk moth invasion, all the examined *Larix* forest types had been dominated by forest species in the dwarf shrub-herb layer: *Vaccinium vitis-idaea, Limnas stelleri, Pyrola incarnata*, etc. The insignificant presence of meadow herbs was explained by constant fires, cattle grazing and recreational pressure cffccts.

An examination of silk moth affected forests indicates the drastic changes in growth conditions induced by the vermin browsing of *Larix* needles (Table 5.2).

**Table 5.2** Change in growth conditions in silk moth affected *Larix* forests

| *Larix* forest | Light intensity, thousand luxe | Ground thawing depth (cm) | Surface air temperature, °C | Soil temperature, °C | | | | | | |
|---|---|---|---|---|---|---|---|---|---|---|
| | | | | Depth (cm) | | | | | | |
| | | | | 5 | 10 | 20 | 30 | 40 | 50 | 80 |
| Intact | 11.8 ± 8.5 | 60–80 | 24.3 | 13.7 | 10.2 | 6.9 | 4.9 | 3.9 | 3.0 | 0.6 |
| Silk moth affected | 26.8 ± 4.2 | 110 | 24.6 | 16.3 | 13.1 | 9.2 | 7.5 | 6.4 | 5.0 | 0.7 |

Under local conditions, complete loss of needles in a tree may happen when wintered old-aged caterpillars in critical numbers of 300–500 specimens per a tree (depending on its size) are present. In that case the whole amount of needles will enter the litter compartment (as a caterpillar metabolism product) within 1–1.5 months. Higher population density of the vermin speeds up the needle utilization. From that point of view, a mass outbreak of silk moths has a positive effect on the functioning of the forest biocoenosis, but its negative consequences for forest ecosystems and forestry are more significant.

Change in floristic composition in the dwarf shrub-herb layer occurs in several stages. Within 1–2 years after a silk moth pest a gradual increase in species number and cover of eurytopic plants is observed (*Fragaria orientalis, Artemisia tanacetifolia, Vicia cracca*). As a rule, these species are common in intact forests though their abundance is low and they occupy the forest edges. In affected forests, due to change in light intensity and temperature, these plants penetrate deep into the forest, featuring high cover abundance values. For instance, in the pre-Amga and pre-Lena uluses an extremely high production of *Fragaria orientalis* fruits was observed, and this eventually decreased again when the forest entered into its next stage.

The next stage of vegetation cover development is characterized by the abundant appearance of typical meadow plants, such as *Hordeum brevisibulatum, Alopecurus arundenaceus, Calamagrostis langsdorffii (Poaceae), Carex globularis (Cyperaceae), Geum allepicum, Potentilla anserina (Rosaceae), Ranunculus monophylla, Ranunculus propinquus (Ranunculaceae), Tephroseris palustris (Asteraceae),* etc. (Fig. 5.3). Numerous fallen dead and strongly affected trees are observed too. The Cajander larch (*Larix cajanderi*) has shown a high resistance to needle browsing. Needle browsing of up to 50–70% during two subsequent summer seasons usually did not yield a mass kill of the trees, or only trees growing at forest

**Fig. 5.3** Vigorous development of the dwarf shrub-herb layer in the silk moth affected *Larix* forest

edges died due to a high concentration of vermin there. Later, a gradual recovery of the dwarf shrub-herb layer with species that are typical for the forest is observed, similar to the way the forest recovers after ground fires.

The relatively high survival capacity of *Larix* stands even following enormous affection impacts facilitates a rapid forest recovery. Hence, the following processes related to permafrost are of special concern. Examination of the silk moth affected forests in the Tattinsky forestry has revealed a dangerous process of an increase in water content in soils which points to the thawing of icy grounds (ice complex). It was induced by excessive sun heat which originally was detained by the tree canopy. Accordingly, we may assume that the disturbance of the temperature conditions in silk moth-affected forests growing in the area of an ice complex may lead to a more extensive thermokarst development, which is characteristic for Central Yakutia. This is not typical for South Siberia and the Far East, however, as they lack perennially frozen grounds with ice complexes. Thus, the large-scale mass outbreaks of the Siberian silk moth represent a threat for permafrost taiga landscapes of Central Yakutia (Fig. 5.4).

Apart from a higher fire risk in the silk moth-affected forests, these forests also represent the best conditions for the development of stem vermin – beetles (long-horned beetles, bark beetles, jewel beetles) and wood-wasps which damage surrounding healthy trees. As a result, the areas of affected forests increase.

In the Siberian silk moth pestholes the larvae and mature specimens of another vermin may occur simultaneously – *Cosmotriche lunicera* Esp. The swarming period of this insect, characterized by a one-year reproductive cycle, starts in the second decade of July to early August. The over-wintered caterpillars appear in a *Larix* canopy at the period of needle flush. Since this species does not produce pestholes of its own and is characterized by less numerous populations, its negative role in Yakutian forests is considered to be less significant then that of the Siberian silk moth.

Another common pest species in forests of the south-western and central regions of the Republic of Sakha (Yakutia) is a Siberian white-toothed moth (*Dasychira albodentata* Brem.) characterized by one year reproductive cycle. There are no records of mass reproduction centres for this species in Yakutia, though the caterpillars of the Siberian white-toothed moth regularly occur in Central Yakutia in *Larix* crowns affected by the Siberian silk moth. They comprise 10–40% of the total needle-eating pest density.

The *Larix* forests of Yakutia (from Verkhoyansk through the central regions) are also the habitat of a widespread casebearer moth species *Coleophora daurica* Flkv. Covering large areas, the epiphytotic pestholes of this species were recorded in the forests of Central Yakutia in the early 1970s. A close study of its biology was conducted by Ammosov (1972, 1975). The species inhabits trees growing along roads with intensive traffic or surrounding alases and settlements. Thus the population density of this insect probably is influenced by anthropogenic activity. The insect belongs to the *Coleophoridae* family and is characterized by small dimensions (wing spread up to 9.5 mm). The caterpillars (up to 6 mm long) live inside the needles, yielding a light-green, later brown needle colour and the needle eventually

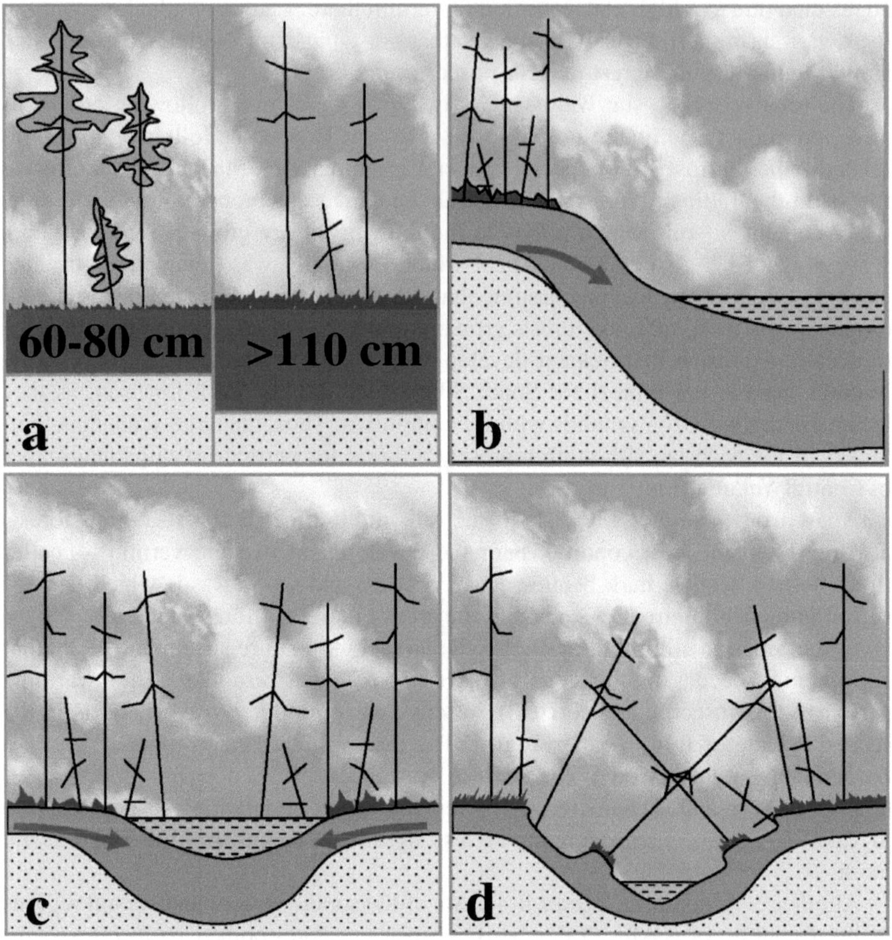


**Fig. 5.4** Change in depth of the seasonally thawed permafrost layer in the Siberian silk moth affected *Larix* forests (**a**) and possible modifications in permafrost landscapes: lake formation in depressions (**b**); initial thermokarst processes (**c**), and thermoerosions on slopes (**d**)

dies. High population density of the insects in a tree crown may result in weakening of the assimilative apparatus which has its effects on growth rates and fruiting of the affected trees, i.e. the linear growth rates of shoots are reduced, and strong suppression of trees is possible (Pleshanov et al. 1978). The population density of this pest may reach tens of millions per hectare. The pestholes recorded in the early 1970s seem permanent and presently still remain. Such a reiterating annual needle browsing may lead to considerable weakening of the tree stand.

The larch budmoth (*Zeiraphera diniana* Gn.) is able to create pestholes over vast territories of the Siberian taiga (Florov 1961; Pleshanov 1972). In Yakutia it occurs in the south-western and central regions. Thus, in 1966 a pesthole of the budmoth, as large as 2 million ha, was recorded in the ripening *Larix cajanderi-Vaccinium vitis idaea*-green moss forest in the Middle Nuyua River basin (Ammosov 1971d). The swarming period of this species lasts from mid July through the end of the second decade of August. The number of larvae in the trees of 3–5 growth classes varied from 24 to 323. Needle damage totaled 99.2% in 100%-affected trees. For pupation, a caterpillar climbs down from the tree top to the litter layer. The pesthole of the budmoth remained active for 1 year. In 1967 the population of the budmoth amounted to single specimens, and the pesthole moved to the territory of the Irkutsk Region, where during previous outbreaks the population density of caterpillars of various ages of this species in ripe and overripe forests reached several tens of thousand specimens per a tree (Kazachinskaya and Kondakov 1964). Having emerged, the caterpillars lodge in a growing needle whorl in search of food. Later, the older caterpillars crop the needles leaving only stipes. The caterpillars tie the damaged needles up with a web and thus disturb their normal functioning. Every larva is stated to damage 300–500 needles during its development period (Pleshanov 1972). Thus, a high population density of caterpillars of the larch budmoth results in severe damages and dying of needles. Besides, a tree top becomes contaminated with web and caterpillar excrements. All this reduces growth rates and fruiting of *Larix*.

Occupying more xeric landscapes, the *Pinus sylvestris* forests of Central Yakutia are less exposed to the negative influence of massing species of needle-eating insects. The pine resin-gall moth *(Retinia resinella* L.) may have a certain effect on the growth rate and fruiting of *Pinus sylvestris*. It affects mainly young growth, though is also common in mature trees growing on forest edges. Trees in the 1–2 age classes are most affected (Petrenko 1965; Ammosov 1967). Damaged trees possess false resin galls containing moths inside. Newly emerged caterpillars start feeding on the needles. Later they penetrate young shoots, forming a false resin gall under the bark at the place where the insect penetrated. The next year the caterpillars penetrate deeper inside the shoot, where they shed their skin and pupate in a resin excrescence. Having spent two winters, the moths appear in June. As a result of the harmful effects of the pine resin-gall moth, *Pinus* trunks become bent, which may lead to misshapen trees.

According to Ammosov (1975), the *deciduous forests* of Central and South Yakutia are characterized by a rather rich fauna of dendro- and thamnophilous insects feeding on birches, aspens, willows, and shrubs. Many of these phytophages feature stable large population sizes, thus actively participating in the substance-energy turnover. There are several species able to produce mass outbreaks.

Black-veined White (*Aporia crataegi* L.) is widely distributed in the middle taiga zone of Yakutia. Its larvae affect *Crataegus dahurica* and *Vaccinium uliginosum.* Mass outbreaks were recorded in 1966–1967 in forests of the Middle Nyuya valley on *Vaccinium uliginosum* thickets, which were completely defoliated (Ammosov 1971c). Only partial recovery was observed the next year, while most plants died. It is obvious that the Black-veined White has a negative effect on berry production and harvest rates. In the central regions of Yakutia the Black-veined White is known as a dangerous pest of *Crataegus* (Ammosov 1967; Petrenko 1965). It prefers *Crataegus* shrubberies growing in open places in the Middle Lena valley. The butterflies lay eggs in piles on the underside of a leaf. The caterpillars live in colonies and do insignificant harm during the first year of their life. They spend the winter in a common nest made of web and fixed openly on branches. Over-wintered caterpillars leave their nests in May and spread over the plant feeding on leaves and flowers.

*Nymphalis xanthomelas* Esp. is also common for the middle taiga zone of Yakutia and severely damages willow shrubberies along the river valleys. The first mass outbreak was recorded by Ammosov in 1967–1968 (Ammosov 1971a). In the pesthole (the Middle Nyuya valley) the willows were completely defoliated. Each shrub contained up to 500 caterpillars browsing leaves up to the central vein. In July, every linear meter of branches bore 3–135 pupae. The average weight of each living pupa was 716 mg. The largest pesthole stretched to 2.5 km, while the areas of other mass outbreak centres along the Nyuya River and on islands were much smaller. There is no doubt that the impact caused by *Nymphalis xanthomelas* is considerable. However, due to their high regenerative capacity, the willows survive and recover their foliage within 1.5 months.

Thus, while the *Lepidoptera* in Yakutia contain about 80 species, the phyllophagous *Lepidoptera* comprise only an insignificant amount of species, but they are able to cause a considerable impact on the Yakutian forest ecosystems, as they are mainly widespread insects. At the very beginning the mass outbreak in a forest develops more or less unnoticed. The pesthole appearance is thought to be preceded by a series of biological and climatic factors which break the resistance and vitality of the trees. Later, as a pesthole expands, the insects attack rather viable trees.

## 5.2 Stem Damaging Insects

**A.I. Averensky**

A group of pest insects with secondary impact are the xylophagous stem pests, though wood-consuming insects comprise not only vermin. On the contrary, most xylophagous species have a positive effect on forest functioning. By feeding on wood and decomposing wood with participation of microflora and wood-destroying fungi, they play a very important sanitary role enriching the soil with organic matter. Xylophagia is the most ancient food specialization of insects. In the course of evolution one group of these insects has diverged to feeding on newly dead or living plant tissues. No doubt, the beetles were among those insects (Rozhkov 1981) representing now the prevailing xylophagous group.

Wood remains in a forest serve long as a food source for numerous living organisms, including insects. Thanks to the vital activity of the latter the organic matter of the forests participates into the general natural cycling. Without the mechanical activity of the insects the decaying process would last for decades (Mamaev 1977). During the period of decay the composition of the xylophilous invertebrate communities undergo changes, but all the organisms are connected in a chain of events. This change is called "succession".

Following the classification of the stages of natural wood decay by Mamaev (1960, 1977), based on literature data and his own studies, each successional stage is named by the Latin name of prevailing (both in number and composition) invertebrate families. Five stages are recognized: for bark – the scolytid, cerambycid, pyrochroid, formicid and lumbricid stages; for wood – the lymexylonid, cerambycid, lucanid, formicid and lumbricid stages.

The first attempts to characterize the stages of insect-induced wood destruction in Yakutia were made by Averensky (1987a,b). The severe climatic conditions of the region determine a poor species assemblage of xylophages. Owing to this, the classification by Mamaev is not useful in Yakutia, since some characteristic families do not occur there or are insignificantly represented. Thus, the following stages are distinguished for Yakutian conifers: for bark – the scolytid, cerambycid, formicid and lumbricid stages; for wood – the cerambycid, formicid and lumbricid stages. The pioneer pest species of the *Scolytidea* and *Cerambycidae* are best represented and studied. Together with these xylophages the whole complex of companion xylophilous insects (saprophages, parasites, predators) develop under barks and in wood. The composition of such complexes depends on the tree species, moisture conditions, and type of wood rot. The companion insect complex is represented by 140 species in South Yakutia and by 102 species in the central regions of Yakutia. These insects serve as food for many birds, which are an essential link maintaining the balance in the ecological chain.

There are plenty of data on resistibility of conifer species to affection by xylophages in adjacent territories (Isaev 1966; Isaev and Girs 1975; Isaev and Petrenko 1968; Isaev et al. 1974; Rozhkov and Byalaya 1966; Rozhkov 1981, etc.).

As for Yakutia, 55 species affect the forests of the middle taiga. In a healthy forest the xylophilous insects inhabit individual ill, or overripe, or otherwise weakened trees. The mass reproduction centres of xylophages develop in cutting areas and along the roads for timber transportation. From these centres the insects radiate to the adjoining tree stands. Large-area pestholes are formed in forests suppressed by various factors: windfall, fires of various ages and dead-standing trees.

We have already mentioned the phyllophage resistibility of *Larix*, the major forest species. This tree is also well adapted to trunk vermin. *Larix* is characterized by a thick bark containing high concentrations of tannins, while its wood is rich in resins and gum known for its protective and repellent properties (Rozhkov 1981). However, since larch wood is an important resource used in building constructions, hydroconstructions, manufacturing of cross-ties, posts, etc., the deterioration of the technical qualities of larch wood caused by xylophages yields significant losses to the national economy.

As said before, pestholes of stem vermin develop in cuttings, burned areas with weakened and dying trees, as well as within the mass outbreak centres of needle- and leaf-eating insects. The results of studies on stem pests in their mass outbreak centres are discussed below.

*Post-fire forests*. A characteristic feature of the forests of Yakutia, as for the whole of Siberia, is their regular damage by fires (Averensky 1979). The permanent presence of stem vermin in tree stands makes them especially susceptible for the impact of the regular fires. Trees do not necessarily die after fires. However, an attack of stem pests considerably aggravates the effect of fires. Suppressed and dying after fires, the trees serve as centres of mass reproduction and reserves of stem pests threaten the adjoining tree stands. That is why an underestimation of the role of xylophages in the forest ecosystems may lead to significant losses in forestry. The primary role in damaging the post-fire tree stands is played by the pioneer insect groups. The composition of such groups depends on the time since the fire took place, and on characteristic features, such as the intensity of the fire, etc.

A study on the origin and development of stem vermin pestholes in post-fire forests was conducted in the south-western regions of Yakutia (Ammosov and Averensky 1971, Averensky 1979; Zabelin and Averensky 1974). The investigation was carried out in *Larix* and *Pinus sylvestris* forests, both on artificial test plots and in the natural post-fire environment. The fauna of stem pests inhabiting burned areas consisted of 42 species, of which only 8 are able to form pestholes (*Monochamus sutor* L., *Monochamus urussovi* Fisch., *Rhagium inquisitor rugipenne* Rtt., *Ips subelongatus* Motsch., *Orthotomicus saturalis* Gyll., *Phaenops guttulata* Gebl., *Stephanopachys linearis* Kug., *Hylobius albosparsus* Boh.). The rest of the species do locally a bit of harm or develop on dead wood.

The species assemblages of stem pest groups are different for *Larix* and *Pinus sylvestris* forests. Pine forests feature colonies of *Acanthocinus carinulatus* Gebl. and *Ips acuminatus* Gyll. In mixed post-fire forests the groups are mixed and an assemblage of pioneer and derivative groups is a most dangerous, complex stem vermin. After a fire, a tree usually starts dying from its butt-end, while pests settle in the butt or locally in some other part of a tree.

If a strong fire takes place in early spring or summer affecting ripe or ripening *Larix* forests mixed with *Pinus sylvestris*, the pests may invade trees the same year. Again pest settlement starts from the butt-end. *Stephanopachys linearis, Orthotomicus saturalis, Ips subelongatus,* as well as *Ips sexdentatus* Boern. preferring a thick pine bark, are very active. In the case study the latter preferred young pine growth and affected 22.2% of the fire-affected trees. The colonies were recorded up to a height of 0.8 m from the ground with increasing density from 0.02 to 1.1 nests per $dm^2$.

In the second half of summer *Phaenops guttulata* and *Monochamus sutor* start invading weakened trees. The year after the fire the derivative pest group is formed, including all the pest species, causing physiological and material harm. They populate trees starting from the ground upwards with a gradual increase in population density. For instance, the density of nests of the Asian larch bark beetle (*Ips subelongatus)* increased from 0.8 to 5.0 per $dm^2$ in a thick larch bark, while of *Orthotomicus saturalis* it increased up to 3.8 per $dm^2$.

The final ecological group is formed only in the third year after the fire. It comprises mainly material wood pests of the *Cerambycidae, Siricidae* families, and *Trypodendron spp*. Later this group is gradually replaced by the xylophages consuming dead wood (larvae of *Callidium, Acmaeops, Clitus*, etc.).

In case of severe fire damage resulting in tree death, the trees are invaded by stem pests populating the whole tree or the middle part of the trunk. The maximum concentration of xylophages is observed in the transitional bark zone in the middle part of the trunks. The leading pioneer species are *Orthotomicus saturalis* and *Ips subelongatus* with rare occurrence of *Melanophila acuminata* Debeer, *Phaenops guttulata, Monochamus spp.*, and *Tetropium gracilicorne* Rtt. Pest population densities and their distribution increase at high rates in fire centres. *Phaenops spp.* are light-demanding and occupy the best-lit parts of trunks around burns. In the case study their number of larvae reached 70 specimens per linear meter of a trunk. The derivative and final xylophage groups are formed in the second year after the fire. They are composed moistly of cerambycids, the activity of their larvae leading to final death of the trees. The density of cerambycids larvae per linear meter usually exceeds the permissible state standards for commercial timber quality by 2–3 times. Such population densities of these material wood pests occur in the second-third year after ground fires. By that time up to 93.9–97.8% of the fire-affected trees are inhabited by xylophages, while the post-fire areas in ripe and ripening forests still serve as a reserve for stem vermin for the forth-fifth years after fires. They may represent a threat to the adjoining tree stands due to the ability of xylophages to rapidly increase their number if trees get weak. In the 6–15-year old fire-sites the xylophages are concentrated only in individual dying trees, just as in a healthy forest. All this evidently shows that fire-caused losses in forests are aggravated owing to the intensive activity of stem vermin.

*Silk moth affected forests.* Their role in the formation of pestholes of stem damaging insects is still subject to study. Only the epiphytotic pestholes of the silk moth in the *Larix* forests of Central Yakutia that took place in 1999–2000 at over 500,000 ha has drawn the attention of specialists.

The start of a stem pesthole and its faunistic composition in silk moth-affected forests depend on the degree of tree damage (primary or secondary cropping, complete or partial crown damage, etc.) and the type of pesthole (temporary or permanent, sporadic or continuous). In the examined silk moth pestholes near settlements and roads the dying and already dead *Larix* trees proved to be a favourable food resource for stem vermin. Tree dying was provoked by complete crown defoliation by the silk moth larvae of various generations, aggravated by weather conditions, such as a drought of several years. This resulted in the absence of the primary ecological group of xylophages (the scolytid stage); the trees died without participation of the stem pests. The trees which still had bark became inhabited by long-horn beetle larvae. *Sirex ermak* Sem. was recorded here and there on individual dead trees. Live trees showed full (affecting the whole tree) and local invasion types by *Ancylocheira strigosa, Monochamus spp.* and *Acanthocunus carinulatus.* The fresh phloem of butt-ends contained the larvae of *Rhagium inquisitor rugipenne* and *Hylobius albosparsus.*

The primary ecological group of stem pests, mainly composed of *Orthotomicus saturalis* and *Ips subelongatus*, occasionally occurred on some trees. As a whole, the faunal composition of the stem vermin fauna the Central Yakutian forests affected by the silk moth was rather poor. The records comprise only 14 xylophage species (Averensky 2007).

As in the post-fire communities, the rates of increase of the stem pest population densities in the forests affected by the silk moth larvae depend on resources of those pests in the surrounding forests. When they are absent, the distribution and population densities of the xylophages increase gradually. They first settle in separate dying trees. Having increased in population density, the pests spread to rather viable trees. Thus, the silk moth-affected *Larix* forests of Central Yakutia represent a threat to surrounding healthy tree stands serving as a source of stem pest activity.

*Cuttings* also represent mass reproduction centres of xylophages thanks to remaining cut trunks, wood chips, stumps, as well as mechanically damaged trees serving as a perfect habitat for the pest insects. The stem vermin fauna in the cuttings of the timber-industry in the southern regions of Yakutia is represented by 34 of the 140 xylophagous species, 12 species being potentially dangerous and 6 of them being able to produce mass outbreaks. These insects are not strictly confined to cuttings, and the species composition of the group depends on tree species, bark thickness (thin, transitional, thick), size of wood chips and other remaining material (branches, butt-ends). The cut trunks and wood remains long retain fresh phloem and this serves as a favourable environment for stem pest development.

Branch phloem and pine crowns with a thin bark dry fast during the first years after cutting and are primarily exposed to xylophage activity. Thin twigs of *Pinus sylvestris* are inhabited by *Pityogenes ircutensis* Egg. and rarely by *Ips acuminatus*, while thick pine branches are occupied by *Ips acuminatus, Phaenops cyanea* F., *Ancylochera strigosa* Gebl. and *Chrysobothris chrysostigma* L., and sometimes by the larvae of *Monochamus galloprovincialis pistor* Germ. *Ips sexdentatus* prefers cut trunks and fallen trees. The density of female passages of the Asian larch bark beetle *(Ips subelongatus)* reaches its maximum values (6–12 per $dm^2$) in the zones of thick and transitional (medium-thick) bark. In the case study the average number of the Asian larch bark beetle per tree was 120–400 specimens with estimated reproduction rates of 500–3700 beetles per stem. The prevailing stem pest species are represented by *Monochamus spp.* and the *Buprestidae* family. The outer layers of logs contained 132 larvae of buprestids and 40 various-aged larch bark beetles per linear meter. This several times exceeded the permissible state standards for commercial timber.

The pine stumps are inhabited by *Ips subelongatus*, *Orthotomicus saturalis,* and *Trypodendron lineatum* Ol. Numerous larvae of *Hylobius albosparsus* and sometimes of *Rhagium inquisitor rugipenne* occur on fresh or slightly dry phloem. *Larix* branches and twigs are usually resistant to xylophage activity, while the zones of transitional and thick bark of logs and fallen *Larix* trunks are most damageable. No doubt, the highest impact is caused by technical pests, by the larvae of *Monochamus sutor* and *M. urussovi*. *Larix* stumps retain fresh phloem for a long time. Under the bark of stumps, near the saw cut area, the female passages of *Ips subelongatus* and *Orthotomicus saturalis* occur. With time, the drooping phloem of butt-ends and thick roots become invaded by the larvae of *Rhagium inquisitor* and *Hylobius albosparsus*, which spend two winters there. The piles of logs, during the first year of their summer storage, are actively occupied by stem pests; the upper layers as well as the lower layers of the logs most exposed to light represent the most favourable habitats for cerambycids. Later, *Urocerus gigas taiganus* Bens. and *Tryptodendron lineatum* appear there.

Spruce forest does not undergo commercial cutting, so, here xylophage samples were collected from individually cut and fallen trees. The thin spruce bark results in a rapid drying up of the cut trunks. This explains the simultaneous occupation of a whole trunk by stem vermin. In the case study colonies of *Pityogenes chalcographus* L. and *Polygraphus polygraphus* L. dominated under the bark and in branches, with *Carphoborus teplouchovi* Spess. distributed all over the trunk up to the thick bark zone, while the female passages of *Ips subelongatus* and *Orthotomicus saturalis* occurred under thick bark and in a butt-end, with the Asian larch bark beetle having a nest density 0.9–1.0 per $dm^2$. The estimated pest population density was 3,684 specimens per a tree for a local micropesthole.

Thus, cuttings represent the most favourable place for stem pests development and serve as their reserve for further spread to adjoining trees in case of weakening or other factors mentioned above. However, the number of vermin in spruce cuttings is not high due to the fact that the wood dries up rather rapidly.

## 5.3 Role of Insects In Herbaceous Plant Communities

**Yu.V. Ermakova**

*Orthoptera* represent one of the major groups of herbivorous animals in herbaceous ecosystems in Yakutia. They play a significant role in the substance-energy cycle, linking the secondary consumers and producers. Even during years of small population sizes they consume a considerable amount of plant mass. During assimilation it is enriched with active microelements from the alimentary tract. Thus, the insect excrements contain a considerable amount of nutrients and stimulants for plant growth. This has great significance for northern regions characterized by slow microbiological processes in soils. The total amount of microorganisms in the excrements is higher than in surrounding soils. After the excrements have quickly decomposed, the nutrients enter the soil being available for plants (Stebaev 1968). Besides, leaf browsing by the orthopterous insects activates plant growth (Lachininsky et al. 2002). Several morphotypes are recognized for *Orthoptera* in relation to the soil-vegetation cover (Pravdin 1978). Each type has its own morphometric indices, which in complex with feeding type, body colour and locomotion peculiarities, reflect the species' adaptation to a habitat represented by a certain vegetation type. The most widespread orthopterous group in Yakutia comprises grass inhabiting insects, which from time to time may go down to the ground, mainly for egg depositing. They possess a slim body and extremities. The typical representative of this group is *Chorthippus albomarginatus* De Geer. Most specialized species lay eggs evenly in a grass turf. The second group inhabits broad-leaved forbs, while the third group is adapted for life on the soil surface and under a herbaceous canopy. They are characterized by a high and lengthened body, developed wings, motley colour and a mixed feeding type. The fourth group comprises soil inhabitants in thin vegetation. They are stumpy insects with a wide pronotum, feed on grasses and forbs, often preferring wormwoods. There is a special insect group inhabiting shrubs and trees. In Yakutia it is represented only by *Ognevia longipennis* Shir.

By feeding type, the majority of the orthopterous insects of Yakutia belong to the oligophages, i.e. they use plants belonging to one family or to a certain group of families. For instance, the diet of the *Chorthippus* genus may consist of grasses, sedges and juncaceous species (Pshenitsyna 1987). Grasshoppers are carnivorous though they may feed on plants too, and the *Tetrigidae* species are saprophages. Yakutia typically knows a very high pressure on vegetation caused by *Orthoptera* owing to the low productivity of herbs, which varies from 0.4 to 1.8 tons/ha in various meadow types (Skryabin and Karavaev 1991). Yakutia does not feature gregarious *Acrididae*. Solitary *Orthoptera* form multi-specific communities confined to certain vegetation types. Some *Acrididae* species, under favorable conditions, can synchronically increase their population densities yielding mass outbreaks. Population densities may reach thousands of specimens per sq m, or hundreds of kilograms of biomass per hectare (Karelina 1971). In such years *Orthoptera* are able to eliminate all above-ground vegetation within a significant area. They have a strong impact on hayfields and pastures and sometimes move to cereal crops. In Yakutia the mass outbreaks of *Acrididae* have a cyclical character, usually occurring in dry years (Karelina 1971).

The orthopterous insects are usually confined to open herbaceous landscapes. The more complex the biotope's structure the more diverse the insect group that inhabits it. The distribution model of a certain species within a biotope depends on the structure of its vegetation cover. The vegetation's layering determines the vertical distribution of the orthopterous morphological types.

*Wet sedge-grass meadows.* The sedge-grass meadows are widespread both in river valleys and in watersheds in the taiga. They are inhabited by the orthopterous communities composed of meadow-bog species: *Stetophyma grossum* L., *Chorthippus montanus* Charp., *Omocestus viridulus* L,. *Tetrix subulata* L., and *T. bipunctata* L., as well as meadow species *Chorthippus parallelus* Zett. and *C. fallax* Zub. Some variation in species composition is possible depending on the combination of plant associations represented in a biotope. In Central Yakutia the orthopterous communities of wet meadows consist of 8–9 species. Besides the major species, in the lake-side meadows the following species may occur: *Bicolorana roeselii* Hagen. and *Metrioptera brachyptera* L.; on tussock *Carex-Calamagrostis* meadows: *Podismopsis genicularis* Shir., *Podismopsis jacuta* Mir. and *Zubovskya koeppeni* Zub. In West and South Yakutia the number of species rises to 11–15 including *Podismopsis poppiusi* Mir., *Chrysochraon dispar* Germ. and *Bicolorana bicolor* Phil. (Karelina 1971, 1974; Potapova and Ermakova 2001), the two latter species being characteristic only for these two regions and not occurring in the rest of the territory.

*Mesic forb-grass meadows.* There are only 2–3 orthopterous species inhabiting the floodplain meadows of Central Yakutia, and their population size is also small. The communities living in forb-grass meadows of above-floodplain terraces are more diverse, comprising 7–8 species with prevalence of *Chorthippus albomarginatus, Omocestus haemorrhoidalis* Charp. and *Aeropus sibiricus* L. All three are concerned as severe agricultural pests. *Metrioptera brachyptera* and *Chorthippus fallax* occur in considerably lower numbers. *Chorthippus intermedius* B.-Bien. and *Stenobotrus lineatus* Panz. occur sporadically. In more mesic meadows the communities are enriched by *Chorthippus parallelus, Bicolorana roeselii* and *Decticus verrucivorus* L. The drier meadows are often inhabited by species prefering steppe vegetation: *Gampsocleis sedakovi* F. d. W., *Montana montana* Koll. and *Glyptobotrus biguttulus* L. The richest communities (up to 15 species) of forb-grass meadows are found in West and South Yakutia. Low-grass meadows in the valleys of small rivers are inhabited by *Melanoplus frigidus* Bohem, *Podismopsis poppiusi*, *Zubovskya koeppeni,*and *Bicolorana roeselii*. In the meadows with higher and denser grass stands they are joined by *Chorthippus montanus, Omocestus viridulus*, *Chrysochraon dispar, Metrioptera brachyptera, Bicolorana bicolor, Chorthippus intermedius,* as well as by meadow species characteristic for Central Yakutia. Further south *Chorthippus apricarius* L. and *Gomphocerus rufus* L. appear.

*Stepped meadows* are formed on above-floodplain terraces, in dry, treeless hollows on watersheds. In Central Yakutia vast areas are occupied by secondary stepped meadows and meadow steppes appearing as a result of forest timber cutting and in former residential areas. The basis of orthopterous communities of stepped meadows may consist of the following species: *Aeropedellus variegatus variegatus* F. d. W., *Bryodema tuberculatum* F., *Glyptobotrus biguttulus, Montana montana* and *Gampsocleis sedakovi*. Also meadow and meadow steppe species *Chorthippus albomarginatus, Aeropus sibiricus, Omocestus haemorrhoidalis, Euthystira brachyptera* and *Chorthippus fallax* may occur in various combinations. The number of species in a community reaches 7–9.

*True steppes*. The steppe vegetation of Yakutia has a relic character. The investigations on the orthopterous fauna of relic steppes were conducted only in Central Yakutia. The steppe *Orthoptera* communities also include species that are characteristic for meadow steppes and stepped meadows. The only species strictly confined to relic steppes is *Celes skalozubovi* Adel. The number of species may vary depending on a type of the steppe landform. For example, there are only 4 orthopterous species inhabiting residential area vicinities. Decrease of anthropogenic pressure yields rises of up to 9–11 species. There are no data on the structure of the steppe communities of Southwestern Yakutia. The communities of Northeastern Yakutia may include *Aeropus sibiricus, Glyptobotrus biguttulus* and *Bryodema tuberculatum* F., these species being characteristic also for the steppes of Central Yakutia. Also the meadow species *Melanoplus frigidus,* and the species inhabiting mountainous regions *Podismopsis gelida* Mir., *Primnoa polaris* Mir. and *Aeropedellus variegatus borealis* Mistsh may occur.

*Forests*. *Orthoptera* inhabit primarily forest edges, though some of them occur under a tree canopy as well. The vegetation of forest edges is characterized by a rather rich floristic composition comprising both meadow steppe, forest and meadow plant species. Correspondingly, the composition of orthopterous insects living there is also diverse in species and morphological types. Species number ranges within 10–15. Most typical species are related to broad-leaved forbs: *Primnoa primnoa* F. d. W., *Podisma pedestris* and *Arcyptera fusca* Pall. Also meadow, meadow steppe and steppe species co-exist in the forest edge vegetation: *Metrioptera brachyptera, Chorthippus intermedius, Stenobotrus lineatus, Aeropus sibiricus, Omocestus haemorrhoidalis, Euthystira brachyptera Chorthippus fallax, Montana montana, Gampsocleis sedakovi, Glyptobotrus biguttulus* and *Bryodema tuberculatum*. Under the canopy of *Larix* and mixed forests *Podismopsis jacuta* and two *Tetrigidae* species occur.

## References

Ammosov YuN (1967) Cheshuekrylye – vrediteli sosnovykh nasazhdeniy zelenoy zony Yakutska. In: Lyubite i okhranyaite prirodu Yakutii. Proceedings of 4th Republican Meeting on Nature Conservation in Yakutia. Knigoizdat, Yakutsk: 156–162 – *The Lepidoptera damaging the pine forests of the green zone of Yakutsk* (in Russian)

Ammosov YuN (1971a) Cherno-zheltaya vanessa (*Vanessa xanthomelas Esp.*) – vreditel ivnyakov Yakutii. In: Ammosov YuN (ed) Vrednye nasekomye i gelminty Yakutii. Knigoizdat, Yakutsk – *Vanessa xanthomelas Esp., the vermin of willow shrubberies of Yakutia* (in Russian)

Ammosov YuN (1971b) K voprosu o massovom razmnozhenii sibirskogo shelkopryada (*Dendrolimus superans sibiricus Tschetv.*) v Tsentralnoy Yakutii. Biologicheskie resursy sushi Severa Dalnego Vostoka 2: 241–246 – *On the question of outbreak of the Siberian silk moth (Dendrolimus superans sibiricus Tschetv.) in Central Yakutia* (in Russian)

Ammosov YuN (1971c) Nasekomye – vrediteli listyev i derevyev, kustarnikov i kustarnichkov Yugo-Zapadnoy Yakutii. In: Ammosov YuN (ed) Vrednye nasekomye i gelminty Yakutii. Knigoizdat, Yakutsk – *The insects damaging needles and leaves of trees, shrubs and dwarf shrubs of South-West Yakutia* (in Russian)

Ammosov YuN (1971d) O vspyshke massovogo razmnozheniya seroy listvennichnoy listovertki (*Zeiraphera diniana* Gn.) v Yuzhnoy Yakutii. In: Scherbakov IP (ed) Okhrana prirody Yakutii. Irkutsk – *On the mass outbreak of the larch budmoth (Zeiraphera diniana Gn.) in South Yakutia* (in Russian)

Ammosov YuN (1972) Cheshuekrylye – potrebiteli listyev derevyev, kustarnikov i kustarnichkov Tsentralnoy i Yuzhnoy Yakutii. In: Ammosov YuN (ed) Fauna i ekologiya nasekomykh Yakutii. Izd-vo YaF SO AN SSSR, Yakutsk – *The Lepidoptera consuming tree, shrub and dwarf shrub leaves in Central and South Yakutia* (in Russian)

Ammosov YuN (1975) Listvennichnaya chekhlonoska daurskaya *Coleophora dahurica Flkv. (Lepidoptera, Coleophoridae)* v Tsentralnoy Yakutii. In: Ammosov YuN (ed) Nasekomye sredney taigi. Izd-vo YaF SO AN SSSR, Yakutsk – *The casebearer moth species Coleophora dahurica Flkv. (Lepidoptera, Coleophoridae) in Central Yakutia* (in Russian)

Ammosov YuN (1978) Sibirskiy shelkopryad (*Dendrolimus superans sibiricus Tschetv.*) v Tsentralnoy Yakutii. In: Rozhkov AS (ed) Khvoynye derevia i nasekomye dednrofagi. Irkutsk – *Siberian silk moth (Dendrolimus superans sibiricus Tschetv.) in Central Yakutia* (in Russian)

Ammosov YuN, Averensky AI (1971) Nasekomye – vrediteli lesov Yuzhnoy Yakutii. In: Scherbakov IP (ed) Okhrana prirody Yakutii. Irkutsk – *The insect pests of the forests of South Yakutia* (in Russian)

Averensky AI (1979) Stvolovye vrediteli na garyakh v lesakh Yuzhnoy Yakutii. In: Scherbakov IP et al. (eds) Lesnye pozhary v Yakutii i ikh vliyanie na prirodu lesa. Nauka, Novosibirsk – *The stem pests in post-fire forests of South Yakutia* (in Russian)

Averensky AI (1987a) Nasekomye – pervichnye razrushiteli drevesiny i kory khvoinykh porod Yakutii. In: Scherbakov IP (ed) Ekologo-biologicheskie osnovy lesovodstvennykh mer v

Yakutii. Izd-vo YaF SO AN SSSR, Yakutsk – *Primary wood and bark destroying insects of conifer species of Yakutia* (in Russian)

Averensky AI (1987b) Kompleksy nasekomykh – razrushiteley drevesiny i kory beryozy v Yakutii. In: Ekologia i geografia chlenistonogikh Sibiri. Proceedings of 6th Meeting of Entomologists of Siberia. Nauka, Novosibirsk: 126–127 – *Insect complexes destroying birch wood and bark in Yakutia* (in Russian)

Averensky AI (2007) Osnovnye resultaty issledovaniy stvolovykh vrediteley na vyrubkakh v lesakh Yugo-Zapadnoy i Tsentralnoy Yakutii. In: Proceedings of the Conference dedicated to 70th Anniversary of PA Timofeyev. Izd-vo YaGU, Yakutsk: 139–145 – *Basic outcomes of investigations on stem damaging insects in cuttings in the forests of South-West and Central Yakutia* (in Russian)

Florov DN (1961) Ocherki istorii izucheniya vrednoy entomofauny taigi Vostochnoy Sibiri. Trudy Vostochno-Sibirskogo filiala, series Biologiya 30 – *Historical studies of investigation of harmful taiga entomofauna of Eastern Siberia* (in Russian)

Isaev AS (1966) Stvolovye vrediteli listvennitsy daurskoy. Nauka, Moscow – *Stem pests of Larix dahurica* (in Russian)

Isaev AS, Girs GI (1975) Vzaimodeystviye dereva i nasekomykh-ksilofagov (na primere listvennitsy sibirskoy). Nauka, Novosibirsk – *Interaction of a tree and xylophagous insects (case study of Larix sibirica)* (in Russian)

Isaev AS, Petrenko ES (1968) Biogeotsenoticheskiye osobennosti dinamiki chislennosti stvolovykh vrediteley. Lesovedeniye 3: 56–65 – *Biogeocoenotic peculiarities of population density dynamics of stem vermin* (in Russian)

Isaev AS, Khlebopros RG, Kondakov YuP (1974) Zakonomernosti dinamiki chislennosti lesnykh nasekomykh. Lesovedeniye 3: 27–42 – *Natural principles of number dynamics of forest insects* (in Russian)

Karelina RI (1971) Saranchovye Viluya i uscherb, prichinyaemy imi selskomu khozyaistvu. In: Okhrana prirody Yakutii. Proceedings of 5th Republican Meeting on Nature Conservation in Yakutia. Irkutsk: 118–125 – *The acridids of the Viluy River basin and their harmful effect on agriculture* (in Russian)

Karelina RI (1974) K faune pryamokrylykh (*Orthoptera*) Yuzhnoy Yakutii. Trudy Vsesoyuznogo entomologicheskogo obschestva Volume 57. Nauka, Leningrad: 112–122 – *On the fauna of Orthoptera of South Yakutia* (in Russian)

Kazachinskaya TP, Kondakov YuP (1964) Glavneyshiye vrednye nasekomye listvennichnykh lesov Krasnoyarskogo kraya. Trudy SIBTI 39: 297–310 – *Major pest insects of the larch forests of Krasnoyarsk Territory* (in Russian)

Kolomiets NG (1957) Sibirskiy shelkopryad – vreditel ravninnoy taigi. In: Trudy po lesnomu khozyaistvu Volume 3. Izd-vo Zapadno-Sibirskogo filiala AN SSSR, Novosibirsk: 61–76 – *Siberian silk moth is a vermin of the plain taiga* (in Russian)

Lachininsky AV, Sergeyev MG, Childebaev MK, Chernyakhovsky ME, Lockwood JA, Kambulin VE, Gapparov FA (2002) Saranchovye Kazakhstana, Sredney Azii i sopredelnykh territoriy. Association for Applied Acridology International and University of Wyoming, Laramie, Wyoming – *The acridids of Kazakhstan, Central Asia, and adjacent territories* (in Russian)

Mamaev BM (1960) Zoologicheskaya otsenka stadiy estestvennogo razrusheniya drevesiny. Izvestiya AN SSSR, series Biology 4: 61–617 – *Zoological assessment of stages of natural decomposition of wood* (in Russian)

Mamaev BM (1977) Biologiya nasekomykh – razrushiteley drevesiny. In: Entomologiya (Itogi nauki i techniki VINITI) Volume 3. Nauka, Moscow – *Biology of wood damaging insects* (in Russian)

Petrenko ES (1965) Nasecomye – vrediteli lesov Yakutii. Nauka, Moscow – *Insect pests of the forests of Yakutia* (in Russian)

Pleshanov AS (1972) Aspecty vzaimnoy adaptatsii listvennitsy i seroy listvennichnoy listovertki. In: Rozhkov AS (ed) Anatomicheskiye, gistokhimicheskiye i biokhimicheskiye preobrazovaniya u listvennitsy pri povrezhdenii nasekomymi. Nauka, Irkutsk – *Aspects of co-adaptation of larch and larch budmoth* (in Russian)

Pleshanov AS, Scherbatyuk AS, Orekhova TP, Epova VI (1978) Osobennosti povrezhdeniya listvennitsy v ochagakh listvennichnoy chekhlonoski daurskoy. In: Rozhkov AS (ed) Khvoynye dereviya i nasekomye-dendrofagi. Irkutsk – *Peculiarities of larch damage in the casebearer moth Coleophora daurica Flkv. pestholes* (in Russian)

Potapova NK, Ermakova YuV (2001) Comparative analysis of *Orthoptera* fauna on two different plots in the Middle Lena valley // The role of permafrost ecosystems in global climate change. Yakutsk: 108–111.

Pravdin FN (1978) Ekologicheskaya geografiya nasekomykh Sredney Azii. Ortopteroidy. Nauka, Moscow – *Ecological geography of the insects of Central Asia. Orthopteroids* (in Russian)

Pshenitsyna LB (1987) Pischevaya izbiratelnost saranchovykh v svyazi s ikh vozdeistviem na stepnye fitotsenozy. Abstract of Thesis for a Candidate Degree. Novosibirsk – *Food selectivity of the acridids in relation to their effect on steppe phytocoenoses* (in Russian)

Rozhkov AS (1963) Sibirskiy shelkopryad. Izd-vo AN SSSR, Moscow – *Siberian silk moth* (in Russian)

Rozhkov AS (1965) Massovoe razmnozhenie sibirskogo shelkopryada i mery borby s nim. Nauka, Moscow – *Mass reproduction of Siberian silk moth and its control measures* (in Russian)

Rozhkov AS (1981) Derevo i nasekomoye. Nauka, Novosibirsk – *A tree and an insect* (in Russian)

Rozhkov AS, Byalaya IV (1966) Vrediteli stvola. In: Rozhkov AS (ed) Vrediteli listvennitsy sibirskoy. Nauka, Moscow – *Stem pests* (in Russian)

Skryabin SZ, Karavaev MN (1991) Zelyony pokrov Yakutii. Knizhnoe izd-vo, Yakutsk – *Green cover of Yakutia* (in Russian)

Stebaev IV (1968) Kharakteristika nadpochvennogo i napochvennogo zoomikrobiologicheskikh kompleksov stepnykh landshaftov Zapadnoy i Srednei Sibiri. Zoologichesky zhurnal 47 (5): 661–675 – *Characteristics of above-ground and on-ground zoomicrobiological complexes of the steppe landscapes of Western and Central Siberia* (in Russian)

Vinokurov NN, Isaev AP (2002) Sibirskiy shelkopryad v Yakutii. In: Nauka i technika v Yakutii: 53–56 – *Siberian silk moth in Yakutia* (in Russian)

Vinokurov NN, Isaev AP, Potapova NK, Nogovitsyna SN (2001) O vspyshke massovogo razmnozheniya sibirskogo shelkopryada v Tsentralnoy Yakutii. Nauka i obrazovaniye 1: 65–68 – *On the mass reproduction peak of the Siberian silk moth in Central Yakutia* (in Russian)

Zabelin OF, Averensky AI (1974) Rezultaty pervogo goda eksperimentalnykh vyzhiganiy. Biologicheskie problemy Severa 5: 180–184 – *The results of the first year after test burnings* (in Russian)

# Chapter 6
# Structural and Functional Peculiarities of the Plants of Yakutia

**T.Ch. Maximov, A.V. Kononov, K.A. Petrov, and B.I. Ivanov**

K.A. Timiryazev emphasized the great role of the green plant, more in particular chlorophyll, in the life of our planet, when he wrote: "The chlorophyll grain is that point, that focus in world space in which the living force of the solar beam transforms into chemical potential". The green plant is the only laboratory in the living world that accumulates solar energy. Assimilating inorganic substances ($CO_2$, $H_2O$ and the elements of mineral nutrition) and solar energy, the green plant preserves it in the form of potential chemical energy in numerous organic substances formed in the process of photosynthesis. During this activity plants constantly produce huge amounts of oxygen and take up $CO_2$. Thus, green plants are the most important agent creating favorable conditions for life on Earth.

## 6.1 Modern Trends of Climatic Changes In Yakutia

In the twentieth century unprecedented climate warming occurred in the world, at least, for the last millennium (Houghton et al. 2001; Solomon et al. 2007): at a global scale the average annual temperature rose by 0.6 °C. An especially quick rise has been observed since the 1970s. Most climatologists presently ascribe this to an increase in the concentration of greenhouse gases, and first of all, carbon dioxide.

The global warming observed in the twentieth century is apparent in all regions of Russia. According to Climate Change Bulletin the warming in all Russia, from the end of nineteenth to the end twentieth century, manifested itself as an overall rise in the mean annual air temperature by approximately 10 °C for this period. After the 1970s the warming trend markedly increased.

Kondratyev et al. (2003) supposed that during the last century the rise in average annual surface air temperature was between 0.3 and 0.6 °C, and since the 1950s this increased by 0.73 °C. The temperature rise is related to unusual properties of an

T.Ch. Maximov (✉)
Institute for biological problems of the cryolithozone, 41 Lenin Ave., Yakutsk, 677980, Russia
e-mail: t.c.maximov@ibpc.ysn.ru

E.I. Troeva et al. (eds.), *The Far North: Plant Biodiversity and Ecology of Yakutia*, Plant and Vegetation 3, DOI 10.1007/978-90-481-3774-9_6,

intensive El-Nino activity and an unprecedented temperature increase in the Indian Ocean (Ross et al. 1999).

Data from meteorological and geocryological stations of the former USSR show a trend of climate warming in most regions of the North during the last three decades, which in turn causes a degradation of the cryolithozone (Israel et al. 1999). As a result, the depth of permafrost occurrence has increased in the Canadian Arctic and Alaska followed by warmer winters as well as nights (Woodward and Smith 1995).

Experts believe that the climate warming is due to the greenhouse effect and is partly caused by mankind burning huge amounts of coal and petroleum and at the same time cutting forests – main stores of $CO_2$, and thus, the carbon content in the atmosphere steadily increases.

Forests have an important role in the carbon sequestration worldwide, as trees store huge amounts of carbon.

In the 1970s climatologists (Manabe and Stouffer 1995; Budyko 1974; and others) raised the issue of possible global warming within a short period of time of ca. 50–100 years. They pointed to continuous glacier thawing and the data from oceanic stations that indicated an increase in carbon dioxide concentration in the atmosphere.

Gavrilova (1993) made an attempt to contrast the temperature curves of the permafrost areas of the two continents of the Northern Hemisphere – in Eurasia and North America. In doing so she used a model of paleoclimatic climate reconstruction at double carbon dioxide concentration, as accepted in the joint Soviet-American report (Budyko et al. 1991). It is assumed that the most drastic warming will take place in the northern parts of the continents, i.e. at high latitudes, particularly in wintertime. Thus, if the warming amounts to 2 °C then the average January temperature at N 40–80° will rise by 2–12 °C; in case it amounts to 4 °C the rise is expected to be 6–20 °C. The warming in summertime will be inconspicuous (2–3 times as less, i.e. a rise of 4–6 °C). Since the winter at high latitudes is longer than the summer, this will affect the annual temperatures which in turn will affect the condition of the permafrost.

For instance, in Yakutsk (Central Yakutia is a classic example of permafrost development) the average temperatures will be as follows: in the mid-twentieth century – January: –43 °C, July: 18.5 °C, annual: –10.5 °C; in the mid-twenty-first century, if the warming amounts to 4 °C, – January: –28 °C, July: 23.5 °C, annual: –1.5 °C.

In the Arctic, in case of 2 °C warming, the temperature will increase by 8–10 °C, in the areas of continuous and discontinuous permafrost on the continents – by 5–6 °C and 4–5 °C, respectively; at 4 °C warming the temperature in the Arctic will increase by 14 °C with a gradual decrease to the south.

According to the average annual temperatures the southern border of permafrost (at the 0 °C isotherm) both in Eurasia and North America can shift to the north: by 5° latitude at 2 °C warming and by almost 10° latitude at 4 °C warming.

Theoretical data are certainly a matter of hot debates. However, multi-year data from meteorological stations prove the validity of such a prognosis. In particular, the readings for the last 20–30 years coincide well with the preliminary calculations for the end of the twentieth century. By 1980 the global temperature rose by 0.5–0.7 °C as compared to the end of the nineteenth century, and by 2000 it increased by 1 °C as predicted.

The "continental" stations also confirm the warming process taking place in the eastern part of the Arctic in Yakutia. Thus, a gradual temperature rise has been registered in the last century at the famous "Cold Poles" of Eurasia in Verkhoyansk and Oymyakon (Fig. 6.1, Verkhoyansk station). While at the end of the nineteenth century the average January temperature was –51 °C (the absolute minimum of –67.8 °C was registered on January 15, 1885 in Verkhoyansk), in the 1990s it was up to –40 °C. The annual temperature for the last 50 years has risen by 1.5 °C, and over the entire century by 2 °C.

The same situation has been observed in continental Central Yakutia (Fig. 6.1, Yakutsk station). Over the last century the winter temperatures have risen by 10 °C and by 7 °C for the last 50 years. The summer temperatures remain relatively stable (±1 °C). However, since winter is longer than summer here, it affects the annual temperature considerably. In the nineteenth century the average annual temperature was –11.5 °C but in the 1990s it rose to –9 °C, for some years being –7 °C.

Accordingly, an analysis of the meteorological data for the last 100–150 years in Siberian and Central Asia permafrost regions has revealed the following: the twentieth century, throughout the northern part of Asia, was warmer than the nineteenth century. So, over 100 years, starting from the end of the nineteenth century, the winter temperature in Eastern Siberia has increased by 10 °C. The average annual temperature for the last century has raised everywhere by 2.0–3.5 °C (Gavrilova 2007).

The degree of climate change depends on geographical factors. A temperature rise is most drastic under continental climate conditions and least in maritime climates. So, climate warming is milder in the Western arctic of Yakutia.

Despite localization, a clear latitudinal distribution is observed. Thus, the warming is 1.5 times stronger at high latitudes (Central and Eastern Yakutia) than in Southern Yakutia. The significant climatic changes occurring in the regions of the North raised the problem of their influence on the basic components of the natural environment of the Sakha Republic (Yakutia). One of these components is the continuous permafrost occurring all over the territory of the Republic of Sakha (Yakutia). Climate warming causes an increase in the seasonally thawing depth and in temperature of the upper cryolithic layers that may result in the activation of destructive cryogenic processes (thermoerosion, thermokarst, landslips, quicksands, icing) and also in a partial loss of the bearing capacity of construction pile bases. There is a real increased danger of destruction of existing engineering constructions (Skachkov 2003).

Analysis of the data obtained from observations at the station Chabyda for 1981–2001 shows no increasing tendencies for the seasonally thawing depth of characteristic terrain types (shallow valley, hillside). It is due to the fact that major factors determining the variability in the seasonal thawing capacity do not vary

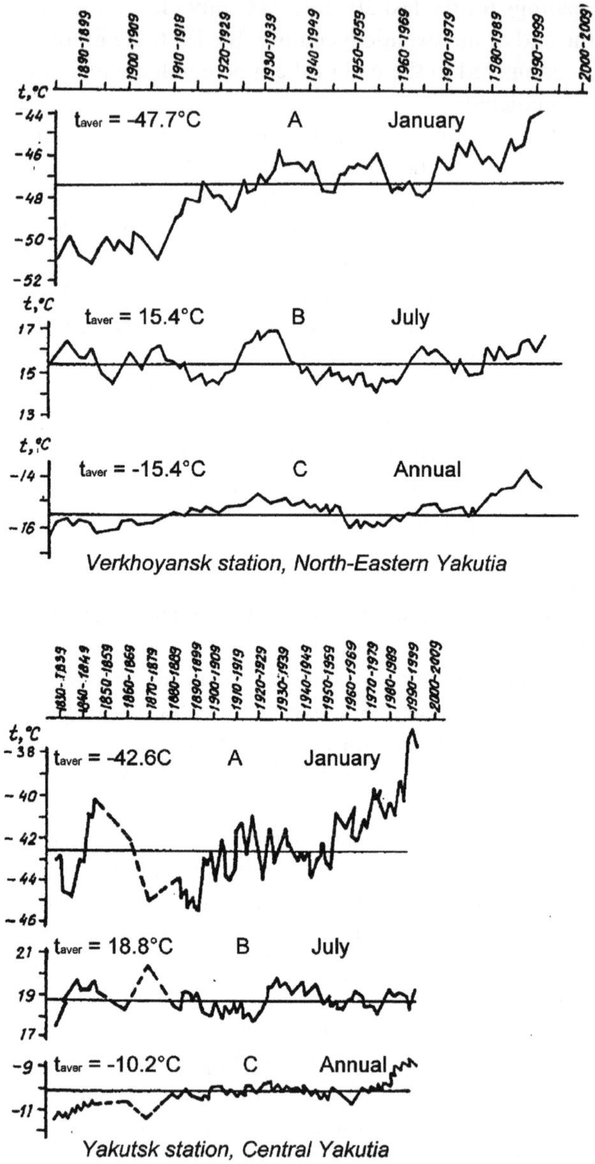


**Fig. 6.1** Sliding 10-year average temperatures, °C. Northeastern and Central Yakutia

sufficiently, i.e. the sum of air temperatures for the warm season hardly increased, and atmospheric precipitation in the summer period tended to decrease (Skachkov 2003).

In 1989–2001 the active soil layer, the cryogenic phenomena and the upper horizons of continuous permafrost layer (down to 10–15 m deep) were monitored by

P.P. Gavrilyev in the taiga, alas and valley landscapes, both natural and cultivated, in various geographic regions of Yakutia. The author clearly noticed the reaction to current climate change: a degradation of the surface cryolithozone layer down to 10–25 m (temperature rise by 1–2 °C at the surface stratum of the annual heat circulation; warming of the top of the permafrost, permafrost thawing from above, activation of destructive cryogenic processes).

The rate and scale of changes of geological and cryogenic conditions in various regions (in the north of Alaska, Western Siberia, Central Yakutia, etc.) are very different (Duchkov and Balobaev 2001; Israel et al. 2002; Gavrilyev 2003; Romanovsky and Osterkamp 2001; Anisimov and Belolutskaya 2002).

According to Desyatkin (2003), an effective factor of the moisture expenditure is a prolonged duration of the warm period due to the effect of global climate warming. Earlier arrivals of spring and advanced autumn periods compared to 1960s resulted in longer warm periods, presently by 20–25 days.

Pozdnyakov (1986) calculated that the daily moisture need in the larch forest is 1.16–1.38 mm. Our research showed that in meadows this value varies from 1.0 to 4.2 mm depending on the phase of vegetation development (Desyatkin 2006). At this rate of evaporation additional loss of moisture due to a prolonged warm period can be predicted to equal 25–35 mm in the forest and 20–100 mm in the meadows. The research showed that a possible excessive moisture loss under global warming can lead to the drying out of the active layer. If the active layer reaches the ice wedge horizon it can result in thermokarst processes.

The global warming, registered as from the second half of the twentieth century, has already strongly influenced the Earth's biota (Gruza et al. 2001; Houghton et al. 2001). The average temperature of the planet's surface, which is expected to rise further, will impact more biological processes in the biosphere and the ecological situation on the planet as a whole.

The resulting changes in species numbers in Yakutia are shown in Table 6.1. An increase in species numbers is caused by widening of the inventory work, as well as by the larger scope of the investigations and the coverage of more systematic groups.

Researches show that many of the species of animals and plants invaded the territory of Yakutia only during the last few decades, presumably due to the climate warming. The likelihood of this presumption may be proven by the expansion of the natural habitats of many species that existed in the region earlier (Labutin and Germogenov 1990; Vinokurov 2002).

Climate change can lead to a sharp increase of agricultural and forest vermins, first of all, insects and rodents. The latter are considered to be preservers and agents of different infectious and parasitic diseases (such as tularaemia, leptospirosis, alveococcosis, echinococcosis and others) (Solomonov 2003).

Climate change may lead to a faster rate of needle dropping in the taiga. While a more rapid decay of fallen leaves results in the enrichment of soil by humus, a faster rate of conifer needle dropping allows a better penetration of sun rays onto the soil and causes melting of the icy ground and intensification of the thermokarst processes (Isaev 2001; Vinokurov and Isaev 2002).

**Table 6.1** Number of plant and animal species in Yakutia (according to Solomonov 2003; Zakharova et al. 2002; Labutin and Germogenov 1990; Vinokurov 2002)

| Groups of organisms | Year | | | |
|---|---|---|---|---|
| | 1935 | 1965 | 1995 | 2000 |
| Fungi | | | 241 | >500 |
| Plants: | | | | |
| vascular | 1190 | 1560 | 1839 | 1916 |
| cryptogams including: | 577 | 1836 | 3609 | 4058 |
| bryophytes | 181 | 236 | 444 | 517 |
| lichens | 42 | 300 | 550 | 705 |
| algae | 354 | 1300 | 2615 | 2836 |
| Animals: | | | | |
| insects | 600 | 1100 | 4000 | 4300 |
| fish (species and forms) | 36 | 53 | 53 | 53 |
| amphibians | 2 | 2 | 4 | 5 |
| reptiles | 2 | 2 | 2 | 2 |
| birds | 138 | 250 | 280 | 291 |
| mammals | 37 | 60 | 63 | 75 |

The death of forests over vast territories of the medium taiga can bring great changes in the structure and functioning of present ecosystems. At the same time, there is the danger of a simultaneous degradation of steppe and meadow-steppe ecosystems within the polar regions and valleys of the Lena and Amga Rivers. There will be serious changes in taiga-alas, subarctic and mountain ecosystems. First of all, the changes will refer to the whole range of animal and plant species of the ecosystems. In its turn, this can cause global changes in the structure and functioning of all the trophic levels that provide the biotic circulation: green plants – producers, their consumers of the 1st and the 2nd order, and in the diversified guild of reducers. Significant thermokarst processes will occur in the entire North. They will be connected with the thawing of rock ice complexes, which are related to continuous permafrost and are widespread both in the forest and tundra zones. The consequent release of great amounts of water along with a thawing of the Arctic Ocean's ice layers will bring floods and the formation of a great number of new lakes. There can be changes in the whole complex of soil-forming processes. Moreover, it will be accompanied by human influence on the environment, both negative and positive. The negative effects include: continuous environment pollution, irrational use of natural resources, anthropogenous degradation of agricultural lands and natural ecosystems. The positive impact, referring to the preservation of biodiversity and stable ecosystems, will be directed towards the creation of a system of territories that will be under special protection, the creation of gene banks, the development of new cultivated plant and animal species; and it will include a transition towards stable development. Unfortunately, it is supposed that in the near future the negative anthropogenous impact will predominate, and this will strengthen the negative effect of global climate warming on nature of the North.

## 6.2 Seasonal Dynamics of Plant Growth and Development

In Yakutia, the commonly recognized continental cold pole of the northern hemisphere, about 2000 higher plant species grow, and that itself is an amazing phenomenon. The adaptive properties of plants to the climatic conditions of the region are unique. The conditions are stipulated by the combination of permafrost, high solar radiation activity and moisture deficit in the soil and atmosphere, especially during the first half of the vegetation season, as well as by a cold prolonged spring, a short summer with significant diurnal temperature amplitudes, a short autumn with rapid changes in temperatures, and late-spring and early-autumn frosts.

The growth and development of plants under natural conditions are regulated by the cumulative effect of various factors of the environment, some of which favor growth processes, while others inhibit them. It is very difficult to evaluate the influence of each factor separately, because a factor taken separately affects the plant differently than in a combination with others.

The factor most strongly inhibiting the growth of annual plants is moisture deficit in the air and soil, especially at the first half of the vegetation season, while late-spring and early-autumn frosts confine the growing season of most cultivated plants.

Investigations of exposed root systems of wild vegetation have shown that different plant species evolved different ways of adaptations to peculiar conditions of the soil environment in the zone with cold soils that contradict, amongst others, the theory of "physiological dryness of cold soils". Apart from plants forming exclusively surface roots, a number of graminoids has been discovered that have root systems penetrating deep and systematically in the permafrost horizons with temperatures of 0.5–0.8 °C.

Roots of trees, shrubs and especially grasses, go deep, to 100, 150, 200 cm, depending on the physico-chemical and agrophysical properties of the soil that are conditioned by insignificant temperature differences in the upper soil horizons of the cryolithozone. In this they differ clearly from the effects of temperatures of similar horizons in more southern areas. At the same time, under a dry climate, plants "suffer" from a moisture deficit and in search for water grow their roots deep down to the moist horizons just above the frozen zone where the soil temperatures are equal or close to zero.

It is obvious that the general character of the development and depth growth of underground organs of plants in the North differs not much from those of plants in the regions with warmer soils. Local wild and cultivated plants with a long evolution cannot undo their tolerance to soil drought, and are drought-resistant and have a short vegetation period.

The first introductive experiments studying the peculiarities of the growth and development of local and introduced shrub species in the cryolithozone were conducted at the Chochur-Muran Experimental Biological Station, that was transformed to the Botanical Garden in 1962. These experiments embraced in total around five thousand plants, specimens and forms of trees and shrubs attributable to 389 species, 80 genera and 35 families, of which 222 species belonging to 70 genera and 29 families were successfully introduced. The main focus was devoted to phenological observations and ecophysiological studies of the water regime, growth periods, development and dormancy of local and introduced trees and shrubs in relation to their frost resistance under the conditions of the winter period in Central Yakutia.

Local trees and shrubs in Central Yakutia begin to become active at relatively low air temperatures, for example at +3 °C in open places, when under the forest canopy, at the depth of the main root mass, the soil temperature still equals –3 to –5 °C (this is in the period May 14–27). Introduced trees and shrubs behave differently. They start their vegetation period 10–15 days later than local species.

During the first years of introduction of the non-indigenous woody plants, originating from habitats that were considerably different from the soil and climatic conditions of Central Yakutia, many partly or completely die being not able to adapt. Some grow as shrubs as a result of the frost. For instance, the Ussurian pear (*Pyrus ussuriensis*), a rather frost-resistant species from Eastern Asia reaching 15 m in height, did not grow up to more than 1 m under local Yakutian conditions. A North-American ash species, *Fraxinus pennsylvanica* Marsh, 2–5 m high in its native land, every year freezes down to the level of its root collar and develops as a deliquescent shrub.

Multi-year observations on the growth processes of woody shrubs in Central Yakutia have shown that the beginning of shoot growth in local species occurs during the first days of June when the mean air temperature is 10–15 °C, while that of introduced species occurs at the end of the second decade of that month at 13–16 °C. Consequently the shoot growth of local plants finishes earlier (in the second decade of July) compared to aliens (in the second decade of August).

These observations on the dependence of the growth processes of woody shrub species on environmental factors show that the local species adapted to the climatic conditions during the vegetation period in Central Yakutia, in the course of a long evolution, by shortening the calendar periods of the phenological stages and shoot growth between spring to autumn and this gives them, after having ceased their growth, more time to prepare for winter.

## 6.3 Ecological-Physiological and Biochemical Adaptation of Plants to Low Temperatures

A characteristic feature of the seasonal dynamics of the growth and development of evergreen plants in the cryolithozone during the vegetation period is that their cessation of growth and entering deep physiological dormancy occurs simultaneously with a sharp decrease in photosynthetic activity at the second half of August or the beginning of September. It coincides with the period of maximal soil respiration intensity and the greatest depth of the seasonally-thawing soil layer.

Numerous perennial herbaceous species grow actively not only in the second half of summer, but also in autumn when air temperatures are low but still above zero, as a result of earlier growth delay and arrested development because of unfavorable factors (e.g. being long flooded, frosts, mechanical damage by herbivorous animals, mowing off).

The temperature at which woody plants enter deep physiological dormancy and perennial herbaceous plants cease their vegetative growth is taken as the average diurnal transition from +5 to 0 °C. This autumn period with low positive air temperatures down to 0 °C lasts about two weeks in Central Yakutia (September 19–October 3) and in North–Western Yakutia (September 7–22).

Under these conditions, characteristically we see a substantial rise of the cell's tolerance to low temperatures, i.e. the ability to adapt by reconstructing its metabolism, to pass through a hardening process. The latter occurs in two consequent phases, and the second one – hardening by temperatures below zero – is possible only after exposure to low positive temperatures (the first phase).

The most favorable weather conditions for increasing the thermal stability of woody shrubs and herbaceous plants that still can grow in autumn in Yakutia occur during the first phase of hardening (at the end of summer and in autumn). The dominant meteorological elements for these conditions are an abundance of clear sunny days, necessary for photosynthesis, and cool nights, slowing down the expenditure of carbohydrates for respiration. About 10–15% of the annual precipitation falls in this period, mainly as snow. The period with weak (up to –5 °C) and moderate (–10 to –15 °C) frosts, suitable for passing the second phase, lasts 25–30 days. Precisely in this period the property to withstand the first low temperatures gradually forms. Many perennial herbaceous plants exposed to hardening remain green till deep in autumn and go under snow in such a state. This is due to the fact that, due to the absence of recurrent warmings, unfavorable factors in the cryolithozone at the beginning of the winter season are minimized, such as damping-off and upheaving of plants which commonly occur in regions with a mild climate.

According to modern concepts the primary role in the defense of the light-harvesting complex (LHC) of the photosynthetic apparatus (PSA) against the negative effects of adverse environmental factors belongs to oxygen-containing carotenoids. We have shown that, the content of xanthophylls (mainly lutein and zeaxanthin) in the needles of evergreen conifers (*Pinus sylvestris, Pinus sibirica, Picea obovata, Juniperus davurica*) and in the leaves (organs) of herbaceous plants (*Equisetum variegatum, Psathyrostachys juncea, Elymus sibiricus, Bromopsis inermis, Elytrigia repens*), growing in Central and North–Western Yakutia, characteristically increases before the first phase of hardening. Such a change in the concentration of LHC PSA yellow pigments in the plant cell occurs at the expense of both decreasing free water content and synthesis of carotenoids. This gives a possibility to terminate the processes of hardening and attain PSA integrity of needles of evergreen species, and prevents chlorophyll dissociation in the leaves of perennial herbaceous plants going under snow in the green state at negative air and soil temperatures.

Furthermore, for the protection from negative temperature impacts, plants increase their frost-resistance thus adapting at the biochemical level, by:

1. Accumulation of sugars possessing osmoregulatory stress-protective action. It is known that the share of easily-mobilized monosaccharides in local plants reaches up to 50–56% in autumn.
2. Accumulation of membrane lipids. Judging from a high content of essential polyunsaturated fatty acids in blood plasma and adipose tissues of the Yakutian horse, we suppose that the leaves of herbaceous plants that grow in autumn are rich in fatty acids.

## 6.4 Photosynthetic Activity of Conifers

A number of works are dealing with the ecophysiological features of larch photosynthesis in Siberia (Maximov et al. 1994, 1995, 2004, 2005a, b; Tabuchi et al. 1994; Hollinger et al. 1995; Schulze et al. 1995; Arneth et al. 1996; Koike et al. 1996, 1998; Vygodskaya et al. 1997; Fujita et al. 1998; Saito et al. 2002; Suzuki et al. 2002; Maximov et al. 2005a, b; Maximov and Ivanov 2003, 2005; and others). These works sufficiently considered the questions of the daily and diurnal dynamics of photosynthesis as well as the influence of environmental factors. However, peculiarities of the seasonal and inter-seasonal dynamics of photosynthesis are discussed insufficiently in these papers because of occasional character of the field observations.

In this chapter we analyze multi-year inter-seasonal variability of photosynthetic activity of Cajander's larch (*Larix cajanderi*), the main forest-forming tree species in Central Yakutia.

Research presented here was conducted between 1998 and 2006 in the territory of the scientific station "Spasskaya Pad" of the Institute for Biological Problems of the Cryolithozone SB RAS in a mature 180-year-old larch forest growing on relatively moist loamy soil with sandy loam inclusions.

### *6.4.1 Diurnal and Seasonal Dynamics of Photosynthesis*

#### 6.4.1.1 Diurnal Dynamics of Photosynthesis

The first large-scale daily ecophysiological studies on photosynthesis of *Larix gmelinii* in Yakutia were conducted under the guidance of professor E.-D. Schulze (Germany) on the Aldan plateau (60°51′ N, 128°16′ E; 155 km south–west of Yakutsk). Publications by Vygodskaya et al. (1997) and other specialists who worked there (Schulze et al. 1995; Hollinger et al. 1995; Arneth et al. 1996) represent a comprehensive series of work dealing with many aspects of photosynthesis, water regime and ecosystem flux of relatively young (125-year-old) larch forest. The studies were done during two weeks in the dry year of 1993 (precipitation sum for June–August was 88 mm).

Our multi-year studies showed (Fig. 6.2) that the daily rate of apparent (net) photosynthesis ($A_{net}$) of *Larix cajanderi,* irrespective of moisture conditions in any

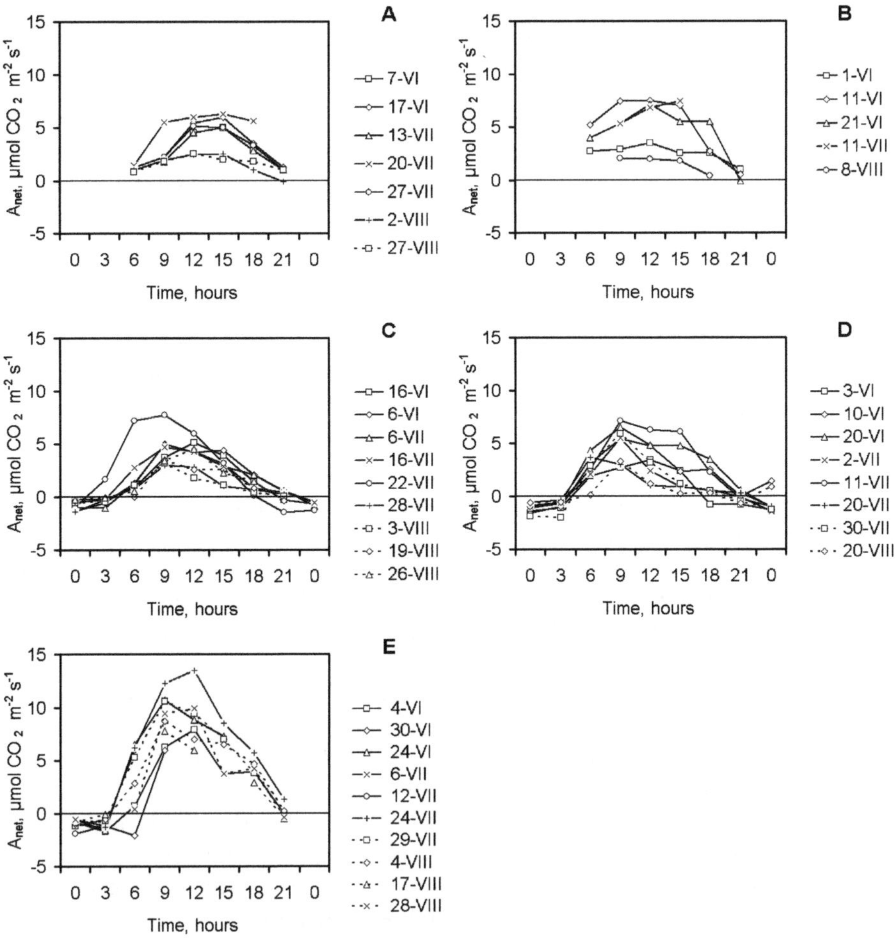


**Fig. 6.2** Diurnal dynamics of apparent photosynthetic rate of *Larix cajanderi* in years with a different hydrothermal regime (1998–2003) Notes. 2002 (**d**): dry year; 1998 (**a**) and 2001 (**c**): extremely dry years; 1999 (**b**) and 2003 (**e**): moist years

particular year, increased from early morning (4–6 h) to afternoon, then gradually decreased, down to negative values (as from about 21 h). Midday depression, characteristic for cultivated plants, was not found in the larch. All this points to a fine adaptation of the tree to a dry climate.

In the beginning of the vegetation period the highest $A_{net}$ value is observed usually before noon, and as from the mid-season the peak shifts to about 9 A.M. By the end of the vegetation period (end of July to August) the $A_{net}$ peak comes back to before noon. However, this occurs not always and depends mainly on air temperature and water vapor deficit (Vygodskaya et al. 1997; Fujita et al. 1998). For most of the summer *Larix* photosynthesized for 15–16 h (from 5–6 to 20–22 h) per day, and a high $A_{net}$ level was maintained for only 2–3 h, which is also apparent from eddy-covariance data over the whole canopy (Vygodskaya et al. 1997; Maximov et al. 2004).

Hollinger et al. (1995) demonstrated a close fit of data on diurnal $A_{net}$ of middle-aged larch forest obtained by different methods, using eddy-covariance over the forest and gas-exchange measurements of separate branches scaled to forest canopy level ($A_{net}$ was 5 μmol $m^{-2}$ $c^{-1}$ in both cases). Later, similar data were shown for a mature larch forest as well (van der Molen et al. 2003).

Vygodskaya et al. (1997) reported the following average values of daytime gas exchange in the canopy of a larch forest: 0.217 and 0.165 mol $m^{-2}{}_{soil}$ $day^{-1}$ obtained by a gas analyzer and the eddy-covariance method, respectively. The difference is believed to be negligible, especially considering the high aerodynamic roughness of the canopy. According to our data obtained earlier (1993–1994) by a gas analyzer, the daily assimilation of $CO_2$ by larch was 0.38 mol $m^{-2}{}_{needle}$ $day^{-1}$ (Maximov et al. 1995). This noteworthy difference probably is a result of soil respiration and scaling questions as well as different area measures (soil and needle): the leaf area index is about 2.0 $m^2$ per $m^2$ of soil.

Analysis of the diurnal carbon dioxide balance of *Larix cajanderi* in Yakutia was measured during extremely dry (2001), dry (2002) and moist (2003) years (Table 6.2). The results showed diurnal assimilation of *Larix cajanderi* in a moist year to equal on average 332.6 mmol $CO_2$ $m^{-2}$ $day^{-1}$, and in an extremely dry year it was 139.5. This goes beyond the values observed for other larch species: from 170 to 220 mmol $CO_2$ $m^{-2}$ $day^{-1}$ (Benecke et al. 1981; Vygodskaya et al. 1997). The maximum, 496 mmol $CO_2$ $m^{-2}$ $day^{-1}$, was observed at the end of July in a humid year. Converting this to 1 ha of forest, it is up to 220 kg $CO_2$ $day^{-1}$ or 61 kg C $ha^{-1}$ $day^{-1}$. Thus, the diurnal consumption of $CO_2$ by *Larix cajanderi* in moist years exceeds approximately 2.4 times the magnitude in dry and extremely dry years. On a diurnal base the $CO_2$ assimilation does not differ much between the years (150.7 and 139.5 mmol $m^{-2}$ $day^{-1}$ in 2001 and 2002 respectively).

$CO_2$ expenses on night respiration in *Larix cajanderi* were almost the same during the years of the study, independently of the moisture content (from 15.6 to 24.3 mmol $m^{-2}$ $night^{-1}$). Nightime respiratory expenses were higher in dry and extremely dry years, within 10.9–16.1% of the daytime-assimilated carbon dioxide. In moist years this parameter was, as a rule, low and did not exceed 5%. For comparison: cultivated plants in the North spend up to 7% of their daytime assimilation for night respiration (Maximov 1989).

Average total daily $A_{net}$ of *Larix cajanderi* was highest in moist years (6.2 μmol $m^{-2}$ $s^{-1}$) and least in extremely dry years (2.5 μmol $m^{-2}$ $s^{-1}$). Significant differences between dry and extremely dry years were not found (2.8 and 2.5 μmol $m^{-2}$ $s^{-1}$ respectively). It is noteworthy that the physiological reaction of the investigated larch to drought is weaker than to humidity. In what are typical dry and extremely dry years for Yakutia $A_{net}$ differs 1.1 times while between dry and moist it differs 2.3 times. This once again testifies to the good adaptation and tolerance of *Larix cajanderi* to moisture deficit. Preliminary results of the studies conducted in Yakutia and Japan on stomatal conductance of the forest canopy clearly indicate that the

**Table 6.2** Daily $CO_2$-exchange parameters of various larch species in different vegetation periods

| Year and weather characteristic | $CO_2$ assimilated, mmol $CO_2$ $m^{-2}$ $day^{-1}$ | $CO_2$ released, mmol $CO_2$ $m^{-2}$ $night^{-1}$ | Daily $CO_2$ balance, mmol $CO_2$ $m^{-2}$ $day^{-1}$ | Night respiration,% | Mean diurnal assimilation, μmol $CO_2$ $m^{-2}$ $s^{-1}$ | Mean nocturnal emission, μmol $CO_2$ $m^{-2}$ $s^{-1}$ |
|---|---|---|---|---|---|---|
| 2001, extremely dry | 139.5 | 15.2 | 124.3 | 10.9 | 2.5 | 0.5 |
| 2002, dry | 150.7 | 24.3 | 126.4 | 16.1 | 2.7 | 0.8 |
| 2003, wet | 332.6 | 16.2 | 316.4 | 4.9 | 6.2 | 0.5 |
| 1993, dry Vygodskaya et al. (1997) | 220 | – | – | – | – | – |
| Benecke et al. (1981) | 175 | – | – | – | – | – |

physiological response of different plant species to dry conditions is the same, i.e. there is a common physiological mechanism of plant tolerance to drought for most tree species.

#### 6.4.1.2 Seasonal Dynamics of Photosynthesis

The main peculiarity in the seasonal dynamics of forest photosynthesis in the cryolithozone is that the lower photosynthetic activity of both herbaceous and woody shrub vegetation falls in the periods of maximum soil respiration and the formation of the active soil layer (end August to early September). Because of the short vegetation period the shrubs begin to prepare for winter as from August and strongly decrease their photosynthetic rates towards dormancy. The main bulk of carbon dioxide evolved as a result of soil respiration remains partially unused by the vegetation (Ivanov et al. 1993, 1994, 1996; Maximov et al. 1996). The fact that northern perennial plants earlier enter dormancy has long been known. However, its effect, in relation to the date of maximum soil respiration falling in the autumn-winter period, on the enrichment of the atmosphere by carbon dioxide was not discussed in the literature available to us.

According to our data (Petrov et al. 1995, 1999) perennial plants in the cryolithozone begin to prepare for the long winter as from the mid vegetation period. To do so they quickly terminate growth processes, timely enter a state of deep dormancy and long autumn hardening. For example, preliminary bud dormancy of some woody shrubs sets in during the first decade of July, and the state of deep dormancy during the first half of August. It has not been reported before that phenolic inhibitors of a stilbenic nature (pinosylvin-(E)-5-(2phenilethenil)-1,3-benzoldiol), which are strong growth suppressors (in biotests), have been extracted and identified from latent buds of alder (*Dushekia fruticosa*). The range of physiologically active concentrations of pinosylvin and methyl ether inhibiting cell distention in a biotest is $4.4 \times 10^{-5}$–$1.1 \times 10^{-4}$ mol, i.e. their inhibiting action is 50 times weaker than abscisic acid and 100 times stronger than known phenolic inhibitors. All these, undoubtedly, accelerate the course of the physiological processes.

The vegetation period of plants usually lasts from June to early September inclusive, varying somewhat depending on weather conditions. In arid and cold springs and in humid years the period of stable $CO_2$ absorption by the ecosystem lengthens by about a week. For example, for the period of studies considered the total duration of *Larix cajanderi* photosynthesis during the growing season was 97 days in an extremely dry year and 104 days in a moist one.

The time of needle expansion and its potential photosynthetic activity at the beginning of the vegetation period are mainly determined by temperature conditions. As our multi-year studies in Yakutia showed the transition of photosynthesis from "blinking" to a stable state is observed at the middle or the end of the third decade of May, when the temperature at 10 cm in the soil becomes higher than 7 °C and that of the air exceeds 10 °C (Fujita et al. 1998; van der Molen et al. 2003). At that time the soil temperature at depths of 60–80 cm is still below zero. Similar data are provided by the work of Schulze et al. (1995).

According to our data a high $A_{net}$ in the larch is initiated immediately after the start of needle growth on the shoots of the previous years (third decade of May to first decade of June). Under favorable hydrothermic conditions (abundance of atmospheric precipitation and high temperature) at the initial stages of growth and development, the high photosynthetic rate of *Larix cajanderi* may be maintained in July as well (Fig. 6.3); in dry years it is restricted, as a rule, by one month (June) only. At the same time, the maximum magnitude of $A_{net}$ in humid years is higher in July than in June.

There is a direct dependence of the photosynthetic activity of *Larix cajanderi* on precipitation (Fig. 6.3). The correlation coefficient ($r^2$) between the photosynthetic rate of larch and the amount of precipitation (June–August) in dry and extremely dry periods was 0.33, and in humid years it increased till 0.77. The exception was 2000 when the curve of the seasonal photosynthesis rate of larch in that dry year had a form (with maximum in July) typical for a humid year (Fig. 6.4).

As our data show, in 1999 the total amount of soil moisture at 0–120 cm increased from 320 mm (5 June) to 416 mm (6 September) due to abundant rainfall. In 2000 a total content of soil moisture (489 mm) was observed already at the beginning of summer (1 June) and it decreased later till 307 mm (5 September). It can be assumed that despite the photosynthetic activity of *Larix cajanderi* in early summer, dry 2000 was affected by the autumn soil moisture reserves of the previous year.

The results of multi-year studies revealed that 1 ha of *Larix cajanderi* forest in extremely dry years assimilates up to 5.4, and in moist years up to 14.4 t $CO_2$

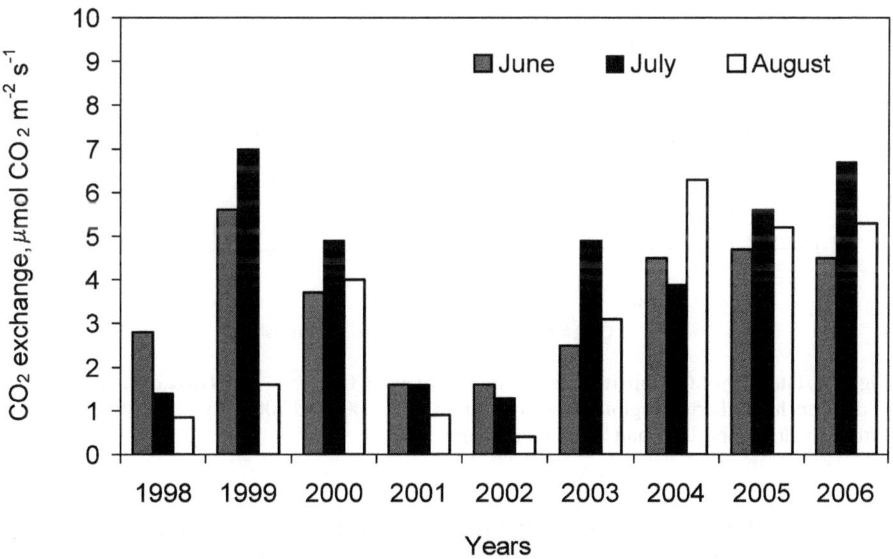


**Fig. 6.3** Mean monthly intensity of the apparent photosynthesis (μmol $CO_2$ $m^{-2}$ $s^{-1}$) of *Larix cajanderi* in years of different hydrothermal regime (1998–2006) Note: 1998 and 2001: extremely dry years; 2000, 2002, 2004: dry years; 1999, 2003, 2005, 2006: humid years

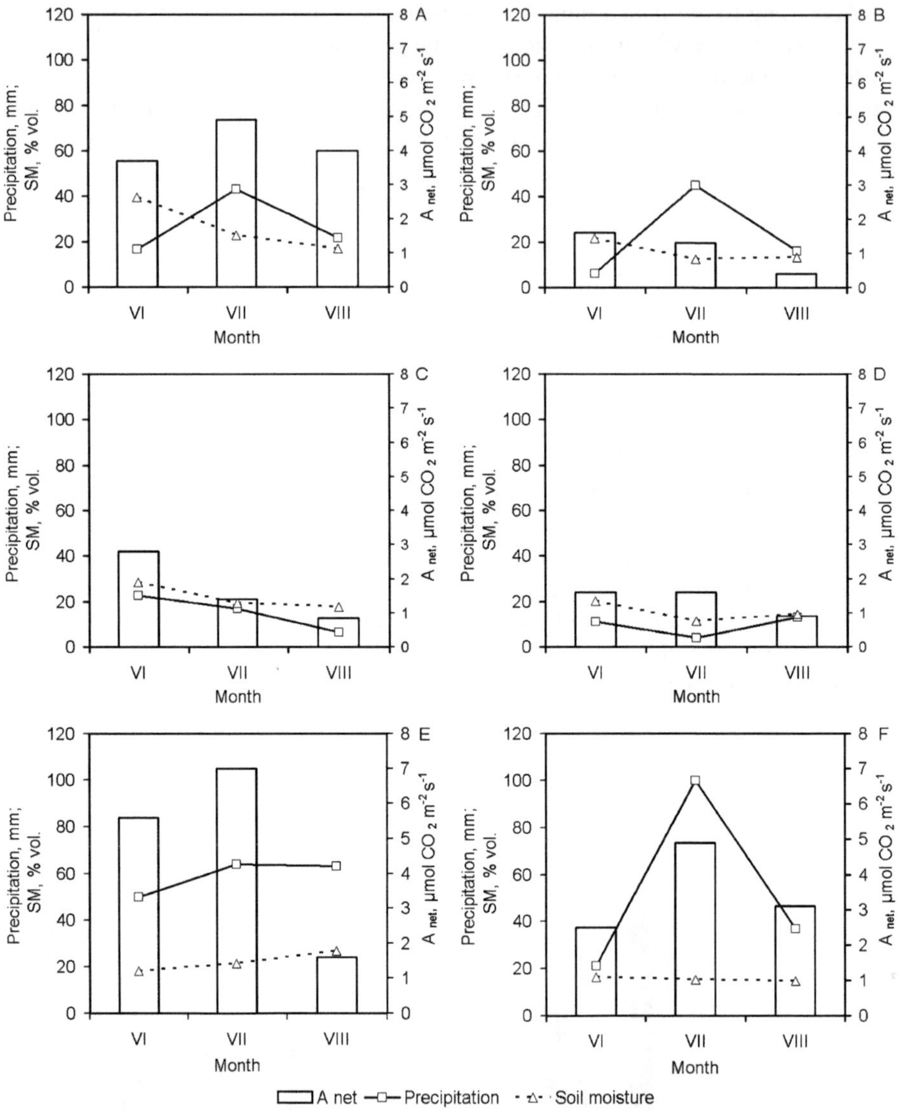


**Fig. 6.4** Intensity of the apparent photosynthesis (μmol $CO_2$ $m^{-2}$ $s^{-1}$) of *Larix cajanderi* in years of different hydrothermal regime and soil moisture Note. 2000 and 2002: dry years; 1998 and 2001: extremely dry years; 1999 and 2003: moist years

$ha^{-1}$ $season^{-1}$. The latter value is 2.5 times higher than in dry periods. Converted to carbon this is 1.5 (2001) and 4.0 t $CO_2$ $ha^{-1}$ $season^{-1}$ (2003).

The moisture regime of the permafrost soils of Yakutia was studied by many researchers since the 60 s of last century (Elovskaya and Musich 1969; Pozdnyakov 1986; Savvinov 1976; Desyatkin 2003 and others). It has been shown that the dynamics of permafrost soil thawing is principally determined by its water-physical properties as well as by type of forest, the degree of development of its wood canopy, the undergrowth, its live cover and ground litter. According to our data, by the end of July the soil layer thaws down to 90 cm, and at the beginning of August down to 120 cm (Sugimoto et al. 2002). It was shown by us for the first time that most of the soil moisture is available for plants in the period when the activity of photosynthesis and stem growth are already at their phase of decrease.

A preferable use of rain water in moist years by Yakutian trees, and of melted ground water in dry years was demonstrated by us in studies using stable isotopes (Sugimoto et al. 2002). This tells about the high adaptation of woody species to the use of atmospheric and soil moisture which gives them a main advantage under conditions of cold soils, and deficit of moisture and nutrition elements. The relatively high level of photosynthesis under the conditions of Yakutia is maintained generally by atmospheric precipitation and, in the case of drought, by the moisture accumulated during the preceding autumn and spring (a marked rise in photosynthesis at the beginning of the vegetation period). In other periods of growth and development (July, August) the capillary rise of melt water from deeper soil layers probably maintains growth processes and photosynthesis. But the effect from this is much weaker than the use of atmospheric moisture. Undoubtedly, a low and stable level of the xylem water potential (–1.5 to –2.0 MPa) to a substantial degree promotes increased consumption and suction of water by the root system from the lower soil horizons (Maximov et al. 1996). Moreover, an enlargement of the transpiration surface in the presence of a well developed root system also favors an effective water use by plants.

It is worthy of note that the conditions of water assimilation under low temperatures is studied insufficiently. Earlier Goncharik (1962), summarizing the data available, underlined that most favorable for plants is the di- and trihydrol structure of water peculiar to low temperatures but not monohydrol, which dominates at high temperatures and is absent in the ice phase. Many researchers believe that under the short vegetation period in the North the quick development of plants in spring and their cold resistance are connected with the intakc of cold water with low content of monohydrols (Ugarov 1974). At low temperatures and frosts the structure of water changes to a favorable one for plants (Pozdnyakov 1963, 1986).

Let us sum up the interesting species-linked peculiarities of the larch. The tree is a mesophyte and able to grow, like pine, in forest-steppe as well as to move up to the higher latitudes of forest-tundra. Distinguished are also its physiological and adaptive functions as regards deciduousness and the period of seed formation. Flowering, insemination and seed ripening occur during one summer season for all the larch species in Yakutia (*Larix gmelinii* and *Larix cajanderi*). However, ripe seeds of *Larix gmelinii* drop off at the end of spring to early summer of the next year (Karpel and Medvedeva 1977), while seeds of *Larix cajanderi* are released in the autumn of their ripening year. At that, the preservation of seeds wintering in the soil is much higher. Autumn seeds of *Larix cajanderi*, germinating under humid conditions in early summer, have a higher chance to give offspring. All these features are partly predetermined by the differences in one of the main physiological functions, the carbon assimilation of *Larix cajanderi* and the diurnal and seasonal dynamics therein.

## *6.4.2 The Ratio of Photosynthesis and Dark Respiration of Plants*

Photosynthesis and respiration are key elements of the plant productive process. Assessment of the quantitative ratio of photosynthesis and respiration is of great importance as comparison of these processes enables us to determine the balance of carbon assimilation and expenditure. Factual material accumulated to date on the quantitative ratio of the components of carbon dioxide gas exchange depending on environmental factors is insufficient and contradictory.

The larch in Siberia excels in its high specific productivity of photosynthesis – 2.7 times higher compared with pine and 3.9 times compared with spruce (Ivanov and Kossovich 1932; Scherbatyuk et al. 1991). Scherbatyuk et al. (1991) showed that the larch differs among conifers (pine and spruce) by a high rate of both photosynthetic activity and respiration in light. Photosynthesis and emission of $CO_2$ in light are closely related with each other, and plants with a high photosynthetic rate are characterized by a high light respiration as well (Laisk 1977; Bykov 1983). During the whole vegetation period the dark respiration of spruce and pine is always higher than the light one, while larch shows a higher light respiration. Larch species use about 60% of assimilated carbon for respiration: nearly a quarter goes to needle respiration, 13–16% to branches, approximately the same to stem and roots. As a consequence, building-up biomass takes about 40% of the assimilated carbon (Scherbatyuk et al. 1991).

During the first international researches conducted by us jointly with Tabuchi et al. (1994) rather low (for Yakutia) values of $A_{net}$ (3.1–4.4 mg $CO_2$ $dm^{-2}$ $h^{-1}$) were observed in 60-year-old larch trees. The measurements were done at the end of July in the extremely dry (33 mm of precipitation for the warm period) year of 1993 in a mature mixed larch-pine forest with dominance of pine. Later we showed that the ability to fix carbon dioxide is much lower in young larch trees compared to pine in mixed forests under drought conditions. $A_{net}$ here was comparable with that of shaded leaves of a mature tree (approximately 160-year-old), the activity of which usually is half that of illuminated leaves (Larcher 1995). As Schulze et al. (1995) notes, Siberian forests, having mainly mature wood stands, differ from the European ones by a high level of annual carbon dioxide exchange due to their age.

Under the conditions of an artificial climate chamber we, together with T. Koike (Japan), obtained very high $A_{net}$ values of larch seedlings, about 30 μmol $CO_2$ $m^{-2}$ $s^{-1}$, at gradually increasing the carbon dioxide concentration up to 1500 ppm. We also established that at doubled $CO_2$ concentration and the air temperature raised by 4 °C, all the main woody species of Yakutia (*Larix cajanderi, Pinus sylvestris,*

*Betula pendula*) have similar values and trends of $A_{net}$ independent of growth conditions (Koike et al. 2000). There is reason to suggest that high values of $A_{net}$ are one of the common adaptive physiological features of Yakutian tree species growing under the conditions of a dry climate and a short vegetation period. It is also known that the rate of photosynthesis depends not only on environmental conditions but also on the structure and physiology of the plants themselves and their organs (Mokronosov 1983; Tselniker et al. 1990). This is well demonstrated by the work of Vygodskaya et al. (1997) for *Larix gmelinii*. These researchers showed that $A_{net}$ was by 40% suppressed as a result of structural and physiological variation, generally by nitrogen distribution in the leaves within the crown. Besides, $A_{net}$ decreased due to low light intensities in the morning and evening (by 12%), and low air humidity (by 75%). That the high changeability in the $A_{net}$ of trees depends on the morphophysiological peculiarities of woody plants was shown by Tselniker et al. (1990).

Quite interesting results of photosynthesis studies adjusted for sunflecks are given in Saito et al. (2002). This important condition is not usually taken into consideration when assessing the carbon balance, although the contribution of penetrating subcanopy flecks may vary from 10 to 80% of incoming radiation, while their total participation occupies less than 10% of diurnal time (Chazdon 1988). According to some data, carbon sequestration due to sunflecks in tropical forests makes up 30–65% of the daily accumulation (Pearcy and Calkin 1983; Pearcy 1987), but it is much less in the temperate zone, e.g., 6–19% in deciduous forests (Weber et al. 1985). Considering these data as well as the relatively high $A_{net}$ of *Larix cajanderi* compared with European species, we may expect a "blinking" carbon capacity for the boreal forest as a whole, and for Yakutia in particular, as it reaches not higher than 25–30% of the total diurnal build-up (Maksimov 2003).

According to our data, maximum values of $A_{net}$ in mature larch trees strongly vary during the vegetation period depending on weather conditions. So, in dry and extremely dry years (1998, 2001, 2002) maximum $A_{net}$ was 6.3–7.5 $\mu$mol $CO_2$ $m^{-2}$ $s^{-1}$, and in humid years (1999, 2003, 2005, 2006) 7.5–13.5 $\mu$mol $CO_2$ $m^{-2}$ $s^{-1}$ (Table 6.3). The significant difference in $A_{net}$ (1.6 times on average) between moist

**Table 6.3** Maximum values of the apparent photosynthesis of *Larix cajanderi* in years of different hydrothermal regime

| Year | Precipitation sum, mm (June–August) | $A_{net}$, $\mu$mol $CO_2$ $m^{-2}{}_{needle}$ $s^{-1}$ |
|---|---|---|
| 1998 | 62.4 | 6.3 (21 June, 11 h) |
| 1999 | 167 | 7.5 (11 June, 11 h) |
| 2001 | 68 | 7.5 (22 July, 11 h) |
| 2002 | 89 | 7.1 (11 July, 8 h) |
| 2003 | 243 | 13.5 (24 July, 11 h) |
| 2004 | 97 | 10.4 (12 August, 12 h) |
| 2005 | 216 | 11.2 (4 August, 15 h) |
| 2006 | 245 | 11.3 (15 July, 12 h) |

**Table 6.4** Apparent photosynthesis of *Larix spp.* in different regions

| Region | Species | $A_{net}$, $\mu$mol $CO_2$ $m^{-2}$ $s^{-1}$ | References |
|---|---|---|---|
| Europe | *Larix decidua* | 3.0 | Benecke et al.(1981) |
| Japan | *Larix leptolepis* | 3.0 | Matyssek and Schulze (1987) |
| North America | *Larix laricina* | 3.0 | Matyssek and Schulze (1987) |
| Moscow region | *Larix decidua* | 25.0 ± 3,1[a] | Malkina (1995) |
| Central Siberia | *Larix gmelinii* | 7.8–11,5 | Koike et al.(1996) |
| Eastern Siberia (Irkutsk region) | *Larix sibirica* | 22.0[a] | Scherbatyuk et al. (1991) |
| Eastern Siberia (Central Yakutia) | *Larix gmelinii* | 8.6 –10.4 | Hollinger et al. (1995); Schulze et al. (1995); Vygodskaya et al. (1997) |
| | *Larix cajanderi* | 4.4–13.5 | Our data |

[a] $A_{net}$ recalculated for estimates based on dry weight, mg $CO_2$ $g^{-1}$ $h^{-1}$.

and dry years is clearly traced. The highest $A_{net}$ of *Larix cajanderi* (13 $\mu$mol $CO_2$ $m^{-2}$ $s^{-1}$) in favorable, moist years was 1.3 times higher than the photosynthetic activity of larches in Southern Yakutia (Vygodskaya et al. 1997), and Central Siberia (Koike et al. 1996), and 4 times higher than that of European, Japanese and American species (Table 6.4).

According to Vygodskaya et al. (1997), the high $A_{net}$ levels of Yakutian larch species, in contrast to others, are connected with a high stomatal conductance and an elevated transpiration rate. The latter is necessary for the normal activity of Yakutian plants under dry conditions in order to prevent overheating and heat shock of the leaf (Stepanov 1976; Maximov 1989 and others).

The results of our studies on $A_{net}$ showed a good concordance with the data of Woodward and Smith (1995) on the biome-wide distribution of maximum $A_{net}$. So, $A_{net}$ values for woody plants of high latitudes, theoretically assessed by them, are between 12.6 and 15.1 $\mu$mol $CO_2$ $m^{-2}$ $s^{-1}$, and experimentally obtained values are 7.6–10.1. They pointed to the high spatial and temporal changeability of larch photosynthesis and the high potential to assimilate carbon dioxide under conditions of extreme deficit of moisture and mineral nutrition. High $A_{net}$ values of *Larix cajanderi* were established both in field and laboratory environments under near optimal conditions of $CO_2$, light and temperature. All this indicates that the environmental factors listed are not limiting the photosynthesis of the Yakutian population of this species. The data obtained are in a good accordance with the results of modeling the global climate warming, in which the provision of plants with water and mineral nutrition is of first importance (Koike et al. 1996; Maximov and Koike 2001; and others).

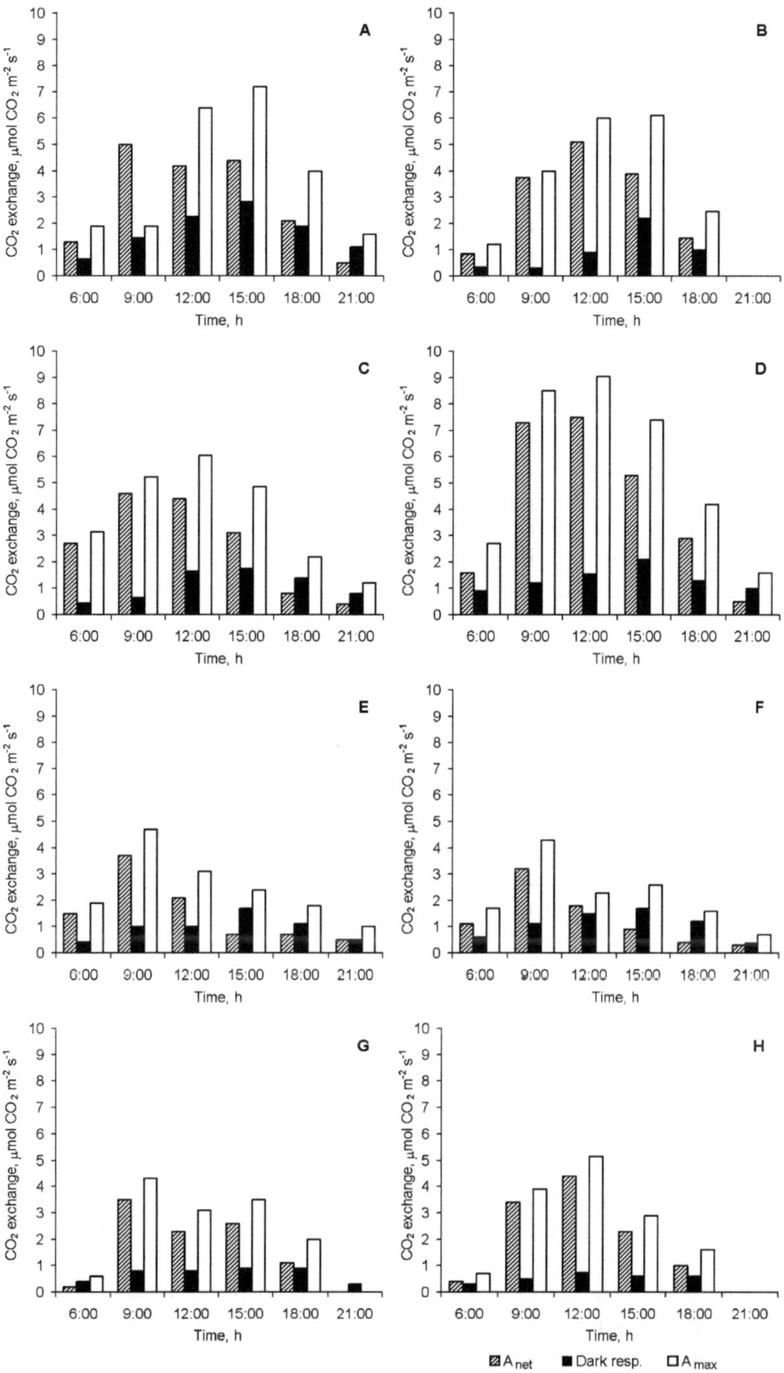


**Fig. 6.5** Ratio of the components of carbon dioxide gas exchange of *Larix cajanderi* (μmol $CO_2$ $m^{-2}$ $s^{-1}$) in extremely dry 2001 (**a**: June 6; **b**: June 16; **c**: July 16; **d**: July 22; **e**: July 28; **f**: August 2; **g**: August 18; **h**: August 25)

The components of carbon dioxide exchange of the larch ($A_{net}$ and dark respiration, $R_{dark}$) differ in years with a different hydrothermal regime. The ratio of $R_{dark}$ to $A_{net}$ reaches its maximum usually at the beginning of the vegetation period (Fig. 6.5). Naturally, the plants need much energy for the growth of new shoots and needles at this time of the year (Zabuga and Scherbatyuk 1982). As shoots grow and the $A_{net}$ increases, the $R_{dark}$ of the larch slightly drops in July and stabilizes in August. An increase in the $R_{dark}$: $A_{net}$ ratio in August is timed to the period of moisture deficit in the air and soil. In a dry year the growth of new shoots is strongly suppressed, especially in July. Therefore, energetic expenses in this period are lowered (Sherbatyuk 1976; Suvorova 2006). According to our multi-year observations, the growth of new shoots of *Larix cajanderi* lasts 1.5 months, i.e. approximately till the end of the second decade of July.

For the vegetation period the dark respiratory expenses of larch needles varied from 22 to 57% of the maximal $A_{net}$ ($A_{max}$) depending on the temperature of the environment and the moisture deficit. The average value of respiratory expenses during this period was 36.4% of the $A_{max}$. It is clear that such a ratio between the $A_{max}$ and $R_{dark}$ testifies to a good adaptation of *Larix cajanderi* to extremely dry conditions of the habitat. In other regions where conifers grow (Siberia as a whole) the $R_{dark}$ is 60% of the $A_{max}$ (Malkina 1995; Scherbatyuk et al. 1991). Comparing light ($R_{light}$) and dark ($R_{dark}$) respiration it becomes clear that the $R_{dark}$ of pine and spruce during the vegetation period is higher than the $R_{light}$ while for larch it is vice versa. A high level of photosynthesis is usually accompanied by a high $R_{light}$ (Laisk 1977; Bykov 1983). Thus, the larch species considered may have low $R_{dark}$ and high $R_{light}$ values at high values of photosynthesis.

During some periods of plant growth and development the $R_{dark}$ may exceed the $A_{net}$ during the day. This is principally observed at hours with a strained hydrothermal regime, at midday and in the afternoon (from 12 to 18 h). As the temperature of the environment increases, the $R_{dark}$ rises as well (up to 3 μmol $CO_2$ $m^{-2}$ $s^{-1}$). The longer the period of plant development, the more biomass is invested in shoot growth and the longer the $R_{dark}$ dominates over the $A_{net}$ (Scherbatyuk et al. 1991).

## 6.5 Parameters of Carbon and Water Cycles In the Forest

Preliminary studies using the eddy-correlation method, basically in Central Siberia, estimate that pine forests form a sink of between 0.5 and 2.5 t C $ha^{-1}$ $year^{-1}$. The annual carbon sequestration in the phytomass of all Russian forests is assessed at $0.27 \pm 0.03$ Gt C $year^{-1}$. Approximations from atmospheric modeling suggest the carbon sink capacity to be 1.5 Gt C $year^{-1}$ for Northern Asia. The latter includes all the changes in land-use during a 10-year period and is based on measurements of atmospheric $CO_2$. Recently, an approximately neutral carbon balance for boreal Eurasia was calculated. These results, obtained using inverse modeling methods, poorly correspond with the observations, and we undoubtedly need much more information on the carbon balance of the Eurasian boreal forests to define more exactly the prerequisites used a priori in the inverse studies.

Almost 65% of the Siberian forests are growing in the permafrost zone, and the forests of the Far East make up 45% of all Siberian forests. The vegetation and soil of Siberian forest ecosystems contain 74 and 249 Gt C, respectively (Dixon et al. 1994). According to our estimates, carbon stocks in the soils of permafrost forest and tundra ecosystems of Yakutia amount to 17 Gt (altogether 126 million ha of forest area and 37 million ha of tundra). That is about 25% of the total carbon stock in the forest soils of the Russian Federation. This carbon has been accumulated during centuries, and a rapid climate change may release huge amounts during a relatively short period, thus making a source rather than a sink of Russia. The total stock of terrestrial phytomass carbon of forests, tundra and meadows of Yakutia is 2.2–4.5 Gt C, including 0.053 Gt C in tundra and meadows.

Long-term (1993–2006) studies were conducted at the scientific forest station “Spasskaya Pad” of the Institute for Biological Problems of the Cryolithozone SB RAS (Central Yakutia, 62°15′ N, 129°37′ E, 30 km north of Yakutsk). The study site is a typical mature 180-year-old larch forest with cowberry cover (*Laricetum limnasozo-vacciniosum*), growing in relatively moist loamy soil with sandy loam inclusions. The average and maximal tree height is 17 and 27.7 m, respectively. Mean tree diameter is 18.5 cm, leaf area index of the tree stand 2.6, and density 876 trees per ha.

For the normal analysis of the net carbon dioxide gas-exchange (NEE) of the ecosystem we used both closed and open eddy-correlation systems at 34 m height. The eddy-correlation system consists of an infrared gas analyzer Licor (USA), an ultrasound anemometer Gill (USA), and a control and data recording unit from Campbell (USA). Soil respiration also was measured using two systems, an automatic one with 4 dark chambers, measuring every 15 min (PP Systems, UK) and a manual EGM-4 (PP Systems, UK). Transect measurements of soil respiration intensity were done along a wind rose (north–west and south–east) every 50 m at a distance of 500 m from the tower.

Under the conditions of the taiga of Central Yakutia woody plants have a rather low biomass of photosynthesizing organs. So, the larch needle mass in Yakutia (1.68 t per ha) is twice as small as in countries with a humid climate. The low needle biomass with a low assimilating leaf area index (up to 2.0) and a short period of photosynthetic activity determines a low primary productivity of the larch, the main forest-forming species of Yakutia (NPP $= 3.1 \pm 0.3$ t $ha^{-1}$ $year^{-1}$). Our long-term (1996–2004) observations showed that during about a 100-day vegetative period (late May to early September) the permafrost forest ecosystems are a sink for carbon dioxide with a maximum absorbing capacity of 22 kg $CO_2$ $ha^{-1}$ $day^{-1}$ or 6.1 kg C $ha^{-1}$ $day^{-1}$. The daily productivity of larch photosynthesis in humid years is 2.5 larger than in dry years, and produces on average $332.6 \pm 45$ mmol $CO_2$ $m^{-2}$ $day^{-1}$ against 139.5 in extremely dry years. This corresponds also to the values measured for other taiga larch forests (*Laricetum vacciniosum*): from 170 to 220 mmol $CO_2$ $m^{-2}$ $day^{-1}$ (Vygodskaya et al. 1997; Benecke et al. 1981).

The seasonal maximum of the photosynthetic activity of the forest canopy in dry years is observed in June, and in moist years in July. During the growing season the woody plants of Yakutia consume 1.5 to 4.0 t C per season. Night time respiration expenses amount to 10.9 and 16.1% from the daytime $CO_2$ assimilation in dry and extremely dry years, respectively. The dark respiration costs of larch for the vegetation period make up 22–57% of the maximum photosynthesis.

The average soil respiration intensity during the growing season reaches 2.2–6.9 kg C $ha^{-1}$ $day^{-1}$, which is characteristic for all Siberian forests, but 3 times less than that of Europe and North America. The maximum soil respiration of permafrost forest in Central Yakutia is observed from late August to early September (up to 10 kg $CO_2$ $ha^{-1}$ $h^{-1}$). The annual soil $CO_2$ emission amounts to an average of $4.5 \pm 0.6$ t C $ha^{-1}$ $year^{-1}$.

The results of multi-year studies into the components of carbon dioxide exchange in permafrost forest ecosystems are summarized in Fig. 6.6.

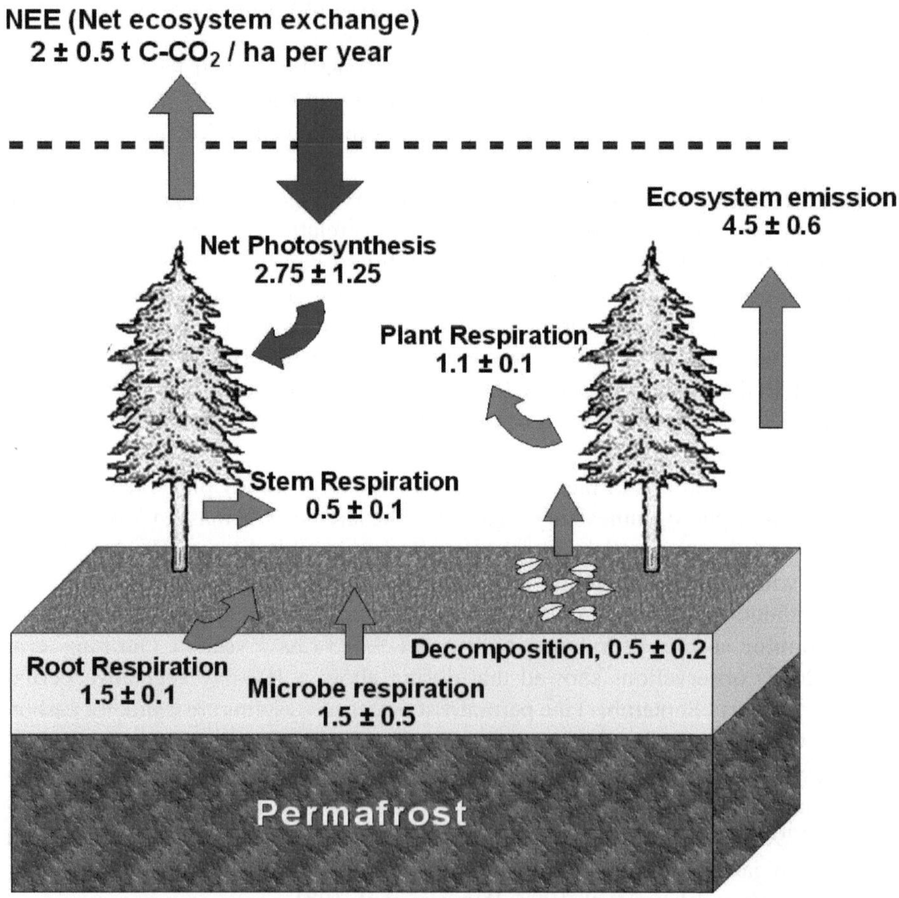


**Fig. 6.6** Annual carbon balance of larch forests in the cryolithozone

The inter-annual variation of net ecosystem exchange (NEE) in the permafrost zone amounts to 1.7–2.4 t C $ha^{-1}$ $year^{-1}$, which results in upper limits of the annual sequestering capacity of 450–617 million t C $year^{-1}$ at the total area of 257.1 million ha of these forests in Russia. The Far East larch forests of Siberia accumulate annually 0.4–1.0 Gt C and that is comparable with the values for European and tropical forests (Shimel et al. 2001). However, this does not include the emission from forest fires. According to an estimate of Isaev et al. (1995) the carbon flux related to post-fire loss of Russian forests amounts to 0.053–0.058 Gt C $year^{-1}$. With the aid of satellite observations the post-fire $CO_2$ emission from this territory is estimated up to 0.14 Gt (van der Werf et al. 2004). Therefore, the net biome productivity (NBP) of Siberian larch forests may be assessed approximately within 0.26–0.86 mlrd t of carbon.

The Proposal of the Russian Federation to the UNO Convention on Climate Change of 1995 estimates the total annual forest carbon sink of Russia to be 0.16 Gt, which is much lower than the values obtained by us and foreign researches. The contribution of Siberian forests (east of the Ural) to this sink we assume to be 55–62%. Annual sequestration of permafrost Siberian larch forests comes almost to half of the overall sink of Russian forests (55%), and soil emission makes up about 27% (Table 6.5).

Parallel measurements of water balance parameters show that the rate of evapotranspiration over a dry larch canopy amounts to 1.5–2.2 mm $day^{-1}$ from May to August, the maximum being 2.9 mm $day^{-1}$ in early July. Total larch evapotranspiration from multi-year-frozen soils made 151–258 mm $yr^{-1}$ (Ohta et al. 2001; Dolman et al. 2004). Canopy evapotranspiration was estimated as 35–50% from the total one. Moisture interception is 15–30% of the total precipitation amount at the open site. Overall evapotranspiration usually exceeded precipitation amount, and the deficit was compensated by melt above-permafrost waters and melt snow water. A schematic model of the annual water budget in a larch forest of the cryolithozone (Fig. 6.7) was compiled by us on the basis of multi-year studies, which testifies to increasing moisture deficit under the conditions of permafrost soils warming and use of moisture accumulated in the soils (up to 50 mm). Considering the significant decrease in carbon dioxide accumulation in dry years and a high frequency of forest fires in the same years, it may be surely stated that there is a noteworthy change in

**Table 6.5** Annual carbon budget of Russia and Siberia, Gt C $year^{-1}$

| | |
|---|---|
| Assimilation by vegetation | |
| Larch forests of Siberia | 0.4–1.0 |
| Larch forests of Yakutia | 0.2–0.4 |
| Emission | |
| Larch forests of Siberia | 0.6–0.9 |
| Larch forests of Yakutia | 0.3–0.4 |
| Annual net gas-exchange, NEE | |
| Larch forests of Siberia | 0.45 |
| Larch forests of Yakutia | 0.18 |

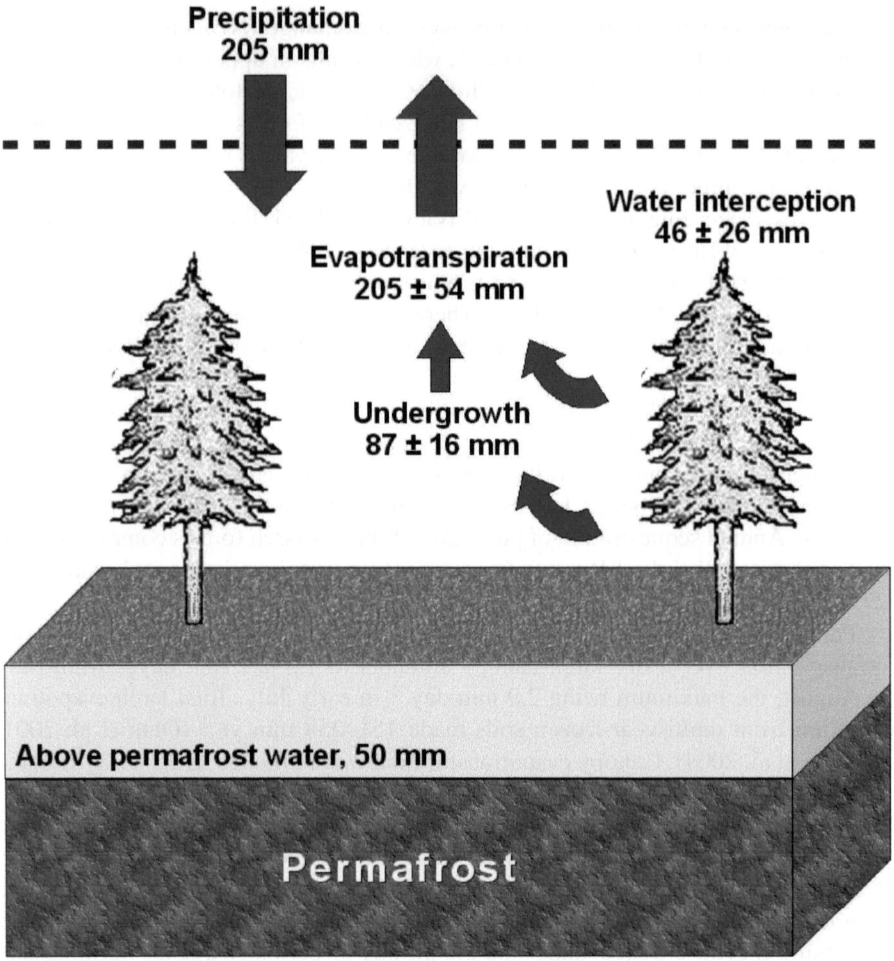


**Fig. 6.7** Annual water balance of larch forests in the cryolithozone

the water and carbon cycle balance, from accumulation to emission, for the forest zone of Yakutia under climate warming.

## 6.6 Response of Yakutian Tree Populations to Possible Climate Change

Enrichment of the atmosphere with carbon dioxide is a serious ecological and political problem because of negative consequences of the greenhouse gas effect disturbing the life-support system of mankind. According to forecasts of scientists, doubling of $CO_2$ concentration in the atmosphere will occur in the mid-twenty-first century, and that results in an increase in the planetary air temperature by $3 \pm 1.5\,°C$ (Manabe and Wetherald 1980; Gavrilova 1993).

Data from meteorological stations show that in most regions of the North there is a clear trend of climate warming for the last three decades (Israel et al. 1999; Ivanov 2003). Two scenarios are discussed in relation to possible climate warming: firstly, the warming will not be accompanied by increased precipitation and will result in total aridization of territories; secondly, a compensatory increase in the amount of atmospheric precipitation will occur. In the first case the amount of precipitation in Central Yakutia will increase insignificantly, from 0 to 30 mm per year, while in northern tundra regions it will exceed multi-year averages by 60 mm (Zavelskaya et al. 1993).

Climate change necessarily must affect the vegetation as its distribution areas and the boundary positions of natural zones depend on the temperature and moisture regimes. A shift in the position of the natural zones will affect the biological productivity of permafrost ecosystems. Therefore the problem arises of the assessment of possible changes and their prediction. Some researchers think that the larch will move into tundra at a rate of about 80 m $y^{-1}$ (Huntley and Birks 1983; Keuk 1989; Velichko et al. 1990). For other woody species the rate may equal to 200–300 m $y^{-1}$, and for pioneer species it may be over 500 m $y^{-1}$ (Budyko 1989). A question arises concerning the possible extinction of some species of herbaceous plants. A movement to the North of $C_4$-plants with their specific features in morphology, physiology and biochemistry is expected. At present the world scientific community intensively discusses the problem of the rate of plant adaptation to the changing growth conditions.

Modeling and experimental studies into the effect of climate warming on the production process of woody plants in Yakutian and Japanese populations were conducted in 1996–1998 in the Forest and Forest Products Research Institute (Sapporo, Japan). One year old seedlings of Yakutian and Japanese populations of larch, pine and birch were grown in plastic pots at a 21 h photoperiod in a liquid nutrient medium (Hyponex with N:P:K = 1:2:1) at a mixing rate of 140 mg N $L^{-1}$ $week^{-1}$. Temperature conditions for the growth and development of the plants were selected on the base of an analysis of maximum and minimum air temperatures for June–August in Central Yakutia and Northern Japan (26/12°C air temperature at day/night time). The experimental plants were grown under a temperature that was 4 °C warmer (30/16°C air temperature at day/night time) and doubled $CO_2$ concentration in the atmosphere (700 ppm), which corresponds to the climatic scenario that is predicted for the mid- to end-twenty-first century (Manabe and Wetherald 1980).

### *6.6.1 Growth and Development*

The process of adaptation of the Yakutian trees to drier climate conditions was expressed in a decreasing size and thickness of the assimilating surface. This

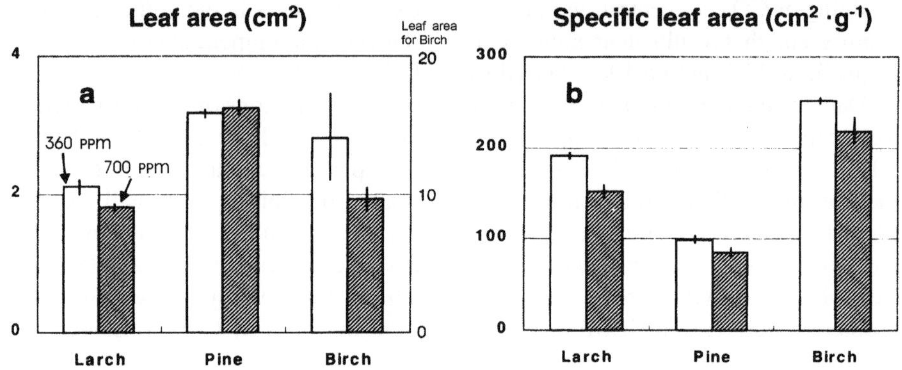


**Fig. 6.8** Influence of doubled $CO_2$ concentration and elevated temperature (+4 °C) on the growth processes of Yakutian trees

trend became especially apparent in the growth and development of birches, both Yakutian and Japanese populations, when modeling the conditions of possible climate warming (Fig. 6.8a).

Specific leaf areas of the trees under elevated temperature and doubled $CO_2$ content also increased as a result of morphological changes of the leaf apparatus towards decreasing thickness (Fig. 6.8b). A reduction of chlorophyll and nitrogen content in the leaves was observed as well.

### *6.6.2 Water Exchange*

In intensity of transpiration Yakutian trees surpassed Japanese ones by, on average, 1.8 times at lowered stomatal conductance and high photosynthetic activity. Low stomatal conductance, favoring economical water use per unit of dry matter synthesized, is an adaptive feature of Yakutian trees. It was formed in the process of a long evolution under conditions of an intensely continental dry climate.

At climate warming (higher temperature treatment experiment) an intensification of the stomatal conductance and transpiration rate of the trees, compared to presently existing growth conditions (control treatment), were observed and, therefore, an increasing water use efficiency (Fig. 6.10).

### *6.6.3 Photosynthesis and Respiration*

Quantum yield of photosynthesis at doubled $CO_2$ concentration and temperature elevated by 4 °C strongly dropped compared to the control (Fig. 6.9) in both conifers and deciduous trees. There was a marked inhibition of this parameter in deciduous trees (birch and larch) but not in the evergreen pine.

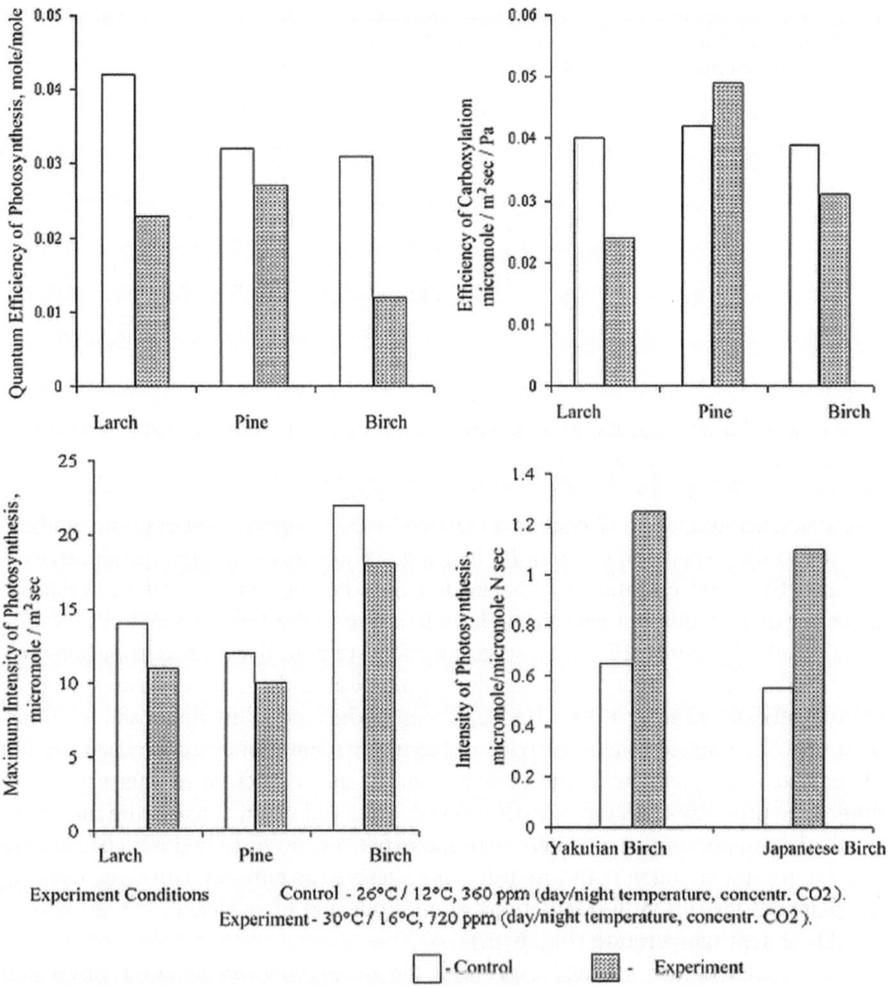


**Fig. 6.9** Photosynthetic activity of coniferous and deciduous trees of Yakutia at possible temperature rise and doubling of $CO_2$ concentration of the atmosphere

The carboxylation efficiency under the possible climate change (temperature rise and doubled $CO_2$) uniformly decreases in comparison with the control for birch and larch, while it increases for pine.

Maximum values of leaf $A_{net}$ intensity of the Yakutian trees in the control were almost 3 times as much as those of the Japanese. The high photosynthetic rate of the Yakutian woody plants compared to the Japanese trees is one of the adaptive features of the former compensating the short growing season. Under the simulated growing conditions the $A_{net}$ of Yakutian and Japanese trees decreased on average 1.6 and 1.4 times compared with the controls.

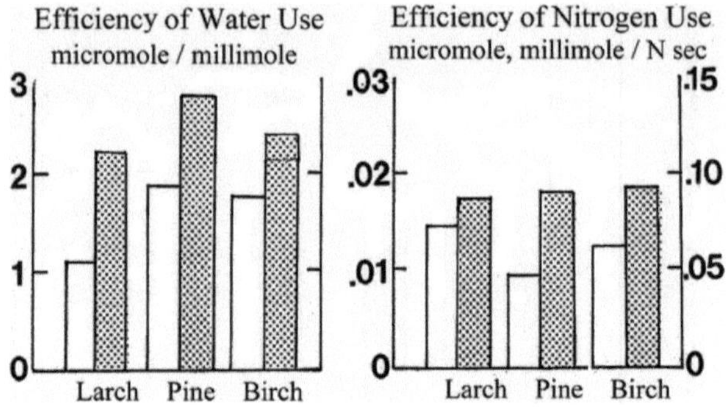


**Fig. 6.10** Water and nitrogen use efficiency of Yakutian trees at possible climate warming. The conditions of experiments and notes are the same as in Fig. 6.9

It is well known that the production process of plants may be restricted both by the photosynthetic leaf rate and the growth function. So, the elevation of the carbon dioxide concentration up to 1000–5000 ppm resulted in leaf "overfeeding", excessive starch accumulation (Acock and Hartwell 1985), suppression of photosynthesis and led to the expansion of the total leaf area. Thus, a rise in plant productivity at elevated $CO_2$ concentration is reached not only through photosynthesis enhancement per unit of area but also through an increase in the leaf area and assimilating potential (Mokronosov 1983). Some researchers suppose that a raise in carbon dioxide concentration affects first the sink magnitude and then, as a consequence, the photosynthesis (Kramer 1981). Hence, in our studies modeling the possible climate warming, the photosynthetic activity of the experimental plants decreased per unit of leaf area compared with the controls due to the restriction of this process by epigenetic processes. A contrary pattern was observed when considering the photosynthetic rate per unit of nitrogen: here the experimental trees had a higher activity of PSA in comparison with the control. This attests to an enhanced nitrogen usage by the plants of the different populations (coniferous and broad-leaved) at an increase of $CO_2$ and air temperature (Fig. 6.10).

The results of our studies, especially the observed drop in chlorophyll content, nitrogen concentration in the leaves and enhanced nitrogen use, are in a good agreement with published data on other woody species under elevated $CO_2$ and air temperature conditions (Bazzaz 1990; Eamus and Jarvis 1989). Sage et al. (1989) showed that the photosynthetic intensity in five plant species was limited by the nitrogen level in the leaves while it increased at elevated $CO_2$ concentrations. After two years of growing under a high carbon dioxide concentration the depression of the photosynthetic rate in *Pinus taeda* was not caused by the deactivation of Rubisco but by a decreased content of nitrogen and Rubisco (Tissue et al. 1993). A similar feature was observed for Alaskan perennials (Oechel and Strain 1985; Tissue and Oechel 1987).

The photosynthetic intensity is strongly affected by the donor-acceptor relationships. Under conditions of enriched carbon dioxide the depression in the photosynthetic rate is closely related to the deficit of phosphates that ultimately influences the transport of assimilates (Mokronosov 1983; Makino 1994; Sage et al. 1989; Sawada and Usuda 1992). Principally, the growth and development of annual plants at a high $CO_2$ concentration is controlled by the level and demand of nutrients (McConnaughay et al. 1993a, b). So, an enlarged input of nitrogen fertilizers changes the course of adaptation of plants to the conditions of enriched $CO_2$, and promotes a high photosynthetic rate to maintain the growth of pine seedlings (Tissue et al. 1993). The same results were obtained by Petterson et al. (1993). The shortage of mineral nutrition of plants in a $CO_2$-enriched atmosphere was accompanied by a significant raise of the biomass of the root system, generally due to its elongation (Kondrasheva et al. 1993; Acock and Hartwell 1985).

### *6.6.4 Respiratory Ability of Leaves Under Increased Temperature and Carbon Dioxide In the Atmosphere*

The respiratory ability of birch leaves in all the treatments of the experiment was most strongly affected in the Yakutian plants. The explanation is that under the conditions of water and mineral nutrition deficit the Yakutian trees spent more assimilates on supporting a unit of biomass than the Japanese plants do (Maximov 1989). The suppression of the respiration intensity in the leaves of the Yakutian trees under a complex influence of elevated temperature and doubled $CO_2$ concentration indicated that temperature is not the dominating factor limiting productivity of forests. An insufficient provision with moisture and mineral nutrition remains the limiting environmental factor for the Yakutian plants.

Additionally, the results of our studies showed a uniform decrease in the quantum yield of photosynthesis, carboxylation efficiency, photosynthetic rate and dark respiration, per unit of leaf area and for both Yakutian and Japanese tree populations, under possible climate warming, due to a restriction of the plant productivity caused by epigenetic processes. Along with nitrogen and water use efficiency, a uniform increase in photosynthetic activity of the trees per unit of nitrogen was noticed. A direct dependence of the activity of the photosynthesizing leaf on its nitrogen content indicates that under the predicted climate change photosynthesis will accelerate or, to the contrary, be reduced.

Obviously climate warming will enhance the activity of soil microorganisms decomposing forest litter. Some researchers think that the thawing of the permafrost will cause a wide distribution of pedoturbations and mixing of soil mass (Karpachevsky 1993). In that case some fractions of the humus and peat horizons, rich in organogens, may be buried into the lower part of the soil profile. This will conserve the organic matter and withdraw the main organogens (nitrogen, carbon) from circulation.

Unfortunately, at present there are no results of long-term studies on the response of most natural ecosystems to an increase in $CO_2$ concentration, and

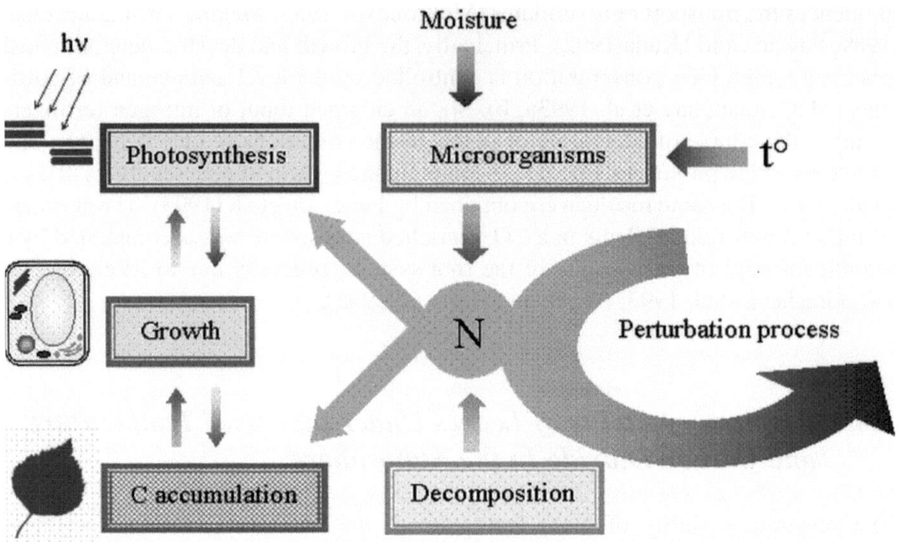


**Fig. 6.11** Hypothetic scheme of physiological responses of woody plants in the cryolithozone to climate warming

there is no proof that a rise of carbon dioxide concentration would lead to a raised net-assimilation of carbon in ecosystems in which woody species dominate (Kondrasheva et al. 1993). Short-term experiments in laboratories and phytotrons showed that doubled $CO_2$ concentration results in a twofold rise of photosynthesis, a decrease in dark and light respiration, a raise of water use efficiency, an optimization of soil nutrition, an increased growth of root systems (Drake et al. 1989; Oechel and Reichers 1987; Tissue and Oechel 1987), and a delay in the ageing of leaves (Melillo et al. 1990). In contrast, long-term exposition of woody plants to an atmosphere with doubled $CO_2$ content showed an inhibition of photosynthesis and plant growth (Veretennikov 1987, 1992).

On the basis of obtained data we constructed a hypothetic scheme of the influence of possible climate warming and pedoturbation processes on the growth and photosynthetic processes of woody plants in the cryolithozone (Fig. 6.11).

## 6.7 Conclusion

Photosynthesis of plants under the dry conditions of the perennially frozen grounds of Yakutia has been studied for a long time. As far back as the 50–60s the first ecological-geographical studies on plant physiology on cold soils were conducted under the guidance of V.P. Dadykin (Blagoveschensk–Yakutsk–Tiksi) (Dadykin and Grigoryeva 1954). It has been shown that the plants of Yakutia, growing under the specific light conditions with a domination of orange-red wavelengths, assimilate more solar radiation during the day. Unfortunately, these experiments were not continued and only at the end of 70s ecological-physiological studies using modern methods were started up again in Yakutia.

New investigations by Yakutian plant physiologists have shown that the character of the physiological-biochemical processes in relation to the formation and development of tolerance to the action of extreme factors (drought, salinity, low temperatures, lack of nitrogen-phosphorus nutrition) depends on individual, breed and species features of cereals. A close connection was established between productivity on the one hand and growth and development periods, energetic exchange and water status in the cell on the other.

Homeostatic growth and development of the woody plants in Yakutia during the short vegetation period is provided by high levels of physiological processes (photosynthesis and transpiration) at relatively low dark and night respiratory expenses for growth and maintenance. A high variability of photosynthesis and dark respiration of the woody plants testifies to an excellent adaptation of *Larix cajanderi* to the peculiar climatic conditions of the cryolithozone.

The high photosynthetic activity of woody plants is determined by an adequate availability of water in the soil and the atmosphere. In particular, the precipitation should be sufficient during most of the growing season. Capillary soil water from above the permafrost horizons, supported by a high stomatal conductance and low xylem potential of the plant, is necessary for the maintenance of a high plant activity when moisture provision are critical in the months of July and August.

It has been shown that a special feature of the water regime of Yakutian soils is that the permafrost restricts the moisture cycle to the melted layer. The dynamics of the hydrothermal regime is in many respects determined by the continuous process of soil thawing, which stops only at the beginning of winter. This process is closely linked to the specific characteristics of the vegetation.

At climate warming the direction of pedoturbation processes, directly affecting the cycle of the main organogens in the ecosystem, will be the dominant factor leading to an increase in the productivity of the forests of the cryolithozone. The production process of Yakutian tree populations under conditions of climate warming will be generally limited by endogenous factors, especially stomatal conductance, as well as by exogenous ones, i.e. the provision of plants with water and mineral organogens, especially with nitrogen.

## References

Acock B, Hartwell AJ (1985) Crop response to elevated carbon dioxide concentrations. In: Direct effects of increasing carbon dioxide on vegetation. Duke University, Durham

Anisimov OA, Belolutskaya MA (2002) Otsenka vliyaniya izmeneniya klimata i degradatsii vechnoy merzloty na infrastrukturu v severnykh regionakh Rossii. Meteorologiya I gydrologiya 6: 15–22 – *Estimation of effect of climate change and permafrost degradation to infrastructure in northern regions of Russia* (in Russian)

Arneth A, Kelliher FM, Bauer G et al. (1996) Environmental regulation of xylem sap flow and total conductance of *Larix gmelinii* trees in eastern Siberia. Tree Physiology 16: 247–255

Bazzaz FA (1990) The response of natural ecosystems to the rising global $CO_2$ levels. Ann Rev Ecol Syst 21: 167–196

Benecke U, Schulze E-D, Matyssek R, Havnarek WM (1981) Environmental control of $CO_2$-assimilation and leaf conductance in *Larix decidua* Mill. I. A comparison of contrasting natural environments. Oecologia 50: 54–61

Budyko MI (1974) Izmeneiye klimata. Hydrometeoizdat, Leningrad: 1974 – *Climate Change* (in Russian)

Budyko MI (1989) Parnikovy effect, izmeneniye klimata i ekosistem. Gidrometeoizdat, Leningrad – *Greenhouse effect, change of climate and ecosystem* (in Russian)

Budyko MI, Israel YuA, McCracken MS, Hecht AD (eds) (1991) Predstoyascheye izmeneiye klimata. Sovmestny sovetsko-amerikansky otchyot o climate i ego izmeneniyakh. Gidrometeoizdat, Leningrad – *Forthcoming climate changes. Joint Soviet-American report on climate and its changes* (in Russian)

Bykov OD (1983) Sootnosheniye fotosinteza i dykhaniya v $CO_2$ gazoobmene na svetu u listyev $C_3$-rasteniy v zavisimosti ot temperatury. Physiologiya rasteniy 30(4) 629–636 – *Photosynthesis/respiration rate in on-light $CO_2$ gas exchange in $C_3$-plants depending on temperature* (in Russian)

Chazdon RL (1988) Sunflecks and their importance to forest understory plants. Adv Ecol Res 18: 1–63

Dadykin VP, Grigoryeva VG (1954) O fotosinteze u rastenij Zapoliaryia pri kruglosutochnom osveshchenii. Doklady AN SSSR 80(2): 114–118 – *On photosynthesis of trans-polar plants under round-the-clock illumination* (in Russian)

Desyatkin RV (2003) Soil water conditions and the dynamics of alas water. In: Ivanov BI (ed) The review of conditions and tendencies of climate change in Yakutia. Proceedings of International Conference. Yakutian branch of the Siberian Division of the Russian Academy of Sciences Publishing House, Yakutsk: 36–47

Desyatkin RV (2006) Pochvoobrazovaniye v termokarstovykh kotlovinakh – alasakh kriolitozony. Abstract of Thesis for a Doctor Degree. Ulan-Ude – *Pedogenesis in thermokarst depressions – alases of cryolithozone* (in Russian)

Dixon RK, Brown S, Houghton RA et al. (1994) Carbon pools and flux of global forest ecosystems. Science 263: 185–190

Dolman AJ, Maximov TCh, Moors EJ et al. (2004) net ecosystem exchange of carbon dioxide and water of Far Eastern Siberia larch (*Larix cajanderi*) on permafrost. Biogeosciences 1:133–146

Drake BG et al. (1989) The effect of elevated atmospheric $CO_2$ on $C_3$ and $C_4$ vegetation on Chesapeake Bay. In: Proceedings of Symposium on the physiological ecology of aquatic plants. Aarhus Publ, Denmark: 1–22

Duchkov AD, Balobaev VT (2001) Evolyutsiya teplovogo i fazovogo sostoyaniya kriolitozony Sibiri. In: Globalnye izmeneniya pririodnoy sredy. Izd-vo SB RAS, Novosibirsk – *Evolution of heat and phase condition of cryolithozone in Siberia* (In Russian).

Eamus D, Jarvis PG (1989) The direct effects of increase in the global atmospheric $CO_2$ concentration on natural and commercial temperate trees and forest. Ad Eco Res 19: 1–55

Elovskaya LG, Musich NI (1969) Yakutskaya ASSR. In: Sokolov AV (ed) Agrokhimicheskaya kharakteristika pochv SSSR/Vostochnaya Sibir. Nauka, Moscow – *The Yakut ASSR* (in Russian)

Fujita N, Yanagisawa N, Sugimoto A (1998) Domination of an East-Siberian taiga around Yakutsk by *Larix gmelinii*, a deciduous conifer, supported by leafing phenology, photosynthetic characteristics and water use efficiency. In: Activity report of GAME-Siberia 14: 81–84

Gavrilova MK (1993) Climate and Permafrost. Permafrost and Periglacial Processes 4(2). John Wiley and Sons, Ltd, Chichester, England: 99–111

Gavrilova MK (2007) Climate change in permafrost regions of Russia during the 20th century. In: Ivanov BI, Maximov TCh(eds) Influence of Climatic and Ecological Changes on Permafrost Ecosystems. YD Publishing House SD RAS, Yakutsk: 9–15.

Gavrilyev PP (2003) Response of ice-rich permafrost terrain in the North to global warming. In: Ivanov BI, Maximov TCh (eds) Influence of Climatic and Ecological Changes on Permafrost Ecosystems. YD Publishing House SD RAS, Yakutsk: 47–52.

Goncharik MN (1962) Vliyaniye ekologicheskikh usloviy na fotosintez kulturnykh rasteniy. Izd-vo AN BSSR, Minsk – *Influence of ecological conditions on photosynthesis of cultivated plants* (in Russian)

Gruza GV, Bardin MYu, Rankova EYa et al. (2001) Ob izmeneniyi temperatury vozdukha i atmosfernykh osadkov na territorii Rossii v 20 veke. In: Izrael YuA (ed) Sostoyaniye i kompleksny monitoring prirodnoy sredy i klimata. Nauka, Moscow – *On Change of Air Temperature and Atmospheric Precipitation in the Territory of Russia in 20th Century* (in Russian)

Hollinger DY, Kelliher FM, Schulze E-D et al. (1995) Initial assessment of multi-scale measures of $CO_2$ and $H_2O$ flux in the Siberian taiga. J Biogeogr 22: 425–431

Houghton JT, Ding Y, Griggs DJ et al. (ed) (2001) IPCC Reports. Climate Change 2001: The Scientific Basis. Contribution of Working Group I to the Third Assessment Report of the Intergovernmental Panel on Climate Change. Cambridge University Press, Cambridge, United Kingdom and New York, NY, USA

Huntley B, Birks HYB (1983) An atlas of Past and Present Pollen Maps for Europe: 0-13000 Years Ago. Cambridge University Press, Cambridge

Isaev AS, Korovin GN, Sukhikh VI et al. (1995) Ekologicheskie problemy pogloscheniya uglekislogo gaza posredstvom lesovosstanovleniya i lesorazvedeniya v Rossiyi. Tsentr ekologicheskoy politiki, Moscow – *Ecological problems of carbon dioxide assimilation by means of forest recovery and forestation in Russia* (in Russian)

Isaev AS (ed) (2001) Bezopasnost Rossii. Regionalnye problemy bezopasnosti. Krasnoyarsky Krai. MGF 'Znaniye', Moscow – *Security of Russia. Regional problems of security. Krasnoyarsk Territories* (in Russian)

Israel YA, Pavlov AV, Anokhin YA (1999) Analiz tekuschego i ozhidaemogo izmeneniya klimata i kriolotozony v Rossiyskoy Arktike. Meteorologiya i gydrologiya 3: 18–27 – *Analysis of current and expected change of climate and cryolithozone in the Arctic of Russia* (in Russian)

Israel YuA, Nazarov IM, Nakhutin AI et al. (2002) Vklad Rossii v izmeneniye kontsentratsii parnikovykh gazov v atmosphere. Meteorologiya i gydrologiya 5: 17–26 – *Contribution of Russia to change of greenhouse gases concentration in atmosphere* (in Russian)

Ivanov BI (ed) 2003. The review of conditions and tendencies of climate change in Yakutia. Proceedings of International Conference. YB of the SD RAS Publishing House, Yakutsk

Ivanov BI, Maximov TCh, Desyatkin RV (1993) On role of cryolithozone soils in global warming climate. In: International Congress on Arctic terrestrial ecosystems. Oplday: 67

Ivanov BI, Maximov TCh, Maximov AP, Kononov AV (1994) Interim Report of Joint Research between FFPRI and YIB «Carbon storage and carbon dioxide budget in Forest Ecosystem». Sapporo

Ivanov BI, Maximov TCh, Kononov AV, Maximov AP (1996) Carbon balance in permafrost forest ecosystems. In: Northern knowledge serves Northern needs. Abstracts of International Conference. Yakutsk: 33

Ivanov LA, Kossovich NL (1932) O rabote assimilyatsionnogo apparata drevesnykh porod. Botanicheskiy zhurnal 17(1): 3–17 – *On work of assimilation apparatus of tree species* (in Russian)

Karpachevsky LO (1993) Rol rasteniy i globalnykh izmeneniy klimata v evolyutsii pochv. Pochvovedenie 9: 20–27 – *The role of plants and global climate changes in the evolution of soils* (in Russian)

Karpel BA, Medvedeva NC (1977) Plodonosheniye listvennitsy daurskoy v Yakutii. Nauka, Novosibirsk – *Fruitage of Daurian larch in Yakutia* (in Russian)

Keuk G (1989) Zuwachsuntersuchungen in Zusammen hang mit den gegenwartigen Waldschaden in Baben Wurtemberg. In: Air pollut and forest decline. Birmensdorf. 1989

Koike T, Lei TT, Tabuchi R, Maximov TCh et al. (1996) Comparison between the photosynthetic capacity of Siberian and Japanese white birch raised under elevated $CO_2$ and temperature. Tree Physiol 16: 381–386

Koike T, Mori S, Matsuura Y et al. (1998) Shoot growth and photosynthetic characteristics in larch and spruce affected by temperature of the contrasting north- and south-facing slopes in eastern Siberia. In: Proceedings of 7th Joint Siberian Permafrost Studies. Yakutsk: 3–12

Koike T, Yazaki K, Funada R et al. (2000) Photosynthetic characteristics of Dahurian Larch, Scotch Pine and White Birch seedlings native to Eastern Siberia raised under elevated $CO_2$. Eurasian J For Res 1: 31–37

Kondrasheva NYu, Kobak KI, Turchinovich IE (1993) Vozmozhnye reaktsii nazemnoy rastitelnosti na uvelichenii konysentratsii $CO_2$ v atmosphere i globalnoye potepleniye. Lesovedenie 4: 71–75 – *Possible reactions of terrestrial vegetation to an increase in $CO_2$ concentration in the atmosphere and global warming* (in Russian)

Kondratyev KYa, Krapivin VF, Varotsos CA (2003) Global Carbon Cycle and Climate Change Springer/PRAXIS, Chichester, UK

Kramer PJ (1981) Carbon dioxide concentration, photosynthesis and dry matter production. Bioscience 31: 29–33

Labutin YuV, Germogenov NI (1990) Ptitsy Yakutii: sovremennye dannye po sostavu i rasprostraneniyu. Preprint. Izd-vo SO RAN, Novosibirsk – *Birds of Yakutia: the modern data on structure and distribution* (in Russian)

Laisk AKh (1977) Kinetika fotosinteza i fotodykhaniya $C_3$-rasteniy. Nauka, Moscow – *Kinetics of photosynthesis and photorespiration in C3-plants* (in Russian)

Larcher W (1995) Physiological Plant Ecology: Ecophysiology and Stress Physiology of Functional Groups. Springer-Verlag, Berlin, Heidelberg

Makino A (1994) Biochemistry of $C_3$-photosynthesis in high $CO_2$. J Plant Res 107: 79–84

Malkina IS (1995) Obmen $CO_2$ molodykh derevyev listvennitsy. Lesovedeniye 5: 59–66 – *$CO_2$-exchange in young larch trees* (in Russian)

Manabe S, Stouffer RJ (1995) Simulation of abrupt climate change induced by freshwater input to the North Atlantic Ocean. Nature 378: 165–167.

Manabe S, Wetherald KJ (1980) On role distribution of climate change resulting from on increase in $CO_2$-content of the atmosphere. J Atmos Sci 37: 99–118.

Maksimov AP (2003) A brief review of larch net-photosynthesis investigations in Central Yakutia. In: Ivanov BI, Maximov TCh (eds) Influence of climatic and ecological changes on permafrost ecosystems. Proceedings of International Conference. YD Publishing House SD RAS, Yakutsk: 87–100

Maximov TCh (1989) Ekologo-fiziologicheskiye issledovaniya fotosinteza yachmenya v usloviyakh Yakutii. Abstract of Thesis for a Candidate Degree. Yakutsk – Ecological-physiological studies of barley photosynthesis under conditions of Yakutia (in Russian)

Maximov TCh, Ivanov BI (2003) The development of international studies of the regional and global carbon cycle in Yakutia permafrost ecosystems. In: Ivanov BI (ed) The review of conditions and tendencies of climate changes in Yakutia. Proceedings of International Conference. YD Publishing House SD RAS, Yakutsk: 34–43

Maximov TCh, Ivanov BI (2005) Monitoring sostoyaniya merzlotnykh ekosistem: 'Spasskaya Pad', Yakutsk. Sibirskiy ekologicheskiy zhurnal 12(4): 777–781 – *Monitoring of permafrost ecosystems condition: Spasskaya Pad, Yakutsk* (in Russian)

Maximov TCh, Koike T (2001) Physiological aspects of adaptation of different populations of trees under expected climate warming. In: Ivanov BI, Maximov TCh (eds) The role of permafrost ecosystems in global climate change. Proceedings of International Conference. YD Publishing House SD RAS, Yakutsk, 2001: 158–166

Maximov TCh, Ivanov BI, Maximov AP, Kononov AV (1994) Interim Report of Joint Research between FFPRI and YIB "Carbon storage and carbon dioxide budget in forest ecosystem". FFPRI, Sapporo

Maximov TCh, Kononov AV, Koike T (1995) Photosynthetic activity of woody plants of Yakutia. In: Symposium on Joint Permafrost Studies between Japan and Russia in 1992–1994. Yakutsk: 24–30

Maximov TCh, Maximov AP, Kononov AV (1996) Balance of carbon dioxide and water in permafrost ecosystems of Yakutia. Proceedings of the Third International Study Conference on GAWEX in Asia and GAME. Cheju: 104–111

Maximov TCh, Dolman AJ, van der Molen MK et al. (2004) The regional and global carbon scales of permafrost-dominated forest ecosystems. In: Maximov TCh, Dolman AJ, Ohta T (eds) C/$H_2O$/Energy balance and climate over boreal regions with special emphasis on eastern Eurasia. Proceedings of International Semi-Open Work-shop. YD Publishing House SD RAS, Yakutsk: 91–94.

Maximov TCh, Dolman AJ, Moors EJ et al. (2005a) Parametry uglerodnogo i vodnogo obmena v lesnykh ekosistemakh. Otchyoty RAN 408(8): 684–686 – *Parameters of carbon and water exchange in forest ecosystems* (in Russian)

Maximov TCh, Maksimov AP, Kononov AV et al. (2005b) Ekologo-fiziologicheskiye osobennosti fotosinteza listvennitsy *Larix cajanderi* v usloviyakh mnogoletney merzloty Yakutii. Lesovedeniye 6: 3–10 – *Ecological-physiological peculiarities of photosynthesis of larch (Larix cajanderi) under conditions of permafrost in Yakutia* (In Russian)

Matyssek R, Schulze E-D (1987) Heterosis in hybrid larch (*Larix decidua* x *leptolepis*). I. The role of leaf characteristics. Trees 1: 219–224

McConnaughay KDM, Berntson GM, Bazzaz FA (1993a) Limitations to $CO_2$-induced growth enhancement in pot studies. Oecologia 94: 550–557

McConnaughay KDM, Berntson GM, Bazzaz FA (1993b) Plant responses to carbon dioxide. Nature 361: 24

Melillo JM et al. (1990) Effect on ecosystem. In: Climate change. The IPCC scientific assessment. Cambridge University Press, Cambridge

Mokronosov AT (1983) Fotosinteticheskaya funktsiya i tselostnost rastitelnogo organizma. 42[d] Timiryazev Readings. Nauka, Moscow – *Photosynthetic function and integrity of the plant organism* – (in Russian)

Oechel W, Reichers GH (1987) Response of a tundra ecosystem to elevated atmospheric carbon dioxide. In: DOE (US Department of Energy) report. Washington DC

Oechel W, Strain BR (1985) Native species response to increasing carbon dioxide concentration. In: Directs effects of increasing carbon dioxide on vegetation. Duke University, Durham

Ohta T, Hiyama T, Tanaka H et al. (2001) Seasonal variation in the energy and water exchanges above and below a larch forest in eastern Siberia. Hydrol Process 15:1459–1476

Pearcy RW (1987) Photosynthetic gas exchange responses of Australian tropical forest trees in canopy, gap and understorymicro-environments. Funct Ecol 1: 169–178

Pearcy RW, Calkin H (1983) Carbon dioxide exchange of $C_3$ and $C_4$ tree species in the understory of a Hawaiian forest. Oecologia 58: 19–25

Petrov KA, Kononov AV, Okhlopkova ZhM, Maximov TChr (1995). O gipermetabolicheskikh svoistvakh ingibitorov rosta pochek *Duschekia fruticosa* L. In: Ecological and genetic studies in Yakutia. Proceedings of Regional Conference. Yakutsk: 74 – On hypometabolic properties of growth inhibitors of *Duschekia fruticosa* L. buds (in Russian)

Petrov KA, Kononov AV, Maximov TChr (1999) O blokiruyuschem deistviyi ingibitorov rosta pochek *Duschekia fruticosa* L na transpiratsiyu izolirovannykh organov rasteniy. In: Proceedings of 4th Congress of plant physiologists of Russia. Moscow: 662 – *On blocking action of growth inhibitors of Duschekia fruticosa L. buds to transpiration of isolated plants organs* (in Russian)

Petterson R, McDonald AJS, Stadenberg I (1993) Response of small birch plants (*Betula pendula* Roth.) to elevated $CO_2$ and nitrogen supply. Plant Cell Environ 16: 1115–1121.

Pozdnyakov LK (1963) Listvennitsa Yakutii. Manuscript of Thesis for a Doctor Degree. Krasnoyarsk – *Larch in Yakutia* (in Russian)

Pozdnyakov LK (1986) Merzlotnoye lesovedeniye. Nauka, Novosibirsk – *Silvics on frozen grounds* (in Russian)

Romanovsky VE, Osterkamp TE (2001) Sistema monitoringa mnogoletney merzloty na Akyaske: Structura i resultaty. Kriosfera Zemli 5 (4): 59–68 – *Permafrost monitoring system in Alaska: structure and results* (in Russian)

Ross DJ, Kelliher FM, Tate KR (1999) Microbiological processes in relation to carbon, nitrogen and temperature regimes in litter and in a sandy mineral soil from a Central Siberia *Pinus silvestris* L. forest. Soil Boil Biochem 31: 757–767

Sage RF, Sharkey TD, Seemann JR (1989) Acclimation of photosynthesis to elevated $CO_2$ in five $C_3$ species. Plant Physiol 89: 590–596

Saito H, Yamamuro K, Tsuno Y et al. (2002) Spatial variations of light intensity and photosynthetic properties within a *Larix gmelinii* tree crown in eastern Siberia. Proceedings of the 10th Symposium on Joint Siberian Permafrost Studies between Japan and Russia in 2001. Tsukuba: 7–14

Savvinov DD (1976) Gidrotermicheskiy rezhim pochv v zone mnogoletney merzloty. Nauka, Novosibirsk – *Hydrothermic regime of soils in the zone of permafrost* (in Russian)

Sawada S, Usuda H (1992) The role of inorganic phosphate in activation of RuBPcase activity in response to changes in source. In: Murata N (ed) Sink balance in research in photosynthesis IV.

Scherbatyuk AS (1976) Rezhim $CO_2$ i fotosinteticheskaya aktivnost osnovnykh vskhodov. In: Biofizicheskiye i sistematicheskiye issledovaniya v lesnoy fitotsenologii. Collection of articles. Petrozavodsk: 112–113 – *Carbon dioxide regime and photosynthetic activity of main saplings* (in Russian)

Scherbatyuk AS, Rusakova LV, Suvorova GG, Yankova LS (1991) Uglekislotny gazoobmen khvoinykh Predbaikalya. Nauka, Novosibirsk – *Carbon-dioxide exchange of the Prebaikalia conifers* (in Russian)

Schimel D, House J, Hibbard K et al. (2001) Recent patterns and mechanisms of carbon exchange by terrestrial ecosystems. Nature 414: 169–172

Schulze E-D, Schulze W, Kelliher FM et al. (1995) Aboveground biomass and nitrogen nutrition in a chronosequence of pristine Dahurian larix stands in Eastern Siberia. Can J For Res 25: 943–960

Skachkov YuB (2003) Temperature monitoring of the high levels of continuous permafrost of Central Yakutia Water. In: Ivanov BI (ed) The review of conditions and tendencies of climate change in Yakutia. Proceedings of International Conference. Yakutian branch of the Siberian Division of the Russian Academy of Sciences Publishing House, Yakutsk: 16–25

Solomon S, Qin D, Manning M et al. (ed) (2007) Climate Change: The Physical Science Basis. Summary for Policymakers. Contribution of Working Group I to the Fourth Assessment Report of the Intergovernmental Panel on Climate Change. Cambridge University Press, Cambridge, United Kingdom and New York, NY, USA

Solomonov NG (2003) Bioecologic consequences of climate warming in permafrost regions. In: Ivanov BI (ed) The review of conditions and tendencies of climate change in Yakutia. Proceedings of International Conference. Yakutian branch of the Siberian Division of the Russian Academy of Sciences Publishing House, Yakutsk: 44–51

Stepanov GN (1976) Vodny rezhim yachmenya v usloviyakh Tsentralnoy Yakutii. Abstract of Thesis for a Doctor Degree. Leningrad – *Water regime of barley under conditions of Central Yakutia* (in Russian)

Sugimoto A, Yanagisawa N, Naito D et al. (2002) Importance of permafrost as a source of water for plants in east Siberia taiga. Ecol Res J 17: 493–503

Suvorova GG (2006) Fotosinteticheskaya aktivnost khvoinykh derevyev v usloviyakh yuga sredney Sibiri. Abstract of Thesis for a Doctor Degree. Irkutsk – *Photosynthetic activity of conifer trees under conditions of Southern Central Siberia* (in Russian)

Suzuki M, Saito H, Iijima H et al. (2002) Photosynthetic and stomatal responses to air vapour deficit within a larch canopy in east Siberia. In: Proceedings of the 10th Symposium on Joint Siberian Permafrost Studies between Japan and Russia in 2001. Tsukuba: 15–19

Tabuchi R, Koike T, Maximov TCh et al. (1994) Gas exchange measurements on some major tree species in Siberian permafrost region in summer. In: Interim report on joint research project "Carbon storage and carbon dioxide budget in forest ecosystems" between Japan and Russia. FFPRI, Sapporo

Tissue DT, Oechel WC (1987) Response of *Eriophorum vaginatum* to elevated $CO_2$ and temperature in the Alaskan arctic tundra. Ecology 68(2): 400–410

Tissue DT, Thomas RB, Strain BR (1993) Long-term effects of elevated $CO_2$ and nutrients on photosynthesis and Rubisco in Loblolly pine seedlings. Plant Cell Environ 16: 859–865

Tselniker YuL, Malkina IS, Yakshina AM (1990) Vertikalny gradient dykhaniya stvolov eli, duba i beryozy. Lesovedeniye 4: 11–18 – *Vertical gradient of stem respiration of spruce, oak and birch* (in Russian)

Ugarov GS (1974) Osobennosti vodnogo rezhima rasteniy pri nizkikh polozhitelnykh temperaturakh. In: Biologicheskiye problemy Severa. Abstracts of All-Union Symposium. Yakutsk: 153–158 – *Peculiarities of water regime of plants under low positive temperatures* (in Russian)

van der Molen MK, Dolman AJ, Maximov TC, Maximov AP (2003) Measurements of $CO_2$-exchange between a larch forest and the atmosphere in spasskaya pad: past results and future plans. In: Influence of climatic and ecological changes on permafrost ecosystems. Yakutsk: 26–32

van der Werf G, Randerson J, Collatz GJ et al. (2004) Continental-scale partitioning of fire emission during the 1997 to 2001 El Nino/La Nina Period. Science 303: 73–76

Velichko AA, Grichuk VP, Zelikson EM, Borisova OK (1990) K otsenke dinamiki borealnykh lesov pri antropogennom poteplenii klimata. In: Severnye lesa: sostoianiye, dinamika, antropogennoe vozdeystvie. Proceedings of International Symposium. Goskomles SSSR, Moscow: 19–29 – *On estimation of boreal forests dynamics during anthropogenically induced climate warming* (in Russian)

Veretennikov AV (1987) Fiziologiya rasteniy s osnovami biokhimii. Voronezhskiy Universitet, Voronezh – *Plant physiology with the basics of biochemistry* (in Russian)

Veretennikov AV (1992) Gazoobmen i rost drevesnykh rasteniy v usloviyakh povyshennogo soderzhaniya CO2 v atmosphere. In: Gazoobmen rasteniy v posevakh i prirodnykh fitotsenozakh. Proceedings of Conference. Syktyvkar: 14–15 – *Gas exchange and growth of tree plants under conditions of elevated* $CO_2$ *content in atmosphere* (in Russian)

Vinokurov NN (2002) Fauna of arthropods of Yakutia. In: Institute biologicheskikh problem kriolitozony. 50 let. Izd-vo DNPO MNiPO RS (Ya), Yakutsk: 64–67

Vinokurov NN, Isaev AP (2002) Sibirskiy shelkopryad v Yakutii. Nauka i tekhnika v Yakutii: 2(3): 53–56 – *The Siberian silk moth in Yakutia* (in Russian)

Vygodskaya NN, Milyukova I, Varlagin A et al. (1997) Leaf conductance and $CO_2$ assimilation of *Larix gmelinii* growing in an eastern Siberian boreal forest. Tree Physiol 17: 607–615

Weber JA, Jurik TW, Tenhunen JD, Gates DM (1985) Analysis of gas exchange in seedlings of *Acer saccharum*: integration of field and laboratory studies. Oecologia 65: 338–347

Woodward FI, Smith TM (1995) Predictions and measurements of the maximum photosynthetic rate, $A_{max}$, at the global scale. In: Schulze E-D, Caldwell MM (eds) Ecophysiology of photosynthesis. Springer-Verlag, Berlin

Zabuga GA, Scherbatyuk AS (1982) Ekologiya sosny obyknovennoy v lesostepi Predbaikalya. Ecologiya 5: 76–78 – *Ecology of Scotch pine in forest-steppe Prebaikalia* (in Russian)

Zakharova VI, Karpov NS, Remigailo PA et al. (2002) Istoriya botanicheskikh issledovaniy v Yakutii. In: Institute biologicheskikh problem kriolitozony. 50 let. – Yakutsk: Izd-vo DNPO MNiPO RS (Ya). – *The history of botanical studies in Yakutia* (In Russian)

Zavelskaya AA, Zukert NV, Polyakova EYu, Pryazhnikov AA (1993) Prognoz vliyaniya izmeneniy klimata na borealnye lesa Rossii. Lesovedenie 3: 16–24 – *A forecast of climate change influence on the boreal forests of Russia* (in Russian)

# Chapter 7
# Nature Conservation Status and Its Prospects

**B.Z. Borisov, M.M. Cherosov, I.A. Fedorov, P.S. Egorova, and P.A. Pavlova**

## 7.1 Natural Reserves Network: Present Situation and Prospects

**B.Z. Borisov and M.M. Cherosov**

The territory of Yakutia occupies a significant part of North–eastern Russia. Rigorous climatic conditions have formed unique plant communities growing on perennially frozen grounds. Winter conditions lasting for nearly half of the year, differences in temperature reaching up to 100°C, and low precipitation values have resulted in a low species diversity and low phytomass as the main characteristic features of the Yakutian vegetation. Within the territory of Yakutia many distribution areas of tree, shrub, herb and other plant species are limited. Under such extreme-threshold conditions any influence, most of all anthropogenic effects, easily damage one or all biotic components, and may lead to the destruction of a whole ecosystem (Kryuchkov 1987). Anthropogenically damaged plant communities recover very slowly. Complete renewal of vegetation may take 100 or 200 years. Hence, permafrost may interfere in this process changing the direction of the recovery course. Accordingly, northern taiga landscapes can be substituted by marshlands or thermokarst lakes. Thus, a landscape can be changed unrecognizably, even if a man would not interfere with natural processes.

In 1992 at the UN Conference in Rio-de-Janeiro (Brasil) it was decided that it is necessary to provide a stable development of human society. This topic is of special urgency at the higher latitudes of the Northern Hemisphere (including Yakutia) which are characterized by extremely vulnerable natural ecosystems.

Up to the mid twentieth century Yakutia was considered a natural "sanctuary" of Russia due to the small population and the low rates of industrial development. Wide-scale development of the North, the Arctic and space started almost simultaneously, in the 1950s. This became possible thanks to the scientific-technical revolution which began at that time. Today it has been proved that development

B.Z. Borisov (✉)
Institute for biological problems of the cryolithozone, 41 Lenin Ave., Yakutsk, 677980, Russia
e-mail: bzborisov@mail.ru

E.I. Troeva et al. (eds.), *The Far North: Plant Biodiversity and Ecology of Yakutia*, Plant and Vegetation 3, DOI 10.1007/978-90-481-3774-9_7,

of the North is impossible by applying the methods and machinery suitable for southern regions, due to the ecological vulnerability of the North (Kryuchkov 1987).

Yakutia is an important ecological and climatic node of the Arctic, which, when getting untied, threatens with negative consequences of a global scale. Due to the shortage of biologically active temperatures the diverse and fragile water, tundra and taiga ecosystems feature slow self-recovery and self-cleaning rates and are easily overstressed. Since they are based on perennially frozen grounds in an ultracontinental climate the possible behaviour of the permafrost under climate change may yield serious consequences.

The weak resistance of the northern ecosystems to external factors is explained as follows: First of all, the northern ecosystems are rather young natural formations from an evolutionary point of view. For instance, the absolute age of the basic soil types of the cryolithozone does not exceed 15,000 years. Secondly, as they thrive under extreme bioclimatic conditions, the northern ecosystems are characterized by a low energy capacity. Thirdly, they have a plain structure which is easily exposed to degradation. And fourthly, the northern ecosystems are characterized by low biological productivity rates and strictly seasonal functioning.

All the abovementioned regional peculiarities of the northern ecosystems have proved to be primary factors for accelerated degradation of the northern biosphere under conditions of industrial development.

Among the factors of severe ecological impact in the territory of Yakutia anthropogenic activity prevails in destabilizing fundamental ecological processes and functions. In a number of regions of the Republic the irrational use of nature has resulted in negative changes in the structure and functioning of both water and terrestrial ecosystems and in a reduction of biological diversity. Industry and agriculture represent the main threats to plant communities and species of Yakutia.

The prior role no doubt belongs to the mining industry, which forms the basis of economical development and welfare of the Republic, though the developmental concept of this industry hardly took ecological points of view into account. Diamond, tin, coal and gas deposits have been exploited for 50 years. This vigorous mining activity has yielded 300 thousand ha of disturbed lands, of which 25,552 ha are wastes of gold mining only.

Intensified agricultural development without consideration of the ecological capacity of the northern ecosystems has also led to a significant degradation of the natural landscapes. Compared to the 1940s, the present productivity of the agricultural lands has decreased 2.5 times. The total area of pastures, hayfields and arable lands has been reduced from 2,815 thousand ha in 1988 to 1,745.8 thousand ha in 1992, leaving just degraded and discarded lands. The most critical situation can be observed in the agricultural regions of the Lena-Amga Interfluve. Misuse of arable lands under conditions of permafrost leads to their secondary salinization making them permanently non-serviceable.

Other urgent ecological problems are the gradual loss of lichens from the vegetation cover of the tundra and deforestation in the taiga zone. Quantification of the main factors negatively affecting the reindeer pastures shows the following: 80% decrease due to overgrazing and irrational use of pastures; 10% due to fires; and 5%

due to technogenic factors. Deforestation is caused by uncontrollable forest fires, irrational timber felling, direct and indirect effects of the mining industry, and high rates of urbanization.

All the abovementioned factors especially affect the valley complexes which are the habitats of the highest biodiversity of both fauna and flora of Yakutia.

As a result of all changes that take place in nature and due to human activity, many taxa of the Yakutian flora presently have a threatened ecological status. The Red Data Book of Yakutia (Dolinin et al. 2000), that was compiled before similar books for the Russian Federation were published, includes 337 species of vascular plants, 13 moss species, 7 lichens and 10 fungi species. It points out that the territory of Yakutia contains the northern boundaries of the distribution areas of many boreal species. This means that such species grow there under conditions of stress and any change in climate or any severing human activity may cause complete disappearance of the Yakutian populations.

The most effective way to preserve the biological diversity is a territorial nature protection network. This issue was first raised in 1939 by V.N. Makarov and V.N. Skalon, who independently proposed four sanctuaries to be established in 1941–1942. World War II thwarted the plans putting them off for an uncertain period. Later, in 1958 a plan for the sanctuary network in the USSR was developed, including the offer to found a sanctuary in the Verkhoyansk Range. Unfortunately, these plans neither were ever fulfilled. In 1963, the famous Russian specialist in forest science I.P. Scherbakov proposed a plan for the foundation of three sanctuaries in the territory of Yakutia: "Ust–Lensky", "Tokkinsky" and "Kitchansky". The plan was approved by the Government and two more sanctuaries, "Viluysky" and "Yano–Indigirsky", were added, and the latter was supposed to include all the Arctic islands. In the early 1980s the work on land allotment for the sanctuaries started. Unfortunately, the obstinate opposition of major land users, represented by mining and agricultural enterprises, resulted in the establishment of only two sanctuaries, "Ust–Lensky" and "Olyokminsky". They were founded in 1984 and 1985, respectively. Besides, three State Reserves were established in 1982–1986, "Chaigurgino", "Bolshoye Tokko", and "Ust–Viluysky". Also 15 more reserves of a regional level were established (Solomonov 2000; Borisov 2000).

As a result, by 1993 two sanctuaries and 18 reserves were realized in Yakutia, covering 110,000 km$^2$ or 3.5% of the territory of the Republic. That size of nature reserve (NR) network did not meet the nature conservation requirements that imply the preservation of biodiversity of the region and of the habitats of plants, animals and man (Solomonov 1988).

In 1993 the renewed Government of Yakutia ordered a new long-term plan for the development of a NR network that would meet the claims of ecologists. That project "Development of NR network in the Republic of Sakha (Yakutia) for the nearest and remote future" was worked out by N.G. Solomonov, N.I. Germogenov and Ya.L. Volpert. It implied the allotment of 12.2% of the territory of Yakutia for NR by 2010 by establishing 65 reserves at various levels. The geobotanical map (Andreyev et al. 1987) was used for selection of the natural territories. This system, in turn, would serve as a basis for development of a more expanded NR network based on a basin approach. In 1994, relying on this project, the President of Yakutia

M.E. Nikolaev issued the decree "On measures of nature reserve development", and the following normative documents were drawn up: "Regulations on the reserved territories of the Republic of Sakha (Yakutia)" and "The law on nature reserves of the Republic of Sakha (Yakutia)". They resulted in the expansion of the NR network of up to 660,000 km$^2$ (21.3% of the territory of Yakutia) by 2000.

At present, the vast NR network consists of 155 objects. They are the state sanctuaries, natural parks, resource reserves, and protected landscapes, covering in total 782,000 km$^2$, or 25% of the territory of the Republic. Not all the existing reserves have a botanical purpose, many being designated in the first place for the protection of animals, i.e. they cover the habitats of rare and vulnerable bird and mammal species, as well as densely populated habitats of game species. Comprehensive biotic protection is provided only in the State Sanctuaries ("Ust–Lensky" and "Olyokminsky") and Natural Parks ("Lena Pillars", "Ust–Viluysky", "Momsky", "Kolymsky"). So how effective can this whole network be for the protection of the plant diversity of Yakutia in practice?

To answer this question we anlaysed the geobotanical map, the map of endemic and rare plants, the map of unique and rare plant communities, and the map of the NR network. The results of superimposing the vegetation map at 1:5,000,000 (Matveyev et al. 1989) and the map of the NR network is given in Table 7.1.

**Table 7.1** Representation of main landscape-geobotanical zones in the modern NR network of Yakutia

| Landscape-geobotanical zone | Area, km$^2$ | Covered by the NR network, km$^2$ | Protection (%) |
|---|---|---|---|
| *Arctic vegetation* | | | |
| Arctic deserts and semi-deserts | 251 | 251 | 100.0 |
| Arctic tundra | 44,904 | 39,089 | 87.0 |
| Subarctic tundra | 139,156 | 39,646 | 28.5 |
| Tundra bogs and wetland complexes | 78,787 | 30,977 | 39.3 |
| Stony deserts and mountainous tundra | 417,202 | 88,719 | 21.3 |
| Vegetation of river valleys and sea coasts | 55,332 | 38,380 | 69.4 |
| *Boreal vegetation* | | | |
| Near-tundra sparse larch forests | 98,274 | 20,710 | 21.0 |
| Northern taiga sparse larch forests | 492,364 | 117,919 | 23.9 |
| Middle taiga forests | 769,274 | 180,386 | 23.4 |
| Mountainous forests | 714,181 | 162,337 | 22.7 |
| *Pinus pumila* shrubberies | 44,265 | 11,096 | 25.0 |
| Boreal bogs | 33,035 | 8,899 | 26.9 |
| River valley vegetation | 155,801 | 35,946 | 23.0 |
| Psammophytic vegetation | 903 | 0 | 0 |

All the geobotanical units are rather well represented in the NR network, except for the territories with psammophytic vegetation. The best examples of this vegetation type (*tukulan*) are located in non-protected sites in the Viluy River basin. *Tukulans* is also found in the Near-Lena region, within the territory of the Natural Park "Lena Pillars". However, they are not depicted on the geobotanical map.

On the one hand, the NR network covers practically all landscape-geobotanical zones, on the other hand, as has been mentioned before, the primary purpose of this network is protection of populations of game species. The protection of plant communities and individual populations of rare and endemic plant species is not a primary goal of the nature reserves. Superimposed maps of the populations of rare plants and of the NR network clearly illustrate this problem (Fig. 7.1).

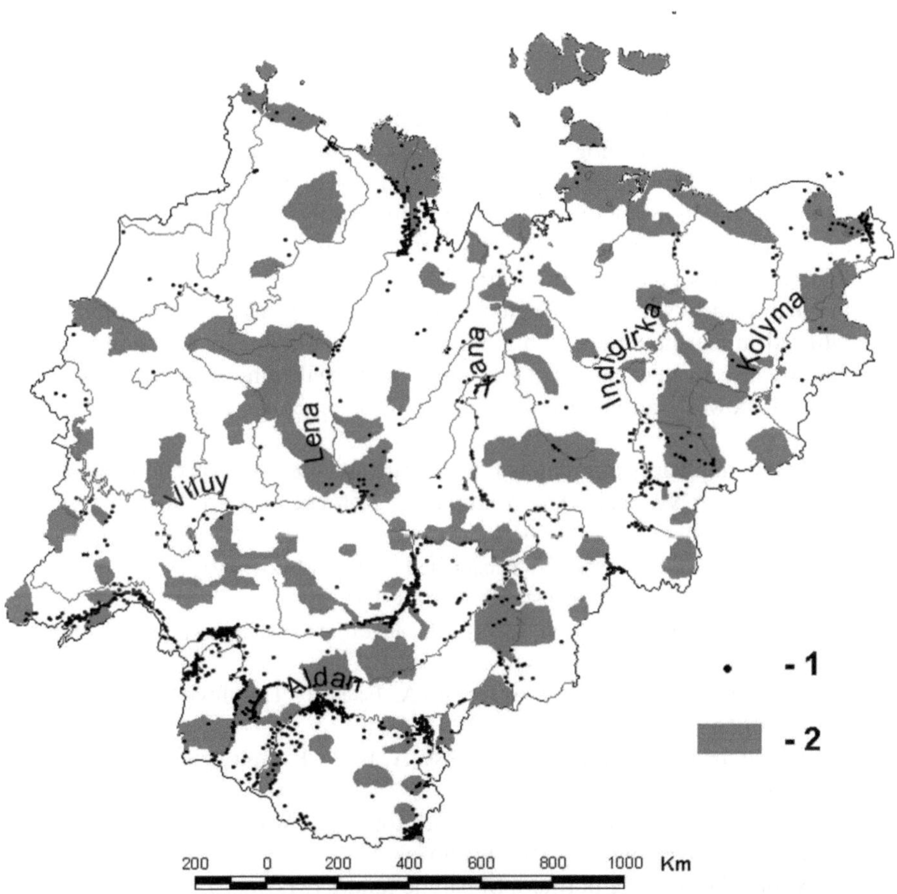


**Fig. 7.1** NR network of Yakutia and localities of occurrence of populations of higher vascular plants listed in the Red Data book of the Republic of Sakha (Yakutia): 1 – locations where the rare plant populations are observed; 2 – NR network

The localities of occurrence of 150 vascular plants of the 337 listed in the Red Data book of Yakutia are not covered by territorial protection. Three of these species belong to Category I of the IUCN classification. They are: *Anoplocaryum helenae, Clintonia udensis,* and *Sorbocotoneaster pozdnjakovii.*

The modern concept of biotical conservation implies the protection of rare and endangered plant and animal species. This is reflected in a series of Red Data Books at various levels: international, federal, and regional (including the Red Data Book of Yakutia). At the same time, presently a new concept is recognized, stating that a plant species can be successfully preserved only within the community where it grows. A plant community is an environment in which the evolution of a species takes place, and where the complex coenotic mutual relations with all the elements of a phytosystem occurs. Thus, it and can be preserved only within this system.

This approach was taken by Ukranian botanists, who have compiled the Green Book of Ukraine (Shelyag–Sosonko 1987). Later on, the concept of conservation of communities and ecosystems was formulated in the UN's convention on conservation of biological diversity.

The issue of publishing a Green Book of Siberia was raised by specialists of the Laboratory of ecology and geobotany of the Central Siberian Botanical Garden, Siberian Branch of the Russian Academy of Sciences, and was actively supported by Siberian botanists. This Book includes 196 communities of primary concern (Koropachinsky 1996). Yakutian botanists have distinguished 14 types of plant communities that are not covered by the NR network of Yakutia (Fig. 7.2). As the map clearly shows, the majority of unprotected unique plant communities are situated in the mountainous regions of South Yakutia. The situation is indeed serious because in that part of the republic industrial mining activities, the construction of power stations, as well as laying of pipelines and electric lines are rigorous.

The Yakutian plant communities that are rare and require protection and that are not covered by the NR network of Yakutia are (Koropachinsky 1996):

1. *Dryas-Kobresia* lichen tundra (described by K. Volotovsky);
2. *Salix cardiophylla*-forb forest (K. Volotovsky);
3. *Salix spp.*-shrub-*Equisetum* forest (A. Isaev, V. Perfilyeva, K. Volotovsky, A. Boichenko, A. Protopopov);
4. *Populus suaveolens-Equisetum* forest (A. Isaev, P. Timofeyev, A. Boichenko, A. Protopopov);
5. *Chosenia arbutifolia*-forb+grass forest (A. Isaev, P. Timofeyev, A. Boichenko, A. Protopopov);
6. *Betula+Populus suaveolens+Picea*-forb forest (K. Volotovsky);
7. Grass-*Carex* meadow bogs (S. Mironova)
8. Polygonal bogs of the northern Subarctic tundra (M. Boch);
9. Petrophytic communities (K. Volotovsky);
10. Hummocky green moss-dwarf shrub hypoarctic tundra communities (A. Egorova).

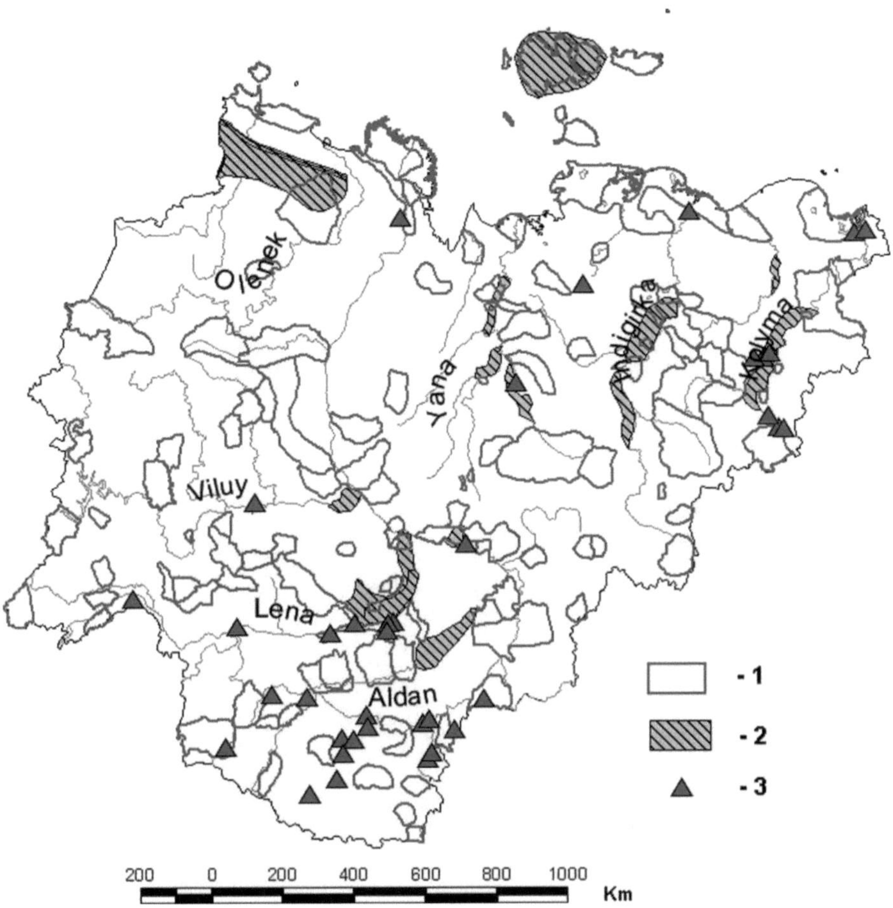


**Fig. 7.2** Rare plant communities listed in the Green Book of Siberia 1 –NR network, 2–3 – Yakutian plant communities that are rare and require protection

Generally, the issue of special botanical reserves is very topical for Yakutia, especially for its central, most populated part. The meadow and steppe communities are most vulnerable as they are actively used for agricultural purposes. In our opinion, protection of botanical objects would be most efficient when covered by small nature reserves in the most urbanized areas. Areas with a high human population density together cover a relatively small area, 75.200 $km^2$ or 2.4% of total territory of Yakutia (Fig. 7.3). Since land allotment questions are very thorny in such areas, the size of such nature reserves can not be large. Preliminary estimates are that an optimal area for geobotanical reserves may total 100–1000 ha. Such reserves are essential for the conservation of both plant, invertebrate, small birds and mammal populations.

The increased amount of personal transport during recent decades has resulted in an increased recreational load. Uncontrolled recreation strongly affects highly

**Fig. 7.3** The most urbanized areas of Yakutia with population densities 60–600 persons per 10 $km^2$ (according to Matveyev et al. 1989)

decorative, as well as food and medicinal plant species. The most striking example is the very attractive *Lilium pensylvanicum*, which used to be a very common species for Central Yakutia. Nowadays it has been exterminated near residential areas and roads.

Special anxiety is caused by the extensive plans for the industrial development of South and West Yakutia. Construction of the railroad, oil and gas pipelines may yield contamination of natural ecosystems by invader species. Oil spill caused by pipeline breakage may seriously affect aquatic and riparian vegetation. The metallurgical works that are to be constructed in South Yakutia may produce "acid rains" that might greatly change the floral composition of vast territories.

Global climate warming also has a great influence on the northern ecosystems, especially on the vegetation of higher (the tundra and Arctic) latitudes and mountainous regions. In Central Yakutia the process of taiga being substituted by forest steppe has already accelerated. This process has a positive effect on the one hand, increasing the flora and fauna richness. On the other hand, all the consequences of this global phenomenon are still unclear.

The NR network should undergo restructuring for a better functionality. In our opinion, the following should be taken into consideration:

1. The areas of the northern natural reserves (covering the Arctic vegetation) should be reduced to a size that would be enough for protection of the animal populations. At the same time, the protected territories northwards of the "Ust–Lensky" Sanctuary should be expanded, covering the mountain masses (the Kharaulakh and Chekanovsky Ranges);
2. In the boreal zone with promising mining activity, the areas of a number of NRs should be decreased to the size that would be optimal for the protection of both boreal vegetation and animal populations inhabiting that area.
3. New NRs for the protection of rare plant species and communities should be established. We offer to consider the following regions for this purpose:

   North–East Yakutia:
   - The Upper Indigirka steppe zone with a high diversity of the relic steppe communities of the Late Pleistocene;
   - The Verkhoyansk steppe zone with similar objects;
   - The Verkhoyansk mountainous zone with rare mountain-steppe and alpine plant species.

   North–West Yakutia:
   - The Middle Olenyok zone covering the valley and watershed areas that are rich in both steppe and boreal plants.

   South–West Yakutia:
   - The Upper Lena taiga-valley zone with elements of dark coniferous forest;
   - The Middle Lena taiga-valley zone with similar objects;
   - The Olyokminsk zone, covering the Lena River valley hosting the rich diversity of species subject to severe anthropogenic activity;
   - The Olyokma–Chara mountain-valley zone with a unique complex of southern steppe petrophytic communities.

   South Yakutia:
   - The Tokko zone with distinct landscapes and a rich floral diversity. Two NRs are situated close to this region though do not cover even one third of its area.

   Central Yakutia:
   - The alas-valley complex covering unique landscapes that developed as a result of permafrost action.

Its unique flora, well adapted to the severe conditions of Yakutia, and its rare plant communities (including steppe vegetation) that are not covered by the network of protected territories, should impel botanists nowadays to contribute into the improvement of the modern NR network of Yakutia, being the largest region of Russia that plays a significant role in Global Change.

Very soon Yakutia may become a territory with expanded anthropogenic landscapes. Tens of large-scale constructions will have an impact on nature as a whole, and on the plant and animal world in particular. From a "sanctuary", Yakutia will turn into a medium-industrial region, having frequent relations with other regions and countries. To expect that all the above-mentioned examples of human activity will not have a significant environmental effect means to be a "naive" optimist. The protection of the botanical inheritance of Yakutia is a matter of today.

## 7.2 The Role of the Yakut Botanical Garden In Conservation of Rare and Endangered Plants

**I.A. Fedorov, P.S. Egorova, and P.A. Pavlova**

As has been mentioned before, South Yakutia is a promising region for rigorous technical activity. The Upper Lena and Aldan Floristic Regions, situated in South Yakutia, are characterized by a rich vegetation, with many endemic and rare plants. Many species grow there at the north–eastern borders of their distribution areas: *Epipactis helleborine*, *Epipogium aphyllum*, *Anagallidium dichotomum, Tussilago farfara,* etc. South Yakutian endemics with a very small distribution area are *Saxifraga lactea, Aconogonon amgense*, *Thermopsis lanceolata subsp. jacutica, Rumex jacutensis*, while *Artemisia remotiloba*, *Papaver anjuicum,* and others are endemic to North East Siberia. It is obvious that the industrial development will threaten the populations of these plants with complete extinction.

In the light of increased technogenic activity the main target of the Yakut Botanical Garden (YBG) is the establishment of a collection of living rare and endangered plant species of Yakutia ex situ. This encompasses the following tasks: (1) Search and collection of genetic resources of rare and endangered wild plant species from the regions of rigorous anthropogenic activity. (2) Study of their biological peculiarities and evaluation of their tolerance to conditions of cultivation.

Since its foundation (1962), the YBG has collected rare and endangered plants (Krotova et al. 1972; Krotova and Yarina 1977; Andreyev 1981; Danilova 1993, 1999, 2005; Danilova et al. 2005). Its collection of herbaceous plants includes 45 rare and endangered species of 20 families (Table 7.2). The list was compiled according to the classification of rare and endangered plant species adopted by the IUCN (Dolinin et al. 2000). All species are represented by 3–4 specimens from habitats with various ecological conditions. This allows to cover as much genetic variability of the species as possible. The period of flowering determines whether

**Table 7.2** Collection of rare species in the Yakut Botanical Garden

| Species name | Category of rarity | Flowering period |
|---|---|---|
| *Aconogonon amgense* | III a | Late summer |
| *Adonis apennina* | II | Late spring |
| *Aquilegia glandulosa* | III d | Late spring |
| *Aquilegia sibirica* | II | Late spring |
| *Artemisia obtusiloba subsp. martjanovii* | III c | Late summer |
| *Artemisia remotiloba* | III b | Late summer |
| *Bergenia crassifolia* | IV | Late spring |
| *Callianthemum isopyroides* | III c | Spring |
| *Centaurea scabiosa* | VI | Mid summer |
| *Cleistogenes squarrosa* | III d | Mid summer |
| *Cypripedium macranthon* | II | Early summer |
| *Dactylorhiza fuchsii* | III d | Does not flower |
| *Delphinium grandiflorum* | II | Mid summer |
| *Dendranthema arcticum subsp. polare* | III | Does not flower |
| *Festuca komarovii* | III c | Mid summer |
| *Gagea pauciflora* | III c | Late spring – early summer |
| *Hemerocallis lilio-asphodelus* | III d | Mid summer |
| *Iris laevigata* | II | Early summer |
| *I. orientalis* | II | Mid summer |
| *Krascheninnikovia lenensis* | I | Mid summer |
| *Leucanthemum pilosiusculum* | II | Early summer |
| *L. vulgare* | III | Mid summer |
| *Lilium pensylvanicum* | II | Early summer |
| *Mentha dahurica* | III d | Late summer |
| *Nyphar pumila* | III d | Does not flower |
| *Omalotheca sylvaticum* | IV | Mid summer |
| *Paeonia anomala* | II | Early summer |
| *Phlojodicarpus sibiricus* | II | Early summer |
| *Ph. villosus* | II | Early summer |
| *Polygala sibirica* | III d | Early summer |
| *Polygonatum odoratum* | III d | Early summer |
| *Potentilla tollii* | III a | Late spring |
| *Pulsatilla ajanensis* | III d | Spring |
| *Pulsatilla turczaninovii* | III d | Spring |
| *Redowskia sofhiifolia* | I | Late spring |
| *Rhodiola borealis* | II | Early summer |
| *Rh. quadrifida* | III c | Early summer |
| *Rh. rosea* | IV | Early summer |
| *Rhododendron aureum* | II | Mid summer |
| *Rumex jacutensis* | III b | Early summer |
| *Scutellaria baicalensis* | IV | Mid summer |
| *Thermopsis lanceolata subsp. jacutica* | II | Early summer |
| *Trollius asiaticus* | II | Early summer |
| *Viola dactyloides* | III c | Late spring |
| *V. patrinii* | III c | Late spring |

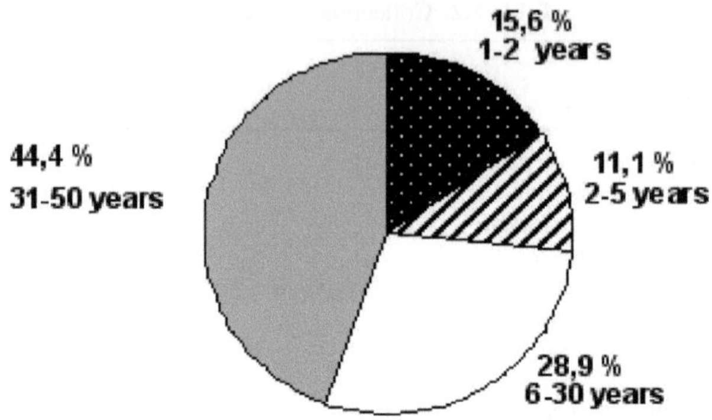


**Fig. 7.4** Duration of growth of the introduced species in the YBG

the cultivated species has time to produce ripe seeds or not, i.e. the plants blooming in spring or early summer are able to bear fruits for seed propagation, while late blooming species have no time for seed ripening. Species that flower in mid summer are able to produce ripe seeds in most cases.

The rare species from the Central Yakutian Floristic Region have been studied for 20–40 years; the species of the Aldan FR for a much shorter time. Most (33 species) have passed the cultivation tests in 1970s–1980s and have been grown for already many years (Fig. 7.4).

The tolerance to cultivation of the species was tested according to the 3-point scale (high tolerant, tolerant and low tolerant) developed by Danilova (1993). It revealed that most of tested rare species (33 of 44) are tolerant or highly tolerant to cultivation. They are characterized by annual fruiting with formation of full-grown seeds, as well as by seed and vegetative propagation. Besides, these species have a high resistibility to diseases and pests. Low tolerant species are the *Orchidaceae* and some other species from different families (*Redowskia sophiifolia, Rhodiola borealis, R. quadrifida,*). They do not form ripe seeds. This group also includes species that require a long time before they grow properly (e.g. *Rumex jacutensis, Rhodiola borealis*).

## References

Andreyev VN (ed) (1981) Dikorastuschiye travy Yakutii v culture. Nauka, Novosibirsk – Wild herbs of Yakutia in culture (in Russian)

Andreyev VN, Galaktionova TF, Perfilyeva VI, Scherbakov IP (1987) Osnovnye osobennosti rastitelnogo pokrova Yakutskoy ASSR. Izd-vo YaF SO AN SSSR, Yakutsk – *Basic features of vegetation cover of the Yakutian ASSR* (in Russian)

Borisov BZ (2000) Sozdaniye seti OOPT Prilenskogo regiona. Abstract of Thesis for a Candidate Degree. Yakutsk – *Development of the Natural Reserves network on the Near-Lena region* (in Russian)

Danilova NS (1993) Introduktsiya mnogoletnikh travyanistykh rasteniy Yakutii. Izd-vo YaNTs SO RAN, Yakutsk – *Introduction of perennial herb plants of Yakutia* (in Russian)

Danilova NS (1999) Lukovichnye geofity v kulture. Izd-vo YaGU, Yakutsk – *Bulbous geophytes in culture* (in Russian)

Danilova NS (ed) (2005) Raznoobrazie rastitelnogo mira Yakutii. Izd-vo SO RAN, Novosibirsk – *Diversity of the vegetation world of Yakutia* (in Russian)

Danilova NS, Borisova SZ, Ivanova NS (2005) Biologiya okhranyaemykh rasteniy Tsentralnoy Yakutii. Izd-vo YaNTs SO RAN, Yakutsk – *Biology of protected plants of Central Yakutia* (in Russian)

Dolinin IN et al. (ed) (2000) Krasnaya kniga Respubliki Sakha (Yakutia) Tom 1: Redkie i nakhodyaschiesya pod ugrozoy ischeznoveniya vidy rasteniy i gribov. NIPK "Sakhapoligrafizdat", Yakutsk – *The Red Data Book of the Republic of Sakha (Yakutia) Vol 1: Rare and endangered species of plants and fungi* (in Russian)

Koropachinsky IYu (ed) (1996) Zelyonaya kniga Sibiri: redkiye i nuzhdayuschiesya v okrane rastitelnye soobschestva. Nauka, Novosibirsk – *Green Book of Siberia: Plant communities that are rare and require protection* (in Russian)

Krotova ZE, Yarina OA (1977) Introduktsiya dekorativnykh rasteniy v usloviyakh Krainego Severa. Nauka, Novosibirsk – *Introduction of ornamental plants under conditions of the Far North* (in Russian)

Krotova ZE, Yarina OA, Govorina TP (1972) Perspektivnye dlya ozeleneniya dikorastuschiye rasteniya Yakutii. In: Pochvennye i botanicheskiye issledovaniya v Yakutii. Collection of articles. Yakutskoye knizhnoye izd-vo, Yakutsk: 154–160 – *Wild plants of Yakutia promising for landscape gardening* (in Russian)

Kryuchkov VV (1987) Sever na grani tysyacheletiy. Mysl, Moscow – *The North at the border of millenniums* (in Russian)

Matveyev IA, Nikolaev ME, Sivtsev TD et al. (eds) (1989) Atlas selskogo khozyaistva Yakutskoy ASSR. GUGK SSSR, Moscow – The atlas of agriculture of the Yakut Autonomous Soviet Socialist Republic (in Russian)

Shelyag–Sosonko YuR (ed) (1987) Zelyonaya kniga Ukrainskoy SSR: redkiye, ischezayuschiye i tipichnye, nuzhdayuschiesya v okhrane rastitelnye soobschestva. Naukova dumka, Kiev – *Green Book of the Ukrainian SSR: Plant communities that are rare, endangered and typical, and require protection* (in Russian)

Solomonov NG (1988) Sistema osobo okhranyaemykh territoriy v Respublike Sakha (Yakutia). Sibirsky ekologichesky zhurnal 3(4): 219–224 – *The Natural Reserves network in the Republic of Sakha (Yakutia)* (in Russian)

Solomonov NG (2000) Problemy sokhraneniya bioraznoobraziya v Respublike Sakha (Yakutia). Nauka i obrazovaniye 1: 135–139 – *Problems of biodiversity conservation in the Republic of Sakha (Yakutia)* (in Russian)

# Colour Plates

**Photo 1** *Duedya* is a first stage of alas formation. A caldron lake develops as a result of thermokarst processes (photo by N. Bosikov)

**Photo 2** *Tympy*, the second stage of alas formation. The lake becomes larger, lake shores start developing (photo by N. Bosikov)

E.I. Troeva et al. (eds.), *The Far North: Plant Biodiversity and Ecology of Yakutia*, Plant and Vegetation 3, DOI 10.1007/978-90-481-3774-9,

**Photo 3** The Myuryu alas as an example of a mature alas. The sides are represented by a *baidjarakh* complex (photo by N. Bosikov)

**Photo 4** A typical landscape for the arctic tundra (the Lena River Delta) (photo by Jan Dirks)

**Photo 5** The subarctic tussock tundra with *Eriophorum vaginatum* (photo by L. Kuznetsova)

**Photo 6** *Larix cajanderi* trees at the northern limits of the taiga zone (the Lower Kolyma River basin) (photo by A. Isaev)

**Photo 7** Northern taiga with *Larix cajanderi* (vicinity of the Chekurovka settlement, the Lower Lena River basin) (photo by L. Kuznetsova)

**Photo 8** Vast tracts of the lowland middle taiga (vicinity of Yakutsk, Central Yakutia) (photo by A. Isaev)

**Photo 9** The *Larix sibirica* taiga in South Yakutia (the Lower Vitim River basin) (photo by A. Isaev)

**Photo 10** "Ship timber", a highly productive *Pinus sylvestris* forest in South-West Yakutia (vicinities of the Vitim settlement) (photo by A. Isaev)

**Photo 11** Medallion tundra: a complex of dwarf shrub and *Alectoria* mountain tundra with patches of bare detritus (Mount Evota in the West Yangi Range, South Yakutia) (photo by L. Kuznetsova)

**Photo 12** The subalpine belt with *Pinus pumila* in autumn (the Central Verkhoyansk Range) (photo by E. Nikolin)

**Photo 13** The crooked forest of *Betula ermannii ssp. lanata* in the subalpine belt of mountains in South Yakutia (the Udokan Range) (photo by L. Kuznetsova)

**Photo 14** *Abies sibirica* is a constituent part of the mountain dark coniferous taiga in the Upper Aldan River basin (photo by E. Troeva)

**Photo 15** Valley complex of the Indigirka River (North-East Yakutia) (photo by V. Sergeyev)

**Photo 16** *Chosenia arbutifolia* forest on the Endybal River floodplain (central part of the Verkhoyansk Region) (photo by A. Isaev)

**Photo 17** *Tukulan*, the northern desert (the Lower Viluy River, Central Yakutia) (photo by L. Kuznetsova)

**Photo 18** Icing in mid-summer on the Echenka River (the Upper Indigirka River basin) (photo by E. Korolyuk)

**Photo 19** A complex of swamped tussock *Carex-Calamagrostis* meadow and *yernik* (shrubberies) is characteristic for depressions (vicinity of the Berdigestyakh settlement, Central Yakutia) (photo by E. Troeva)

**Photo 20** Steppe on a south-facing slope of the Adycha River valley (the Yana River basin, North-East Yakutia) (photo by E. Troeva)

**Photo 21** *Stipa+Artemisia* steppe in forest-steppe landscapes of Central Yakutia (the Middle Lena River basin) (photo by A. Isaev)

**Photo 22** *Ribes fragrans* and *Potentilla asperrima* are the characteristic species for acid bedrock landscapes (the Oymyakon region) (photo by E. Nikolin)

**Photo 23** Dead *Larix cajanderi* forest after invasion of the Siberian silk moth (vicinity of the Chychymakh settlement, the Middle Amga River basin, Central Yakutia) (photo by A. Isaev)

**Photo 24** *Chamaenerion angustifolium* stage of post-fire succession (the Neleger locality, vicinity of Yakutsk) (photo by A. Isaev)

**Photo 25** Grass stage of post-fire succession (vicinity of the Chychymakh settlement, the Middle Amga River basin, Central Yakutia) (photo by A. Isaev)

**Photo 26** Initial stage of technogenic succession on mine dump (South Yakutia). Derived community *Chamaenerion angustifolium* (photo by S. Mironova)

**Photo 27** Fruits of *Sorbocotoneaster pozdnjakovii* Pojark., the spontaneous generic hybrid occurring only in the Middle Aldan River basin (South Yakutia) (photo by L. Kuznetsova)

**Photo 28** *Redowskia sophiifolia* Cham. et Schlecht. is observed only in the territory of the Natural Park "the Lena Pillars" (Central Yakutia ) (photo by E. Nikolin)

**Photo 29** *Aconogonon amgense (*Michal. et V. Perf.) Tzvel. is an endemic species of the Upper Amga River basin (photo by A. Isaev)

**Photo 30** *Potentilla tollii* Trautv., the endemic species of the Yana River basin, after flowering (photo by E. Troeva)

**Photo 31** Steppe community with *Krascheninnikovia lenensis (*Kumin.) Tzvel. (*Chenopodiaceae*). In Yakutia this species is found only on the right bank of the Middle Lena River on calcareous bedrock (photo by E. Troeva)

**Photo 32** *Hericium coralloides* (Fr.) Pers. is a rare fungi species listed in the Red Data Book of Russia. In Yakutia it occurs only in the South (photo by L. Mikhalyova)

# Index

MIX
Papier aus verantwortungsvollen Quellen
Paper from responsible sources
FSC® C105338

If you have any concerns about our products,
you can contact us on
**ProductSafety@springernature.com**

In case Publisher is established outside the EU,
the EU authorized representative is:
**Springer Nature Customer Service Center GmbH**
**Europaplatz 3, 69115 Heidelberg, Germany**

Printed by Libri Plureos GmbH
in Hamburg, Germany